Maple V
Library Reference Manual

Bruce W. Char   Keith O. Geddes   Gaston H. Gonnet
Benton L. Leong   Michael B. Monagan   Stephen M. Watt

Maple V
Library Reference Manual

Springer-Verlag

New York Berlin Heidelberg London Paris
Tokyo Hong Kong Barcelona Budapest

Bruce W. Char
Department of Mathematics
  and Computer Science
Drexel University
Philadelphia, PA 19104
U.S.A.

Keith O. Geddes
Department of Computer Science
University of Waterloo
Waterloo, ON
Canada N2L 3G1

Gaston H. Gonnet
Department Informatik
ETH Zentrum
8092 Zürich
Switzerland

Benton Leong
Symbolic Computation Group
University of Waterloo
Waterloo, ON
Canada N2L 3G1

Michael B. Monagan
Department Informatik
ETH Zentrum
8092 Zürich
Switzerland

Stephen M. Watt
IBM Thomas J. Watson
  Research Center
P.O. Box 218
Yorktown Heights, NY 10598
U.S.A.

Maple is a registered trademark of Waterloo Maple Software.

Library of Congress Cataloging-in-Publication Data
Maple V Library reference manual /
    Bruce Char . . . [et al.] 1st ed.
        p.   cm.
      Includes bibliographical references and index.
      1. Maple (Computer program)   I. Char, Bruce W.
    QA155.7E4M36 1991
    510.'285'5369dc20   91-20181

Printed on acid-free paper.

Photocomposed copy prepared from the authors' TeX files.
Printed and bound by R.R. Donnelley & Sons, Harrisonburg, VA.
Printed in the United States of America.

9 8 7 6 5 4 3 2   **(Corrected second printing, 1992)**

ISBN 0-387-97592-6 Springer-Verlag New York Berlin Heidelberg
ISBN 3-540-97592-6 Springer-Verlag Berlin Heidelberg New York

Preface

The design and implementation of the Maple system is an on-going project of the Symbolic Computation Group at the University of Waterloo in Ontario, Canada. This manual corresponds with version V (roman numeral five) of the Maple system. The on-line *help* subsystem can be invoked from within a Maple session to view documentation on specific topics. In particular, the command `?updates` points the user to documentation updates for each new version of Maple.

The Maple project was first conceived in the autumn of 1980, growing out of discussions on the state of symbolic computation at the University of Waterloo. The authors wish to acknowledge many fruitful discussions with colleagues at the University of Waterloo, particularly Morven Gentleman, Michael Malcolm, and Frank Tompa. It was recognized in these discussions that none of the locally-available systems for symbolic computation provided the facilities that should be expected for symbolic computation in modern computing environments. We concluded that since the basic design decisions for the then-current symbolic systems such as ALTRAN, CAMAL, REDUCE, and MACSYMA were based on 1960's computing technology, it would be wise to design a new system "from scratch". Thus we could take advantage of the software engineering technology which had become available in recent years, as well as drawing from the lessons of experience.

Maple's basic features (elementary data structures, input/output, arithmetic with numbers, and elementary simplification) are coded in a systems programming language for efficiency. For users, there is a high-level language with a modern syntax more suitable for describing algebraic algorithms. An important property of Maple is that most of the algebraic facilities in the system are implemented using the high-level user language. The basic system, or kernel, is sufficiently compact and efficient to be practical for use in a shared environment or on personal computers that have as little as two megabytes of main memory. Library functions are loaded into the system as required, adding to the system such facilities as polynomial factorization, equation solving, indefinite integration, and matrix manipulation functions. The modularity of this design allows users to consume computer resources proportional to the algebraic facilities actually being used.

The system kernel is written in macros which can be translated by a locally-developed macro processor (called Margay) into versions of the kernel in the C programming language for various different operating systems. Operating systems currently supporting a Maple implementation include UNIX and various UNIX-like systems, 386 DOS, Macintosh Finder, DEC VMS, IBM VM/CMS and Amiga DOS.

Acknowledgments

This manual has benefited from comments given to us by students, faculty, and staff at the University of Waterloo, and by users at sites throughout the world. These "friends of Maple" are too numerous to list here. We hope they will feel that their assistance has been acknowledged when they find their suggestions incorporated into the present edition. For their assistance in the production of this manual we would like to thank Kate Atherley, Greg Fee, Krishna Gopinathan, Blair Madore, Liyuan Qiao, Marc Rybowicz, Katy Simonsen, and Hans Ziemann.

The Maple project has received support from various sources, including the Academic Development Fund of the University of Waterloo, the Natural Sciences and Engineering Research Council of Canada, Digital Equipment Corporation, the Sloan Foundation, and the Information Technology Research Centre of Ontario.

Contents

1

The Maple Library

1.1 Introduction

This manual describes all the functions available in the Maple library. The library is where the code for over 95% of Maple's mathematical knowledge and expertise resides. The library routines are available for your inspection, study, and when appropriate, modification and extension. Each of these routines is written in the Maple programming language. Through the library, the Maple system and the algorithms that it uses can be dynamically extended and customized.

Every Maple library routine is described in detail in this manual. Each description includes a brief description of the function, the parameters it uses, and further details about the function. Also included are extensive examples of the use of each routine. One of the easiest ways to become familiar with a new routine is to examine some typical uses of the routine.

All of the descriptions that are printed in this volume can also be produced on-line from within a Maple session. Once Maple has been invoked, the command ?*topic* will show you the documentation for the topic. For example, `?linalg` will produce the same documentation for the `linalg` (Linear Algebra) package of routines as is printed in this volume. With a few formatting exceptions, all of the descriptions available in this volume can be accessed verbatim on-line.

The electronic form of these descriptions does have one advantage over the printed form. With many versions of Maple, especially for workstations, the documentation appears as a separate window on the screen. Electronic cutting and pasting operations allow you to easily copy an on-line example of a function's use and paste it into your main Maple worksheet. You may then execute the example immediately or you may edit the example and then execute it. This process is an easy way to explore the capabilities of new Maple functions that you wish to learn.

This volume accompanies the *Maple V Language Reference Manual*, which describes the Maple language and the Maple system itself. *First Leaves: A Tutorial Introduction to Maple V* provides an introduction to Maple for the beginning user.

1.2 Description of the Maple Library

Categories of Library Functions

Maple's library functions fall into four categories: 1) "builtin" functions internal to the Maple system, 2) demand-loaded library functions, 3) miscellaneous library functions which are not automatically loaded on demand, and 4) packages of functions which are normally loaded via the `with` command.

Functions in the first category are coded in the Maple kernel using the system implementation language to achieve efficiency and in some cases to achieve functionality that would not be available via user-level Maple code. Functions in the second category are specified by Maple code in the Maple system library, and their names are initially assigned as unevaluated `readlib`-defined functions (see Section 7.9 of the *Maple V Language Reference Manual*). Functions in the third category are specified by Maple code and they cannot be used without being explicitly loaded by the user, using the `readlib` function. Packages of functions are specified by Maple code and may be loaded via the `with` library function, as described below.

The distinction between the first two categories will be essentially irrelevant to Maple users. These functions are grouped together and listed alphabetically in Section 2.1. For a specific function 'f', the user can easily determine which of the two categories it belongs to by entering Maple and using the command `eval(f, 1);`. This command evaluates f to just one level; this allows us to peek into the contents of the variable f without evaluating it fully.

For functions in the first category, the result will be a procedure definition with '**options builtin**' specified and with an integer for the procedure body, as in

```
proc () options builtin; 51 end
```

This indicates that this function is built into the Maple kernel and is coded in a system implementation language and not in the Maple programming language. Builtin functions cannot be inspected by the user since only Maple language routines can be meaningfully displayed. The number of builtin functions in the library is miniscule compared to the rest of the library, which can be inspected.

For functions in the second category, the result of `eval(f, 1);` will be

```
readlib('f')
```

This shows that the function f is defined in the library. By noting that f is `readlib`-defined, we mean that the initial value of the variable f is the unevaluated `readlib` call shown. As soon as f is evaluated or used normally (such as being part of a function call), the library Maple-level code for f is automatically loaded and assigned to f. Such automatic demand-loading of code for the majority of Maple functions gives the user instant and transparent access to the library. Demand-loading provides the advantage of requiring space only for code that is actually used.

Section 2.2 presents the miscellaneous library functions which are available in the Maple library but which must be explicitly loaded before they can be used. Since these functions are not `readlib`-defined, before using one of them you must load it explicitly through a command such as

```
readlib(f);
```

This reads the procedure definition for f from the appropriate place in the library and then assigns the procedure to f.

Chapters 3 through 7 present the various Maple packages which are currently defined.

Inspecting Maple Library Routines

With the exception of the builtin functions, the Maple source code for each routine may be displayed. To look at the code for a function **f**, use the command

```
print(f);
```

The function **f** is evaluated normally before being printed. If it had not been loaded previously and is **readlib**-defined, the code for **f** will be loaded. If the function **f** is one of the miscellaneous library functions in category 3 and therefore is not **readlib**-defined, you will need to explicitly load the code for **f** before printing it.

The code displayed for **f** may be in either an abbreviated form or a full display form. The abbreviated form simply displays the function header for **f**, followed by ... for the deleted body, followed by **end**. The full display form shows the entire definition of the procedure in a nicely indented fashion. A user interface setting allows the user to choose how procedures are to be displayed. The command

```
interface( verboseproc=setting );
```

is used to set **verboseproc** to 0, 1, or 2. By default, the value for **verboseproc** is 1. The value of 1 prints user-defined procedures in full display form and library procedures in abbreviated form. A setting of 2 displays both in full display form. A setting of 0 displays both in abbreviated form.

The code that is displayed is formatted by Maple from the internal code that is loaded for a procedure. Since Maple source level comments are not maintained in the internal code, you will not see these nor will you see the code as it was originally formatted in the Maple source files. All of the original, commented Maple source code for the library can be ordered by contacting:

Waterloo Maple Software
160 Columbia Street West
Waterloo, Ontario
Canada N2L 3L3
Telephone: (519) 747-2373
E-mail: wmsi@daisy.uwaterloo.ca

Argument Evaluation Rules

For all functions, all arguments are evaluated (from left to right) before being passed to the procedure, with the following six exceptions: **assigned**, **evaln**, **eval**, **evalhf**, **seq**, and **userinfo**. For the functions **assigned** and **evaln**, the arguments are evaluated to a name. For the functions **eval** and **evalhf**, no evaluation of arguments is performed prior to invoking the procedure, which then performs the specified type of evaluation. The function **seq** is evaluated using **for**-loop semantics, meaning that the index of sequence-formation (appearing in the second argument) is first evaluated to a name, and this name is assigned a specified value prior to each evaluation of the first argument. Finally, the function **userinfo** uses delayed evaluation of arguments beyond the first two in order to recognize occurrences of **print** and **lprint** for appropriate handling.

The Package Concept

The concept of a *package* in Maple permits the user to define an entire collection of new functions with just one Maple command, namely the `with` command. In addition to defining new user-level functions, packages provide two other services. First, they provide the necessary support functions to tie these routines properly into the rest of the system. This includes such facilities as type definitions and rules for expansion of mathematical expressions involving newly-defined functions. Secondly, they provide convenient short names for each of the routines in the package. In some cases, the effect is to redefine a name which is already used for some other purpose in the Maple session (or a name which is pre-defined by Maple). A warning is issued whenever such a redefinition of a name occurs via the `with` command.

Maple packages load very quickly because, in effect, loading a package requires nothing more than defining the locations of the new functions in the library. No code is loaded, but each of the member functions of a package becomes `readlib`-defined. For example, execution of the command

```
with(linalg);
```

causes a table (named `linalg`) to be loaded. The indices of the table are the various function names, and the table entry for each index is the `readlib` definition of the corresponding code for the function. For example, `linalg[transpose]` has the value

```
readlib(`linalg/transpose`)
```

Furthermore, the `with` command has the side-effect of assigning to each function name itself (such as the name `transpose`) the corresponding `readlib` definition. Hence, if the package is loaded via `with` then the simple function name `transpose` may be used; otherwise, the long-form notation `linalg[transpose]` would have to be used as the function name. As soon as the `transpose` function is first called, the code for this routine is loaded into memory.

1.3   Format of Library Function Descriptions

Each of the function descriptions which follow have the following sections: a one-line description, prototypes showing how the function is called, information about what types of arguments are expected, further information about the function, some examples of the function's use, and finally a list of related functions or topics. In the descriptions of the function's parameters, italicized names are prototypes and may be replaced by an expression of the appropriate type according to the restrictions noted in the PARAMETERS section. (Further conditions for the parameters may be described in the SYNOPSIS section.) Parameters shown in a `typewriter` font in the CALLING SEQUENCE section are parameters which are to be used verbatim. Some parameters may be described as being optional and may therefore be omitted when the function is called.

2
Main Routines

2.1 Standard Library Functions

2.1.1 abs – absolute value of real or complex argument
Synopsis:
- The syntax is `abs(e)` for an expression `e`. Automatic simplifications will be applied to this function.

2.1.2 Ai, Bi – Airy wave functions
Synopsis:
- The Airy wave functions, `Ai` and `Bi`, are linearly independent solutions for w in the equation $w'' - wz = 0$. More precisely:

$$Ai(z) = c_1 f(z) - c_2 g(z)$$

$$Bi(z) = \sqrt{3}[c_1 f(z) + c_2 g(z)]$$

- $f(z)$ and $g(z)$ are defined as:

$$f(z) = \sum_{k=0}^{\infty} 3^k (\frac{1}{3})_k \frac{z^{3k}}{(3k)!}$$

$$g(z) = \sum_{k=0}^{\infty} 3^k (\frac{2}{3})_k \frac{z^{3k+1}}{(3k+1)!}$$

where

$$(\alpha + \frac{1}{3})_0 = 1$$

and

$$3^k(\alpha + \frac{1}{3})_k = (3\alpha + 1)(3\alpha + 4)\ldots(3\alpha + 3k - 2)$$

for arbitrary α and positive integers k.

- The constants c_1 and c_2 are defined as

$$c_1 = \frac{3^{-2/3}}{\Gamma(2/3)}$$

$$c_2 = \frac{3^{-1/3}}{\Gamma(1/3)}$$

2.1.3   alias – define an abbreviation or denotation

Calling Sequence:

    alias(*e1*, *e2*, ..., *eN*)

Parameters:

    e1, *e2*, ..., *eN*   –   zero or more equations

Synopsis:

- Mathematics is full of special notations and abbreviations. These notations are typically encountered in written material as statements like "let J denote the Bessel function of the first kind" or "let alpha denote a root of the polynomial x^3-2". The purpose of the **alias** facility is to allow the user to state such abbreviations for the longer unique names that Maple uses and, more generally, to give names to arbitrary expressions.

- To define F as an alias for fibonacci, use **alias(F=fibonacci)**. To redefine F as an alias for hypergeom simply use **alias(F=hypergeom)**. To remove this alias for F use **alias(F=F)**.

- Aliases work as follows. Consider defining **alias(J=BesselJ)**, and then entering J(0,-x). On input the expression J(0,-x) is transformed to **BesselJ(0,-x)**. Maple then evaluates and simplifies this expression as usual. In this example Maple returns **BesselJ(0,x)**. Finally, on output, **BesselJ(0,x)** is replaced by J(0,x).

- The arguments to **alias** are equations. When **alias** is called, the equations are evaluated from left to right but are not subjected to any existing aliases. This means that you cannot define one alias in terms of another. Next, the aliases are defined. Finally, a sequence of all existing aliases is returned as the function value.

- The following improvements for Maple V should be noted. An alias may be defined for any Maple object except a numerical constant. You may assign to a variable by assigning to its alias. The global symbol I in Maple has been defined to be an alias for `sqrt(-1)`. Hence if you want to see the complex constant displayed as j then you would use `alias(I=I,j=sqrt(-1))`. Parameters and local variables are not affected by aliases. Thus it is possible to have a local variable or parameter called I.

Examples:

```
> alias(C=binomial);                    ⟶              I, C

> C(4,2);                               ⟶                6

> C(n,m);                               ⟶             C(n, m)

> alias(F=F(x)):
  alias(Fx=diff(F(x),x)):
  diff(F,x);                            ⟶               Fx

> diff(Fx,x);                           ⟶                d
                                                      ---- Fx
                                                       dx

> has(Fx,x);                            ⟶              true

> alias(alpha=RootOf(x^2-2)):
  factor(x^4-4,alpha);                  ⟶                                2
                                          (x + alpha) (x - alpha) (x  + 2)

> alias(f=g);                           ⟶        I, C, F, Fx, alpha, f

> f:=sin;                               ⟶            f := sin

> g(Pi);                                ⟶                0
```

SEE ALSO: `subs`, `macro`

2.1.4   allvalues – evaluate all possible values of expressions involving RootOfs

Calling Sequence:

>   allvalues(*expr*)
>   allvalues(*expr*, ´d´)

Parameters:

>   *expr* –   any expression, table, list, or set of expressions
>   ´d´   –   the character ´d´

Synopsis:

- The most common application of **allvalues** is to evaluate expressions returned from **solve** involving RootOfs.

- Typically, a RootOf represents more than one value. Thus, expressions involving RootOfs will generally evaluate to more than one value or expression. The procedure **allvalues** will return, in an expression sequence, all such values (or expressions) generated by the combinations of different values of the RootOfs.

- The procedure `allvalues` will attempt to evaluate expressions exactly using solve. The roots of nth degree polynomial equations where $n \leq 4$ can be obtained exactly. Where roots cannot be obtained exactly, `allvalues` will use `fsolve` to obtain a numerical solution. In this case, no symbolic constants can be used in the particular `RootOf` argument.

- The procedure `allvalues` will recognize `&RootOf` as an inert `RootOf` operator. It will be treated in exactly the same manner as `RootOf`.

- The optional second parameter ´d´ is used to specify that `RootOf`s of the same equation represent the same value and they should not be evaluated independently of one another.

- Nested `RootOf`s are supported by `allvalues`.

Examples:
```
> readlib(allvalues):
> expr := RootOf(_Z^2-1) + 1/RootOf(_Z^4-1)^2;
                              2                 1
              expr := RootOf(_Z  - 1) + ----------------
                                                  4    2
                                          RootOf(_Z   - 1)
> allvalues(expr);
                    2, 2, 0, 0, 0, 0, -2, -2
> allvalues(2*RootOf(_Z^3-1)-1);
                           1/2           1/2
                 1, - 2 + I 3   , - 2 - I 3
> allvalues(sin(&RootOf(_Z^2-a^2))*RootOf(_Z^2-1));
                 sin(a), - sin(a), sin(- a), - sin(- a)
> allvalues(RootOf(_Z^2-1)-1/RootOf(_Z^2-1),d);
                           0, 0
```

SEE ALSO: `solve`, `fsolve`, `RootOf`, `evala`

2.1.5   anames – sequence of assigned names

Calling Sequence:
    `anames( );`

Synopsis:
- The procedure `anames` returns an expression sequence of names that are currently assigned values other than their own name.

SEE ALSO: `assigned`, `unames`

2.1.6 arc-trigonometric functions

Synopsis:

- The following function names are used:
  arcsin, arccos, arctan, arccsc, arcsec, arccot.

- Automatic simplifications will be applied to these functions.

2.1.7 arc-hyperbolic functions

Synopsis:

- The following function names are used:
  arcsinh, arccosh, arctanh, arccsch, arcsech, arccoth.

- Automatic simplifications will be applied to these functions.

2.1.8   array – create an array

Calling Sequence:

array(*indexfcn, bounds, list*)

Parameters:

indexfcn	–	(optional) an indexing function
bounds	–	(optional) sequence of ranges
list	–	(optional) list of initial values

Synopsis:

- An array is a specialization of a table, with zero or more specified dimensions, where each dimension is an integer range. The result of executing the **array** function is to create an array. For example, `V := array(1..10)` creates a one dimensional array (a Maple vector) of length 10 but with no explicit entries. The command `A := array(1..m,1..n)` creates a two dimensional array (a Maple matrix) with **m** rows and **n** columns.

- All parameters to the array function are optional and may appear in any order. The *bounds* parameter is a sequence of integer ranges which must appear consecutively. If the *bounds* are not specified then they are deduced from the *list* of initial values.

- The *indexfcn* can be a procedure or a name specifying how indexing is to be performed - see **indexfnc** for more information. The built-in indexing functions are symmetric, antisymmetric, sparse, diagonal, and identity. If *indexfcn* is not specified, then "ordinary" indexing is used.

- The *list* of initial values may be a list of equations (cf. tables), or a list of values (one-dimensional), or a nested list of lists (row-by-row).

- The **map** function can be used to apply a function to each entry of an array. For example, **map** (`simplify`, A) simplifies each entry of the array A.

- Arrays have special evaluation rules (like procedures) so that if the name `A` has been assigned an array then `A` evaluates to the name `A` and `eval(A)` yields the actual array structure.

- The `op` function can be used to pick apart an array structure. Specifically `op(eval(A))` yields *indexfcn*, *bounds*, and *entries* where *entries* is a list of equations corresponding to the explicit entries in the array (cf. entries).

Examples:
```
> v := array(1..4):
  for i to 3 do v[i] := i^2 od:
  print(v);                           ⟶              [ 1, 4, 9, v[4] ]

> v[2];                               ⟶                     4

> v[0];                               ⟶    Error, index outside of bounds
> A := array(1..2,1..2):
  A[1,2] := x:
  A[1,1];                             ⟶                   A[1, 1]

> A[1,2];                             ⟶                     x

> print(A);                           ⟶            [ A[1, 1]      x    ]
                                                   [                   ]
                                                   [ A[2, 1]   A[2, 2] ]

> A := array( symmetric, 1..2,1..2, [ [1,x]
  , [x,x^2] ] ):
  op(1,eval(A));                      ⟶                  symmetric

> op(2,eval(A));                      ⟶                1 .. 2, 1 .. 2

> op(3,eval(A));                      ⟶                     2
                                          [(2, 2) = x , (1, 2) = x, (1, 1) = 1]

> map(diff,A,x);                      ⟶                 [ 0   1 ]
                                                        [       ]
                                                        [ 1  2 x ]
```

SEE ALSO: `table`, `matrix`, `vector`, `linalg`, `print`, `map`, `indices`, `entries`

2.1.9   **assemble** – **assemble a sequence of addresses into an object**
         disassemble – **break an object into its component addresses**
         pointto – **obtain the expression pointed to by an address**
         addressof – **obtain the address which points to an expression**

Calling Sequence:
       assemble(*addrseq*)
       disassemble(*ptr*)
       pointto(*ptr*)
       addressof(*expr*)

Parameters:

addrseq	–	sequence of integer addresses, the first integer representing the type of the object
ptr	–	integer pointer
expr	–	Maple expression

Synopsis:

- This collection of four functions is known as the "hackware package" in Maple. These functions allow access to the internal representations of Maple objects and to the addresses pointing to them.

- The user should become familiar with the internal representation of Maple objects before using this collection of functions. See the Maple Language Reference Manual for details.

- Extreme care is required since it is very easy to cause various types of system errors through the use of these functions.

- The functions **assemble** and **disassemble** are a complementary pair of functions:

```
assemble( addrseq )  -->  ptr
disassemble( ptr )   -->  addrseq
```

- The first function, **assemble**, assembles a sequence of integer addresses into a Maple object and returns an integer pointer *ptr* to the object. The second function, **disassemble**, looks at the Maple object pointed to by the integer pointer *ptr*, disassembles the object into its component parts, and returns the sequence of addresses of the component parts. The first integer in *addrseq* represents the type of the object.

- The functions **pointto** and **addressof** are a complementary pair of functions:

```
pointto( ptr )     -->  expr
addressof( expr )  -->  ptr
```

- The function **pointto** takes an integer pointer *ptr* and returns the Maple expression to which it points. The **addressof** function takes a Maple expression and returns the integer address which points to it.

- An object whose only reference is through an **addressof()** reference could be collected by garbage collection. To avoid this, either make sure that all the objects for which you compute **addressof()** are assigned to variables or delay garbage collection (see **gc**).

Examples:

```
> f := 3*x^2+5;
```
$\longrightarrow$
$$f := 3\,x^2 + 5$$

```
> a := addressof(f);
```
$\longrightarrow$ $\quad$ a := 268536196

```
> pointto(a);
```
$\longrightarrow$
$$3 x^2 + 5$$

```
> d := disassemble(a);
```
$\longrightarrow$ d := 11, 268562996, 268520600, 268520632, 268520568

```
> pointto(d[2]);
```
$\longrightarrow$
$$x^2$$

```
> pointto(d[3]);
```
$\longrightarrow$ 3

```
> pointto(d[4]);
```
$\longrightarrow$ 5

```
> pointto(d[5]);
```
$\longrightarrow$ 1

```
> g := assemble(d[1],d[2],d[3]);
```
$\longrightarrow$ g := 268563944

```
> pointto(g);
```
$\longrightarrow$
$$3 x^2$$

2.1.10   assign – perform assignments

Calling Sequence:
> assign(a, b)
> assign(a = b)
> assign(t)

Parameters:

> a – a name
> b – any expression
> t – a list or set of equations

Synopsis:

- For the cases **assign**(a, b) and **assign**(a = b) where a is a name, the assignment $a := b$ is made and **NULL** is returned. The arguments to **assign** are evaluated. Hence if a is already assigned the name b, then **assign**(a,2) would assign b the value 2. Note that b is any valid expression, with one exception: it cannot be an expression sequence containing more than one element.

- If the argument is a list or set of equations, then **assign** is applied to each equation in the list or set.

- One use of this function is to apply it to a set of equations returned by the **solve** function when it is desired to assign the solution values to the variables.

Examples:
```
> assign(a,2);
  a;
```
$\longrightarrow$ 2

```
> assign(b=99);
  assign(c=a);
  a,b,c;
```
$\longrightarrow$ 2, 99, 2

```
> s := solve( {x+y=1, 2*x+y=3}, {x,y} );        ⟶          s := {y = -1, x = 2}
>
  assign(s);
  x,y;                                           ⟶                2, -1
```

SEE ALSO: unassign

2.1.11 assigned – check if a name is assigned

Calling Sequence:
> assigned(*n*)

Parameters:
> *n* – name, subscripted name, or function call

Synopsis:

• The **assigned** function returns true if *n* has a value other than its own name, and returns false otherwise.

• This function is one of the exceptions to the normal evaluation rule for arguments of a function. The argument to **assigned** will only be evaluated as a name (see the **evaln** function) rather than fully evaluated.

• The definition of **assigned** for array/table subscripts is

```
        assigned( A[i] )  =  evalb( A[i] <> evaln(A[i]) )
```

• The definition of **assigned** for function calls is

```
        assigned( f(x) )  =  evalb( f(x) <> evaln(f(x)) )
```

Examples:
```
> a := table(symmetric);         ⟶         a := table(symmetric,[])
> a[1,2] := x;                    ⟶              a[1, 2] := x
> assigned(a[1,2]);              ⟶                  true
> assigned(a[2,1]);              ⟶                  true
> assigned(a[1,1]);              ⟶                  false
> assigned(abs(x));             ⟶                  false
> assigned(abs(2));             ⟶                  true
```

SEE ALSO: anames, unames, evaln

2.1.12   asympt – asymptotic expansion

Calling Sequences:

  asympt(f, x)

  asympt(f, x, n)

Parameters:

  f – an algebraic expression in x

  x – a name

  n – positive integer (truncation order)

Synopsis:

- The function **asympt** computes the asymptotic expansion of f with respect to the variable x (as x approaches infinity).

- The third argument n specifies the truncation order of the series expansion. If no third argument is given, the value of the global variable `Order` (default `Order = 6`) is used.

- Specifically, **asympt** is defined in terms of the series function

```
subs( x=1/x, series( subs(x=1/x, f), x=0, n ) );
```

However, the expression returned will be in sum-of-products form rather than in the series form.

Examples:

```
> asympt(x/(1-x-x^2),x);
                       1       2       3       5        1
         - 1/x +  ----  -  ----  +  ----  -  ----  + O(----)
                    2       3       4       5        6
                   x       x       x       x        x
> asympt(n!/sqrt(2*Pi),n,3);
                   1/2       1           1            1
                  n    +  -------  +  --------  + O(----)
                              1/2        3/2         5/2
                          12 n      288 n          n
                 -----------------------------------
                                  n
                           (1/n)   exp(n)
> asympt(exp(x^2)*(1-erf(x)),x);
                   1            1             3          1
                -------  -  ----------  +  ----------  + O(----)
                   1/2         1/2  3         1/2  5       7
                 Pi    x     2 Pi    x      4 Pi    x      x
> asympt(sqrt(Pi/2)*BesselJ(0,x),x,3);
     sin(x + 1/4 Pi)         cos(x + 1/4 Pi)          sin(x + 1/4 Pi)        1
   --------------- - 1/8 --------------- - 9/128 --------------- + O(----)
        1/2                    3/2                     5/2             7/2
       x                      x                       x              x
```

SEE ALSO: `limit`, `series`

2.1.13   bernoulli – Bernoulli numbers and polynomials

Calling Sequence:

    `bernoulli(`n`)`

    `bernoulli(`n`, `x`)`

Parameters:

    n – a non-negative integer

    x – an expression

Synopsis:

- The `bernoulli` function computes the nth Bernoulli number, or the nth Bernoulli polynomial, in the expression x. The nth Bernoulli number B(n) is defined by the exponential generating function:

    `t/(exp(t)-1) = sum( B(n)/n!*t^n, n=0..infinity )`

Examples:

`> bernoulli(4);`	$\longrightarrow$	$-1/30$
`> bernoulli(4,x);`	$\longrightarrow$	$- 1/30 + x^2 + x^4 - 2 x^3$
`> bernoulli(4,1/2);`	$\longrightarrow$	$7/240$

SEE ALSO: `euler`

2.1.14   BesselI, BesselK – Bessel functions

Synopsis:

- These Bessel functions are solutions to

$$z^2\frac{d^2w}{dz^2} + z\frac{dw}{dz} - (z^2 + v^2)w = 0$$

- `BesselI` is a Bessel function of the first kind usually denoted

$$\texttt{BesselI}(v, z) = I_v(z)$$

- `BesselK` is a Bessel function of the second kind usually denoted

$$\texttt{BesselK}(v, z) = K_v(z)$$

2.1.15   BesselJ, BesselY – Bessel functions

Synopsis:

- These Bessel functions are solutions to

$$z^2 \frac{d^2w}{dz^2} + z \frac{dw}{dz} + (z^2 - v^2)w = 0$$

- `BesselJ` is a Bessel function of the first kind usually denoted

$$\texttt{BesselJ}(v, z) = J_v(z)$$

- `BesselY` is a Bessel function of the second kind usually denoted

$$\texttt{BesselY}(v, z) = Y_v(z)$$

2.1.16   Beta(x,y) – the Beta function

Synopsis:

- The Beta function has syntax `Beta(x,y)` is defined as follows:

$$Beta(x, y) = \frac{\text{GAMMA}(x) \times \text{GAMMA}(y)}{\text{GAMMA}(x + y)}$$

2.1.17   binomial – binomial coefficients

Calling Sequence:

>   binomial(n, r)

Parameters:

> n, r –   expressions

Synopsis:

- The binomial function computes the binomial coefficients. If the arguments are both positive integers with $0 \le r \le n$, then `binomial(`n`, `r`)` = $n!/r!/(n-r)!$ which is the number of ways of choosing r objects from n distinct objects.

- If n and r are integers which do not satisfy $0 \le r \le n$, or n and r are rationals or floats, then the more general definition `binomial(`n`, `r`)` = `GAMMA(`n`+1)/(GAMMA(`r`+1)/GAMMA(`n`-`r`+1)` is used. Otherwise, for symbolic arguments, some simplifications such as `binomial(`n`, 1)` = n can be made, but typically, `binomial` returns unevaluated.

Examples:

```
> binomial(4, 2);
```

$\longrightarrow$ 6

```
> binomial(2, 1/2);              ⟶                    16
                                                     ----
                                                     3 Pi
> binomial(2.1, 2);             ⟶                1.155000000
> binomial(n, 2);               ⟶                binomial(n, 2)
> expand(");                    ⟶                 1/2 (n - 1) n
```

SEE ALSO: `combinat`, `combinat[multinomial]`

2.1.18 cat – concatenating expressions

Calling Sequence:

   cat(*a*, *b*, *c*, ...)

Parameters:

   a, b, c, etc. –   any expressions

Synopsis:

- The `cat` function is commonly used to concatenate strings together. The arguments are concatenated to form either a string or an object of type `` `.´ ``.

- Note that the result of `cat(`*a, b, c*`)` is equivalent to doing

    `` `` `` . a . b . c

SEE ALSO: `concat`, `substring`, `length`

2.1.19 chebyshev – Chebyshev series expansion

Calling Sequence:

   chebyshev(*expr*, *eq/nm*, *eps*)

Parameters:

   expr   –   an algebraic expression or a procedure

   eq/nm –   an equation x = a..b or a name x

   eps   –   (optional) a numeric value

Synopsis:

- This function computes the Chebyshev series expansion of *expr*, with respect to the variable x on the interval `a..b`, valid to accuracy *eps*.

- If *eq/nm* evaluates to a name x then the equation x = -1..1 is assumed.

- If the third argument *eps* is present then it specifies the desired accuracy; otherwise, the value used is *eps* = 10^(-Digits). It is an error to specify *eps* less than 10^(-Digits).

- The expression or procedure *expr* must evaluate to a numerical value when x takes on a numerical value. Moreover, it must represent a function which is analytic in a region surrounding the interval a..b.

- The resulting series is expressed in terms of the Chebyshev polynomials T(k,x), k = 0, 1, 2, ... with floating-point series coefficients. Conversion to ordinary polynomial form can be accomplished by simply loading the Chebyshev polynomial definitions via with(orthopoly,T).

- The series computed is the "infinite" Chebyshev series, truncated by dropping all terms with coefficients smaller than *eps* multiplied by the largest coefficient.

- Note: The name T used in representing the Chebyshev polynomials is a global name, so the user must ensure that this name has no previous value.

Examples:
```
> Digits := 5:
> chebyshev(cos(x), x);
    .76520 T(0, x) - .22981 T(2, x) + .0049533 T(4, x) - .000041877 T(6, x)

> chebyshev(exp(x), x=0..1, .001);
     1.7534 T(0, 2 x - 1) + .85040 T(1, 2 x - 1) + .10521 T(2, 2 x - 1)

          + .0087222 T(3, 2 x - 1)
```

SEE ALSO: orthopoly, series, taylor

2.1.20   chrem – Chinese Remainder Algorithm

Calling Sequence:

> chrem(*u*, *m*)

Parameters:

> *u*   –   the list [u0,u1,...,un] of evaluations
>
> *m*   –   the list of moduli [m0,m1,...,mn]

Synopsis:

- If *u* is a list of integers, chrem(*u*,*m*) computes the unique positive integer a such that a mod m0 = u0, a mod m1 = u1, ..., a mod mn = um, and a < m0 m1 ... mn. The moduli m0, m1, ..., mn must be pairwise relatively prime.

- If *u* is a list of polynomials, chrem is applied to the coefficients of each term in [u0, u1, ..., um].

Examples:
```
> chrem( [1,2], [5,7] );              ⟶              16
> chrem( [3*x,2*y+x], [5,7] );        ⟶         30 y + 8 x
```

2.1.21   Ci – the cosine integral

Synopsis:

- The cosine integral is defined as follows:

$$Ci(x) = \gamma + ln(x) + \int_0^x \frac{\cos t - 1}{t} dt$$

2.1.22   coeff – extract a coefficient of a polynomial

Calling Sequence:

```
coeff(p,x)
coeff(p,x,n)
coeff(p,x^n)
```

Parameters:

p – a polynomial in x

x – the variable (an expression)

n – (optional) an integer

Synopsis:

- The `coeff` function extracts the coefficient of x^n in the polynomial p. Note that input expression p must be `collected` in x. Use the function `collect(p,x)` prior to calling `coeff`, if necessary.

- If the third argument is omitted, it is determined by looking at the second argument. Thus `coeff(p,x^n)` is equivalent to `coeff(p,x,n)` for $n <> 0$.

- The cases of the second argument being a number or a product are disallowed since they do not make sense.

- The related functions `lcoeff`, `tcoeff` and `coeffs` extract the leading coefficient, trailing coefficient and all the coefficients of p in x respectively.

Examples:

```
> p := 2*x^2 + 3*y^3 - 5:
  coeff(p,x,2);                    ⟶                    2

> coeff(p,x^2);                    ⟶                    2

> coeff(p,x,0);                    ⟶                    3
                                                    3 y  - 5

> q := 3*a*(x+1)^2 + sin(a)*x^2*y - y^2*x +
    x - a:
  coeff(q,x);                      ⟶    Error, unable to compute coeff
> q := collect(q,x);               ⟶                              2
                                        q := (sin(a) y + 3 a) x
```

$$+ (6\, a - y^2 + 1)\, x + 2\, a$$

```
> coeff(q,x);
```
$\longrightarrow$
$$6\, a - y^2 + 1$$

```
> coeff(p,x*y,2);
```
$\longrightarrow$ Error, wrong number (or type) of parame\
ters in function coeff;

SEE ALSO: collect, lcoeff, tcoeff, coeffs, sort

2.1.23 coeffs – extract all coefficients of a multivariate polynomial

Calling Sequence:

coeffs(p, x, $'t'$);

Parameters:

p – multivariate polynomial

x – (optional) indeterminate or list/set of indeterminates

t – (optional) name

Synopsis:

- The coeffs function returns an expression sequence of all the coefficients of the polynomial p with respect to the indeterminate(s) x.

- If x is not specified, *coeffs* computes the coefficients with respect to all the indeterminates of p (see the indets function). If a third argument t is specified (call by name), it is assigned an expression sequence of the terms of p. There is a one-to-one correspondence between the coefficients and the terms of p.

- Note that p must be collected with respect to the appropriate indeterminates.

Examples:

```
> s := 3*v^2*y^2+2*v*y^3;
```
$\longrightarrow$
$$s := 3\, v^2\, y^2 + 2\, v\, y^3$$

```
> coeffs( s );
```
$\longrightarrow$
$$3,\, 2$$

```
> coeffs( s, v, 't' );
```
$\longrightarrow$
$$2\, y^3,\, 3\, y^2$$

```
> t;
```
$\longrightarrow$
$$v,\, v^2$$

SEE ALSO: collect, coeff, tcoeff, lcoeff, indets

2.1.24   collect – collect coefficients of like powers

Calling Sequence:
    collect(a, x)
    collect(a, x, *form*, *func*)

Parameters:

a	–	an expression
x	–	an indeterminate, or a list or set of indeterminates
form	–	(optional) name
func	–	(optional) procedure

Synopsis:

- The `collect` function views a as a polynomial in x and collects all the coefficients with the same power of x .

- The second argument x can be a single indeterminate (univariate case) or a list or set of indeterminates x_1, x_2, ..., x_n (multivariate case) . The indeterminates can be names or unevaluated function calls but not sums or products.

- Two forms for the result are available. The form is specified by the third argument. It may be the name `recursive` (the default) or the name `distributed`.

- The recursive form is obtained by first collecting the coefficients in x_1, then for each coefficient in x_1, collecting the coefficients in x_2 and so on. The distributed form is the form obtained by collecting the coefficients of $x_1 \hat{\ } e_1 * x_2 \hat{\ } e_2 * ... * x_n \hat{\ } e_n$ together.

- Note that if x is an indeterminate (univariate case), the recursive and distributed forms yield the same result.

- A function may be specified as an optional fourth argument. It is applied to the coefficients of the collected result. Often `normal` or `factor` will be used.

- Note that the resulting polynomial is not necessarily sorted. To sort a polynomial see the `sort` function.

Examples:
```
> collect( (x+1)*(x+2), x );
```
$\longrightarrow$
$$x^2 + 3\,x + 2$$

```
> collect( y*(sin(x)+1)+sin(x), sin(x) );
```
$\longrightarrow$
$$(y + 1)\,\sin(x) + y$$

```
>
  p := x*y+z*x*y+y*x^2-z*y*x^2+x+z*x:
  collect( p, x );
```
$\longrightarrow$
$$(y - z\,y)\,x^2 + (y + z\,y + 1 + z)\,x$$

```
> collect( p, [x,y] );
```
$\longrightarrow$
$$(1 - z)\,y\,x^2 + ((1 + z)\,y + 1 + z)\,x$$

```
> collect(p,[x,y],distributed);
```
$\longrightarrow$
$$(1 + z)\ x + (1 + z)\ x\ y + (1 - z)\ y\ x^2$$

```
> int(-x^2/exp(x),x);
```
$\longrightarrow$
$$\frac{x^2}{\exp(x)} + 2\ \frac{x}{\exp(x)} + \frac{2}{\exp(x)}$$

```
> collect(",exp(x));
```
$\longrightarrow$
$$\frac{x^2 + 2\ x + 2}{\exp(x)}$$

SEE ALSO: `sort`

2.1.25   combine – combine terms into a single term

Calling Sequence:
>   combine(*f*)
>   combine(*f*, *n*)

Parameters:
>   *f* – any expression
>   *n* – a name

Synopsis:

- The `combine` function applies transformations which combine terms in sums, products, and powers into a single term. This function is applied recursively to the components of lists, sets, and relations; that is, *f* and *n* may be lists (or sets) of expressions and names, respectively.

- For many functions, the transformations applied by `combine` are the inverse of the transformations that are applied by expand. For example, consider the well-known identity

```
    sin(a+b) = sin(a)*cos(b) + cos(a)*sin(b)
```

- The `expand` function applies the identity from left to right whereas the combine function does the reverse.

- Subexpressions involving `Int`, `Sum`, and `Limit` are combined into one expression where possible using linearity; that is, `c1*f(a,range) + c2*f(b,range)` ==> `f(c1*a+c2*b,range)`.

- A specific set of transformations is obtained by specifying a second (optional) argument *n* (a name) which is one of the following: `exp`, `ln`, `power`, `trig`, `Psi`. For additional information and examples about the transformations applied by each of these, see `combine[`*n*`]`.

Examples:

```
> combine(Int(x,x=a..b)-Int(x^2,x=a..b));    ⟶
```

$$\int_{a}^{b} x - x^2 \; dx$$

```
> combine(Limit(x,x=a)*Limit(x^2,x=a)+c);    ⟶
```

$$\lim_{x \to a} x^3 + c$$

```
> combine(4*sin(x)^3,trig);                  ⟶        - sin(3 x) + 3 sin(x)
> combine(exp(x)^2*exp(y),exp);              ⟶           exp(2 x + y)
> combine(2*ln(x)-ln(y),ln);                 ⟶
```

$$\ln\left(\frac{x^2}{y}\right)$$

```
> combine((x^a)^2,power);                    ⟶
```

$$x^{(2\,a)}$$

```
> combine(Psi(-x)+Psi(x),Psi);               ⟶    2 Psi(x) + Pi cot(Pi x) + 1/x
```

SEE ALSO: expand, factor, student

2.1.26   combine/exp – combine exponentials

Calling Sequence:

    combine(*f*, exp)

Parameters:

    f – any expression

Synopsis:

- Expressions involving exponentials are combined by applying the following transformations:

```
        exp(x)*exp(y)    ==>   exp(x+y)

        exp(x)^y         ==>   exp(x*y)

        exp(x+n*ln(y))   ==>   y^n*exp(x)   where n is an integer
```

Examples:

```
> combine(exp(x)*exp(-x),exp);               ⟶                1
> combine(exp(5)^2*exp(3),exp);              ⟶              exp(13)
```

```
> combine(exp(x+3*ln(y)),exp);          ⟶                    3
                                                           y  exp(x)

> combine(exp(x+3/2*ln(y)),exp);        ⟶              exp(x + 3/2 ln(y))
```

2.1.27   combine/ln – combine logarithmic terms

Calling Sequence:

   combine(*f*, ln)

Parameters:

   f – any expression

Synopsis:

- Expressions involving logarithms are combined by applying the following transformations:

```
        a*ln(x)      ==>   ln(x^a)
        ln(x)+ln(y)  ==>   ln(x*y)
```

   where it should be noted that the first of the above transformations is not applied if *a* is a logarithmic term.

Examples:

```
> combine(a*ln(x+1),ln);                ⟶                          a
                                                            ln((x + 1) )

> combine(a*ln(x)+3*ln(x),ln);          ⟶                     a 3
                                                            ln(x  x )

> combine(a*ln(x)*ln(y),ln);            ⟶              a ln(x) ln(y)

> combine(3*ln(2)-2*ln(3),ln);          ⟶                 ln(8/9)

> combine(ln(1+x)+5-ln(1-x),ln);        ⟶                      x + 1
                                                        5 + ln(-----)
                                                              1 - x
```

2.1.28   combine/power – combine terms with powers

Calling Sequence:

   combine(*f*, power);

Parameters:

   f – any expression

Synopsis:

- Expressions involving powers are combined by applying the following transformations:

```
x^y*x^z          ==>   x^(y+z)
(x^y)^z          ==>   x^(y*z)
exp(x)*exp(y)    ==>   exp(x+y)
exp(x)^y         ==>   exp(x*y)
sqrt(-a)         ==>   I*sqrt(a)
a^n*b^n          ==>   (a*b)^n
```

for arbitrary x, y, z, integers a, $b > 1$, and rational n.

Examples:

```
> combine(x^3*x^(m-3),power);
```
$\longrightarrow$
$$x^m$$

```
> combine((3^n)^m*3^n,power);
```
$\longrightarrow$
$$3^{(n\,m\,+\,n)}$$

```
> combine(exp(x)^7*exp(y),power);
```
$\longrightarrow$
$$\exp(7\,x\,+\,y)$$

```
> combine(2^(1/2)*3^(1/2),power);
```
$\longrightarrow$
$$6^{1/2}$$

```
> combine((-2)^(1/2)*3^(1/2),power);
```
$\longrightarrow$
$$I\,6^{1/2}$$

2.1.29 combine/Psi – combine Psi functions

Calling Sequence:

    combine(f, Psi);

Parameters:

   f – any expression

Synopsis:

- Expressions involving `Psi` and its derivatives are combined by applying the recurrence and reflection formulae.

- The following identities are used:

```
Psi(1+x)    =   Psi(x) + 1/x
Psi(1-x)    =   Psi(x) + Pi*cot(Pi*x)
Psi(n,x+1)  =   Psi(n,x) + (-1)^n * n! / x^(n+1)
Psi(n,1-x)  =   (-1)^n * ( Psi(n,x) + diff(Pi*cot(Pi*x),x$n) )
```

Examples:

```
> combine(Psi(x)+Psi(1+x),Psi);
```
$\longrightarrow$
            2 Psi(1 + x) - 1/x

```
> combine(Psi(1-x)-Psi(x)+Psi(y),Psi);
```
$\longrightarrow$
            Pi cot(Pi x) + Psi(y)

```
> combine(Psi(2,x-1/2)-Psi(2,x+1/2),Psi);     ⟶
```

$$- \frac{2}{(x - 1/2)^3}$$

2.1.30   combine/trig – combine trigonometric terms

Calling Sequence:

   `combine(f, trig);`

Parameters:

   f –   any expression

Synopsis:

- Products and powers of trigonometric terms involving sin, cos, sinh and cosh are combined into a sum of trigonometric terms by repeated application of the transformations

```
        sin(a)*sin(b)   ==>   1/2*cos(a-b) - 1/2*cos(a+b)
        sin(a)*cos(b)   ==>   1/2*sin(a-b) + 1/2*sin(a+b)
        cos(a)*cos(b)   ==>   1/2*cos(a-b) + 1/2*cos(a+b)
```

- where `sin(a)^2` and `cos(b)^2` are special cases of the above. The form of the result is a sum of trigonometric terms whose arguments are integral linear combinations of the original arguments.

- An important special case is when the input is a polynomial in `sin(x)` and `cos(x)` over a field, in which case the result is a canonical form; namely,

```
        sum( a[i]*sin(i*x), i=-n..n ) + sum( b[i]*cos(i*x), i=-n..n )
```

- where `a[i]`, `b[i]` are in the field and `n` is bounded by the total degree of the input polynomial in `sin(x)` and `cos(x)`.

Examples:

```
> combine(sin(x)^2,trig);              ⟶          1/2 - 1/2 cos(2 x)
> combine(sinh(x)*cosh(x),trig);       ⟶              1/2 sinh(2 x)
> combine(2*sin(x)*cos(y),trig);       ⟶          sin(x + y) + sin(x - y)
> f := sin(x)^2*cos(x) + 3*cos(x)^3 + 2*
  sin(x)*cos(x);                       ⟶                  2              3
                                             f := sin(x)  cos(x) + 3 cos(x)

                                                   + 2 sin(x) cos(x)

> combine(f,trig);                     ⟶      5/2 cos(x) + 1/2 cos(3 x) + sin(2 x)
> f := 512*sin(x)^5*cos(x)^5;          ⟶                        5       5
                                             f := 512 sin(x)  cos(x)

> combine(f,trig);                     ⟶      sin(10 x) + 10 sin(2 x) - 5 sin(6 x)
```

2.1.31   compoly – determine a possible composition of a polynomial

Calling Sequence:

    compoly(r, x)

Parameters:

   r – a polynomial

   x – a variable upon which the composition will be made

Synopsis:

- The function compoly returns a pair p(x), x=q(x) such that subs(x=q(x), p(x)) is equal to r, the input polynomial. If such a pair cannot be found, it returns FAIL. p(x) and q(x) are non-linear polynomials and q(x) has a low-degree in x greater or equal to 1.

- Note that the composition may not be unique.

Examples:

```
> readlib(compoly):
  compoly( x^6-9*x^5+27*x^4-27*x^3-2*x^2*y+
  6*x*y+1, x );
```
$\longrightarrow$
$$x^3 - 2\,x\,y + 1,\ x = x^2 - 3\,x$$

```
> compoly( x^4-3*x^3-x+5, x );
```
$\longrightarrow$
$$\text{FAIL}$$

2.1.32   Content – inert content function
Primpart – inert primitive part function

Calling Sequence:

    Content(a, x, ´pp´)
    Primpart(a, x, ´co´)

Parameters:

   a – multivariate polynomial in x

   x – (optional) name or set or list of names

   pp – (optional) unevaluated name

   co – (optional) unevaluated name

Synopsis:

- Content and Primpart are "placeholders" for the content and primitive part of a polynomial over a coefficient domain. They are used in conjunction with mod and evala as described below. See content and primpart for more information.

- The calls Content(a, x) mod p and Primpart(a, x) mod p compute the content and primitive part of a respectively modulo the prime integer p. The argument a must be a multivariate polynomial over the rationals or over a finite field specified by RootOfs.

- The calls evala(Content(a,x)) and evala(Primpart(a,x)) compute the content and primitive part of a respectively over a coefficient domain which may include algebraic numbers. The polynomial a must be a multivariate polynomial with algebraic number coefficients specified by RootOfs.

- The optional arguments $'pp'$ and $'co'$ are assigned a/Content(a) and a/Primpart(a) respectively, computed over the appropriate coefficient domain.

Examples:

> Content(x*(y+4)+y^2+4,x) mod 5;	$\longrightarrow$	y + 4
> Primpart(x*(y+4)+y^2+4,x) mod 5;	$\longrightarrow$	y + x + 1
> a := 5*x^3+3*y^2;	$\longrightarrow$	$a := 5\,x^3 + 3\,y^2$
> Content(a,x) mod 11;	$\longrightarrow$	1
> Primpart(a,x,'c1') mod 11;	$\longrightarrow$	$x^3 + 5\,y^2$
> c1;	$\longrightarrow$	5

SEE ALSO: content, primpart, mod, evala, RootOf

2.1.33   content – content of a multivariate polynomial
           primpart – primpart of a multivariate polynomial

Calling Sequence:

    content(a, x, $'pp'$)
    primpart(a, x, $'co'$)

Parameters:

    a – multivariate polynomial in x

    x – (optional) name or set or list of names

    pp – (optional) unevaluated name

    co – (optional) unevaluated name

Synopsis:

- If a is a multivariate polynomial with integer coefficients, **content** returns the content of a with respect to x, thus returning the greatest common divisor of the coefficients of a with respect to the indeterminate(s) x. The indeterminate(s) x can be a name, a list, or a set of names.

- The third argument pp, if present, will be assigned the primitive part of a, namely a divided by the content of a.

- If the coefficients of a in x are rational functions then the content computed will be such that the primitive part is a multivariate polynomial over the integers whose content is 1.

- Similarly, `primpart` returns $a/$`content`(a,x). The third argument co, if present, will be assigned the content. Note: Whereas the `sign` is removed from the content, it is not removed from the primitive part.

Examples:

```
> content( 3 - 3*x, x );                    ⟶           3

> content( 3*x*y + 6*y^2, x );              ⟶          3 y

> content( 3*x*y + 6*y^2, [x,y]);           ⟶           3

> content( -4*x*y+6*y^2, x,'pp');           ⟶          2 y

> pp;                                       ⟶        3 y - 2 x

> content( x/a - 1/2, x, 'pp' );            ⟶           1
                                                       ---
                                                       2 a

> pp;                                       ⟶        2 x - a

> primpart( -4*x*y + 6*y^2, x );            ⟶        3 y - 2 x

> primpart( x/a - 1/2, x );                 ⟶        2 x - a
```

SEE ALSO: `icontent`, `gcd`, `coeffs`, `Content`

2.1.34   convert – convert an expression to a different form

Calling Sequence:

   `convert(`*expr, form, arg3, ...*`)`

Parameters:

expr	–	any expression
form	–	a name
arg3, ...	–	(optional) other arguments

Synopsis:

- The `convert` function is used to convert an expression from one form to another. Some of the conversions are data-type conversions, for example `convert([x,y], set)` Others are form conversions, for example `convert(x^3-3*x^2+7*x+9,horner,x)` yields `(((x^3)*x+7)*x)+9`.

- The types of known conversions are (the second argument *form* must be one of these):

`+`	`*`	D	array	base	binary
confrac	decimal	degrees	diff	double	eqnlist
equality	exp	expln	expsincos	factorial	float
fraction	GAMMA	hex	horner	hostfile	hypergeom
lessthan	lessequal	list	listlist	ln	matrix
metric	mod2	multiset	name	octal	parfrac
polar	polynom	radians	radical	rational	ratpoly
RootOf	series	set	sincos	sqrfree	tan
vector					

- Further information is available under **convert**[*form*] where *form* is one of the forms from the above list.

- A user can make his own conversions known to the **convert** function by defining a Maple procedure in the following way. If the procedure `convert/f` is defined, then the function call convert(a,f,x,y,...) will invoke `convert/f`(a,x,y,...);

Examples:
```
> convert( 9, binary );                    ⟶           1001

> convert( [$(1..10)], `+` );              ⟶            55

> convert( 1.23456, fraction );            ⟶           3858
                                                       ----
                                                       3125

> convert( (x^2+1)/x, parfrac,x);          ⟶         x + 1/x
```

SEE ALSO: type, operators[binary]

2.1.35   convert/`+` – convert to sum
convert/`*` – convert to product

Calling Sequence:
>   convert(*expr*, `+`)
>   convert(*expr*, `*`)

Synopsis:
- The "convert to sum" call sums all the operands of *expr*.

- The "convert to product" call multiplies all the operands of *expr*.

Examples:
```
> convert(sin(x)*(y-x),`+`);               ⟶       sin(x) + y - x

> convert(sin(x)+y-x,`*`);                 ⟶        - sin(x) y x

> convert(f(a,b,c),`+`);                   ⟶         a + b + c
```

2.1.36 convert/array – convert to array

Calling Sequence:

    convert(*object*, array, *bounds*, *indexfcn*)

Parameters:

object	–	a table, list or array
bounds	–	(optional) dimensions of the new array
indexfcn	–	(optional) indexing function of new array

Synopsis:

- The convert/array function converts a table, list, or array to an array. The bounds or indexing function can be specified as additional parameters.

Examples:

```
> convert([[1,2],[3,4]],array,0..1,-1..0)
  ;
```
$\longrightarrow$
```
array(0 .. 1,-1 .. 0,, [
    (0, -1) = 1
    (0, 0) = 2
    (1, -1) = 3
    (1, 0) = 4
])
```

```
> t := table([(1,1)=1,(1,2)=3,(2,2)=4]):
  convert(t,array,symmetric);
```
$\longrightarrow$
```
          [ 1   3 ]
          [       ]
          [ 3   4 ]
```

```
> convert(",array,3..4,3..4);
```
$\longrightarrow$
```
array(3 .. 4,3 .. 4,, [
    (3, 3) = 1
    (3, 4) = 3
    (4, 3) = 3
    (4, 4) = 4
])
```

SEE ALSO: array

2.1.37 convert/base – convert between bases

Calling Sequence:

    convert(*n*, base, *beta*)
    convert(*l*, base, *alpha*, *beta*)

Parameters:

n	–	a decimal number
l	–	a list of digits
alpha	–	(optional) base of input

    beta   –   base of output

Synopsis:

- In the first case, `convert/base` converts the base 10 number n to base *beta*.

- In the second case, it converts the list of digits l in base *alpha* to base `beta`. The list of digits is interpreted as the number

$$\texttt{sum(l[k]*alpha\^{}(k-1), k=1..nops(l))}$$

- It is assumed that $0 \leq l[k] < alpha$ and that l is positive.

- The number is returned as a list of digits in base *beta*.

Examples:		
`> convert(17,base,3);`	$\longrightarrow$	`[2, 2, 1]`
`> convert([2,2,1],base,3,10);`	$\longrightarrow$	`[7, 1]`
`> convert(1000,base,60);`	$\longrightarrow$	`[40, 16]`

SEE ALSO: `convert[binary]`, `convert[octal]`, `convert[hex]`, `convert[decimal]`

2.1.38   convert/binary – convert to binary form

Calling Sequence:
    `convert(n, binary)`

Parameters:
    n – a decimal number

Synopsis:

- The function "convert/binary" converts a decimal number n to a binary number. The number may be either positive or negative, and may be either an integer or a floating point number. In the case of a floating point number, an optional third argument determines the total number of digits of precision in the answer (the default being Digits).

- The binary number is returned as a base 10 number consisting of the digits 1 and 0 only.

Examples:		
`> convert(123,binary);`	$\longrightarrow$	`1111011`
`> convert(-5,binary);`	$\longrightarrow$	`-101`

SEE ALSO: `convert[base]`, `convert[decimal]`, `convert[hex]`, `convert[octal]`

2.1.39   convert/confrac – convert to continued-fraction form

Calling Sequence:

```
convert(expr, confrac )
convert(expr, confrac, maxit)
convert(expr, confrac, 'cvgts' )
convert(expr, confrac, maxit, 'cvgts')
convert(expr, confrac, var)
convert(expr, confrac, var, order)
```

Synopsis:

- Convert to confrac converts a number, series, rational function, or other algebraic expression to a continued-fraction approximation.

- If *expr* is numeric then **maxit** (optional) is the maximum number of partial quotients to be computed, and **cvgts** (optional) will be assigned a list of the convergents. A list of the partial quotients is returned as the function value.

- If *expr* is a series and no additional arguments are specified, a continued-fraction approximation (to the order of the series) is computed. It is equivalent to either an (n,n) or (n,n-1) Pade approximant (depending on the parity of the order). By specifying `subdiagonal` as an optional third argument, the continued-fraction computed will be equivalent to a (n,n) or (n-1,n) Pade approximant.

- If *expr* is a **ratpoly** (quotient of polynomials) in x, the calling sequence is convert(*expr*, confrac, x). The rational form is converted into its associated continued-fraction form as required for efficient evaluation of numerical subroutines.

- If *expr* is any other algebraic expression, the third argument specifies a variable and (optionally) the fourth argument specifies order. The series function is applied to the arguments to obtain a series and then case series applies.

- Otherwise, `convert/confrac` is applied to each component of a non-algebraic structure.

Examples:

```
> convert(2.3,confrac);                    ⟶              [2, 3, 3]
> convert(21/13,confrac,cvgts);            ⟶          [1, 1, 1, 1, 1, 2]
> cvgts;                                   ⟶
                                                                    21
                                                   [1, 2, 3/2, 5/3, 8/5, ----]
                                                                    13
> convert(exp(x),confrac,x);               ⟶
                                                               x
                                               1 + -------------------------
                                                                 x
                                                   1 + ---------------------
                                                                 x
                                                       - 2 + ---------------
                                                                 x
                                                           - 3 + ---------
                                                                 2 + 1/5 x
```

```
> r := (1+1/2*x+1/12*x^2) / (1-1/2*x+1/12*
  x^2):
  convert(r,confrac,x);
```
$$\longrightarrow \qquad 1 + \cfrac{12}{x - 6 + \cfrac{12}{x}}$$

SEE ALSO: `convert[ratpoly]`

2.1.40   convert/D – convert expressions involving derivatives to use the D operator

Calling Sequence:

convert(*expr*, D)

Synopsis:

- This function converts expressions involving derivatives, in the diff notation, to the D operator notation.

Examples:

```
> f := diff(y(x), x$2):
  convert(f, D);
```
$$\longrightarrow \qquad D^{(2)}(y)(x)$$

SEE ALSO: `convert[diff]`

2.1.41   convert/decimal – convert to base 10

Calling Sequence:

convert(*n*, decimal, binary)
convert(*n*, decimal, octal)
convert(*string*, decimal, hex)

Parameters:

n   –   an integer
string   –   a hexadecimal number

Synopsis:

- The function `convert/decimal` converts a binary, octal, or hexadecimal number to a decimal number.

- The binary or octal number is given as an integer; the hexadecimal number is given as a string.

Examples:

```
> convert(101,decimal,binary);
```
$$\longrightarrow \qquad 5$$

```
> convert(23,decimal,octal);                    ⟶            19
> convert(`1A`,decimal,hex);                     ⟶            26
```

SEE ALSO: `convert[binary]`, `convert[octal]`, `convert[hex]`, `convert[base]`

2.1.42 convert/degrees – convert radians to degrees

Calling Sequence:

    convert(number, degrees)

Synopsis:

- convert/degrees converts an expression from radians to degrees by multiplying it by 180/Pi.

Examples:

```
> convert(Pi/2,degrees);                         ⟶        90 degrees
```

SEE ALSO: `convert[radians]`

2.1.43 convert/diff – convert an expression of the form D(f)(x) to diff(f(x),x)

Calling Sequence:

    convert(*expr*, diff, x);

Parameters:

    x – a variable

Synopsis:

- This function will replace, where possible, the use of the D operator by the use of the the diff function.

Examples:

```
> f := D(y)(x) - a*D(z)(x):
  convert(f, diff, x);
```

$$\left(\frac{d}{dx} y(x) \right) - a \left(\frac{d}{dx} z(x) \right)$$

SEE ALSO: `convert[D]`

2.1.44 convert/double – convert a double precision floating point number from one format to another

Calling Sequence:

    convert(*expr*, double, to1)
    convert(*expr*, double, to1, from1)

Parameters:

> *expr* – a floating point number or hex string
>
> *to1* – format to which the string or float is to be converted
>
> *from1* – (optional) format of the existing string

Synopsis:

- `convert/double` converts a double precision floating point number from one format to another. We support converting between Maple float and IBM, MIPS or VAX "hex" formats.

- If there are three arguments then *to1* must be either IBM, MIPS, or VAX, and *expr* must be a Maple floating point number.

- If there are four arguments then `to1` must be Maple and `from1` must be either IBM, MIPS, or VAX. *expr* must then be a valid hex string in the given format.

Examples:

```
> a := 1.2345:
  b := convert(a,double,mips);              ⟶        b := 3FF3C083126E978D
> convert(b,double,maple,mips);             ⟶           1.234500000
```

SEE ALSO: `convert[hex]`

2.1.45   convert/eqnlist – convert to a list of equations

Calling Sequence:

```
    convert( table, eqnlist )
    convert( array, eqnlist )
    convert( list, eqnlist )
```

Synopsis:

- `convert/eqnlist` will convert a table, array, or list to a list of equations that define it. This will be of the form `[index=element,...]`.

- For a list, the index will be the sequence of operand (`op`) numbers required to get to the element. Any list element which is an equation is incorporated in the result as an `index=element` equation.

Examples:

```
> A := array([x,y,z]):
  convert(A,eqnlist);              ⟶          [1 = x, 2 = y, 3 = z]
> B := [2=x,1=y,z]:
  convert(B,eqnlist);              ⟶          [2 = x, 1 = y, 3 = z]
> C := array([[1,2],[3,4]]):
  convert(C,eqnlist);              ⟶     [(1, 1) = 1, (2, 1) = 3, (1, 2) = 2,
                                               (2, 2) = 4]
> D := [ [4,3=5],(1,2)=7 ]:
  convert(D,eqnlist);              ⟶     [(1, 1) = 4, (1, 3) = 5, (1, 2) = 7]
```

2.1.46   convert/equality – convert relations to equalities
         convert/lessthan – converts relations to `` `<` `` form
         convert/lessequal – converts relations to `` `≤` `` form

Calling Sequence:

    convert(*relation*, equality)
    convert(*relation*, lessthan)
    convert(*relation*, lessequal)

Synopsis:

- convert/equality converts a relation to an equality, by replacing the relation with =.

- convert/lessthan replaces the relation with $<$ and convert/lessequal replaces the relation with $\leq$.

- Note that the inequalities $>$ and $\geq$ are automatically converted to $<$ and $\leq$ by Maple.

Examples:

    > convert(x<y^2,equality); $\longrightarrow$ 2
                                                              x = y

    > convert(x<=sin(x),lessthan); $\longrightarrow$ x < sin(x)

    > convert(x=y,lessequal); $\longrightarrow$ x <= y

SEE ALSO: `simplex[equality]`

2.1.47   convert/exp – convert trig functions to exponentials

Calling Sequence:

    convert(*expr*, exp);

Synopsis:

- convert/exp converts all trigonometric functions appearing in the expression to their corresponding exponential form.

Examples:

    > convert(sin(x),exp); $\longrightarrow$ / | 1   \
                                               - 1/2 |exp(I x) - --------| I
                                                     \   exp(I x)/

    > convert(tanh(x),exp); $\longrightarrow$ 2
                                                   exp(x) - 1

                                                        2
                                                   exp(x) + 1

    > convert(exp(x^2)-2*sinh(x^2),exp); $\longrightarrow$ 1

                                                           2
                                                        exp(x)

SEE ALSO: `convert[expsincos]`, `convert[sincos]`, `convert[ln]`

2.1.48 convert/expln – convert elementary functions to exp and ln

Calling Sequence:

convert(*expr*, expln);

Synopsis:

- convert/expln converts all elementary functions to exp and ln

---

Examples:
```
> f := cos(x)*sin(x):
> convert(f, expln);
             /                  1    \ /                1    \
    - 1/2 |1/2 exp(I x) + ----------| |exp(I x) - --------| I
             \                2 exp(I x)/ \            exp(I x)/
```

SEE ALSO: convert[exp], convert[ln]

2.1.49 convert/expsincos – convert trig functions to exp, sin, cos

Calling Sequence:

convert(*expr*, expsincos);

Synopsis:

- convert/expsincos converts all trigonometric functions appearing in the expression to sin and cos, and converts all hyperbolic trigonometric functions to exponential form.

Examples:
```
> convert(cot(x),expsincos);            ⟶                cos(x)
                                                          ------
                                                          sin(x)

> convert(sinh(x),expsincos);           ⟶                        1
                                                    1/2 exp(x) - --------
                                                                 2 exp(x)
```

SEE ALSO: convert[exp], convert[sincos]

2.1.50 convert/factorial – convert GAMMAs and binomials to factorials

Calling Sequence:

convert(*expr*, factorial)
convert(*expr*, factorial, *indets*)

Synopsis:

- Convert GAMMAs, binomials and multinomial coefficients in an expression to factorials.

- If an indeterminate or set of indeterminates is specified, then only GAMMAs and binomials involving a specified indeterminate will be converted to the factorial function.

Examples:

```
> convert(GAMMA(x),factorial);                    ⟶                    x!
                                                                      ----
                                                                       x

> convert(binomial(m,3),factorial);              ⟶                    m!
                                                               1/6  --------
                                                                    (m - 3)!
```

SEE ALSO: `convert/GAMMA`

2.1.51   convert/float – convert to floating-point

Calling Sequence:

    convert(*expr*, float)

Synopsis:

- `convert/float` converts an expression to floating-point to the precision given by the global variable `Digits`.

- `convert/float` is just a call to evalf.

Examples:

```
> convert(1/2,float);                            ⟶                .5000000000
> convert(Pi,float);                             ⟶                3.141592654
> convert(sin(1),float);                         ⟶                .8414709848
```

SEE ALSO: `Digits`, `evalf`

2.1.52   convert/GAMMA – convert factorials and binomials to GAMMAs

Calling Sequence:

    convert(*expr*, GAMMA)
    convert(*expr*, GAMMA, *indets*)

Synopsis:

- Convert factorials, binomials and multinomial coefficients in an expression to the GAMMA function.

- If an indeterminate or set of indeterminates is specified, then only factorials and binomials involving a specified indeterminate will be converted to the GAMMA function.

Examples:

```
> convert(x!,GAMMA);                             ⟶              GAMMA(x + 1)

> convert(binomial(m,3),GAMMA);                  ⟶              GAMMA(m + 1)
                                                          1/6 ------------
                                                              GAMMA(m - 2)
```

> `convert(x!*y!*z!,GAMMA,{x,y});` $\longrightarrow$ `GAMMA(x + 1) GAMMA(y + 1) z!`

SEE ALSO: `convert/factorial`

2.1.53   convert/hex – convert to hexadecimal form

Calling Sequence:

   `convert( ` *number*`, hex )`

Synopsis:

- `convert/hex` converts a positive decimal integer to a hexadecimal number.

- The hex number is returned as a string consisting of the characters 0 through 9 and A through F.

Examples:

> `convert(123456,hex);` $\longrightarrow$ `1E240`

> `convert(100,hex);` $\longrightarrow$ `64`

SEE ALSO: `convert/base`, `convert/binary`, `convert/decimal`, `convert/octal`
            `convert/double`

2.1.54   convert/horner – convert a polynomial to Horner form

Calling Sequence:

   `convert( ` *poly*`, horner )`
   `convert( ` *poly*`, horner, ` *var* ` )`

Parameters:

   poly   –   polynomial
   var   –   (optional) variable or list/set of variables

Synopsis:

- `convert/horner` writes the polynomial *poly* in the variable(s) *var* in horner or "nested" form.

- If only two arguments are specified, it is equivalent to `convert(`*poly*`, horner, indets(`*poly*`))`

- If the third argument is a list or set of variables, the conversion is applied recursively to the coefficients of *poly* in the first variable. A list allows for control over the order of the conversions.

Examples:

> `convert(x^2+3*x+4,horner,x);` $\longrightarrow$ $4 + (3 + x)\, x$

> `poly := y^2*x^2 + 2*y^2*x + 2*y*x^2 + 4*`
  `y*x + x^2 + 2*x:`
  `convert(poly,horner,x);` $\longrightarrow$ $(2\, y^2 + 4\, y + 2 + (y^2 + 2\, y + 1)\, x)\, x$

```
> convert(poly,horner,[x,y]);
```
$\longrightarrow$ (2 + (4 + 2 y) y + (1 + (2 + y) y) x) x

SEE ALSO: convert

2.1.55   convert/hostfile – convert Maple filename to host filename

Calling Sequence:

convert(*filename*, hostfile)

Synopsis:

- Convert to hostfile converts the Maple filename to the actual filename used on the host system.

- The filename returned will differ from system to system.

Examples:

```
> convert(`[bwchar]/subdir/foo.m`,hostfile)
  ;
```
$\longrightarrow$ [bwchar]/subdir/foo.m
```
> convert(`bwchar/subdir/foo.m`,hostfile)
  ;
```
$\longrightarrow$ bwchar/subdir/foo.m
```
> convert(`foo.m`,hostfile);
```
$\longrightarrow$ foo.m

SEE ALSO: read, save

2.1.56   convert/hypergeom – convert summations to hypergeometrics

Calling Sequence:

convert(*expr*, hypergeom)

Synopsis:

- Converts any call to sum or Sum in an expression into a hypergeometric function, if possible.

- No attempt is made to ensure that the resulting hypergeometric function is convergent or terminates.

Examples:

```
> Sum( 1/k/binomial(k+n,k), k=1..infinity):
> convert( ", hypergeom);
```

$$\frac{\text{hypergeom}([1, 1], [2 + n], 1)}{1 + n}$$

SEE ALSO: hypergeom, simplify[hypergeom]

2.1.57   convert/list – convert to a list

Calling Sequence:

```
convert( table, list )
convert( vector, list )
convert( expression, list )
convert( operand, list, list )
convert( operand, list, `=` )
```

Synopsis:

- `convert/list` will take a table or one dimensional array and make a list of the elements. It will make a list of the operands of an expression.

- With the third operand `list`, `convert/list` will make a list of lists. See convert[listlist] for more information.

- With the third operand `=`, `convert/list` will make an equation list. See convert[eqnlist] for more information.

Examples:
```
> A := array([x,y,z]):
  convert(A,list);                         ⟶              [x, y, z]

> convert(r+s-t,list);                     ⟶              [r, s, - t]

> B := table([(3)=123,(7)=456]):
  convert(B,list);                         ⟶              [123, 456]
```

2.1.58   convert/listlist – convert to a list of lists

Calling Sequence:

```
convert( array, listlist )
convert( list, listlist )
```

Synopsis:

- `convert/listlist` will convert an array or list into a list of lists.

- The list of lists formed from an array is equivalent to the list that would be passed to the `array` function to recreate the original array.

- The list of lists is formed from a list by creating an equation list from the original list, using this list to create an array, and then returning the list of lists that forms the array. Basically, that means the list is viewed as equations defining an array, and then `convert/listlist` is applied to that array.

Examples:
```
> A := array([[1,2],[3,4]]):
  convert(A,listlist);                     ⟶              [[1, 2], [3, 4]]
```

```
> B := [4=12,2=7,1=8,3=6]:
  convert(B,listlist);                        ⟶              [8, 7, 6, 12]
> C := [ [1=6,2=4],(2,1)=9,(2,2)=6 ]:
  convert(C,listlist);                        ⟶              [[6, 4], [9, 6]]
```

SEE ALSO: `array`, `convert[eqnlist]`

2.1.59   convert/ln – convert arctrig functions to logarithms

Calling Sequence:

> `convert( expr, ln );`

Synopsis:

- `convert/ln` converts all inverse trigonometric functions appearing in the expression to their corresponding logarithmic form.

Examples:

```
> convert(arcsin(x),ln);                     ⟶                         2 1/2
                                                         - I ln((1 - x )    + I x)

> convert(arctanh(x),ln);                    ⟶                      1 + x
                                                           1/2 ln(-----)
                                                                   1 - x
```

SEE ALSO: `convert[exp]`

2.1.60   convert/matrix – convert an array or a list of lists to a matrix

Calling Sequence:

> `convert( A, matrix );`

Synopsis:

- `convert/matrix` converts a two dimensional array or a list of lists to a matrix.

- *A* must be either a two dimensional array or a list of lists

Examples:

```
> A := array(0..1,-1..0,[[1,2],[3,4]]):
  convert(A, matrix);                        ⟶              [ 1  2 ]
                                                            [      ]
                                                            [ 3  4 ]
```

SEE ALSO: `convert[array]`, `matrix`, `array`

2.1.61   convert/metric – convert to metric units

Calling Sequence:

    convert(*expr*, metric, imp/US);

Parameters:

    expr   –   any expression (dependent on "form")

    metric –   an unevaluated name

    imp/US –   unevaluated names

Synopsis:

- `convert(` *expr*, metric `)`; will convert imperial measurements to metric measurements.

- If a third parameter, `imp` or `US`, appears the measures specified will be taken to be imperial or U.S., respectively.

- The following units are known to the `convert` function when it is called to convert from a non-metric expression to a metric one.

acre	acres	bu	bushel	bushels	chain
chains	cm	cord	cords	feet	foot
ft	furlong	furlongs	gal	gallon	gallons
Gals	gill	gills	gr	hr	ins
inch	inches	kg	km	Lb	lbs
light_year	light_years	mi	Mile	miles	MPG
MPH	ounce	Ounces	oz	Ozs	pint
pints	pole	poles	pound	pounds	quart
quarts	yard	yards	yd	yds	

Examples:

`> convert(4*ft + 3*inches, metric);`	$\longrightarrow$	1.295400000 m
`> convert(yards, metric);`	$\longrightarrow$	.9144 m
`> convert(5*gallons, metric, imp);`	$\longrightarrow$	22.7298155 lt
`> convert(5*gallons, metric, US);`	$\longrightarrow$	18.92705892 lt

2.1.62   convert/mod2 – convert expression to mod 2 form

Calling Sequence:

    convert(*expr*, mod2)

Synopsis:

- `convert/mod2` reduces an expression modulo 2.

- The expression may contain the following Boolean operators:

  and   not   or

- These are converted to their equivalent modulo 2 representation.

Examples:

`> convert( 7*x^3*y+5, mod2 );`	$\longrightarrow$	x y + 1
`> convert( x^2*y^2+x*y, mod2 );`	$\longrightarrow$	0
`> convert( x and (not y), mod2 );`	$\longrightarrow$	x (1 + y)
`> convert( x or y, mod2 );`	$\longrightarrow$	x + y + x y

SEE ALSO: `msolve`, `logic`, `logic[convert]`

2.1.63   convert/multiset – convert to a multiset

Calling Sequence:

  `convert(f,multiset)`

Parameters:

  f – a table, list, or algebraic expression

Synopsis:

- This utility function converts f to a `multiset`. A multiset is represented in the form `[ [e[1],m[1]], ..., [e[n],m[n]] ]`; it is a list of pairs where each `e[i]` is a value (an expression), and `m[i]` is its multiplicity (an integer).

- If f is a table, each index is interpreted as a value and the associated entry (which should be an integer) is interpreted as its multiplicity.

- If f is an algebraic expression, each factor is interpreted as a value, and the exponent to which this factor is raised (which should be an integer) is interpreted as its multiplicity.

- If f is a list, the number of instances of each entry in the list is interpreted as its multiplicity.

Examples:

`> t := table( [x=1,y=3,z=2] ):` `  convert( t, multiset );`	$\longrightarrow$	`[[y, 3], [x, 1], [z, 2]]`
`> convert( x*y^3*z^2, multiset );`	$\longrightarrow$	`[[x, 1], [y, 3], [z, 2]]`
`> convert([x,y,z,y,z,y],multiset);`	$\longrightarrow$	`[[y, 3], [x, 1], [z, 2]]`

SEE ALSO: `convert[set]`

2.1.64   convert/string – convert an expression to a string (name)

convert/name – a synonym for convert/string

Calling Sequence:

    convert(*expr*, string)
    convert(*expr*, name)

Synopsis:

- convert(expr,string) converts the expression *expr* to a string. In this context, **name** is simply a synonym for **string**.

Examples:

> convert(sin(x^2),string); $\longrightarrow$ sin((x)^(2))

> convert(a+b+c*d,name); $\longrightarrow$ (a)+(b)+((c)*(d))

SEE ALSO: name, type[name]

2.1.65   convert/octal – convert to octal form

Calling Sequence:

    convert(*number*, octal)
    convert(*number*, octal, *precision*)

Synopsis:

- convert/octal converts a decimal number to an octal number. The number may be either positive or negative, and may be either an integer or a floating point number. In the case of a floating point number an optional third argument determines the total number of digits of precision in the answer (the default being Digits).

- The octal number is returned as a base 10 number consisting of the digits 0 through 7.

Examples:

> convert(12345,octal); $\longrightarrow$ 30071

> convert(100,octal); $\longrightarrow$ 144

SEE ALSO: convert[base], convert[binary], convert[decimal], convert[hex]

2.1.66   convert/parfrac – convert to partial fraction form

Calling Sequence:

    convert(*f*, parfrac, *x*)
    convert(*f*, parfrac, *x*, *factored*)

Parameters:

f	–	rational function
x	–	main variable name
factored	–	(optional) `true` or `false`

Synopsis:

- Convert to `parfrac` performs a partial fraction decomposition of the rational function *f* in the variable *x*.

- The last argument `factored` is optional. If specified, its value should be `true` when it is not necessary to apply normal to *f*, and the denominator of *f* is already in the desired factored form. It should be `false` otherwise. The default is `false`.

- This allows for various types of partial fraction decomposition:

default	Maple´s factor routine will be applied to the denominator
square-free	apply square-free factorization to the denominator before invoking `convert/parfrac` with *factored* = `true`
fully factored	factor the denominator into linear factors over its splitting field before invoking `convert/parfrac` with *factored* = `true`.

Examples:

```
> convert(x^2/(x+2), parfrac, x);
```
$$\longrightarrow \quad x - 2 + \frac{4}{x+2}$$

```
> convert(x/(x^2-3*x+2), parfrac, x);
```
$$\longrightarrow \quad -\frac{1}{x-1} + \frac{2}{x-2}$$

```
> convert(x/(x-b)^2, parfrac, x);
```
$$\longrightarrow \quad \frac{b}{(x-b)^2} + \frac{1}{x-b}$$

```
> poly := x^5-2*x^4-2*x^3+4*x^2+x-2:
  f := 36 / convert(poly, sqrfree, x);
```
$$\longrightarrow \quad f := \frac{36}{(x-2)^2 (x^2-1)^2}$$

```
> convert(f, parfrac, x, true);
```
$$\longrightarrow \quad \frac{4}{x-2} - 4\,\frac{x+2}{x^2-1} - 12\,\frac{x+2}{(x^2-1)^2}$$

```
> convert(f, parfrac, x);
```
$$\longrightarrow \quad \frac{4}{x-2} - \frac{9}{(x-1)^2} - \frac{3}{(x+1)^2} - \frac{4}{x+1}$$

SEE ALSO: `convert`, `int`, `Int`

2.1.67 convert/polar – convert to polar form

Calling Sequence:

    convert(*expr*, polar)

Parameters:

  expr – a (complex) expression

Synopsis:

- `convert/polar` converts an expression to its representation in polar coordinates.

- The expression is represented as `polar(r,theta)` where `r` is the modulus and `theta` is the argument of the complex value of the expression.

Examples:

`> convert(3+4*I,polar);`	$\longrightarrow$	`polar(5, arctan(4/3))`
`> convert(3*I,polar);`	$\longrightarrow$	`polar(3, 1/2 Pi)`

SEE ALSO: `evalc`

2.1.68 convert/polynom – convert a series to polynomial form

Calling Sequence:

    convert(*series*, polynom)

Synopsis:

- Convert to polynom converts a Taylor series to a polynomial. If *series* is not a Taylor series then the conversion to sum-of-products form takes place but the result is not of type `polynom`.

Examples:

`> s := series(sin(x),x,5);` $\longrightarrow$
$$s := x - 1/6\ x^3 + O(x^5)$$

`> type(s,polynom);` $\longrightarrow$ `false`

`> p := convert(s,polynom);` $\longrightarrow$
$$p := x - 1/6\ x^3$$

`> type(p,polynom(anything,x));` $\longrightarrow$ `true`

`> t := series(exp(x)/x,x,3);` $\longrightarrow$
$$t := x^{-1} + 1 + 1/2\ x + O(x^2)$$

`> q := convert(t,polynom);` $\longrightarrow$
$$q := 1/x + 1 + 1/2\ x$$

`> type(q,polynom(anything,x));` $\longrightarrow$ `false`

`> type(q,`+`);` $\longrightarrow$ `true`

SEE ALSO: `convert[series]`

2.1.69 convert/radians – convert degrees to radians

Calling Sequence:

    convert(*number*, radians)

Synopsis:

- convert/radians converts an expression from degrees to radians by multiplying it by 180/Pi.

- The expression must be of degree 1 in the indeterminate degrees.

Examples:

> convert(90*degrees,radians); $\longrightarrow$ 1/2 Pi

SEE ALSO: convert[degrees]

2.1.70 convert/radical – convert RootOf to radicals and I

Calling Sequence:

    convert(*expr*, radical)

Synopsis:

- Change occurrences of RootOf(..) to the corresponding radical notation when possible. The "radical notation" for a^(1/2), when sign(a) = -1, will be (sign(a)*a)^(1/2) * I.

- In particular, RootOf(_Z^2+1) is replaced by I.

- In general, if the RootOf argument is a polynomial in _Z with only one term in _Z, say _Z^n, then the RootOf expression is converted to n-th root notation (modulo the introduction of I as noted above).

- For conversion in the opposite direction use convert(*expr*,RootOf).

Examples:

```
> convert( RootOf(_Z^2 + 1), radical );        ⟶              I
> convert( RootOf(_Z^2 - 2), radical );        ⟶              1/2
                                                              2
> convert( RootOf(_Z^2 + _Z + 1), radical)
  ;                                            ⟶               2
                                                    RootOf(_Z  + _Z + 1)
> eval( subs(RootOf=solve, ") );               ⟶           1/2              1/2
                                                  - 1/2 + 1/2 I 3   , - 1/2 - 1/2 I 3
```

SEE ALSO: RootOf, convert[RootOf], solve

2.1.71   convert/rational – convert float to an approximate rational

Calling Sequence:

    convert(*float*, rational)
    convert(*float*, fraction)
    convert(*float*, rational, *digits*)
    convert(*float*, fraction, *digits*)

Synopsis:

- convert/rational converts a floating-point number to an approximate rational number. The meaning of the type **fraction** is identical with **rational** in this context.

- The accuracy of the conversion will depend on the value of the global variable Digits, or the value of *digits* if specified as an integer.

- If the third argument *digits* is the name ´**exact**´ then an exact conversion of *float* to a rational will be performed; thus **Float(f,e)** becomes simply **f*10^e**. Note that exact conversion executes much more quickly than the more sophisticated conversion.

Examples:

> convert(0.125,rational);	$\longrightarrow$	1/8
> convert(2.345,rational);	$\longrightarrow$	469

		200
> convert(0.3333333333,rational);	$\longrightarrow$	1/3
> convert(0.3333333333,rational,exact);	$\longrightarrow$	3333333333

		10000000000
> convert(evalf(Pi),rational);	$\longrightarrow$	104348

		33215
> convert(evalf(Pi),rational,3);	$\longrightarrow$	22/7
> convert(evalf(Pi),rational,exact);	$\longrightarrow$	1570796327

		500000000

SEE ALSO: Digits

2.1.72   convert/ratpoly – convert series to a rational polynomial

Calling Sequence:

    convert(*series*, ratpoly);
    convert(*series*, ratpoly, *numdeg*, *dendeg*);

Synopsis:

- Converts a series to a rational polynomial (rational function). If the first argument is a Taylor or Laurent series then the result is a Pade approximation, and if it is a Chebyshev series then the result is a Chebyshev-Pade approximation.

- The first argument must be either of type `'laurent'` (hence a Laurent series) or else a Chebyshev series (represented as a sum-of products in terms of the basis functions `T(k,x)` for integers `k`).

- If the third and fourth arguments appear, they must be integers specifying the desired degrees of numerator and denominator, respectively. (Note: The actual degrees appearing in the approximant may be less than specified if there exists no approximant of the specified degrees.)

- If the third and fourth arguments are not specified then the degrees of numerator and denominator are chosen to be m and n, respectively, such that `m+n+1 = order( `*series*` )` and either `m=n` or `m=n+1`. (The order of a Chebyshev series is defined to be `d+1` where `d` is the highest-degree term which appears.)

- For the Pade case, two different algorithms are implemented. For the pure univariate case where the coefficients contain no indeterminates, a "fast" algorithm due to Cabay and Choi is used. Otherwise, an algorithm due to Geddes based on fraction-free symmetric Gaussian elimination is used.

Examples:

```
> series(exp(x), x);
```

$$1 + x + 1/2\ x^2 + 1/6\ x^3 + 1/24\ x^4 + 1/120\ x^5 + O(x^6)$$

```
> convert(", ratpoly);
```

$$\frac{1 + 3/5\ x + 3/20\ x^2 + 1/60\ x^3}{1 - 2/5\ x + 1/20\ x^2}$$

```
> Digits := 5:
> chebyshev(cos(x), x);
```

$$.76520\ T(0, x) - .22981\ T(2, x) + .0049533\ T(4, x) - .000041877\ T(6, x)$$

```
> convert(", ratpoly, 2,2);
```

$$\frac{.76025\ T(0, x) - .19673\ T(2, x)}{T(0, x) + .043088\ T(2, x)}$$

2.1.73 convert/RootOf – convert radicals and I to RootOf notation

Calling Sequence:

convert(*expr*, RootOf)

Synopsis:
- Change all occurrences of the algebraic constant I, and all occurrences of radicals (may be algebraic constants or functions), to `RootOf` notation.

- In particular, I is replaced by `RootOf(_Z^2+1)`.

- If the input expression is an unnamed table then the conversion routine is mapped onto the elements of the table.

- To convert `RootOf` notation to I and radicals (where possible) use `convert(`*expr*`, radical)`.

Examples:

```
> convert( (1 + 2^(1/2)) * I, RootOf );
```
$$\longrightarrow \quad (1 + RootOf(\_Z^2 - 2))\,RootOf(\_Z^2 + 1)$$

```
> convert( 234543^(1/5), RootOf);
```
$$\longrightarrow \quad RootOf(\_Z^5 - 234543)$$

```
> convert( (x^2 + 2*x + 1)^(1/2), RootOf )
  ;
```
$$\longrightarrow \quad RootOf(\_Z^2 - x^2 - 2\,x - 1)$$

```
> convert( (x^3 - 6*x^2 - 6*x - 7)^(1/2),
  RootOf );
```
$$\longrightarrow \quad RootOf(\_Z^2 - x^3 + 6\,x^2 + 6\,x + 7)$$

```
> convert( ", radical );
```
$$\longrightarrow \quad (x^3 - 6\,x^2 - 6\,x - 7)^{1/2}$$

SEE ALSO: `RootOf`, `allvalues`, `convert[radical]`

2.1.74   convert/series – convert to series

Calling Sequence:
```
    convert( expr, series )
    convert( expr, series, var )
```

Parameters:
    expr   –   a series or polynomial
    var   –   the series variable

Synopsis:
- `convert/series` converts a polynomial to a series.

- To convert an arbitrary expression to a series, use the series function.

- The parameter *var* is only necessary if *expr* is multivariate.

Examples:
```
> p := x^2+1:
  s := convert(p,series);
```
$$\longrightarrow \quad s := 1 + x^2$$

```
> type(p,series);                    ⟶              false
> type(s,series);                    ⟶              true
```

SEE ALSO: `series`, `convert[polynom]`

2.1.75   convert/set – convert to a set

Calling Sequences:

    convert(*table*, set)

    convert(*array*, set)

    convert(*expr*, set)

Synopsis:

- convert/set will make a set of the entries of a table or array.

- convert/set will make a set of the operands of an expression.

Examples:

```
> A := array(1..2,1..2,[[1,2],[3,4]]):
  convert(A,set);                    ⟶              {1, 2, 3, 4}
> convert([a,b,c,d],set);            ⟶              {a, b, d, c}
> convert(x+y-z,set);                ⟶              {x, y, - z}
```

2.1.76   convert/sincos – convert trig functions to sin, cos, sinh, cosh

Calling Sequence:

    convert(*expr*, sincos);

Synopsis:

- convert/expsin converts all trigonometric functions appearing in the expression to sin and cos, and converts all hyperbolic trigonometric functions to sinh and cosh.

Examples:

```
> convert(cot(x),sincos);           ⟶              cos(x)
                                                    ------
                                                    sin(x)

> convert(tanh(x),sincos);          ⟶              sinh(x)
                                                    -------
                                                    cosh(x)
```

SEE ALSO: `convert[exp]`, `convert[expsincos]`

2.1.77   convert/sqrfree – convert to square-free form

Calling Sequence:

    convert(a, sqrfree, x)

Parameters:

   a –   polynomial

   x –   optional variable name

Synopsis:

- Convert to `sqrfree` performs a square-free factorization of the polynomial a. It is most often called with the third argument x specified. In this case, `content(`a`, `x`)` is first removed (thus making a primitive) before doing the square-free factorization. The resulting factorization is of the form:

```
content(a,x) * p[1]^1 * p[2]^2 * ... * p[k]^k
where  gcd(p[i],p[j]) = 1  for 1 <= i < j <= k
```

- If called with only two arguments, a `complete` square-free factorization will be done. A variable is first chosen for which the above computation is done. Next, the content is made square-free by recursive application of `convert` to `sqrfree`. This is repeated until there are no variables left in the content.

Examples:
```
> convert(x^2+4*x+4,sqrfree,x);                ⟶                          2
                                                                   (x + 2)

> poly := y^2*x^3+2*y^2*x^2+y^2*x+2*y*x^3+
  4*y*x^2+2*y*x:
  convert(poly,sqrfree,x);                      ⟶          2                  2
                                                        (y  + 2 y) x (x + 1)

> convert(poly,sqrfree);                        ⟶          2                  2
                                                        (y  + 2 y) x (x + 1)
```

SEE ALSO: `content`, `gcd`

2.1.78   convert/tan – convert trig function to tan

Calling Sequence:

    convert($expr$, tan);

Synopsis:

- `convert/tan` converts all trigonometric functions appearing in the expression to tan.

Examples:
```
> f := sin(x)*cos(x);                           ⟶              f := sin(x) cos(x)
```

```
> convert(f, tan);
```
$$\longrightarrow \qquad 2\ \frac{\tan(1/2\ x)\ (1 - \tan(1/2\ x)^2)}{(1 + \tan(1/2\ x)^2)^2}$$

SEE ALSO: convert[sincos], convert[expsincos]

2.1.79   convert/trig – convert all exponentials to trigonometric and hyperbolic trigonometric functions

Calling Sequence:

> convert(*expr*, trig);

Synopsis:

- convert/trig converts all exponential functions appearing in the expression to sin, cos, sinh and cosh via Euler's formula. Existing trigonometrics will be combined into a form given by `combine/trig`.

Examples:

```
> convert(1/2*exp(x) + 1/2*exp(-x),trig);   ⟶          cosh(x)

> convert(1/4*exp(x)^2-1/4/exp(x)^2,trig)
  ;                                          ⟶        1/2 sinh(2 x)

> convert(cos(x)^2*sin(x),trig);             ⟶   1/4 sin(3 x) + 1/4 sin(x)
```

SEE ALSO: combine[trig]

2.1.80   convert/vector – convert a list or an array to a Maple vector

Calling Sequence:

> convert(L, vector);

Synopsis:

- convert/vector converts an array or a list to a Maple vector

Examples:

```
> A := array(0..1, -1..0, [[1,2],[3,4]]):
  convert(A, vector);                        ⟶        [ 1, 2, 3, 4 ]

> convert([1,2], vector);                    ⟶           [ 1, 2 ]
```

SEE ALSO: convert[matrix], convert[array]

2.1.81   copy – create a duplicate array or table

Calling Sequence:

    copy(a);

Parameters:

    a – any expression

Synopsis:

- The purpose of the copy function is to create a duplicate table (or array) which can be altered without changing the original table (or array). If **a** is not a table (or array), **a** is returned.

- This is necessary since the statements **s := table(); t := s;** leave both names **s** and **t** evaluating to the same table structure. Hence, unlike other Maple data structures, assignments made via one of the names affect the values associated with the other name as well.

- Note that copy is not recursive. This means that if **a** is a table of tables, the table data structure for **a** is copied but the table structures for the entries of **a** are not copied.

Examples:

`> s[1] := x;`	$\longrightarrow$	`s[1] := x`
`> t := s;`	$\longrightarrow$	`t := s`
`> t[1] := y;`	$\longrightarrow$	`t[1] := y`
`> s[1];`	$\longrightarrow$	`y`
`> u := copy(s);`	$\longrightarrow$	`u := table([`
		`        1 = y`
		`    ])`
`> u[1] := z;`	$\longrightarrow$	`u[1] := z`
`> s[1];`	$\longrightarrow$	`y`

2.1.82   D – Differential operator

Calling Sequence:

    D(*f*)

    D[*i*] (*f*)

Parameters:

    f – expression which can be applied as a function

    i – positive integer or expression or sequence of such

Synopsis:

- Let **f** be a function of one argument. The call D(**f**) computes the derivative of the function **f**. The derivative is a function of one argument such that D(**f**)(**x**) = diff(**f**(**x**), **x**). That is, D(**f**) = unapply(diff(**f**(**x**), **x**), **x**). Thus D is a mapping from unary functions to unary functions.

- Let f be a function of n arguments. The call D[i](f) computes the partial derivative of f with respect to its ith argument. More generally, D[i,j](f) is equivalent to D[i](D[j](f)), and D[](f) = f. Thus D[i] is a mapping from n-ary functions to n-ary functions.

- The argument f must be an algebraic expression which can be treated as a function. It may contain constants, known function names (such as exp, sin), unknown function names (such as f, g), arrow operators (such as x -> x^2), and the arithmetic and functional operators. For example, f+g, f*g, and f@g are valid since (f+g)(x) = f(x)+g(x), (f*g)(x) = f(x)*g(x), and (f@g)(x) = f(g(x)) where f@g denotes functional composition.

- The usual rules for differentiation hold. In addition, it is assumed that partial derivatives commute. Hence D(f+g) = D(f) + D(g), D(f*g) = D(g*f) = D(f)*g + D(g)*f, D(f@g) = D(f)@g * D(g) and so forth.

Examples:

> D(sin);	$\longrightarrow$	cos
> D(exp+ln+Pi+tan);	$\longrightarrow$	$\exp + (a \to 1/a) + 1 + \tan^2$
> D(f);	$\longrightarrow$	D(f)
> D(f@g);	$\longrightarrow$	D(f)@g D(g)
> D[1](f);	$\longrightarrow$	D[1](f)
> D[i,j](f);	$\longrightarrow$	D[i, j](f)
> D[2,1](f)-D[1,2](f);	$\longrightarrow$	0
> f := x -> x^2;	$\longrightarrow$	$f := x \to x^2$
> D(f);	$\longrightarrow$	x -> 2 x
> f := (x,y) -> exp(x*y);	$\longrightarrow$	f := (x,y) -> exp(x y)
> D[](f);	$\longrightarrow$	f
> D[1](f);	$\longrightarrow$	(x,y) -> y exp(x y)
> D[2](f);	$\longrightarrow$	(x,y) -> x exp(x y)
> D[i](D[j,i](f));	$\longrightarrow$	D[i, i, j](f)
> D[1](D[2,1](f));	$\longrightarrow$	$(x,y) \to 2\ y\ \exp(x\ y) + y^2\ x\ \exp(x\ y)$

SEE ALSO: diff, `@`, `@@`, unapply, operators, taylor

2.1.83   define – define characteristics of an operator name

Calling Sequence:

```
define(aa(oper))
define(oper, property1, property2, ...)
```

Parameters:

    oper   –   name of the operator being defined

    aa   –   name of an abstract algebraic object (currently: group and linear)

    property.i –   name of the property of an operator or equation defining a value
                    of the operator or an equation `f(args)=val` defining a special
                    value or a `forall` expression

Synopsis:

- `define` defines the evaluation and simplification rules of an operator. It will create the appropriate entries for such functions as `simplify`, `eval`, `expand`, `diff`, `series`, and `testeq` so that the properties of the operator are realized. All these actions are side-effects, since `define` returns NULL.

- Operators can be used in the infix notation (&-names):

        `a &+ b;`   `a &* b &* c;`   `&A x`

- or in the functional notation (any name):

        `Op(a,b);`   `Op(a);`   `Op(a,b,c)`

Examples:
```
> define(linear(`&L`));
> define(group(G, 0, invG));
> define(`&+`, associative, commutative, identity=0);
> define(Ls, unary, type=[algebraic], Ls(0)=0,
>    forall(x, Ls(exp(x))=Ls(x)+1, Ls(ln(x))=Ls(x)-1));
> define(F, forall(integer(x), F(x) = F(x-1)+F(x-2)), F(0)=0, F(1)=1);
```

See Also: `operators`, `neutral`, `procedures`, `define[linear]`, `define[group]`,
        `define[operator]`, `define[forall]`

2.1.84   define/forall – define a property for an operator

Calling Sequence:

    `define(`*F*`,forall(`*vars*`, `*F(args1)=res1*`, `*F(args2)=res2*`, ...)`

Parameters:

    vars   –   is a name or a typed name, or a list of either names or typed
                 names

    F   –   is the name of the operator being defined

    args.i –   are the arguments on which this operator will be used

    res.i   –   is the desired result of *F(args.i)*

Synopsis:
- The `forall()` expression is a valid argument of a define call, and it provides operational properties of the operator.

- This expression will define special relations which hold for certain arguments of the operator. The expression $F(args.i)=res.i$ will make the operator produce the expression $res.i$ every time it is given the arguments $args.i$. The parameters $args.i$ and $res.i$ can be any Maple expression. The names defined in $vars$ are used as matching arguments which can take any possible value. The substitutions are then done for all possible such values.

- The names defined in $vars$ should appear exactly once in $args.i$. The parameter $vars$ can be a name or a list of names, in which case these names will match any Maple expression. If the names are of the form `type(xxx)`, where `type` is a valid type-name and `xxx` is a name, then the name `xxx` will be matched only if it is of the given type.

Examples:
```
> define(F, forall(x, F(x^2)=F(x), F(1/x)=F(x)));
> define(Ls, forall([integer(y),x], Ls(x^y)=y*Ls(x)));
> define(G, forall(x, G(exp(x+a))=F(x-b)));
```

SEE ALSO: `match[property]`, `match`

2.1.85   define/group – define characteristics of a group operator

Calling Sequence:
> define(group(*OpName*, *Identity*, *Inverse*))

Parameters:

OpName	–	is the name of the binary operator defined in the group
Identity	–	is the identity symbol or value of the group
Inverse	–	is an unary operator which computes the inverse of the elements in the group

Synopsis:
- `define/group` sets up a definition of *OpName* which allows evaluation, simplification and expansion.

- This function also sets up a definition of the inverse.

- The properties of a group are:

  Closure
  Associativity
  Identity on left and right
  Inverse (unary operator)

Examples:
```
> define(group(`&+`, 0, `&-`));
  z &+ 0;                              ⟶                    z

> &- (&- a);                          ⟶                    a

> a &+ (&- a);                        ⟶                    0
```

SEE ALSO: `define[operator]`

2.1.86   define/linear – define characteristics of a linear operator

Calling Sequence:

   `define(linear(L))`

Parameters:

   L – is the name of the linear operator being defined

Synopsis:

- `define/linear` sets up a definition of L which allows simplification and evaluation.

- The properties of a linear operator are:

```
L(K1*a + K2*b) = K1*L(a) + K2*L(b)
L(K3) = K3*L(1)
for any constants K1, K2 and K3
```

Examples:
```
> define(linear(L));
  L(5);                               ⟶                 5 L(1)

> L(3*x+z);                           ⟶              3 L(x) + L(z)

> L(Pi);                              ⟶                Pi L(1)
```

SEE ALSO: `define[operator]`

2.1.87   define/operator – define an operator by a list of properties

Calling Sequence:

   $define(f, \ p1, \ ..., \ pn)$

Parameters:

   f – name of the operator being defined

   $p.i$ – name of the property of an operator or an equation

Synopsis:

- The call `define(f,p1,...,pn)` defines the evaluation and simplification rules of the operator f to have the properties $p1,...,pn$. The name f will be assigned a Maple procedure implementing the given properties. To see the procedure, use the `print` command. Presently the following properties are understood:

- `unary`; The operator is unary.

- `binary`; The operator is binary.

- `associative`; The operator is associative (n-ary), so $f(x,f(y,z)) = f(f(x,y),z) = f(x,y,z)$.

- `commutative` or `symmetric`; The operator f is commutative, so $f(x,y) = f(y,x)$. The arguments will be ordered to guarantee uniqueness.

- `antisymmetric`; The operator f is antisymmetric, so $f(x,y) = -f(y,x)$. The arguments will be ordered to guarantee uniqueness.

- `inverse=`g; Define the inverse (unary) operator of f to be the operator g.

- `identity=`x; The expression x is to be considered the left and right identity of the operator f.

- `zero=`x; Define the "zero" of f to be x, so if any argument of f is x then the result is x.

Examples:
```
> define(`&+`, associative, commutative, identity=0);
> x &+ (y &+ z) &+ (0 &+ x);
                        &+(y, x, x, z)
> define(f, commutative, associative, inverse=g);
> f( g(x), z, f(x,y), y );
                        f(y, y, z)
```

SEE ALSO: `neutral`, `procedures`, `print`

2.1.88   degree – degree of a polynomial
ldegree – low degree of a polynomial

Calling Sequence:

    `degree(`a,x`)`

    `ldegree(`a,x`)`

Parameters:

    a –   any expression

    x –   (optional) indeterminate or a list or set of indeterminates

Synopsis:

- The `degree` and `ldegree` functions are used to determine the highest degree and lowest degree of the polynomial a in the indeterminate(s) x, which is most commonly a single name but may be any indeterminate or a list or set of such.

- The polynomial a may have negative integer exponents in x. Thus **degree** and **ldegree** functions may return a negative or positive integer. If a is not a polynomial in x in this sense, then **FAIL** is returned.

- The polynomial a must be in collected form in order for **degree/ldegree** to return an accurate result. For example, given (x+1)*(x+2) - x^2, **degree** would not detect the cancellation of the leading term, and would incorrectly return a result of 2. Applying the function **collect** or **expand** to the polynomial before calling **degree** avoids this problem.

- If x is a set of indeterminates, the **total** degree/ldegree is computed. If x is a list of indeterminates, then the **vector** degree/ldegree is computed. The vector degree is defined as follows:

```
degree(p,[]) = 0
degree(p,[x1,x2,...]) = degree(p,x1) + degree(lcoeff(p,[x2,...]))
```

- The total degree is then defined as

```
degree(p,{x1,...,xn})
        = maximum degree(p,{x1,...xn}) over the terms in a sum
        = degree(p,[x1,...,xn]) otherwise
```

- Notice that the vector degree is sensitive to the order of the indeterminates, whereas the total degree is not. Finally, if x is not specified, this is short for **degree(a,indets(a))**, meaning that the total degree in all the indeterminates is computed.

Examples:

> degree(2/x^2+5+7*x^3,x);	$\longrightarrow$	3
> ldegree(2/x^2+5+7*x^3,x);	$\longrightarrow$	-2
> degree(x*sin(x),x);	$\longrightarrow$	FAIL
> degree(x*sin(x),sin(x));	$\longrightarrow$	1
> degree((x+1)/(x+2),x);	$\longrightarrow$	FAIL
> degree(x*y^3+x^2,[x,y]);	$\longrightarrow$	2
> degree(x*y^3+x^2,{x,y});	$\longrightarrow$	4
> ldegree(x*y^3+x^2,[x,y]);	$\longrightarrow$	4

SEE ALSO: lcoeff, tcoeff

2.1.89   Det – inert determinant

Calling Sequence:

    Det(A)

Parameters:

    A   –   matrix

Synopsis:

- The Det function is a placeholder for representing the determinant of the matrix A. It is used in conjunction with mod and modp1, which define the coefficient domain as described below.

- The call Det(A) mod p computes the determinant of the matrix A mod p where p is a prime integer and the entries in A are rationals, or in general, multivariate polynomials over a finite field.

- The call modp1(Det(A),p) computes the determinant of the matrix A mod p where p is a prime integer and the entries of A are modp1 polynomials.

Examples:

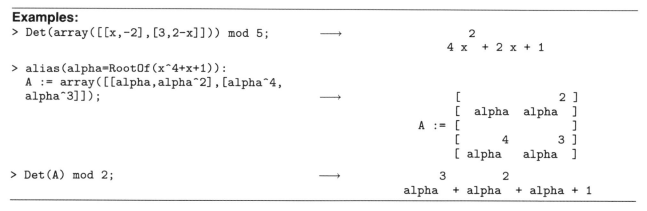

```
> Det(array([[x,-2],[3,2-x]])) mod 5;
```
$$4x^2 + 2x + 1$$

```
> alias(alpha=RootOf(x^4+x+1)):
  A := array([[alpha,alpha^2],[alpha^4,
  alpha^3]]);
```
$$A := \begin{bmatrix} alpha & alpha^2 \\ alpha^4 & alpha^3 \end{bmatrix}$$

```
> Det(A) mod 2;
```
$$alpha^3 + alpha^2 + alpha + 1$$

SEE ALSO: mod, modp1, det

2.1.90   diff or Diff – partial differentiation

Calling Sequence:

    diff(*a, x1, x2, ..., xn*)

    Diff(*a, x1, x2, ..., xn*)

Parameters:

    a   –   an algebraic expression

    x1, ... –   names

Synopsis:

- diff computes the partial derivative of *a* with respect to *x1, x2, ..., xn*, respectively.

- Note that where *n* is greater than 1, the call to diff is the same as diff called recursively. Thus diff(f(x), x, y); is equivalent to the call diff(diff (f(x), x), y);

- `diff` has a user interface that will call the user´s own differentiation functions. This means that if the procedure `diff/f` is defined then the function call `diff(f(x, y, z), y)` will invoke `diff/f`(x,y,z,y) to compute the derivative.

- The sequence operator `$` is useful for forming higher-order derivatives. `diff(f(x),x$4)`, for example, is equivalent to `diff(f(x),x,x,x,x)` and `diff(g(x,y),x$2,y$3)` is equivalent to `diff(g(x,y),x,x,y,y,y)`

- If the derivative cannot be expressed (if the expression is an undefined function), the diff function call itself is returned. (The prettyprinter displays the diff function in a two-dimensional d/dx format.)

- The capitalized function name `Diff` is the inert `diff` function, which simply returns unevaluated. The prettyprinter understands Diff to be equivalent to diff for printing purposes.

- The differential operator `D` is also defined in Maple; see `?D`.

Examples:

```
> diff(sin(x),x);
```
$\longrightarrow$ `cos(x)`

```
> diff(sin(x),y);
```
$\longrightarrow$ `0`

```
> diff(sin(x),x$3);
```
$\longrightarrow$ `- cos(x)`

```
> diff(1/(x+2),x);
```
$\longrightarrow$

```
       1
   - --------
          2
     (x + 2)
```

```
> Diff(exp(x),x);
```
$\longrightarrow$

```
    d
   ---- exp(x)
    dx
```

SEE ALSO: `D`, `int`, `dsolve`, `$`, `seq`

2.1.91   dilog(x) – the dilogarithm integral

Synopsis:

- The dilogarithm integral is defined as follows:

$$dilog(x) = \int_1^x \frac{\ln t}{1 - t} dt$$

2.1.92   discrim – discriminant of a polynomial

Calling Sequence:

     $discrim(p, x)$

Parameters:

p – polynomial in x

x – independent variable

Synopsis:

- If d=degree(p,x) and a=lcoeff(p,x) then the discriminant is

$$(-1)^{\wedge}(d*(d\ 1)/2)*resultant(p,diff(p,x),x)/a$$

Examples:

```
> p := a*x^2 + b*x + c;
```
$\longrightarrow$
$$p := a\ x^2 + b\ x + c$$

```
> discrim(p,x);
```
$\longrightarrow$
$$-4\ a\ c + b^2$$

SEE ALSO: resultant, degree, lcoeff

2.1.93 DistDeg – distinct degree factorization

Calling Sequence:

DistDeg(a, x) mod p

Parameters:

a – univariate polynomial in x

x – name

p – prime integer

Synopsis:

- This function computes the distinct degree factorization of a monic square-free univariate polynomial over a finite field mod p.

- The distinct degree factorization is returned as a list of pairs of the form [[f[1],d[1]],...,[f[m], d[m]]] where a = f[1] * ... * f[m] and each f[k] is a product of deg(f[k])/d[k] irreducible factors of degree d[k].

Examples:

```
> a := x^6+x^5+x^3+x;
```
$$a := x^6 + x^5 + x^3 + x$$

```
> DistDeg(a,x) mod 2;
```
$$[[x^2 + x,\ 1],\ [x^4 + x + 1,\ 4]]$$

```
> Factors(a) mod 2;
```
$$[1,\ [[x,\ 1],\ [x^4 + x + 1,\ 1],\ [x + 1,\ 1]]]$$

SEE ALSO: Sqrfree, Factors, RootOf

2.1.94   Divide – inert divide function

Calling Sequence:

   Divide(a, b, $'q'$)

Parameters:

   a, b –   multivariate polynomials

   q –   optional unevaluated name

Synopsis:

- The `Divide` function is a placeholder for the division of the polynomial a by b . It is used in conjunction with either `mod`, `modp1`, `evala` or `evalgf` as described below.

- The call `Divide`($a,b,'q'$) `mod` p determines whether b divides a modulo p, a prime integer. It returns **true** if the division succeeds and assigns the quotient to q, such that $a=b*q$; otherwise it returns **false**. The polynomials a and b must be multivariate polynomials over the rationals or over a finite field specified by `RootOf`s.

- The call `modp1`(`Divide`($a,b,'q'$),p) does likewise for a and b polynomials in the `modp1` representation, p a prime integer.

- The call `evala`(`Divide`($a,b,'q'$)) does likewise for a and b multivariate polynomials defined over an algebraic number field specified by `RootOf`s.

- The call `evalgf`(`Divide`($a,b,'q'$),p) does likewise for a and b multivariate polynomials over an algebraic number field modulo p defined by `RootOf`s. This function must be explicitly loaded by typing `readlib(evalgf)`.

Examples:
```
> Divide(x^3+x^2+2*x+3, x+2, 'q') mod 5;        ⟶                true

> q;                                             ⟶                  2
                                                                  x  + 4 x + 4

> a := x^2+x-2-RootOf(_Z^2-2):
  b := x-RootOf(_Z^2-2):
  evala(Divide(a, b, 'q'));                      ⟶                true

> q;                                             ⟶                      2
                                                            x + 1 + RootOf(_Z  - 2)
```

SEE ALSO: `mod`, `divide`, `Quo`, `Prem`, `Gcd`

2.1.95   divide – exact polynomial division

Calling Sequence:

   divide(a, b, $'q'$)

Parameters:

a, b	–	polynomials with rational number coefficients
q	–	an unevaluated name

Synopsis:

- `divide` checks if the polynomial b divides a over the rationals. If so, true is returned; otherwise false is returned.

- If the division is successful and there is a third argument $'q'$, then the value of the quotient is assigned to q. In the case of an unsuccessful division the name q will not be affected.

Examples:
```
> divide(x^3-y^3, x-y, 'q');
```
$\longrightarrow$ true
```
> q;
```
$$\longrightarrow \qquad y^2 + x\,y + x^2$$

SEE ALSO: `rem`, `quo`, `gcd`, `Divide`

2.1.96   dsolve – solve ordinary differential equations

Calling Sequence:

     dsolve(*deqns*, *vars*)

     dsolve(*deqns*, *vars*, *option*)

Parameters:

deqns	–	ordinary differential equation in *vars*, or set of equations and/or initial conditions
vars	–	variable or set of variables to be solved for
option	–	one of: `explicit`, `laplace`, `series`, `numeric`

Synopsis:

- `dsolve` is able to find closed-form solutions to many differential equations. The solution is returned either as an equation in `y(x)` and `x` (or whatever variables were specified) or in parametric form `[x=f(_T),y(x)=g(_T)]` where `_T` is the parameter. Any arbitrary constants are represented as `_C1`, `_C2`, ..., `_Cn`.

- The `explicit` option forces the solution to be returned explicitly in terms of the dependent variable, if possible.

- The `laplace` option causes `dsolve` to solve using Laplace transforms. One advantage to using this option is that differential equations may contain the `Dirac` or `Heaviside` functions. These functions are not recognized by the rest of `dsolve`.

- The `series` option causes `dsolve` to solve using a series method. The order of the solution can be specified by setting `Order`.

- The initial conditions must be specified at x=0 if the **laplace** or **series** option is being used. Otherwise, the conditions may be initial or boundary conditions specified at any points. Derivatives in conditions are specified by applying D to the function name, thus the second derivative of y at 0 is given as D(D(y))(0) or (D@@2)(y)(0).

- If the **numeric** option is specified then the differential equations must be specified with initial-value conditions. In this case, a procedure is returned as the result of the **dsolve** function. If this procedure is assigned to the name F, for example, then invoking F(t) for a numeric value of the independent variable t invokes a numerical method to solve the differential equation (or system of equations), yielding the numerical solution at t. The numerical solution is returned as an expression sequence consisting of the value of t, followed by the values of the dependent variables at the point t. For further information, see the topic **dsolve[numeric]**.

Examples:

```
> dsolve(diff(y(x),x$2) - y(x) = 1, y(x));
                    y(x) = - 1 + _C1 exp(x) + _C2 exp(- x)

> dsolve({diff(v(t),t)+2*t=0, v(1)=5}, v(t));
                                        2
                            v(t) = - t  + 6

> dsolve(diff(y(x),x) - t*y(x) = 0, y(x), explicit);
                            y(x) = exp(t x) _C1

> de1 := diff(y(t),t$2) + 5*diff(y(t),t) + 6*y(t) = 0;
                    /  2     \
                    | d      |     / d      \
            de1 := |----- y(t)| + 5 |---- y(t)| + 6 y(t) = 0
                    |  2     |     \ dt     /
                    \ dt     /

> dsolve({de1, y(0)=0, D(y)(0)=1}, y(t), laplace);
                    y(t) = 2 exp(- 5/2 t) sinh(1/2 t)

> sys := diff(y(x),x)=z(x), diff(z(x),x)=y(x):   fcns := {y(x), z(x)}:
> dsolve({sys,y(0)=0,z(0)=1}, fcns);
        {z(x) = 1/2 exp(x) + 1/2 exp(- x), y(x) = 1/2 exp(x) - 1/2 exp(- x)}

> dsolve({sys,y(0)=0,z(0)=1}, fcns, series);
                        2        4      6                    3          5        6
    {z(x) = 1 + 1/2 x  + 1/24 x  + O(x ), y(x) = x + 1/6 x  + 1/120 x  + O(x )}

> F := dsolve({sys,y(0)=0,z(0)=1}, fcns, numeric);
F := proc(x) `dsolve/numeric/result2`(x,269200824,[1,1]) end

> F(0);   F(0.5);   F(1.0);
                                0, 0, 1.

                    .5000000000, .5210953039, 1.127625961

                        1., 1.175201186, 1.543080624

> sys2 := {(D@@2)(y)(x)=y(x), y(0)=1, D(y)(0)=1}:
> s := dsolve(sys2, {y(x)}, numeric);
s := proc(x) `dsolve/numeric/result2`(x,269230164,[2]) end
```

```
> s(1.0);  s(1.5);
```

$$1., \ 2.718281808$$

$$1.500000000, \ 4.481689021$$

SEE ALSO: `diff`, `D`, `convert[diff]`, `convert[D]`, `rsolve`, `laplace`, `series`, `dsolve[numeric]`, `mellin`

2.1.97   dsolve/numeric – numeric solution of ordinary differential equations

Calling Sequence:
> dsolve(*deqns*, *vars*, numeric)

Parameters:

   deqns – set of ordinary differential equations and initial conditions

   vars   – variable or set of variables to be solved for

Synopsis:

- Invoking the `dsolve` function with the `numeric` option yields a procedure as a result. Then the procedure may be invoked for various numeric values of the independent variable, to compute the numeric solution at specific points. The numeric solution is returned as an expression sequence consisting of the value of the independent variable, followed by the values of the dependent variables at that point.

- The method used is a Fehlberg fourth-fifth order Runge-Kutta method. Reference: Subroutine RKF45 appearing in **Computer Methods for Mathematical Computations** by G.E. Forsythe, M.A. Malcolm, and C.B. Moler.

- Default values are chosen for the relative and absolute error tolerances. Specifically, the default value is `10^(-Digits+3)` for both tolerances. These values may be user-specified, by assigning values to the global names _RELERR and _ABSERR. For example,

```
      _RELERR := Float(1,-4);   _ABSERR := Float(1,-Digits+2);
```

- The computation may return with an error message indicating a value of the variable `IFLAG` from the rkf45 procedure. This flag indicates the status of the numerical computation, as specified below.

```
    IFLAG = 2 -- Indicates successful return and is the normal mode
                 for continuing integration.
          =-2 -- A single successful step in the specified direction
                 has been taken.  Normal mode for continuing
                 integration one step at a time.
          = 3 -- Integration was not completed because relative error
                 tolerance was too small.  _RELERR must be increased
```

before continuing.

= 4 -- Integration was not completed because more than
       3000 derivative evaluations were needed. This
       is approximately 500 steps.

= 5 -- Integration was not completed because solution
       vanished making a pure relative error test
       impossible. Must use nonzero _ABSERR to continue.

= 6 -- Integration was not completed because requested
       accuracy could not be achieved using smallest
       allowable stepsize. User must increase the error
       tolerance before continued integration can be
       attempted.

= 7 -- It is likely that rkf45 is inefficient for solving
       this problem. Too much output is restricting the
       natural stepsize choice. Use the one-step integrator
       mode.

= 8 -- Invalid input parameters.

Examples:
```
> f := dsolve({D(y)(x) = y(x), y(0)=1}, y(x), numeric);
f := proc(x) `dsolve/numeric/result2`(x,268629772,[1]) end

> f(0);  f(0.5);  f(1.0);
                        0, 1.

              .5000000000, 1.648721265

                  1., 2.718281810

#
> e := 0.25:  alpha := evalf(Pi/4):
> de1 := { (D@@2)(x)(t) = -alpha^2 * x(t) / (x(t)^2 + y(t)^2)^(3/2),
>   (D@@2)(y)(t) = -alpha^2 * y(t) / (x(t)^2 + y(t)^2)^(3/2) }:
> init1 := { x(0)=1-e, D(x)(0)=0, y(0)=0, D(y)(0)=alpha*sqrt((1+e)/(1-e)) }:
> F := dsolve(de1 union init1, {x(t),y(t)}, numeric);
F := proc(t) `dsolve/numeric/result2`(t,268826428,[2,2]) end

> for t from 0 by 0.5 to 2.0 do F(t) od;
                        0, .75, 0

            .5000000000, .6197679946, .4777913619

                1., .2944174609, .8121784964

          1.500000000, -.1051764913, .9580380433

              2., -.4902999153, .9398749131
```

```
#
> t := 't':
> Digits := 20:
> de2 := { diff(y(t),t$3) - 2*diff(y(t),t$2) + y(t) }:
> init2 := { y(0)=1, D(y)(0)=1, (D@@2)(y)(0)=1 }:
> G := dsolve(de2 union init2, y(t), numeric);
G := proc(t) `dsolve/numeric/result2`(t,269001616,[3]) end

> _RELERR := Float(1,-18);
                                          -17
                        _RELERR := .1*10

> G(1.0);
Error, (in dsolve/numeric/result2) rkf45 returned with IFLAG =, 3
> _RELERR := Float(1,-10);
                                          -9
                        _RELERR := .1*10

> G(1.0);
          1.0000000000000000000, 2.7182818284158535504
```

2.1.98   Ei – the exponential integral

Synopsis:

- The exponential integral is defined as follows:

$$Ei(x) = \int_{-\infty}^{x} \frac{e^t}{t} dt$$

2.1.99   Eigenvals – compute the eigenvalues/vectors of a numeric matrix

Calling Sequence:

   Eigenvals(*A, vecs)*
   Eigenvals(*A, B, vecs)*

Parameters:

   A,B –   square matrices of real or complex numbers

   vecs –   an optional square matrix for the eigenvectors

Synopsis:

- Eigenvals(*A)* returns an array of the eigenvalues of *A*. The eigenvalues are computed by the QR method. The matrix is first balanced and transformed into upper Hessenberg form. Then the eigenvalues (eigenvectors) are computed.

- If the matrices are real and an optional parameter *vecs* is supplied, the eigenvectors are returned in *vecs* in the form of an n by n array such that the ith column is the ith eigenvector corresponding to the ith eigenvalue. If the ith eigenvalue is complex, then the ith column and the (i+1)th column in the eigenvector array are the real and imaginary parts corresponding to the ith eigenvalue.

- If at least one of the entries in matrix A or B is complex and if an optional parameter *vecs* is supplied, then the eigenvector corresponding to the ith eigenvalue is returned in the ith column of *vecs* (an n by n array).

- If the matrix is declared as symmetric then the routine will handle the matrix specially (using a faster algorithm).

- This routine also handles the generalized eigenvalue problem: find eigenvalues and eigenvectors L and X such that AX = LBX where A and B are square matrices of the same dimensions.

- The function Eigenvals itself is inert. To actually compute the eigenvalues and eigenvectors, the user must evaluate the inert function in the floating point domain, by evalf(Eigenvals(*A*)).

Examples:
```
> A := array([[1,2,4],[3,7,2],[5,6,9]]):
> evalf(Eigenvals(A));
                 [ -.8946025434, 13.74788904, 4.146713484 ]
```

SEE ALSO: linalg[eigenvals], linalg[eigenvects]

2.1.100   eqn – produce output suitable for troff/eqn printing

Calling Sequence:
    eqn(*expr*, filename)

Parameters:
    expr   –   any expression
    filename –   (optional) output to a file

Synopsis:
- This function first must be loaded by typing readlib(eqn):

- The eqn function produces output which is suitable for printing the input expression *expr* with a troff/eqn processor. It knows how to format integrals, limits, sums, products and matrices.

- The mathematical format is taken, in general, from the CRC handbook or the Handbook of Mathematical functions.

- The functions sum, product, int, diff, limit, and log are aliased by Sum, Product, Int, Diff, Limit, and Log, so that these can be used to prevent evaluation by Maple.

- User-defined function printing can be interfaced by including a procedure `eqn/*function-name*`.

- Note eqn will not produce .EQ or .EN lines in the output file.

Examples:
```
> readlib(eqn):
> eqn(x^2 + y^2 = z^2);
{{{  "x"   sup 2 }^+^{  "y"   sup 2 }}~~=~~{  "z"   sup 2 }}
# Put this output in the file EqnFile
> eqn(x^2 + y^2 = z^2, EqnFile);
> eqn(Int(1/(x^2+1), x) = int(1/(x^2+1), x));
{{int  { {( {{  "x"   sup 2 }^+^1 } )} sup -1 }~d   "x" }~~=~~{arctan (
 "x" )}}
```

SEE ALSO: `latex`, `fortran`, `C`, `writeto`, `appendto`

2.1.101   erf – error function

Synopsis:
- The error function is defined by

$$\mathrm{erf}(x) = \frac{2}{\sqrt{\pi}} \int_0^x \exp(-t^2)\, dt$$

- Automatic simplifications will be applied to this function.

2.1.102   erfc – the complementary error function

Synopsis:
- The complementary error function is defined by

$$\mathrm{erfc}(x) = 1 - \mathrm{erf}(x) = 1 - \frac{2}{\sqrt{\pi}} \int_0^x e^{-t^2}\, dt$$

Automatic simplifications will be applied to this function.

2.1.103   ERROR – error return from a procedure

Calling Sequence:
> ERROR($expr_1$, $expr_2$, ...)

Parameters:
> $expr_1$, ...   –   expression sequence (possibly null)

Synopsis:
- A call to the ERROR function causes an immediate return to the point where the current procedure was invoked.

- If the procedure was not invoked within the context of **traperror** then control returns to the top level of the Maple system and the following message is printed: "Error, *in procname*" followed by the evaluated expression sequence specified in the call to **ERROR**. (Here, **procname** will be the name with which the procedure was invoked).

- If the procedure was invoked within the context of **traperror**, then the value of the global name **lasterror** is assigned the evaluated expression sequence specified in the call to **ERROR**. In this case, the value of the **traperror** call is also this same expression sequence.

Examples:
```
> f := proc (x) if x<0 then ERROR(`invalid x`, x) else x^(1/2) fi end:
> f(-3);
Error, (in f) invalid x, -3
> traperror(f(-3)):
> if " = lasterror then `error occurred` else `no error occurred` fi;
                              error occurred
```

SEE ALSO: procedures, traperror, procname, return, `lasterror`

2.1.104   euler – Euler numbers and polynomials

Calling Sequence:
>   euler(*n*)
>   euler(*n*, *x*)

Parameters:
>   *n* – a non-negative integer
>   *x* – an expression

Synopsis:
- The function euler computes the *n*th Euler number, or the *n*th Euler polynomial in *x*. The *n*th Euler number $E(n)$ is defined by the exponential generating function:

$$2/(exp(t)+exp(-t)) = sum(E(n)/n!*t^n, n = 0..infinity)$$

Examples:
```
> euler(6);                    ⟶            -61

> euler(4,x);                  ⟶            4       3
                                           x  - 2 x  + x

> euler(4,1/2);               ⟶            5/16
```

SEE ALSO: bernoulli

2.1.105   Eval – Evaluate a polynomial

Calling Sequence:

    Eval(a, $x=n$))

    Eval(a, $\{x1=n1,x2=n2,\dots\}$)

Parameters:

a	–	multivariate polynomial
x, $x1$, $x2$,...	–	names
n, $n1$, $n2$,...	–	evaluation points

Synopsis:

- The `Eval` function is a place holder for polynomial evaluation. The polynomial a is evaluated at $x = n$ ($x1=n1$, $x2=n2$, ... for the multivariate case).

- The call `Eval`($a,x=n$) `mod` p evaluates the polynomial a at $x=n$ modulo p. The polynomial a must be a multivariate polynomial over a finite field.

- The call `modp1(Eval`(a,n)`,`p) evaluates the polynomial a at $x = n$ modulo p where a must be a univariate polynomial in the `modp1` representation, with n an integer and p an integer > 1.

Examples:

```
> Eval(x^2+1,x=3) mod 5;
```
$\longrightarrow$
$$0$$

```
> Eval(x^2+y,{x=3,y=2}) mod 5;
```
$\longrightarrow$
$$1$$

```
> alias(alpha=RootOf(x^4+x+1)):
  a := x^4+alpha*x^2+1;
```
$\longrightarrow$
$$a := x^4 + alpha\ x^2 + 1$$

```
> Eval(a,x=alpha) mod 5;
```
$\longrightarrow$
$$alpha^3 + 4\ alpha$$

SEE ALSO: `subs`, `Powmod`

2.1.106   eval – explicit evaluation

Calling Sequence:

    eval(x)

    eval(x, n)

Parameters:

x	–	an expression
n	–	an integer

Synopsis:

- The normal evaluation rules in Maple are full evaluation for global variables, and one-level evaluation for local variables and parameters. Sometimes the user requires full evaluation or one-level evaluation explicitly. The `eval` function is used for this purpose.

- The call `eval(x)` means fully evaluate the expression x.

- The call `eval(x, 1)` means evaluate the expression x one level. More generally, `eval(x, n)` yields n-level evaluation.

- If `x := y` and `y := 1` in a Maple session, what does the variable x evaluate to? In an interactive session where x and y are global variables, x would evaluate to 1 and we would say that x is "fully evaluated". For one-level evaluation, we would use the command `eval(x, 1)` which would in this case yield y. However, inside a Maple procedure, if x is a local variable or a parameter, then the standard mode is one-level evaluation, so x would evaluate to y. For local variables and parameters, full evaluation is obtained by calling `eval(x)`, yielding 1 in this example.

- In normal Maple usage, the user is unlikely to need to use the `eval` function. The default evaluation rules are usually what is desired.

- One common case where a user may need to use `eval` to force full evaluation is when applying the `subs` function. See `subs` for an example.

SEE ALSO: `subs`, `evalf`, `evalm`, `evaln`, `evalhf`, `evalb`, `evala`, `evalc`, `student[value]`

2.1.107   evala – evaluate in an algebraic number field

Calling Sequence:
    `evala(`*expr*`)`

Parameters:
    expr – an expression or an unevaluated function call

Synopsis:
- If *expr* is an unevaluated function call (such as `evala(Gcd(u,v))`), then the function is performed in the smallest algebraic number field possible. See `help(function)` for more information, where `function` is one of the following:

Content	Divide	Expand	Factor	Gcd	Gcdex	Normal
Prem	Primpart	Quo	Rem	Resultant	Sprem	Sqrfree

- Otherwise, *expr* is returned unchanged, after first checking for dependencies between the `RootOf`s in the expression.

- If a dependency is noticed between `RootOf`s during the computation, then an error occurs, and the dependency is indicated in the error message (this is accessible through the variable `lasterror`).

- An additional argument can be specified for `Factor`. This is an algebraic number, or a set of algebraic numbers, which are to be included in the field over which the factorization is to be done.

Examples:
```
> evala(Factor(x^2-2), RootOf(_Z^2-2));
```
$$(x + RootOf(\_Z^2 - 2)) \; (x - RootOf(\_Z^2 - 2))$$
```
> evala(Quo(x^2 - x + 3, x - RootOf(_Z^2-3), x, 'r'));
```
$$x - 1 + RootOf(\_Z^2 - 3)$$
```
> r;
```
$$6 - RootOf(\_Z^2 - 3)$$

SEE ALSO: `RootOf`, `alias`

2.1.108   evalb – evaluate as a Boolean expression

Calling Sequence:

    `evalb(`x`)`

Parameters:

    x – an expression

Synopsis:

- The purpose of the evalb function is to force evaluation of expressions involving relational operators to **true** or **false** if possible; otherwise, it returns an unevaluated expression.

- Normally expressions containing the relational operators =, <>, <, ≤, >, and ≥ are treated as algebraic equations or inequalities by Maple. However, when passed as arguments to the **evalb** function (or when appearing in a Boolean context in an **if** or **while** statement), they are evaluated to **true** or **false** if possible.

- Note that expressions involving > and ≥ are converted into equivalent expressions involving < and ≤, respectively.

Examples:

`> x = x;`	$\longrightarrow$	`x = x`
`> evalb(x=x);`	$\longrightarrow$	`true`
`> evalb(x=y);`	$\longrightarrow$	`false`
`> evalb(x>y);`	$\longrightarrow$	`y - x < 0`

SEE ALSO: `Boolean`

2.1.109   evalc – evaluate in the complex number field

Calling Sequence:

>   evalc(*expr*)

Parameters:

>   *expr*   –   any expression

Synopsis:

- This function is for evaluating and simplifying complex numbers. evalc attempts to split an expression into its real and imaginary components.

- When possible, the result of evalc will be in the canonical form $expr_1$ + $expr_2$ * I involving the symbol I, which is an alias for sqrt(-1).

- The result of evalc of a complex series will be a series with each coefficient in the above canonical form. When evalc encounters an unevaluated function call (such as f(1+I) where f is not defined) then it will attempt to put the arguments in the above canonical form.

- If e := evalc(*expr*) then the "real part" of e is coeff(e,I,0) and the "imaginary part" is coeff(e,I,1). Alternatively, evalc(Re(*expr*)) and evalc(Im(*expr*)) return the real and imaginary parts of an expression.

- To find the complex conjugate of an expression use evalc(conjugate(*expr*)).

- A complex number may be represented to evalc as polar(r,theta) where r is the modulus and theta is the argument of the number.

Examples:

> evalc((3 + 5*I)*(7 + 4*I));	$\longrightarrow$	1 + 47 I
> evalc(Im((3 + 5*I)*(7 + 4*I)));	$\longrightarrow$	47
> evalc((5 - I)/(1 + 2*I));	$\longrightarrow$	3/5 - 11/5 I
> evalc(2^(1+I));	$\longrightarrow$	2 cos(ln(2)) + 2 sin(ln(2)) I
> evalc(sin(I));	$\longrightarrow$	sinh(1) I
> evalc(exp(I));	$\longrightarrow$	cos(1) + sin(1) I
> evalc(conjugate(exp(I)));	$\longrightarrow$	cos(1) - sin(1) I
> evalc(polar(r,theta));	$\longrightarrow$	r cos(theta) + r sin(theta) I

SEE ALSO: evalc[polar], convert[polar], alias

2.1.110   evalc/polar – polar coordinates

Calling Sequence:

>   polar(*m, t*)

Parameters:
> m, t –   real expressions

Synopsis:
- The polar function is a place holder for specifying polar coordinates. The parameter m is the magnitude of the number and t is the angle.

- A complex value of the form `x+y*I` can be converted to polar coordinates by using the `convert` function. Polar coordinates can be converted back to a complex value by applying the `evalc` function.

Examples:
```
> convert(x+y*I, polar);
```
$$\longrightarrow \quad polar((x^2 + y^2)^{1/2}, arctan(y, x))$$

```
> evalc(polar(x,y));
```
$$\longrightarrow \quad x\ cos(y) + x\ sin(y)\ I$$

SEE ALSO: `convert`, `evalc`

2.1.111   evalf – evaluate using floating-point arithmetic

Calling Sequence:
> `evalf(`*expr*`, n)`

Parameters:
> *expr* –   any expression
>
> *n*   –   (optional) integer specifying number of digits

Synopsis:
- `evalf` evaluates to floating-point numbers expressions (or subexpressions) involving constants such as `Pi`, `E`, `gamma`, and functions such as `exp`, `ln`, `sin`, `arctan`, `cosh`, `GAMMA`, and `erf`. For the complete list of known constants, see `ininames`. For the complete list of known functions, see `inifcns`.

- The accuracy of the result is determined by the value of the global variable `Digits`. By default the results will be computed using 10-digit floating-point arithmetic, since the initial value of `Digits` is 10. A user can change the value of `Digits` to any positive integer.

- If a second parameter, `n`, is present the result will be computed using n-digit floating-point arithmetic.

- `evalf` has an interface for evaluating user-defined constants and functions. For example, if a constant K is required, then the user must define a procedure called `` `evalf/constant/K` ``. Then calls to `evalf(K)` will invoke `` `evalf/constant/K`() ``. For a function example, if the procedure `` `evalf/f` `` is defined then the function call `evalf(f(2.3, 5.7), 20)` will invoke `` `evalf/f`(2.3, 5.7) `` and the global variable `Digits` will be 20 during the `evalf` call.

- If **evalf** is applied to an unevaluated definite integral then numerical integration will be performed (when possible). To invoke numerical integration without first invoking symbolic integration, use the inert form **Int** as in: **evalf(Int(f,x=a..b))** (See **int** for further options for numerical integration.)

- There are currently two floating-point numerical linear algebra routines supported: **Eigenvals** and **Svd**. See **Eigenvals** and **Svd** for more information.

- When evalf is invoked on a hypergeometric function, the internal function `evalf/hypergeom` is called for efficient processing of this case.

Examples:

> evalf(Pi);	$\longrightarrow$	3.141592654
> evalf(5/3*exp(-2)*sin(Pi/4),15);	$\longrightarrow$	.159494160850686
> evalf(cos(1) + sin(1)*I);	$\longrightarrow$	.5403023059 + .8414709848 I

> evalf(3/4*x^2+1/3*x-sqrt(2)); $\longrightarrow$

$$.7500000000 \ x^2 \ + \ .3333333333 \ x$$
$$- \ 1.414213562$$

> int(exp(x^3), x=0..1); $\longrightarrow$

$$\int_{0}^{1} \exp(x^3) \ dx$$

> evalf(");	$\longrightarrow$	1.341904418
> evalf(Int(tan(x),x=0..Pi/4));	$\longrightarrow$	.3465735903

SEE ALSO: **ininames, inifcns, int, Eigenvals, Svd, linalg**

2.1.112   evalhf – evaluate an expression using hardware floating-point

Calling Sequence:

   evalhf(*expr*)

Parameters:

   expr   –   an expression to be evaluated numerically

Synopsis:

- A call to **evalhf** evaluates an expression to a numerical value using the hardware floating-point of the system. The evaluation is done in double precision.

- The only argument of **evalhf** should be an expression which evaluates to a single value. As opposed to **evalf**, no symbolic answers are permitted.

- The argument evaluated by `evalhf` can include function calls, either to standard functions (such as `sin` and `arctan`) or to user-defined functions in Maple (see `evalhf[function]`).

- The `evalhf` function converts all its arguments to hardware floats, computes the answer and converts the answer to a Maple float result. Hence the user never deals with hardware floating-point numbers; these are confined to `evalhf`.

- It is advantageous to do as much computation as possible within a single call to `evalhf`, to avoid the conversion overhead.

- Maple includes `evalhf` for the purpose of gaining speed in numerical computations or for users who wish to use the hardware floating-point system.

- If `Digits` is set equal to (the integer part of) the special value `evalhf(Digits)` (approximately 15 on many 32-bit architectures) then `evalf` and `evalhf` (when both succeed) should produce similar results.

Examples:

```
> evalhf(sin(exp(gamma+2)+ln(cos(Catalan)))
  );                                          ⟶           .09801979012383793
> f := proc(n) if n<2 then n else (n+1)*
  f(n-1)/n fi end:
  evalhf(f(100)+f(10)+f(1));                  ⟶                  57.
> g := proc(x) x^5*ln(x)/(1+x^2) end:
  evalhf(g(2));                               ⟶           4.436141955583650
> Digits := trunc( evalhf(Digits) );         ⟶             Digits := 15
> evalf(g(2));                                ⟶           4.43614195558365
```

SEE ALSO: `evalhf[topic]` for the topics: `functions, arrays, fortran, constants, var, fcnlist, boolean`

2.1.113   evalhf/array – handling of arrays

Synopsis:

- `evalhf` handles only one special structure: arrays. Arrays passed as parameters to `evalhf` should have all their entries evaluating to floats or unassigned.

- Unassigned array entries are treated as 0 by `evalhf`.

- In general, `evalhf` can pass arrays downwards (to its called routines). These arrays can be modified and returned as function values, but cannot be returned as function values at the top level.

- We distinguish two levels of interaction for arrays: (a) top-level, first call to `evalhf` and (b) function-to-function within an `evalhf` call. Thus

```
(a) A := array(1..2,3..10); B := array(....);
    evalhf(f(A) + g(var(B)));
(b) f := proc(..) ..., A:=array();  g(...A,...) ... end;
    evalhf(f(...));
```

- This distinction has to be made because at the top level, arrays are converted to hardware floats, something which is not necessary in between internal **evalhf** calls. This conversion is expensive and should be used judiciously.

- Arrays within **evalhf** are passed by reference, and hence can be modified by any sub-function which uses them.

- Arrays at the top level (a) are normally passed by value. By using the **var**(*array*) construct, arrays are passed by value and result, being converted into hardware floats and after evaluation, converted back.

- The function **array(...)** is valid inside **evalhf** calls and accepts dimension information only. The result is an array of the given dimensions initialized to all zeros. No indexing or initializing information can be given to the function **array**.

- Indexing functions are not handled by **evalhf**.

- There is no mechanism, within **evalhf**, to find the dimension of an array from the object itself.

- Global arrays or global array references are not permitted within **evalhf**.

SEE ALSO: `evalhf[var]`

2.1.114   evalhf/boolean – Boolean expressions

Synopsis:
- Boolean expressions are evaluated according to the usual Maple rules, except that no result can remain unevaluated. Every expression should evaluate to **true** or **false**.

- Some Boolean functions have a special treatment: (a) **type(...,numeric)**, **type(...,float)** and **type(...,rational)** will return **true** regardless of the argument; (b) **type(...,integer)** returns **true** if the float argument represents an exact integer.

- Boolean variables or expressions (outside **if** or **while** statements) will have the values **1.0** (**true**) and **0.0** (**false**).

- Equations and relations in general will be treated as Boolean expressions if encountered outside an "if" or "while" statement.

2.1.115   evalhf/constant – constants known to evalhf

Synopsis:
- The following constants are known to `evalhf`:

E	Pi	gamma	Catalan
false	FLT_RADIX	DBL_MANT_DIG	DBL_DIG
DBL_EPSILON	DBL_MIN_EXP	DBL_MIN	DBL_MIN_10_EXP
DBL_MAX_EXP	DBL_MAX	DBL_MAX_10_EXP	LNMAXFLOAT
true			

- `evalhf(Digits)` returns the approximate number of digits of the hardware floating point system in use, regardless of the actual setting of `Digits`. This constant takes into account the precision returned by the standard mathematical functions.

- Inside `evalhf`, `true` has the value `1.0` and `false` has the value `0.0`.

- The following constants are taken from the ANSI-C definition of floating point parameters of an IEEE-standard floating point system.

- `FLT_RADIX`: The base of the floating point system

- `DBL_MANT_DIG`: Significant base digits

- `DBL_DIG`: Equivalent decimal precision

- `DBL_EPSILON`: Smallest x such that `1.0+x` $<>$ `1.0`

- `DBL_MIN_EXP`: Minimum normalized exponent

- `DBL_MIN`: Minimum normalized positive number

- `DBL_MIN_10_EXP`: Minimum exponent of a representable power of 10

- `DBL_MAX_EXP`: Maximum exponent

- `DBL_MAX`: Maximum number

- `DBL_MAX_10_EXP`: Maximum exponent of a representable power of 10

- `LNMAXFLOAT`: `ln(DBL_MAX)*0.99999`

2.1.116 evalhf/fcnlist – list of functions handled

Synopsis:
- The following is a list of functions handled directly by `evalhf`, consequently generated without using Maple code, merely by calling the corresponding mathematical function from the C library or executing C code.

abs	arccos	arccosh	arccot	arccoth	arccsc	arccsch
arcsec	arcsech	arcsin	arcsinh	arctan	arctan	arctanh
array	cos	cosh	cot	coth	csc	csch
erf	ERROR	evalb	evalf	evalf	evalhf	exp
factorial	Float	GAMMA	iquo	iquo	irem	irem
length	ln	lnGAMMA	lprint	max	max	min
mod	modp	mods	print	RETURN	sec	sech
sin	sinh	sqrt	sum	tan	tanh	time
trunc						

- Some implementations of Maple may be on systems which do not have all the above functions in the C library. This should not cause any differences other than speed and truncation errors. Consult your local C expert to determine which functions could be missing.

- The functions listed above should execute at speeds significantly higher than the Maple-`evalf` equivalents.

2.1.117   evalhf/fortran – and its relation to Fortran

Synopsis:

- Typically a Fortran program or a set of functions can be easily converted into a Maple function which can be executed by `evalhf`. The speed of a function executed by `evalhf` compared to a function compiled in optimized Fortran is on a ratio anywhere between 1:5 to 1:50. Converting Fortran into Maple-`evalhf` is still one or two orders of magnitude faster than running the equivalent code under standard Maple.

- The Whetstone benchmark gives a ratio of 1:35 in favour of compiled Fortran (under a VAX running Unix BSD 4.3).

- The following differences and problems should be observed when converting Fortran into Maple-`evalhf`:

- The only type handled by `evalhf` is floating point (double precision). Integers and Booleans are treated as floats also.

- There is no equivalent of the common or equivalence statements.

- Any Fortran expression which will evaluate over the integers, in particular expressions assigned to integer variables, should be surrounded by the function `trunc()`.

- Array declarations are dynamic with the `array()` function, and not static.

- All variables should be declared as local variables.

- There is no goto statement in Maple. Most Fortran programs using complicated gotos will need to be restructured.

- Fortran may return values through assigned arguments. This will not work under **evalhf**. Arrays with a single element may solve this problem.

- Fortran is very liberal with the array dimensions and will allow a subroutine to work with an array that has a declaration different from the caller. This is not allowed in Maple-**evalhf**; furthermore, arrays can only be passed as a whole, not just by the mention of a single element.

- Returned values in Maple are the last value computed. In Fortran these values are assigned to a variable with the same name as the function.

- There is neither a read nor format statement.

2.1.118   evalhf/function – handling of Maple functions
Synopsis:
- The function **evalhf** recognizes many "system" functions which are made available through the C library. See evalhf[fcnlist] for more information on this topic.

- A Maple function can be executed by **evalhf** subject to the following limitations.

- All the arguments to the function must evaluate to floats or to arrays of floats.

- All the intermediate steps of computation inside the function must evaluate to floats.

- Global variables can be used, but cannot be assigned. Global variables should evaluate to a float.

- Some constants can be used without definition. See **evalhf[constants]** for more information.

- Options are not effective (including option **remember** and option **trace**). High settings of printlevel will not cause tracing of the statements executed under **evalhf**.

- All Maple statements are effective, provided they do not generate a symbolic structure. However, the **for ... in** construct cannot be used. For-loop variables iterate as usual but using floats. Neither **break** nor **next** statements are allowed.

- Arrays are treated specially. See **evalhf[arrays]** for more information.

- The **type** function is treated specially for compatibility with the rest of Maple.

- The **op** function is useless as there are no symbolic structures. In particular, it is ineffective in picking up parts of floating point numbers (as opposed to standard Maple).

- No garbage collection happens during an **evalhf** call. None is needed, as **evalhf** uses almost no memory.

SEE ALSO: evalhf, evalhf[fcnlist], evalhf[constants], evalhf[arrays]

2.1.119   evalhf/var – passing arrays by value/result; var(name)

Synopsis:

- The **var** construct is a place holder for Maple array arguments passed as parameters to functions evaluated by **evalhf**.

- For example,

```
A := array(....);
evalhf(f(var(A)));
```

- The **var** array arguments are converted to hardware floating point. Then **evalhf** evaluates the function with this float array and on return, the arguments are converted back into the Maple array. This achieves an effect similar to **var** arguments in Pascal function calls, but in reality it does call by value/result.

- The **var** mechanism is the only mechanism available that will obtain an array result from a computation done by **evalhf** other than individual calls to compute each array entry.

- This conversion process is only necessary when passing an array from Maple to an **evalhf** function call. Calls within functions already under **evalhf** evaluation do not need the **var(..)** feature.

SEE ALSO: `evalhf[arrays]`

2.1.120   evalm—evaluate a matrix expression

Calling Sequence:

   evalm(*matrix expression*)

Parameters:

   matrix expression – an expression

Synopsis:

- The function **evalm** evaluates an expression involving matrices. It performs any sums, products, or integer powers involving matrices, and will map functions onto matrices.

- Note that Maple may perform simplifications before passing the arguments to **evalm,** and these simplifications may not be valid for matrices. For example, **evalm(A^0)** will return **1,** not the identity matrix.

- Unassigned names will be considered either symbolic matrices or scalars depending on their use in an expression.

- To indicate non-commutative matrix multiplication, use the operator **&***. The matrix product **ABC** may be entered as **A &* B &* C** or as **&*(A,B,C)**, the latter being more efficient. Automatic simplifications such as collecting constants and powers will be applied. Do NOT use the ***** to indicate purely matrix multiplication, as this will result in an error. The operands of **&*** must be matrices (or names) with the exception of 0. Unevaluated matrix products are considered to be matrices. The operator **&*** has the same precedence as the ***** operator.

- Use 0 to denote the matrix or scalar zero. Use **&*()** to denote the matrix identity. It may be convenient to use **alias(Id=&*())**.

- If a sum involves a matrix and a Maple constant, the constant will be considered as a constant multiple of the identity matrix. Hence matrix polynomials can be entered in exactly the same fashion as fully expanded scalar polynomials.

Examples:
```
> alias(Id=&*()):
  S := array([[1,2],[3,4]]):
  T := array([[1,1],[2,-1]]):
  evalm(S+2*T);
```
$\longrightarrow$
$$\begin{array}{cc} [\ 3 & 4\ ] \\ [&] \\ [\ 7 & 2\ ] \end{array}$$

```
> evalm(S^2);
```
$\longrightarrow$
$$\begin{array}{cc} [\ \ 7 & 10\ ] \\ [&] \\ [\ 15 & 22\ ] \end{array}$$

```
> evalm(sin(S));
```
$\longrightarrow$
$$\begin{array}{cc} [\ \sin(1) & \sin(2)\ ] \\ [&] \\ [\ \sin(3) & \sin(4)\ ] \end{array}$$

```
> evalm(S &* T);
```
$\longrightarrow$
$$\begin{array}{cc} [\ \ 5 & -1\ ] \\ [&] \\ [\ 11 & -1\ ] \end{array}$$

```
> evalm(A &* B &* (2*B)-B &* Id);
```
$\longrightarrow$
$$2\ (A\ \&*\ B\ )^2 -\ B$$

```
> evalm(&*(A,B,0));
```
$\longrightarrow$
$$0$$

SEE ALSO: `array`, `matrix`, `linalg`, `alias`, `operators`

2.1.121   evaln – evaluate to a name

Calling Sequence:
     evaln(*expr*)

Parameters:
     expr – an expression

Synopsis:
- The **evaln** function is used to "evaluate to a name" or, more generally, to form an assignable object. The name (assignable object) will not itself be evaluated.

- A typical example of its use would be **divide(a,b,evaln(t[i]))** where i is the index of a **for** loop.

- If the expression is a simple name (such as a string) then the same effect can be achieved by using single quotes on the name.
- `evaln` can be used on names, subscripts, function calls, and concatenations.

Examples:

```
> i := 1;                        ⟶              i := 1
> evaln(i);                      ⟶                 i
> evaln(a.i);                    ⟶                a1
> evaln(a.(1..3));               ⟶            a1, a2, a3
> evaln(t[i]);                   ⟶               t[1]
> evaln(f(i));                   ⟶               f(1)
> divide(x^2,x,evaln(t[i]));     ⟶               true
> t[i];                          ⟶                 x
> evaln(3);                      ⟶     Error,
                                       Illegal use of an object as a name
```

SEE ALSO: `uneval, assigned, `:=``

2.1.122   example – produce an example of a function or type

Calling Sequence:

   example(*f*)

Parameters:

   f – The name of a function or a valid type

Synopsis:

- This is a function designed to illustrate the use of functions and to give executable examples. Once the example is produced it can be executed.
- If `example` is given a type, `example` builds an expression of the given type. The type can be a basic type or a structured type.
- `example` has a large library of examples, but can also create its own random ones for a function with a type definition.

Examples:

```
> example(int);                  ⟶              /
                                                |
                                                |  cos(arctan(x)) dx
                                                |
                                                /

> example(solve);                ⟶       solve(cos(x) + y = 9, x)
> example(name=algebraic);       ⟶          n = (B + 10) A
```

SEE ALSO: `help`

2.1.123   exp – the exponential function

Synopsis:

- The syntax is `exp(x)` for an expression `x`.

- The global constant `E` is understood to represent `exp(1)`.

- Automatic simplifications will be applied to this function.

2.1.124   Expand – inert expand function

Calling Sequence:

Expand(a)

Parameters:

a –   any expression

Synopsis:

- The `Expand` function is a placeholder for representing the `expand` function. It is used in conjunction with `mod`, `evala`, or `evalgf`, which define the coefficient domain as described below.

- The call `evala(Expand(a))` expands products in a where a may contain algebraic extensions defined by `RootOf`s.

- The calls `Expand(a) mod p` and `evalgf(Expand(a),p)` expand products in a over the integers mod p where a may contain algebraic extensions defined by `RootOf`s.

Examples:

```
> Expand( (x+2)^2*(x-2) ) mod 3;
```
$\longrightarrow$
$$x^3 + 2x^2 + 2x + 1$$

```
> alias(alpha=RootOf(x^2-2)):
  evala(Expand( (x+alpha)^2*(x-alpha) ));
```
$\longrightarrow$
$$alpha\ x^2 - 2\ alpha + x^3 - 2x$$

```
> alias(beta=RootOf(x^2+x+1)):
  Expand( (x+beta)^2*(x-beta) ) mod 2;
```
$\longrightarrow$
$$x^3 + x + beta\ x^2 + beta\ x + 1$$

SEE ALSO: `expand`, `evala`, `RootOf`, `Normal`, `mod`

2.1.125   expand – expand an expression

Calling Sequence:

expand($expr$, $expr_1$, $expr_2$, . . ., $expr_n$)

Parameters:

$expr$ –   any algebraic expression

$expr_1$, $expr_2$, ..., $expr_n$ – (optional) expressions

Synopsis:
- The primary application of expand is to distribute products over sums. This is done for all polynomials. For quotients of polynomials, only sums in the numerator are expanded — products and powers are left alone.

- expand also knows how to expand most of the mathematical functions, including `sin`, `cos`, `tan`, `sinh`, `cosh`, `tanh`, `det`, `erf`, `exp`, `factorial`, `GAMMA`, `ln`, `max`, `min`, `Psi`, `binomial`, `sum`, `product`, `int`, `limit`, `bernoulli`, `euler`, `BesselJ`, `BesselY`, `BesselI`, and `BesselK`.

- The optional arguments $expr_1$, $expr_2$, ..., $expr_n$ are used to prevent particular sub-expressions in $expr$ ($expr_1$, $expr_2$, ..., $expr_n$) from being expanded. To prevent all functions from being expanded, do `frontend(expand,[`$expr$`]);`

- expand has an interface that allows the user to make a function known to expand in the following way. If the procedure `` `expand/f` `` is defined, then the call `expand(f(x));` will generate the function call `` `expand/f`(x); ``

Examples:

```
> expand((x+1)*(x+2));
```
$\longrightarrow$
$$x^2 + 3\,x + 2$$

```
> expand((x+1)/(x+2));
```
$\longrightarrow$
$$\frac{x}{x+2} + \frac{1}{x+2}$$

```
> expand(1/(x+1)/x);
```
$\longrightarrow$
$$\frac{1}{(x+1)\,x}$$

```
> expand(sin(x+y));
```
$\longrightarrow$
$$\sin(x)\,\cos(y) + \cos(x)\,\sin(y)$$

```
> expand(cos(2*x));
```
$\longrightarrow$
$$2\,\cos(x)^2 - 1$$

```
> expand(exp(a+ln(b)));
```
$\longrightarrow$
$$\exp(a)\,b$$

```
> expand(ln(x/(1-x)^2));
```
$\longrightarrow$
$$\ln(x) - 2\,\ln(1-x)$$

```
> expand((x+1)*(y+z));
```
$\longrightarrow$
$$x\,y + x\,z + y + z$$

```
> expand((x+1)*(y+z), x+1);
```
$\longrightarrow$
$$(x+1)\,y + (x+1)\,z$$

```
> expand(BesselJ(2,t));
```
$\longrightarrow$
$$2\,\frac{\mathrm{BesselJ}(1,t)}{t} - \mathrm{BesselJ}(0,t)$$

SEE ALSO: `expandoff`, `expandon`, `collect`, `combine`, `factor`, `sort`, `Expand`

2.1.126   Factor – inert factor function

Calling Sequence:

    Factor(a, K)

Parameters:

    a –   multivariate polynomial

    K –   optional specification of the algebraic extension

Synopsis:

- The `Factor` function is a placeholder for representing the factorization of the polynomial a over a unique factorization domain. It is used in conjunction with either `mod`, `evala` or `evalgf`, which define the coefficient domain as described below.

- The call `Factor(a) mod p` computes the factorization of a over the integers modulo p a prime integer. The polynomial a must have rational coefficients or coefficients from a finite field specified by `RootOf`s.

- The call `evala(Factor(a,K))` computes the factorization of the polynomial a over an algebraic number field defined by the extension K, which is specified as a `RootOf` or a set of `RootOf`s. The polynomial a must have algebraic number coefficients.

- The calls `Factor(a,K) mod p` and `evalgf(Factor(a,K),p)` compute the factorization of the polynomial a over the Galois field (finite field) defined by the algebraic extension K over the integers mod p a prime integer. The extension K is specified by a `RootOf` and must be irreducible over the integers mod p. Note: `evalgf` must be explicitly loaded by using `readlib(evalgf)`

Examples:

```
> Factor(x^2+3*x+3) mod 7;                    ⟶          (x + 4) (x + 6)

> alias(alpha=RootOf(x^2-2)):
  evala(Factor(x^2-2,alpha));                 ⟶      (x + alpha) (x - alpha)

> alias(alpha=RootOf(x^2+x+1)):
  Factor(x^3+1,alpha) mod 2;                  ⟶  (x + 1) (x + alpha + 1) (x + alpha)
```

SEE ALSO: `factor`, `Factors`, `Expand`, `mod`, `evala`, `evalgf`, `RootOf`, `alias`

2.1.127   factor – factor a multivariate polynomial

Calling Sequences:

    factor(a)

    factor(a,K)

Parameters:

    a –   an expression

    K –   an algebraic extension

Synopsis:

- The function `factor` computes the factorization of a multivariate polynomial with integer, rational, or algebraic number coefficients.

- If the input is a rational function, then a is first "normalized" (see `normal`) and the numerator and denominator of the resulting expression are then factored. This provides a "fully-factored form" which can be used to simplify in the same way the `normal` function is used. However, it is more expensive to compute than normal.

- If the input a is a list, set, equation, range, series, relation, or function, then `factor` is applied recursively to the components of a.

- The call `factor(a,K)` factors a over the algebraic number field defined by K. K must be a single `RootOf`, a list or set of `RootOf`s, a single radical, or a list or set of radicals.

- If the second argument K is not given, the polynomial is factored over the rationals. Note that any integer content (see first example below) is not factored.

Examples:

```
> factor(6*x^2+18*x-24);
```
$\longrightarrow$
$$6\ (x + 4)\ (x - 1)$$

```
> factor({x^3+y^3 = 1});
```
$\longrightarrow$
$$\{(x + y)\ (x^2 - x\ y + y^2) = 1\}$$

```
> factor((x^3-y^3)/(x^4-y^4));
```
$\longrightarrow$
$$\frac{x^2 + x\ y + y^2}{(x + y)\ (x^2 + y^2)}$$

```
> factor(x^3+5);
```
$\longrightarrow$
$$x^3 + 5$$

```
> factor(x^3+5, 5^(1/3));
```
$\longrightarrow$
$$(x^2 - 5^{1/3}\ x + 5^{2/3})\ (x + 5^{1/3})$$

```
> factor(y^4-2,sqrt(2));
```
$\longrightarrow$
$$(y^2 + 2^{1/2})\ (y^2 - 2^{1/2})$$

```
> alias(alpha = RootOf(x^2-2)):
  factor(y^4-2,alpha);
```
$\longrightarrow$
$$(y^2 + \text{alpha})\ (y^2 - \text{alpha})$$

SEE ALSO: `ifactor`, `Factor`, `factors`, `sqrfree`, `collect`, `galois`, `irreduc`, `roots`

2.1.128 Factors – inert factors function

Calling Sequence:

    Factors(a, K)

Parameters:

a –   multivariate polynomial

K –   optional specification for an algebraic extension

Synopsis:

- The `Factors` function is a placeholder for representing the factorization of the multivariate polynomial a over U, a unique factorization domain. The construct `Factors(a)` produces a data structure of the form `[u,[[f[1],e[1]],...,[f[n],e[n]]]]` such that `a = u * f[1]^e[1] * ... * f[n]^e[n]`, where each `f[i]` is a primitive irreducible polynomial.

- If U is a field then `u` is the leading coefficient of a. Otherwise, `u` is not factored in U.

- The difference between the `Factors` function and the `Factor` function is only the form of the result. The `Factor` function, if defined, returns a Maple sum of products more suitable for interactive display and manipulation.

- The call `Factors(a) mod p` computes the factorization of a over the integers modulo p, a prime integer. The polynomial a must have rational coefficients or coefficients over a finite field specified by `RootOfs`.

- The call `Factors(a,K) mod p` computes the factorization over the finite field defined by K, an algebraic extension of the integers mod p where K is a `RootOf`.

- The call `modp1(Factors(a),p)` computes the factorization of the polynomial a in the `modp1` representation modulo p a prime integer.

Examples:
```
> Factors(2*x^2+6*x+6) mod 7;
                    [2, [[x + 4, 1], [x + 6, 1]]]
> Factors(x^5+1) mod 2;
                        4    3    2
            [1, [[x + 1, 1], [x  + x  + x  + x + 1, 1]]]
> alias(alpha=RootOf(x^2+x+1));
                              I, alpha
> Factors(x^5+1,alpha) mod 2;
        2                          2
    [1, [[x  + (alpha + 1) x + 1, 1], [x  + alpha x + 1, 1], [x + 1, 1]]]
```

SEE ALSO: `factors`, `Factor`, `Irreduc`, `Sqrfree`, `Expand`, `mod`, `modp1`

2.1.129   FresnelC – Fresnel cosine integral

Synopsis:

- The Fresnel cosine integral is defined as follows:

$$\mathtt{FresnelC}(x) = \int_0^x \cos(\frac{\pi}{2}t^2)dt$$

2.1.130   FresnelS – Fresnel sine integral

Synopsis:
- The Fresnel sine integral is defined as follows:

$$\text{FresnelS}(x) = \int_0^x \sin(\frac{\pi}{2}t^2)dt$$

2.1.131   Fresnelf – auxiliary Fresnel function

Synopsis:
- The Fresnelf auxiliary function is defined as

$$\text{Fresnelf}(z) = (\frac{1}{2} - \text{FresnelS}(z)) \cos(\frac{\pi}{2}z^2) - (\frac{1}{2} - \text{FresnelC}(z)) \sin(\frac{\pi}{2}z^2)$$

2.1.132   Fresnelg – auxiliary Fresnel function

Synopsis:
- The Fresnelg function is defined as follows:

$$\text{Fresnelg}(z) = (\frac{1}{2} - \text{FresnelC}(z)) \cos(\frac{\pi}{2}z^2) - (\frac{1}{2} - \text{FresnelS}(z)) \sin(\frac{\pi}{2}z^2)$$

2.1.133   fnormal – floating-point normalization

Calling Sequence:
    fnormal(e)
    fnormal(e, digits)
    fnormal(e, digits, epsilon)

Parameters:
e	–	an algebraic expression, or a list, set, relation, series, or range of algebraic expressions
digits	–	(optional) number of digits for floating-point evaluation (defaults to the value of the global variable Digits)

> $epsilon$ – (optional) error tolerance for "fuzzy zero" (defaults to the value
> `Float(1,-Digits+2)`)

Synopsis:

- The value returned by **fnormal** is equivalent to e under the assumption that all numeric values with magnitude less than $epsilon$ may be considered to be zero.

- If e is a list, set, range, series, equation, or relation, then **fnormal** is applied recursively to the components of e.

Examples:

```
> fnormal(.0000000001);
```
$\longrightarrow$ 0

```
> fnormal(1+.0000000001*I);
```
$\longrightarrow$ $1.$

```
> fnormal({10^(-10),10^(-9),10^(-8)});
```
$\longrightarrow$
$\{0, .1000000000*10^{-7}\}$

SEE ALSO: `simplify`, `evalf`, `evalhf`

2.1.134   fortran – generate Fortran code

Calling Sequence:

> `fortran(s);`

Parameters:

> s – an expression, array of expressions, or list of equations

Synopsis:

- The **fortran** function generates Fortran code for evaluating the input. The input **s** must be either a single algebraic expression, an array of algebraic expressions, or a list of equations of the form **name = algebraic** where the latter is understood to mean a sequence of assignment statements. Currently, **fortran** cannot generate output for Maple procedures.

- By default the output is sent to standard output. An additional argument of the form **filename = foo** can be used to direct the output to the file **foo**.

- If the keyword **optimized** is specified as an additional argument, common subexpression optimization is performed. The result is a sequence of assignment statements in which temporary values are stored in local variables beginning with the letter **t**. The global names **t0, t1, t2, ...** are reserved for use by **fortran** for this purpose.

- The global variable precision can be assigned either single or double (single by default) for single or double precision respectively.

- The Fortran language has a continuation line limit of 19 lines, which if exceeded will result in an error during compilation. For large expressions which exceed this limit, the **fortran** routine will automatically break up the expression. The global names **s0, s1, s2, ...** are reserved for use by **fortran** for this purpose.

Examples:
```
> s := ln(x)+2*ln(x)^2:
  fortran([a = s]);
> fortran(s);
> fortran(s,optimized);
```
$\longrightarrow$
$\longrightarrow$
$\longrightarrow$
```
a = alog(x)+2*alog(x)**2
t0 = alog(x)+2*alog(x)**2
t1 = alog(x)
t2 = t1**2
t4 = t1+2*t2
```

SEE ALSO: optimize, C, eqn, latex

2.1.135   frontend – process general expression into a rational expression

Calling Sequence:
>   frontend(p, x, f)

Parameters:

p – a procedure

x – a list of arguments to p

f – (optional) a list of two sets: first, a set of type names not to be
frozen; second, a set of expressions not to be frozen (default is
$[\{`+`,`*`\},\{\}]$)

Synopsis:

• The purpose of **frontend** is to extend the domain of computation for many of the functions in Maple.

• For example, the procedure used by Maple´s **normal** function is defined to work over the domain of rational functions. Thus in order to handle more general expressions such as expressions involving **sin(x)** or **sqrt(x)** in a reasonable way, **frontend** is employed to "temporarily freeze" all occurrences of such expressions for unique names. This is always valid.

• However, it is important to understand that the zero equivalence property of the normal function is only guaranteed if the sub-expressions which are frozen are algebraically independent.

• Each item in the list x is "frozen". The order in which the "freezing" occurs is as follows.

• If the argument is atomic, it is not frozen. Integers, names, rationals, and floats are considered to be atomic.

• If the argument is of type `+` or `*` or `^` and the exponent is an integer, then freezing is applied recursively.

• If the argument has any subexpression in the set of expressions, then freezing is applied recursively.

• If the argument is one of the types in the set of type names, freezing is applied recursively.

• Otherwise, the expression is substituted for a unique name.

- The procedure p is then evaluated with the "frozen" argument(s). Any frozen names occurring in the result are substituted back for their original subexpressions.

Examples:

```
> a := sin(x)^2 + 2*sin(x) + 1;
```
$\longrightarrow$
$$a := \sin(x)^2 + 2\sin(x) + 1$$

```
> b := sin(x) + 1;
```
$\longrightarrow$
$$b := \sin(x) + 1$$

```
> gcd(a, b);
```
$\longrightarrow$ Error, (in gcd) arguments must be polyn\
omials over the rationals

```
> frontend(gcd, [a, b]);
```
$\longrightarrow$
$$\sin(x) + 1$$

2.1.136   fsolve – solve using floating-point arithmetic

Calling Sequence:

    fsolve(*eqns*, *vars*, *options*);

Parameters:

    eqns   –   an equation or set of equations

    vars   –   (optional) an unknown or set of unknowns

    options –   (optional) parameters controlling solutions

Synopsis:

- The conventions for passing equations and variables, and returning the answers, are the same for fsolve as for solve.

- For a general equation, fsolve attempts to compute a single real root. However, for polynomials it will compute all real (non-complex) roots, although exceptionally ill-conditioned polynomials may cause fsolve to miss some roots.

- To compute all roots of a polynomial over the field of complex numbers, use the **complex** option.

- The options available are:

    complex — Find one root (or all roots, for polynomials) over the complex floating-point numbers.

    fulldigits — This option prevents fsolve from decreasing Digits for intermediate calculations at high settings of Digits. With this option fsolve may escape ill-conditioning problems, but the routine slows down.

    maxsols=n — Find only the n least roots. This option is only meaningful for polynomials, where more than one root is computed.

    interval — Specifically: a..b or x = a..b or {x=a..b, y=c..d, ...}. Search for roots in the given interval only. The ranges are open intervals, i.e. the endpoints are not included in the range.

- Note that an fsolve computation may fail to find a root even though one exists, in which case specifying appropriate range information may result in a successful computation.

Examples:
```
> fsolve( tan(sin(x))=1, x );                      →           .9033391108

> poly := 23*x^5 + 105*x^4 - 10*x^2 + 17*
  x:
  fsolve( poly, x, -1..1 );                        →        0, -.6371813185

> fsolve( poly, x, maxsols=3 );                    →    0, -4.536168981, -.6371813185

> q := 3*x^4 - 16*x^3 - 3*x^2 + 13*x +
  16:
  fsolve(q, x, 1..2);                              →           1.324717957

> fsolve(q, x, 2..5);
  fsolve(q, x, 4..8);                              →           5.333333333

> fsolve(q, x, complex);                           →    - .6623589786 + .5622795121 I,

                                                        - .6623589786 - .5622795121 I,

                                                        1.324717957, 5.333333333

> f := sin(x+y) - exp(x)*y = 0:
  g := x^2 - y = 2:
  fsolve({f,g},{x,y},{x=-1..1,y=-2..0});           →    {x = -.6687012050, y = -1.552838698}
```

SEE ALSO: solve, dsolve, isolve, msolve, rsolve

2.1.137   galois – compute the Galois group of an irreducible polynomial

Calling Sequence:
    galois(f)

Parameters:
    f – an irreducible univariate rational polynomial up to degree 7

Synopsis:
- The galois function returns an expression sequence of three elements. The first is the name of the Galois group (with + if even), the second is its order and the third is a set of generators of the group up to conjugacy. A string represents the disjoint cycle structure of each generator.

- The argument is tested first to see whether it is indeed an irreducible univariate rational polynomial of degree less than 8. An error is given if it is not in the right form.

- The galois procedure is able to print out details of how the Galois group is computed. To see some of this information you must first type infolevel[galois]:=1; before calling galois. To see all of the details type infolevel[galois]:=2; before invoking galois.

Examples:
```
> galois(x^4+x+1);                                 →     S4, 24, {(1 2), (1 2 3 4)}

> galois(t^5-5*t+12);                              →     +D5, 10, {(2 5)(3 4), (1 2 3 4 5)}
```

```
> galois(x^5+2);                         ⟶        F20, 20, {(2 3 5 4), (1 2 3 4 5)}
> galois(x^7+4*x^5-3*x^2+5);             ⟶        S7, 5040, {(1 2), (1 2 3 4 5 6 7)}
```

SEE ALSO: group, GF

2.1.138 GAMMA – the gamma function

Synopsis:
- The GAMMA function is defined by

$$\mathtt{GAMMA}(x) = \Gamma(x) = \int_0^\infty \exp(-t)\, t^{x-1}\, dt$$

- Automatic simplifications will be applied to this function.

2.1.139 gc – garbage collection

Calling Sequence:

gc(n)

Parameters:

n – an (optional) integer

Synopsis:
- Invocation of this function causes a garbage collection to occur. Garbage collection has the effect of deleting all data to which no references are made. It also cleans the remember tables of procedures with an option system or option **builtin** by removing entries that have no other references to them.

- The optional argument n controls the frequency of automatic garbage collections, which will occur (approximately) after every n words used (i.e. 4n bytes used). The current value of this frequency is available as **status[5]**, the fifth entry in the `status` sequence.

- This function returns the value **NULL** and prints a message showing three values:

      bytes used=xxxx, alloc=yyyy, time=zzzz

- where **xxxx** is the total bytes used, **yyyy** is the total bytes actually allocated, and **zzzz** is the total CPU time used (in seconds) since the start of the session.

- If the optional argument n is 0 then the above messages are not printed for subsequent calls to the automatic garbage collector, the frequency of automatic garbage collection is not affected, and a garbage collection does not immediately occur.

SEE ALSO: `status`, words

2.1.140   Gcd – inert gcd function

Calling Sequence:

Gcd(a, b)

Gcd(a, b, $'s'$, $'t'$)

Parameters:

a, b –   multivariate polynomials

s, t –   unevaluated names

Synopsis:

- The Gcd function is a placeholder for representing the greatest common divisor of a and b where a and b are polynomials. If s and t are specified, they are assigned the cofactors. Gcd is used in conjunction with either **mod**, **modp1** or **evala** as described below which define the coefficient domain.

- The call Gcd(a,b) mod p computes the greatest common divisor of a and b modulo p a prime integer. The inputs a and b must be polynomials over the rationals or over a finite field specified by RootOf expressions.

- The call modp1(Gcd(a,b),p) does likewise for a and b, polynomials in the modp1 representation.

- The call evala(Gcd(a,b)) does likewise for a and b multivariate polynomials with algebraic extensions defined by RootOf expressions.

Examples:

> Gcd(x+2,x+3) mod 7;	$\longrightarrow$	1
> Gcd(x^2+3*x+2,x^2+4*x+3,$'s'$,$'t'$) mod 11;	$\longrightarrow$	x + 1
> s, t;	$\longrightarrow$	x + 2, x + 3

SEE ALSO: gcd, mod, evala, RootOf, Gcdex

2.1.141   gcd – greatest common divisor of polynomials
lcm – least common multiple of polynomials

Calling Sequence:

gcd($a,b,'cofa','cofb'$)

lcm($a,b,\ldots$)

Parameters:

a, b –   multivariate polynomials over the rationals

$cofa, cofb$ –   (optional) unevaluated names

Synopsis:

- The gcd function computes the greatest common divisor of two polynomials a and b with rational coefficients.

- The `lcm` function computes the least common multiple of an arbitrary number of polynomials with rational coefficients.

- The optional third argument *cofa* is assigned the cofactor $a/\text{gcd}(a,b)$.

- The optional fourth argument *cofb* is assigned the cofactor $b/\text{gcd}(a,b)$.

Examples:

```
> gcd(x^2-y^2,x^3-y^3);                     ⟶                    y - x

> lcm(x^2-y^2,x^3-y^3);                     ⟶        3    4    4    3
                                                  - y x  + y  - x  + x y

> gcd(6,8,a,b);                             ⟶                  2

> a;                                        ⟶                  3

> b;                                        ⟶                  4
```

SEE ALSO: `gcdex`, `igcd`, `ilcm`, `Gcd`, `numtheory[GIgcd]`

2.1.142   Gcdex – inert gcdex function

Calling Sequence:

Gcdex(a, b, x, ´s´, ´t´)

Parameters:

a, b – multivariate polynomials

x – main variable

s, t – (optional) unevaluated names

Synopsis:

- The `Gcdex` function is a placeholder for the extended Euclidean algorithm applied to a and b which are polynomials in x over a field. `Gcdex` computes g, the greatest common divisor of a and b, which is a monic polynomial in x. Additionally s and t are (if present) assigned polynomials in x such that $a*s + b*t = g$ with $\text{degree}(s,x) < \text{degree}(b,x)$ and $\text{degree}(t,x) < \text{degree}(a,x)$. `Gcdex` is used in conjunction with either **mod** or **evala** as described below, both of which define the coefficient domain.

- The call `Gcdex($a,b,x$,´$s$´,´$t$´)` **mod** p performs the computation modulo p a prime integer. The multivariate polynomials a and b must have rational coefficients or coefficients in a finite field specified by **RootOfs**.

- The call `evala(Gcdex($a,b,x$,´$s$´,´$t$´))` does likewise. The multivariate polynomials a and b must have algebraic number coefficients specified by **RootOfs**.

Examples:

```
> Gcdex(x^2+x+1,x^2-x+1,x,´s´,´t´) mod
  11;                                       ⟶                  1
```

> s, t; $\longrightarrow$ 5 x + 6, 6 x + 6

SEE ALSO: `gcdex`, `Gcd`, `mod`, `evala`, `RootOf`

2.1.143   gcdex – extended Euclidean algorithm for polynomials

Calling Sequence:
    gcdex($A,B,x,$ $'s','t'$)
    gcdex($A,B,C,x,$ $'s','t'$)

Parameters:
    A, B, C, x –   polynomials in the variable x
    s, t –   (optional) unevaluated names

Synopsis:
- If the number of parameters is less than six, **gcdex** applies the extended Euclidean algorithm to compute polynomials s and t such that $s * A + t * B = g$ (where g is the gcd of A and B as univariate polynomials in x) with **degree**(s,x) < **degree**(B,x) and **degree**(t,x) < **degree**(A,x). The gcd is returned as the function value.

- In the case of six parameters, **gcdex** solves the polynomial diophantine equation $s * A + t * B = C$. The input polynomials must satisfy **degree**(C,x) < **degree**(A,x) + **degree**(B,x) and if C is divisible by **gcd**(A,B) then s and t will be polynomials satisfying **degree**(s,x) < **degree**(B,x) and **degree**(t,x) < **degree**(A,x). The **NULL** value is returned as the function value.

- Note that if the input polynomials are multivariate then, in general, s and t will be rational functions in variables other than x.

- In the six-parameter case, if C is not divisible by **gcd**(A,B) then s and t will be nontrivial rational functions in x.

Examples:
> gcdex(x^3-1,x^2-1,x,'s','t'); $\longrightarrow$ - 1 + x

> s,t; $\longrightarrow$ 1, - x

> gcdex(x^2+a,x^2-1,x^2-a,x,'s','t');
    s,t; $\longrightarrow$ $-\dfrac{a-1}{a+1},\ 2\ \dfrac{a}{a+1}$

SEE ALSO: `igcdex`, `gcd`, `Gcdex`

2.1.144   genpoly – generate polynomial from integer n by Z-adic expansion

Calling Sequence:
    genpoly(n, b, x);

Parameters:

n –   integer or polynomial with integer coefficients

b –   integer

x –   variable name

Synopsis:

- `genpoly` computes the unique polynomial `a(`x`)` over `Z[`x`]` from the integer n with coefficients less than $b/2$ in magnitude such that `subs(`x`=`b`,a(`x`))` `=` n.

- This is directly related to b-adic expansion of an integer. If the base-b representation of the integer n is `c0 + c1*`b` + ... + ck*`b`^k` where `ci` are integers modulo b (using symmetric representation) then the polynomial generated is `c0 + c1*`x` + ... + ck*`x`^k` .

- If n is a polynomial with integer coefficients then each integer coefficient is expanded into a polynomial.

Examples:
```
> genpoly( 11, 5, x );
```
$$1 + 2\,x$$
```
> genpoly( 11*y^2-13*y+21, 5, x );
```
$$(1 + 2\,x)\,y^2 + (2 + 2\,x - x^2)\,y + 1 - x + x^2$$

2.1.145   harmonic – the harmonic function

Synopsis:

- The `harmonic` function is defined as

$$\text{harmonic}(n) = \sum_{i=1}^{n} \frac{1}{i} = \Psi(n+1) + \gamma$$

This function is known to `evalf`.

2.1.146   has – test for a specified subexpression

Calling Sequence:

`has( `f`, `x` );`

Parameters:

f –   an expression

x –   an expression or list or set of expressions

Synopsis:
- If f contains the expression x, then the function **has** will return **true**; otherwise it will return **false**. The expression f contains x iff a subexpression of f is equal to x.

- If x is a list or set, then f contains x iff f contains at least one item in x.

Examples:

> has((a+b)^(4/3), a+b);	$\longrightarrow$	true
> has((a+b)^(4/3), a);	$\longrightarrow$	true
> has(a+b+c, a+b);	$\longrightarrow$	false
> has(a+b+c, {c,d,e});	$\longrightarrow$	true
> has(a+b+c, {d,e,f});	$\longrightarrow$	false

SEE ALSO: **member**, **op**

2.1.147   hastype – test for a specified type

Calling Sequence:
> hastype(*expr*, *t*);

Parameters:

    expr – any expression

    t – name of a datatype

Synopsis:
- The Boolean function **hastype** returns **true** if and only if *expr* has any subexpressions of the type t.

Examples:

> hastype((x+y),`+`);	$\longrightarrow$	true
> hastype((x+y)*(1/2*x+y), fraction);	$\longrightarrow$	true
> hastype(x+7*y, fraction);	$\longrightarrow$	false

SEE ALSO: **has**, **type**

2.1.148   help – descriptions of syntax, datatypes, and functions

Calling Sequence:
> ?*topic* or ?*topic,subtopic* or ?*topic*[*subtopic*] or
> help(*topic*) or help(*topic,subtopic*) or help(*topic*[*subtopic*])

Synopsis:

`?intro`	introduction to Maple
`?library`	Maple library functions and procedures
`?index`	list of all help categories
`?index,`*category*	list of help files on specific topics
`?`*topic*	explanation of a specific topic
`?`*topic,subtopic*	explanation of a subtopic under a topic
`?distribution`	for information on how to obtain Maple
`?copyright`	for information about copyrights

- Note 1: The recommended way to invoke help is to use the question mark.

- Note 2: When invoking help using the function call syntax, `help(`*topic*`)`, Maple keywords (reserved words) must be enclosed in backquotes. For example, `help(quit)` causes a syntax error. Use `help(`quit`)` instead. Note that the string delimiter is the backquote (`` ` ``), not the apostrophe (´), nor the double quote ("). When using the question mark syntax for help, no quotes are required.

- Note 3: A command must end with a semicolon, followed by **RETURN** or **ENTER**, before Maple will execute it and display the result. The semicolon can appear on the next line if you forget to end the command with it, but it must appear. There can be multiple commands on one line, separated by semicolons or colons. An exception to this is when a line starts with a question mark in which case help is invoked and no semicolon is required.

- To contact Waterloo Maple Software, see `?distribution`. To contact the authors of Maple, see `?scg`.

SEE ALSO: `keywords`, `quotes`, `colons`, `quit`, `example`, `scg`, `distribution`, `TEXT`

2.1.149   Hermite – compute the Hermite normal form of a matrix mod p
Smith – compute the Smith normal form of a matrix mod p

Calling Sequence:

Hermite(A, x)

Smith(A, x)

Parameters:

A – a rectangular matrix of univariate polynomials in x

x – name

Synopsis:

- `Hermite` and `Smith` are inert forms of `linalg[hermite]` and `linalg[smith]` respectively.

- Hermite and Smith are placeholders for representing the hermite and smith normal forms respectively. They are used in conjunction with mod as is described below.

- Hermite(A, x) mod p computes the Hermite normal form (reduced row echelon form) of an m by n rectangular matrix of univariate polynomials in x over the integers modulo p. The polynomial coefficients must be rational or elements of a finite extension field specified by RootOfs.

- Smith(A, x) mod p computes the Smith normal form of a matrix with univariate polynomial entries in x over the integers modulo p. The coefficients of the polynomial must be either rational or elements of a finite extension field specified by RootOfs.

Examples:

```
> A := array([[1+x, 1+x^2],[1+x^2, 1+x^4]])
  ;
```
$$\longrightarrow \quad A := \begin{bmatrix} & & & 2 \\ 1+x & & 1+x \\ & & & \\ & 2 & & 4 \\ 1+x & & 1+x \end{bmatrix}$$

```
> Hermite(A,x) mod 2;
```
$$\longrightarrow \quad \begin{bmatrix} & & & & & 2 & \\ 1+x & & & & 1+x & \\ & & & & & & \\ & & 4 & 3 & 2 & \\ 0 & & x + x + x + x \end{bmatrix}$$

```
> Smith(A,x) mod 2;
```
$$\longrightarrow \quad \begin{bmatrix} 1+x & & & 0 & \\ & & & & & & \\ & & 4 & 3 & 2 & \\ 0 & & x + x + x + x \end{bmatrix}$$

SEE ALSO: `linalg[hermite]`, `linalg[smith]`, `RootOf`

2.1.150   icontent – integer content of a polynomial

Calling Sequence:
    `icontent( expr );`

Parameters:
    expr   –   an expanded polynomial with rational coefficients

Synopsis:

- `icontent` computes a rational number such that dividing the expanded polynomial by it makes the polynomial primitive over the integers. Thus, for integer coefficients, `icontent` returns the greatest common divisor of the coefficients.

- For rational coefficients, `icontent` computes the least common multiple of the denominators of all fractions in addition to the greatest common divisor of the numerators of the coefficients.

Examples:
```
> icontent( 3*x + 12*y );
```
$\longrightarrow \qquad 3$

> icontent(7/6*y + 21/4*z); $\longrightarrow$ 7/12

SEE ALSO: content

2.1.151   ifactor – integer factorization

Calling Sequence:
>   ifactor(n)
>   ifactor(n, *method*)

Parameters:

n –   integer or a rational

method   –   (optional) name of base method for factoring

Synopsis:

- ifactor returns the complete integer factorization of n.

- The answer is in the form: u * ''(f1)^e1 * ... * ''(fn)^en such that n = u * f1^e1 * ... * fn^en where u equals sign(n), f1, ..., fn are the distinct prime factors of n, and e1, ..., en are their multiplicities (negative in the case of the denominator of a rational).

- The **expand** function may be applied to cause the factors to be multiplied together again.

- If a second parameter is specified, the named method will be used when the front-end code fails to achieve the factorization. By default, the Morrison-Brillhart algorithm is used as the base method. Currently accepted names are:

    'squfof'   - D. Shanks' undocumented square-free factorization;
    'pollard'   - J.M. Pollard's rho method;
    'lenstra'   - Lenstra's elliptic curve method; and
    'easy'   - which does no further work.

- If the 'easy' option is chosen, the result of the ifactor call will be a product of the factors that were easy to compute, and a name _c.m indicating an m-digit composite number that was not factored.

- The pollard base method accepts an additional optional integer: ifactor(n,pollard,k), which increases the efficiency of the method when one of the factors is of the form k*m+1.

Examples:
> ifactor(61); $\longrightarrow$ (61)

> ifactor(60); $\longrightarrow$

$$(2)^2 \ (3) \ (5)$$

> ifactor(-144); $\longrightarrow$

$$- \ (2)^4 \ (3)^2$$

`> expand(");`	$\longrightarrow$	-144
`> ifactor( 60, easy );`	$\longrightarrow$	$(2)^2$ (3) (5)
`> ifactor( 4/11 );`	$\longrightarrow$	$\dfrac{(2)^2}{(11)}$
`> n := 8012940887713968000041:` `  ifactor( n, easy );`	$\longrightarrow$	(13) (457) $\_c19$
`> ifactor( n );`	$\longrightarrow$	(13) (457) (473638939) (2847639359)

SEE ALSO: `ifactors, isprime, factor, type[facint]`

2.1.152   igcd – greatest common divisor of integers
ilcm – least common multiple of integers

Calling Sequence:

    $\text{igcd}(x_1, x_2, \ldots);$
    $\text{ilcm}(x_1, x_2, \ldots);$

Parameters:

    $x_1, x_2, \ldots$ –   any integers

Synopsis:

- The function `igcd` computes the greatest common divisor of an arbitrary number of integers. The function `ilcm` computes the least common multiple of an arbitrary number of integers.

Examples:

`> igcd();`	$\longrightarrow$	0
`> ilcm();`	$\longrightarrow$	1
`> igcd( -10, 6, -8 );`	$\longrightarrow$	2
`> ilcm( -10, 6, -8 );`	$\longrightarrow$	120
`> igcd( a, b );`	$\longrightarrow$	`igcd(a, b)`

SEE ALSO: `gcd, lcm`

2.1.153   igcdex – extended Euclidean algorithm for integers

Calling Sequence:

    $\text{igcdex}(\ a,\ b,\ 's',\ 't'\ )$

Parameters:

     a,b –   integers

     s,t –   (optional) names

Synopsis:

- `igcdex` returns g = `igcd`(a,b), and optionally s and/or t such that

$$g = s\ a + t\ b$$

Examples:

```
> igcdex(2,3,'s','t');
```
                         $\longrightarrow$ 1

```
> s; t;
```
                         $\longrightarrow$ -1

                                                 1

SEE ALSO: `igcd`, `ilcm`, `gcdex`

2.1.154   indets – find indeterminates of an expression

Calling Sequence:

     `indets(`*expr*`)`

     `indets(`*expr*`,`*typename*`)`

Parameters:

     expr   –   any expression

     typename   –   (optional) the name of a type

Synopsis:

- `indets` with only one argument returns a set containing all the indeterminates of *expr*.

- *expr* is viewed as a rational expression (an expression formed by applying only the operations +, –, *, / to some given symbols). Therefore expressions such as `sin(x)`, `f(x,y)`, and `x^(1/2)` are treated as indeterminates.

- Expressions of type `constant` such as `sin(1)`, `f(3,5)`, and `2^(1/2)` are not considered to be indeterminates in the single-argument case.

- If a second argument *typename* is specified then the value returned is a set containing all subexpressions in *expr* which are of type `typename`, including subexpressions which may not have been considered to be indeterminates in the single-argument case.

Examples:

```
> indets( x*y + z/x );
```
                         $\longrightarrow$ {z, y, x}

```
> e := x^(1/2) + exp(x^2) + f(9):
  indets(e);
```
                         $\longrightarrow$ $\{x, x^{1/2}, \exp(x^2)\}$

```
> indets(e,function);
```
$\longrightarrow$
$$\{exp(x^{2}),\ f(9)\}$$

2.1.155   indices – indices of a table or array
entries – entries of a table or array

Calling Sequence:
> indices(*t*)
> entries(*t*)

Parameters:
> *t*   –   a table or array

Synopsis:
- The `indices` and `entries` functions return sequences of the indices (or keys) and the entries respectively of the table or array *t* corresponding to the entries that are explicitly stored.

- The result returned is a sequence of lists in an apparently arbitrary order which cannot be controlled by the user. However, there is a one-to-one correspondence between the result of `indices` and `entries`.

Examples:
```
> t := table([green=gruen, red=rot,
  blue=blau, black=schwarz]):
  indices(t);
```
$\longrightarrow$ `[blue], [red], [black], [green]`
```
> entries(t);
```
$\longrightarrow$ `[blau], [rot], [schwarz], [gruen]`
```
> a := array([[1,2],[3,4]]):
  indices(a);
```
$\longrightarrow$ `[2, 2], [1, 1], [2, 1], [1, 2]`
```
> entries(a);
```
$\longrightarrow$ `[4], [1], [3], [2]`

SEE ALSO: `op`, `array`, `table`

2.1.156   int or Int – definite and indefinite integration

Calling Sequences:

int(*f*, *x*)	Int(*f*, *x*)
int(*f*, *x*=*a*..*b*)	Int(*f*, *x*=*a*..*b*)
int(*f*, *x*=*a*..*b*, 'continuous')	Int(*f*, *x*=*a*..*b*, 'continuous')

Parameters:
> *f*   –   an algebraic expression or a procedure, the integrand
>
> *x*   –   a name
>
> *a, b*   –   interval on which integral is taken

continuous – (optional) suppresses the check for continuity

Synopsis:

- The function `int` computes the indefinite or definite integral of f with respect to the variable x. The name `integrate` is a synonym for `int`.

- Indefinite integration is performed if the second argument x is a name. Note that no constant of integration appears in the result. Definite integration is performed if the second argument is of the form $x=a..b$ where a and b are the endpoints of the interval of integration.

- If Maple cannot find a closed form for the integral, the function call itself is returned. (The prettyprinter displays the `int` function using a stylized integral sign.)

- The capitalized function name `Int` is the inert `int` function, which simply returns unevaluated. The prettyprinter understands `Int` to be equivalent to `int` for printing purposes.

- In the case of a definite integral which returns unevaluated, numerical integration may be invoked by applying `evalf` to the unevaluated integral. To invoke numerical integration without first invoking symbolic integration, use the inert function `Int` as in: `evalf( Int(f, x=a..b) )`.

- When applying numerical integration, the default method can be overridden by using the calling sequence

```
readlib(`evalf/int`):
`evalf/int`(f, x=a..b, Digits, _NCrule);
```

  in which case an adaptive Newton-Cotes rule will be applied. (The default method applies singularity-handling and then uses Clenshaw-Curtis quadrature, which is a Chebyshev series method. Only if this fails does it then invoke adaptive Newton-Cotes.)

- If `f` is a procedure then `int` will return unevaluated; if `evalf` is applied in this case, numerical integration will be performed using an adaptive Newton-Cotes rule.

- For symbolic definite integration, the function `iscont` is invoked to ensure the integrand is continuous before taking limits at the endpoints. This check can be disabled by calling `int` with third argument `continuous` (a global name).

- Note that Maple's `series` function may be invoked on an unevaluated integral to compute a series expansion of the integral (when possible).

- When `int` is applied to a series structure, the internal function `int/series` is invoked to compute the integral efficiently.

Examples:
```
> int( sin(x), x );
```
$$- \cos(x)$$
```
> int( sin(x), x=0..Pi );
```

```
> int( x/(x^3-1), x );
                    2                      1/2                    1/2
    1/3 ln(x - 1) - 1/6 ln(x  + x + 1) + 1/3 3    arctan(1/3 (2 x + 1) 3    )
> int( exp(-x^2), x );
                              1/2
                      1/2 Pi     erf(x)

> int( exp(-x^2)*ln(x), x=0..infinity );
                        1/2                1/2
              - 1/4 Pi     gamma - 1/2 Pi     ln(2)

> int( exp(-x^2)*ln(x), x );
                          /
                          |       2
                          |  exp(- x ) ln(x) dx
                          |
                          /

> series( ", x=0, 4 );
                                              3          5
              (ln(x) - 1) x + (- 1/3 ln(x) + 1/9) x  + O(x )

> int( ", x );
                   2       2       4              4      6
         1/2 ln(x) x  - 3/4 x  - 1/12 x  ln(x) + 7/144 x  + O(x )

> int( exp(-x^2)*ln(x), x=0..1 );
                         1
                         /
                         |       2
                         |  exp(- x ) ln(x) dx
                         |
                         /
                         0

> evalf(");
                            -.9059404763

> int( exp(-x^2)*ln(x)^2, x=0..infinity );
          5/2        1/2     2        1/2                     1/2     2
    1/16 Pi   + 1/8 Pi    gamma + 1/2 Pi   gamma ln(2) + 1/2 Pi    ln(2)

> Int( exp(-x^2)*ln(x)^2, x=0..infinity );
                        infinity
                         /
                         |         2      2
                         |    exp(- x ) ln(x)  dx
                         |
                         /
                         0

> evalf(");
                         1.947522180
```

```
> int( 1/(x+a)^2, x=0..2, 'continuous' );
                    1
             - ----- + 1/a
               2 + a
```

SEE ALSO: `diff`, `evalf`, `series`, `limit`, `iscont`, `student[Int]`, `laplace`, `mellin`

2.1.157   interface – set or query user interface variables

Calling Sequence:

    `interface( `*arg1, arg2,* ` ... )`

Parameters:

    arg.i   –   is either of the form *name=val* or simply *name*

Synopsis:

- The function `interface` is provided as a unique mechanism of communication between Maple and the user interface (called Iris). Specifically, this function is used to set and query all variables which affect the format of the output but do not affect the computation.

- If an argument is of the form *name=val* then this specifies the setting for the named variable.

- If an argument is simply a *name* then the current value of the named variable is retrieved and returned as the function value. It is an error if more than one argument is just a *name*.

- The standard set of interface variables, i.e. the set of variables which should be supported by all types of user interfaces, are:

Name	Values	Explanation
echo	0..4	0 - do not echo under any circumstance; 1 - (default) echo whenever the input or the output is not from/to the terminal, but do not echo as a result of a read statement; 2 - echo whenever the input or the output is not from/to the terminal; 3 - echo only as a result of a read statement; 4 - echo everything. The echo option is superseded by quiet; so if quiet=true, no echo will occur.
indentamount	integer	The number of spaces items are indented if they must be broken across lines. This is also used in formatting procedures.
labelling labeling	true/false	This flag enables the use of the %1, %2, ... labels for subexpressions on output. It defaults to true.
labelwidth	integer	An item must be at least this many characters long to be considered for a %n label. The measurement used is approximate.
prettyprint	true/false	Boolean variable which determines if the output will be in two dimensions. If false, linear printing is used.
prompt	string	The string which is printed when user input is expected.
quiet	true/false	An interface constant which will suppress all auxiliary printing (logo, garbage collection messages, bytes used messages,

and prompt).

screenheight	integer	The height of the screen in characters.
screenwidth	integer	The width of the screen in characters. Must be greater than ten.
plotdevice	string	The name of the plotting device. See `plot[device]` for a list of known devices.
plotoutput	string	The name of a file where the plot output will be stored. See `plot`.
preplot	list(integer)	A list of integers which encodes the sequence to be sent before a plot to initialize the output device. Negative values specify a delay. See `plot[setup]` for values.
postplot	list(integer)	A list of integers which encodes the sequence to be sent after a plot is made to restore the output device. Negative values specify a delay. See `plot[setup]` for values.
terminal	string	Terminal setup for plotting: sets the magic values for plotdevice, preplot, and postplot for the terminals listed under the terminal directory in the library (currently: kd500g, vt240, xtek).
verboseproc	0,1,2	0 - do not print the body of a procedure, print only a skeleton of the form proc(x) ... end ; 1 - (default) print the body of user-defined

procedures but do not print the body of
Maple library procedures;
2 - print the body of all procedures.

version   string   An interface constant. It provides the name
of the interface; it is readable but cannot
be set.

Examples:
```
interface( plotdevice=postscript, plotoutput=myfile );
oldquiet := interface( quiet, quiet=true );
interface( echo=2, prompt=`#-->` );
```

SEE ALSO: `plot`, `plot[device]`, `plot[setup]`, `plotsetup`

2.1.158   Interp – inert interp function

Calling Sequence:
> Interp(x, y, v)
> Interp(x, y)

Parameters:
> x –   list or vector of independent values, x[1],..x[n]
> y –   list or vector of dependent values, y[1],..y[n]
> v –   variable name to be used in polynomial

Synopsis:
- The function `Interp` is a placeholder for polynomial interpolation. It is used in conjunction with `mod` and `modp1`.

- The call `Interp(x,y,v) mod p` computes the polynomial of degree at most `n-1` in the name `v` that interpolates the points `(x[1],y[1])`, `(x[2],y[2])`,..., `(x[n],y[n]) mod p`. The points must be from a finite field.

- The call `modp1(Interp(x,y),p)` computes the interpolation polynomial in the `modp1` representation where `x` and `y` must be lists.

- All the independent values in x must be distinct. In other words, a particular value modulo p must not occur more than once in x.

Examples:
```
> Interp([2,5,6], [9,8,3], x) mod 11;
```
$$\longrightarrow \qquad 8\,x^2 + 6\,x + 9$$

```
> alias(alpha=RootOf(x^4+x+1));
```
$$\longrightarrow \qquad I, \text{ alpha}$$

```
> a := Interp([0,1,alpha],[alpha,alpha^2,
  alpha^3], x) mod 2;
```
$$\longrightarrow \qquad a := x^2 + (alpha^2 + alpha + 1)\ x + alpha$$

SEE ALSO: `interp`, `mod`, `modp1`, `Eval`

2.1.159   interp – polynomial interpolation

Calling Sequence:
>   interp(x, y, v)

Parameters:
> x – list or vector of independent values, x[1],..x[n+1]
> y – list or vector of dependent values, y[1],..y[n+1]
> v – variable to be used in polynomial

Synopsis:
- The function interp computes the polynomial of degree less than or equal to **n** in the variable v which interpolates the points $(x[1],y[1])$, $(x[2],y[2])$, ..., $(x[n+1],y[n+1])$.

- If the same x-value is entered twice, it is an error, whether or not the same y-value is entered; all independent values must be distinct.

SEE ALSO: `Interp`

2.1.160   irem – integer remainder
iquo – integer quotient

Calling Sequence:
>   irem(m,n)   irem($m,n,\,'q'$)
>   iquo(m,n)   iquo($m,n,\,'r'$)

Parameters:
> m, n – any expressions
> q, r – names

Synopsis:
- If m and n are both integers the function **irem** computes the integer remainder of m divided by n. If the third argument is present it will be assigned the quotient. Likewise, **iquo** computes the integer quotient of m divided n and if the third argument is present assigns it the remainder.

- Specifically, if m and n are integers then **irem** returns **r** such that $m = n*q + r$, **abs(r)** < **abs**(n) and **m*r** $\geq$ 0.

- If m and n are not both integers then `irem` remains unevaluated.

Examples:

> irem(23,4,´q´);	$\longrightarrow$	3
> q;	$\longrightarrow$	5
> iquo(23,4,´r´);	$\longrightarrow$	5
> r;	$\longrightarrow$	3
> irem(-23,4);	$\longrightarrow$	-3
> irem(23,-4);	$\longrightarrow$	3
> irem(-23,-4);	$\longrightarrow$	-3
> irem(x,3);	$\longrightarrow$	irem(x, 3)

SEE ALSO: **rem, quo**

2.1.161   Irreduc – inert irreducibility function

Calling Sequence:
> Irreduc(a)
> Irreduc(a,K)

Parameters:
> a – multivariate polynomial
> K – RootOf

Synopsis:

- The `Irreduc` function is a placeholder for testing the irreducibility of the multivariate polynomial a. It is used in conjunction with `mod` and `modp1`.

- Formally, an element a of a commutative ring R is said to be "irreducible" if it is not zero, not a unit, and a = b*c implies either b or c is a unit.

- In this context where R is the ring of polynomials over the integers mod p, which is a finite field, the units are the non-zero constant polynomials. Hence all constant polynomials are not irreducible by this definition.

- The call `Irreduc(`a`) mod p` returns **true** iff a is "irreducible" modulo p. The polynomial a must have rational coefficients or coefficients from a finite field specified by `RootOf` expressions.

- The call `Irreduc(`a,K`) mod p)` returns **true** iff a is "irreducible" modulo p over the the finite field defined by K, an algebraic extension of the integers mod p where K is a `RootOf`.

- The call `modp1(Irreduc(a),p)` returns **true** iff a is "irreducible" modulo p. The polynomial a must be in the `modp1` representation.

Examples:

> Irreduc(2) mod 7;	$\longrightarrow$	false

```
> Irreduc(2*x^2+6*x+6) mod 7;                   false
```
$\longrightarrow$

```
> Irreduc(x^4+x+1) mod 2;                        true
```
$\longrightarrow$

```
> alias(alpha=RootOf(x^4+x+1)):
  Irreduc(x^4+x+1,alpha) mod 2;                  false
```
$\longrightarrow$

```
> Factor(x^4+x+1,alpha) mod 2;
```
$\longrightarrow$

$$(x + \text{alpha}) (x + \text{alpha}^2 + 1)$$

$$(x + \text{alpha} + 1) (x + \text{alpha}^2)$$

SEE ALSO: isprime, irreduc, mod, modp1, RootOf, Factor

2.1.162   irreduc – polynomial irreducibility test

Calling Sequence:
> irreduc(a); or irreduc(a, K);

Parameters:
> a –   a univariate polynomial
>
> K –   (optional) algebraic number field extension

Synopsis:
- The irreduc function tests whether a univariate polynomial over an algebraic number field is irreducible. It returns true if a is irreducible, false otherwise. Note that a constant polynomial by convention is reducible.

- The call irreduc(a); tests for irreducibility over the field implied by the coefficients present; if all the coefficients are rational, then the irreducibility test is over the rationals.

- The call irreduc(a,K); tests for irreducibility over the algebraic number field defined by K. K must be a single RootOf, a list or set of RootOfs, a single radical, or a list or set of radicals.

Examples:
```
> irreduc( 2 );                                  false
```
$\longrightarrow$

```
> irreduc( x^3+5 );                              true
```
$\longrightarrow$

```
> irreduc( x^3+5, 5^(1/3) );                     false
```
$\longrightarrow$

```
> factor( x^3+5, 5^(1/3) );
```
$\longrightarrow$

$$(x^2 - 5^{1/3} x + 5^{2/3}) (x + 5^{1/3})$$

```
> alias(alpha=RootOf(x^3-5)):
  irreduc( x^3+5, alpha );                       false
```
$\longrightarrow$

```
> factor( x^3+5, alpha );
```
$\longrightarrow$

$$(x^2 - \text{alpha} \, x + \text{alpha}^2) (x + \text{alpha})$$

SEE ALSO: isprime, factor, roots, Irreduc, RootOf

2.1.163   isolve – solve equations for integer solutions

Calling Sequence:

    `isolve(`*eqns,* *vars*`)`

Parameters:

    eqns – a set of equations or a single equation

    vars – (optional) set of variables or a variable

Synopsis:

- The procedure `isolve` solves the equations in *eqns* over the integers. It solves for all of the indeterminates occurring in the equations.

- The optional second argument *vars* is used to name global variables that have integer values and occur in the solution, and if there is only one argument, then the global names _N1, _N2, and so forth are used.

- It returns the `NULL` value if there are no integer solutions.

Examples:

`> isolve(3*x-4*y=7);` $\longrightarrow$ `{y = - 7 + 3 _N1, x = - 7 + 4 _N1}`

`> isolve(x^2=3);`

2.1.164   isprime – primality test

Calling Sequence:

    `isprime(`*n,* *iter*`)`

Parameters:

    n   – integer

    iter – (optional) positive integer (default 5)

Synopsis:

- The function isprime is a probabilistic primality testing routine.

- It returns false if n is shown to be composite within *iter* tests and returns true otherwise. If *isprime* returns true, n is "very probably" prime - see Knuth Vol 2, 2nd edition, Section 4.5.4, Algorithm P for a reference.

SEE ALSO: `nextprime`, `prevprime`, `ithprime`

2.1.165   isqrt – integer square root
iroot – integer nth root

Calling Sequence:
>   isqrt(*x*)
>   iroot(*x*, *n*)

Parameters:

> x – an integer
>
> n – an integer

Synopsis:

- The `isqrt` function computes an integer approximation to the square root of x. The approximation is exact for perfect squares, and the error is less than 1 otherwise. Note: if $x < 0$, `isqrt(x)` returns 0.

- The `iroot` function computes an integer approximation to the nth root of x. The approximation is exact for perfect powers, and the error is less than 1 otherwise. Note: if $x < 0$, and n is even, `iroot(x, n)` returns 0.

- The `iroot` function should be defined by the command `readlib(iroot)` before it is used.

Examples:
```
> isqrt(10);
```
$\longrightarrow$ 3
```
> readlib(iroot):
  iroot(100, 3);
```
$\longrightarrow$ 5

SEE ALSO: `sqrt`, `issqr`, `psqrt`, `numtheory[msqrt]`, `numtheory[mroot]`

2.1.166 ithprime – determine the ith prime

Calling Sequence:
>   ithprime(*i*)

Parameters:

> i – positive integer

Synopsis:

- The function `ithprime` returns the ith prime number, where the first prime number is 2.

SEE ALSO: `isprime`, `nextprime`, `prevprime`

2.1.167 laplace – Laplace transform

Calling Sequence:
>   laplace(*expr*, *t*, *s*)

Parameters:

expr	–	expression to be transformed
t	–	variable *expr* is transformed with respect to
s	–	parameter of transform

Synopsis:

- The `laplace` function applies the Laplace transformation to *expr* with respect to *t*.

- Expressions involving exponentials, polynomials, trigonometrics (`sin`, `cos`, `sinh`, `cosh`) with linear arguments, and Bessel functions (`BesselJ`, `BesselI`) with linear arguments can be transformed.

- The `laplace` function also recognizes derivatives (`diff`) and integrals (`int` or `Int`).

- When transforming expressions like `diff(y(t), t, s)`, `laplace` will insert the initial values `y(0)`, `D(y)(0)`, etc. `D(y)(0)` is the value of the first derivative at 0; `D(D(y))(0)` is the second derivative at 0, and so on.

- Both `laplace` and `invlaplace` recognize the Dirac-delta (or unit-impulse) function as `Dirac(t)` and Heaviside's unit step function as `Heaviside(t)`.

Examples:

```
> laplace(t^2+sin(t)=y(t), t, s);
```
$$\frac{2 s^2 + 2 + s^3}{s^3 (s^2 + 1)} = laplace(y(t), t, s)$$

```
> laplace(t^(3/2)-exp(t)+sinh(a*t), t, s);
```
$$3/4 \frac{Pi^{1/2}}{s^{5/2}} - \frac{1}{s - 1} + \frac{a}{s^2 - a^2}$$

```
> laplace(diff(y(t), t$2)-y(t)=sin(a*t), t, s);
```
$$(laplace(y(t), t, s) s - y(0)) s - D(y)(0) - laplace(y(t), t, s) = \frac{a}{s^2 + a^2}$$

```
> laplace(BesselI(0,a*t), t, s);
```
$$\frac{1}{(s^2 - a^2)^{1/2}}$$

SEE ALSO: `invlaplace`, `ztrans`, `dsolve`, `D` , `mellin`, `int`

2.1.168 latex – produce output suitable for latex printing

Calling Sequence:
> latex(*expr*)
> latex(*expr*, *filename*)

Parameters:
> *expr*   –   any expression
>
> *filename* –   (optional) file in which to put the output

Synopsis:
- The `latex` function produces output which is suitable for printing the input *expr* with a LaTeX processor. It knows how to format integrals, limits, sums, products and matrices.

- The mathematical format is taken, in general, from the CRC handbook or the Handbook of Mathematical functions.

- The inert functions `Sum`, `Int`, `Diff`, `Limit` and `Log` can be used instead of `sum`, `int`, `diff`, `limit` and `log` to prevent evaluation by Maple.

- It is possible to extend the abilities of `latex` to format other objects by defining a procedure with the name `` `latex/functionname` ``. The `latex` function will call this procedure when it encounters a function call to *functionname* within *expr*. For more information, see `latex[functions]`.

- Note that the output of the `latex` command does not include the commands to invoke the latex math environment.

Examples:
```
> latex(x^2 + y^2 = z^2);
x^{2}+y^{2}=z^{2}
# Put this output in the file LatexFile
> latex(x^2 + y^2 = z^2, LatexFile);
> latex(Int(1/(x^2+1), x) = int(1/(x^2+1), x));
\int \!\left (x^{2}+1\right )^{-1}{dx}=\arctan(x)
```

SEE ALSO: `eqn`, `fortran`, `C`, `writeto`, `appendto`, `latex[functions]`, `latex[names]`

2.1.169 latex[functions] – how latex formats functions

Synopsis:
- When `latex` processes a Maple object of type `function` (i.e., an unevaluated function call), it checks to see if there exists a procedure by the name of `latex/function_name`, where *function_name* is the name of the function. If such a procedure exists, it is used to format the function call.

- For instance, invoking `latex(Int(exp(x), x=1..3))` causes `latex` to check to see if there is a procedure by the name of `latex/Int`. Since such a procedure exists, the above call to `latex` returns the result of `latex/Int(exp(x), x=1..3)` as its value. This allows LaTeX to produce standard mathematical output in most situations.

- If such a procedure does not exist, `latex` formats the function call by recursively formatting the operands to the function call and inserting parentheses and commas at the appropriate points.

- Maple has pre-defined `latex/` functions for the following constructs:

@	@@	D	Diff	Int	Limit	Log
Sum	abs	binomial	diff	exp	factorial	int
limit	ln	log10	log2	sum		

- Standard mathematical functions such as `sin`, `cos`, and `tan` are converted to `\sin`, `\cos`, and `\tan`. See `latex[names]` for a description of how this is done.

Examples:
```
> latex(Int(exp(x), x));              ⟶    \int \!{e^{x}}{dx}
> latex(Limit(3+y(x), x=0,left));     ⟶    \lim _{x\rightarrow 0^{-}}3+y(x)
```

SEE ALSO: `latex`, `latex[names]`

2.1.170   latex[names] – how latex formats names

Synopsis:

- When `latex` processes a Maple object of type `string` it makes various tests to see if special processing is desirable. This allows LaTeX to produce standard mathematical output in most situations.

- On encountering a string, `latex` first tests to see if the given string refers to an entry in the table `latex/special_names`. If so, that entry's value is used. This is how `I` is converted to `i`, `Pi` to `pi`, and so forth.

- Next, `latex` checks to see if the given string is contained in the set `latex/greek`. If it is, then a backslash is prepended to its value. This allows names like "alpha" and "beta" to be printed as Greek symbols.

- Then, `latex` checks to see if the given string is contained in the set `latex/mathops`. If it is, then a backslash is prepended to its value. This allows function names like "log" and "cos" to be printed correctly without further processing.

- The fonts used by `latex` are controlled by two variables: `latex/csname_font`, which is initially set to `\it`, and `latex/verbatim_font`, which is initially set to `\tt`. The former is used for printing multi-character strings that don´t require back-quoting; the latter is used for those strings that do.

- The `latex` command searches for special function formatting rules before it searches for special name formatting rules. See `latex[functions]` for more information.

Examples:

> latex(3 * Pi / I);	$\longrightarrow$	-3\,\pi \,\sqrt {-1}
> latex(Total_profit*Rate);	$\longrightarrow$	{\it Total_profit}\,{\it Rate}
> latex(GAMMA(x)/gamma+psi/beta);	$\longrightarrow$	{\frac {\Gamma (x)}{\gamma}}+{\frac {\p\
		si}{\beta}}

SEE ALSO: `latex`, `latex[functions]`

2.1.171   lcoeff – leading coefficient of a multivariate polynomial
tcoeff – trailing coefficient of a multivariate polynomial

Calling Sequence:

lcoeff(p) or tcoeff(p)

lcoeff(p, x) or tcoeff(p, x)

lcoeff(p, x, $'t'$) or tcoeff(p, x, $'t'$)

Parameters:

p –   multivariate polynomial

x –   (optional) indeterminate or list or set of indeterminates

$'t'$ –   (optional) unevaluated name

Synopsis:

- The functions `lcoeff` and `tcoeff` return the leading (trailing) coefficient of p with respect to the indeterminate(s) x. If x is not specified, `lcoeff` (`tcoeff`) computes the leading (trailing) coefficient with respect to all the indeterminates of p. If a third argument t is specified ("call by name"), it is assigned the leading (trailing) term of p.

- If x is a single indeterminate, and d is the degree of p in x, then `lcoeff`(p, x) is equivalent to `coeff`(p, x, d). If x is a list or set of indeterminates, lcoeff (tcoeff) computes the leading (trailing) coefficient of p considered as a multivariate polynomial in the variables x.

- Note that p must be `collected` with respect to the appropriate indeterminates before calling `lcoeff` or `tcoeff` – see `coeff` for further details.

Examples:

> s := 3*v^2*w^3*x^4+1;	$\longrightarrow$	$s := 3\,v^2\,w^3\,x^4 + 1$
> lcoeff(s);	$\longrightarrow$	3
> tcoeff(s);	$\longrightarrow$	1
> lcoeff(s, [v,w], 't');	$\longrightarrow$	$3\,x^4$
> t;	$\longrightarrow$	$v^2\,w^3$

SEE ALSO: `collect`, `coeff`, `coeffs`, `indets`, `degree`, `ldegree`

2.1.172   length – length of an object

Calling Sequence:

length(*expr*)

Parameters:

expr   –   any expression

Synopsis:

- If *expr* is an integer, the number of decimal digits is returned (where the length of **zero** is defined to be zero and the length of a negative integer is defined to be the same as the length of the absolute value of the integer).

- If *expr* is a string, the number of characters in *expr* is returned.

- For other objects, the length of each operand of *expr* is computed recursively and added to the number of words used to represent *expr*. This gives a measure of the size of *expr*.

Examples:

> length(0);	$\longrightarrow$	0
> length(11);	$\longrightarrow$	2
> length(14.5);	$\longrightarrow$	7
> length(abc);	$\longrightarrow$	3
> length(x+2*y);	$\longrightarrow$	9

2.1.173   lexorder – test for lexicographical order

Calling Sequence:

lexorder(*name$_1$*, *name$_2$*)

Parameters:

name$_1$, *name$_2$*   –   unevaluated names

Synopsis:

- The function **lexorder** returns **true** if *name$_1$* occurs before *name$_2$* in lexicographical order, or if *name$_1$* is equal to *name$_2$*. Otherwise, it returns **false**.

- The lexicographical order depends in part upon the ordering of the underlying character set, which is system-dependent.

- For names consisting of ordinary letters, lexicographical order is the standard alphabetical order.

Examples:

> lexorder(a, b);	$\longrightarrow$	true
> lexorder(A, a);	$\longrightarrow$	true

`> lexorder(` `` ` a ` ``, a);`	$\longrightarrow$	true
`> lexorder(greatest, great);`	$\longrightarrow$	false
`> lexorder(` `` `*` ``, `` `^` ``);`	$\longrightarrow$	true

2.1.174 lhs – left hand side of an expression
rhs – right hand side of an expression

Calling Sequence:

lhs(*expr*)

rhs(*expr*)

Parameters:

expr – an equation, inequality, relation, or range

Synopsis:

- lhs(*expr*) returns the left hand side of *expr*, which is equivalent to op(1, *expr*).

- rhs(*expr*) returns the right hand side of *expr*, which is equivalent to op(2, *expr*).

Examples:

`> e := y = a*x^2 + b;`	$\longrightarrow$	$e := y = a\,x^2 + b$
`> lhs(e);`	$\longrightarrow$	y
`> rhs(e);`	$\longrightarrow$	$a\,x^2 + b$
`> r := 2..5;`	$\longrightarrow$	$r := 2 .. 5$
`> lhs(r);`	$\longrightarrow$	2
`> rhs(r);`	$\longrightarrow$	5

SEE ALSO: op

2.1.175 limit – calculate limits
Limit – inert form of limit

Calling Sequences:

limit(*f*, *x*=*a*);   limit(*f*, *x*=*a*, *dir*);

Limit(*f*, *x*=*a*);   Limit(*f*, *x*=*a*, *dir*);

Parameters:

f – an algebraic expression

x – a name

a — an algebraic expression (limit point, possibly `infinity`, or `-infinity`

dir — (optional) direction: one of `left`, `right`, `real`, or `complex`

Synopsis:
- The `limit` function attempts to compute the limiting value of f as x approaches a.

- If dir is not specified, the limit is the real bidirectional limit, except in the case where the limit point is `infinity` or `-infinity`, in which case the limit is from the left to `infinity` and from the right to `-infinity`. For help with directional limits see `limit[dirs]`.

- `limit` will assume that any unassigned variables in the expression are real and nonzero (as signum assumes). Thus, `limit(a^2*x, x=infinity)` will return `infinity`.

- The output from limit may be a range (meaning a bounded result) or an algebraic expression, possibly containing infinity. For further help with the return type see `limit[return]`.

- Most limits are resolved by computing series. By increasing the value of the global variable `Order`, `limit` will increase its ability to solve problems with significant cancellation.

- Based on *A New Algorithm for Computing Symbolic Limits Using Generalized Hierarchical Series*, K.O. Geddes and G.H. Gonnet, Symbolic and Algebraic Computation (Proceedings of ISSAC ´88), P. Gianni, (ed.), Lecture Notes in Computer Science, Springer-Verlag, Berlin, 1989, pp 490-495.

- If Maple cannot find a closed form for the limit, the function call itself is returned. (The prettyprinter displays the limit function using a two-dimensional format.)

- The capitalized function name `Limit` is the inert `limit` function, which simply returns unevaluated. The prettyprinter understands `Limit` to be equivalent to `limit` for printing purposes.

Examples:

```
> limit(sin(x)/x, x=0);                    ⟶              1
> limit(exp(x), x=infinity);               ⟶           infinity
> limit(exp(x), x=-infinity);              ⟶              0
> limit(1/x, x=0, real);                   ⟶          undefined
> limit(exp(x^2)*(1-erf(x)), x=infinity);  ⟶              0
> Limit(sin(x), x=0);                      ⟶        Limit  sin(x)
                                                    x -> 0
```

SEE ALSO: `limit[dirs]`, `limit[return]`, `series`, `signum`

2.1.176 limit[dir] – directional limits

Calling Sequence:
    `limit(f, point, dir)`

Synopsis:

- If *dir* is **left** or **right**, the limit is a directional limit, taken from the left or right, respectively. If *dir* is **real**, the limit is the bidirectional real limit. If *dir* is **complex**, the limit is omni-directional, from all complex directions to the point.

- If *dir* is not specified, the limit is the real bidirectional limit, except in the case where the limit point is **infinity** or **-infinity**, in which case the limit is from the left to **infinity** or from the right to **-infinity** respectively.

- If *dir* is **complex**, the limit point **infinity** denotes complex infinity, that is, all infinities in the complex plane. If *dir* is **real**, the limit point **infinity** denotes both positive and negative infinity, and the limit is done bidirectionally. Otherwise, the limit point **infinity** denotes positive infinity, and **-infinity** denotes negative infinity.

Examples:

> limit(exp(x), x=infinity);	$\longrightarrow$	infinity
> limit(exp(x), x=-infinity);	$\longrightarrow$	0
> limit(exp(x), x=infinity,real);	$\longrightarrow$	undefined
> limit(1/x, x=0, right);	$\longrightarrow$	infinity
> limit(1/x, x=0, left);	$\longrightarrow$	- infinity
> limit(1/x, x=0);	$\longrightarrow$	undefined
> limit(1/x, x=0, real);	$\longrightarrow$	undefined
> limit(1/x, x=0, complex);	$\longrightarrow$	infinity
> limit(1/x, x=infinity, real);	$\longrightarrow$	0
> limit(-x, x=infinity, complex);	$\longrightarrow$	infinity
> limit(-x, x=infinity);	$\longrightarrow$	- infinity

SEE ALSO: limit, limit[multi], limit[return]

2.1.177   limit[multi] – multidirectional limits

Calling Sequence:
>   limit(*f*, *points*)
>   limit(*f*, *points*, *dir*)

Parameters:
>   *f*   –   an algebraic expression
>   *points* –   a set of equations of the form **x=a**
>   *dir*   –   (optional) direction

Synopsis:
- Given a set of points as its second argument, limit attempts to compute the limiting value of f in a multi-dimensional space.
- The optional direction *dir* is applied to all points, and consequently each point will be approached from the same direction; from the left or right, bidirectional (real), or complex.
- If the limit depends on the direction approached, undefined is returned.
- A limiting point defined as x=x is ignored.

Examples:
```
> limit((x^2-y^2)/(x^2+y^2), {x=0,y=0});          ⟶                    undefined

> limit(x+1/y, {x=0,y=infinity});                 ⟶                         0

> limit(x*y, {x=0,y=infinity});                   ⟶                    undefined

> limit((sin(x^2)-sin(y^2))/(x-y), {x=0,
  y=0});                                          ⟶
```

$$\lim\left(\frac{\sin(x^2) - \sin(y^2)}{x - y},\ \{x = 0,\ y = 0\}\right)$$

SEE ALSO: limit[dir], limit[return], limit

2.1.178   limit[return] – values returned by limit

Calling Sequence:
   limit(*f*, *point*, *dir*)

Parameters:

f	–	an algebraic expression
point	–	an equation, x=a, where x is a name and a is the limit point
dir	–	(optional) direction

Synopsis:
- The meaning of a returned infinity depends on *dir*. If *dir* is complex, then infinity denotes complex infinity. Otherwise, the result infinity denotes real positive infinity, and -infinity denotes real negative infinity.
- If limit returns a numeric range it means that the value of the limiting expression is known to lie in that range for arguments restricted to some neighborhood of the limit point. It does not necessarily imply that the limiting expression is known to achieve every value infinitely often in this range.
- If the limit is known to be undefined, or each side of a two-sided limit has a different value, or for a multi-dimensional limit the limiting value depends on the direction from which the limit is approached, then undefined is returned.

- If `limit` is unable to evaluate the limit, then it returns unevaluated. Unless the limit is unevaluated, the limit returned is independent of the variable given in *point*.

Examples:

`> limit(exp(x), x=infinity);`	$\longrightarrow$	infinity
`> limit(1/x, x=0, complex);`	$\longrightarrow$	infinity
`> limit(sin(1/x), x=0);`	$\longrightarrow$	-1 .. 1
`> limit(exp(x), x=infinity,real);`	$\longrightarrow$	undefined
`> limit(tan(x), x=infinity);`	$\longrightarrow$	undefined
`> limit(a*x, x=infinity);`	$\longrightarrow$	signum(a) infinity

SEE ALSO: `limit`, `limit[dirs]`, `limit[multi]`

2.1.179   ln, log, log10 – logarithm functions

Synopsis:

- The syntax `ln(x)` for an expression `x` represents the natural logarithm of `x` (logarithm to the base e).

- Automatic simplifications will be applied to the `ln` function.

- The `log10` function gives the logarithm base 10.

- The `log` function initially has the same value as the `ln` function. The user may assign to `log` a function that computes logarithms to some other base.

2.1.180   lprint – linear printing of expression

Calling Sequence:

    `lprint(`*expr*$_1$`, `*expr*$_2$`, ...)`

Parameters:

    expr$_1$, *expr*$_2$, ...   –   any expressions

Synopsis:

- The function `lprint` returns NULL and prints its arguments in a one-dimensional format.

- The expressions, each separated from the others by three blanks, are printed on a single line.

- Expressions returned from `lprint` are in general valid input to Maple.

- Maple will print all expressions using `lprint` if the **interface** variable **prettyprint** is set to `false` (default `true`).

Examples:
```
> lprint(3*x^2,x);
> g := (x^4 - y)/(y^2-3*x);
```
$\longrightarrow$ `3*x**2   x`

$\longrightarrow$

$$g := \frac{x^4 - y}{y^2 - 3\,x}$$

```
> lprint(g);
```
$\longrightarrow$ `(x**4-y)/(y**2-3*x)`

SEE ALSO: `print, printlevel, interface`

2.1.181   macro – define an macro - abbreviation

Calling Sequence:

   $\text{macro}(e_1,\ e_2,\ \ldots,\ e_n)$

Parameters:

   $e_1, e_2, \ldots, e_n$ –   zero or more equations

Synopsis:

- The `macro` facility is a simple abbreviation facility to be used when writing Maple functions and library code.

- The effect of invoking `macro(f = g)` is that Maple when reading input from the terminal or a file will transform all occurrences of `f` on input into `g` except for parameters and local variables.

- The arguments to macro are equations which are neither evaluated nor subjected to any existing macros. This means that you cannot define one macro in terms of another. You can however change the macro definition by doing `macro(f = h)`. The command `macro(f = f)` will remove the macro definition for `f`.

- A macro may be defined for any Maple object, except a numerical constant.

- A typical use of the macro facility is the following. Suppose we are writing a Maple procedure that uses the Maple library routine `combinat['fibonacci']` in several places, and we wish to avoid having to type this long function name every time it is used. We can define the following: `macro(F=combinat['fibonacci'])`, and then the symbol `F` can be used as an abbreviation for `combinat['fibonacci']` throughout the remainder of the code.

Examples:
```
> macro(v = linalg[vandermonde]);
  v([3,2,1]);
```
$\longrightarrow$
```
[ 1   3   9 ]
[           ]
[ 1   2   4 ]
[           ]
[ 1   1   1 ]
```

SEE ALSO: `alias, subs`

2.1.182   map – apply a procedure to each operand of an expression

Calling Sequence:

> map(*fcn*, *expr*, *arg*₂, ..., *arg*ₙ)

map(fcn, $expr$, arg_2, ..., arg_n)

Parameters:

fcn	–	a procedure or a name
expr	–	any expression
*arg*ᵢ	–	(optional) further arguments to *fcn*

Synopsis:

- The `map` function applies *fcn* to the operands of *expr*.

- The ith operand of *expr* is replaced by the result of applying *fcn* to the ith operand. This is done for all the operands of *expr*.

- For a table or array, *fcn* is applied to each element of the table or array.

- If *fcn* takes more than one argument, they are to be specified as additional arguments, arg_2, arg_3,..., arg_n, which are simply passed through as the second, third ,..., n^{th} arguments to *fcn*.

Examples:

```
> map(f, x + y*z);                        ⟶        f(x) + f(y z)
> map(f, y*z);                            ⟶          f(y) f(z)
> map(f, {a,b,c});                        ⟶        {f(a), f(b), f(c)}
> map(x -> x^2, x + y);                   ⟶            2    2
                                                      x  + y
> map(proc(x,y) x^2+y end, [1,2,3,4], 2); ⟶        [3, 6, 11, 18]
```

SEE ALSO: `op`, `proc`, `operators[functional]`, `select`

2.1.183   match – pattern matching

Calling Sequence:

> match(*expr* = *pattern*, *v*, ´*s*´)

Parameters:

expr	–	the expression to be matched
pattern	–	the pattern to match
v	–	the name of the main variable
´*s*´	–	the name of the return argument

Synopsis:

- The `match` function returns `true` if it was able to match *expr* to *pattern*, and `false` otherwise. If the match was successful, *s* is assigned ("call by name") a substitution set such that `subs`(*s*, *pattern*) = *expr*.

- The main variable is a variable which must be matched exactly into the pattern. In other words, the main variable cannot be substituted for any value.

Examples:
```
> match(ln(k)/k^(1/2) = A*ln(k)^P*k^Q, k, 's');
                              true

> s;
                    {P = 1, A = 1, Q = -1/2}

> match(5*x^2-3*x+z*x+y = a*(x+b)^2+c, x, 's');
                              true

> s;
                                                    2
       {a = 5, b = - 3/10 + 1/10 z, c = - 9/20 + 3/10 z - 1/20 z  + y}
```

SEE ALSO: `solve[identity]`, `subs`

2.1.184   max – maximum of numbers
min – minimum of numbers

Calling Sequence:
> $\max(x_1, x_2, \ldots)$
> $\min(\ x_1, x_2, \ldots)$

Parameters:
> $x_1, x_2, \ldots$ – any expressions

Synopsis:
- The functions **max** and **min** return the maximum or minimum respectively of one or more arguments.

- Most often the arguments will be of type **numeric**, being integers, rationals, or floats. However, the functionality is more general, allowing any type of argument for which an unevaluated function call may be returned.

Examples:

`> max(3/2, 1.49);`	$\longrightarrow$	3/2
`> min(3/2, 1.49);`	$\longrightarrow$	1.49
`> max(3/5, evalf(ln(2)), 9/13);`	$\longrightarrow$	.6931471806
`> max(x+1, x+2, y);`	$\longrightarrow$	max(x + 2, y)

SEE ALSO: `maximize`, `minimize`, `extreme`

2.1.185   maxnorm – infinity norm of a polynomial

Calling Sequence:

maxnorm(a)

Parameters:

a – an expanded polynomial

Synopsis:

- Given a polynomial, `maxnorm` computes the coefficient with the largest absolute value.
- The coefficients of the polynomial should be integer, rational, or floating point numbers.

Examples:

```
> maxnorm(x-3*y);
```
$\longrightarrow$

3

SEE ALSO: `norm`, `linalg[norm]`

2.1.186   MeijerG – Special case of the general Meijer G function

Calling Sequence:

MeijerG(m, a, z)

Parameters:

m – an integer greater than 2

a – real expression

z – complex or real expression

Synopsis:

- The general Meijer G function is defined as

```
 mn  /  | a  a  . . . a  \
G    | x |  1  2        p  |
 pq  \  | b  b  . . . b  /
        | 1  2        q
```

- In certain cases, it reduces to the generalized hypergeometric function. However, our special case is a three-argument function defined by:

```
                    m   0 /  |  0  0  . . .  0  \
MeijerG (m,a,z) =  G      | x |                  |
                  m-1  m  \  | -1 -1  . . . a-1 -1 /
```

- It is very useful to express successive derivatives of the incomplete Gamma function `GAMMA`(a, z) with respect to its first parameter a:

$$\frac{d^m\ \mathrm{GAMMA}(a,z)}{da^m} = \ln^m z\ \mathrm{GAMMA}(a,z) + m\,z\sum_{i=0}^{m-1}\ln^{m-i-1}z\ P_i^{m-1}\ \mathrm{MeijerG}\,(3+i,a,z)$$

$$\text{where}\ \ p_j^i = \binom{i}{j}\,j! = \frac{i!}{(i-j)!}$$

- Special cases of this function reduce to simpler special functions. For example, `MeijerG(3, n, z)` for **n** a positive integer reduces to an expression involving the exponential integral `Ei(-z)`.

- Reference: *Evaluation of classes of definite integrals involving elementary functions via differentiation of special functions*, K.O. Geddes, M.L. Glasser, R.A. Moore, and T.C. Scott, AAECC (Applicable Algebra in Engineering, Communication, and Computing) 1, November 1990, pp. 149-165.

Examples:

`> z*MeijerG(2, a, z);`	$\longrightarrow$	`GAMMA(a, z)`
`> z*MeijerG(3, 1, -z);`	$\longrightarrow$	`Ei(z)`

SEE ALSO: `GAMMA`, `Ei`

2.1.187   mellin – Mellin transform

Calling Sequence:

   `mellin(`*expr*`, ` *t*`, ` *s*`)`

Parameters:

  expr – expression to be transformed

  t   – variable *expr* is transformed with respect to

  s   – parameter of transform

Synopsis:

- The function `mellin` applies the Mellin transformation to *expr* with respect to *t*.

- Some expressions involving exponentials, polynomials, algebraic functions, trigonometrics (`sin`, `cos`, `sinh`, `cosh`) or various special functions can be transformed.

- The `mellin` function attempts to reduce the expression according to a set of simplification rules and then tries to match the reduced expression against an internal table of basic Mellin transforms. Entries to this table can be added by using the function `mellintable`.

- The class of expressions which can be treated by `mellin` is limited by some weaknesses of the function `solve/identity`.

Examples:
```
> mellin(a*x^b*exp(-x^(1/4)),x,s);
                    4 a GAMMA(4 s + 4 b)
> mellin(exp(-3*x^2)/(exp(x^2)-1),x,s);
            /                        1     \
     1/2   |Zeta(1/2 s) - 1 - --------| GAMMA(1/2 s)
            |                   (1/2 s)|
            \                  2       /
```

SEE ALSO: `mellintable`, `laplace`, `int`

2.1.188   member – test for membership in a set or list

Calling Sequence:

member(x, s, $'p'$)

Parameters:

x –   any expression

s –   a set or list

$'p'$ –   (optional) an unevaluated name

Synopsis:

- The function **member** determines if x is a member of the set or list s. It returns **true** if so, and **false** otherwise.

- If a third argument $'p'$ is present and **member** yields **true**, then the position of the first x in s will be assigned to p.

Examples:
```
> member(y, {x, y, z});                    ⟶        true
> member(y, {x*y, y*z});                    ⟶        false
> member(x*y, [x*y, w+u, y]);               ⟶        true
> member(w, [x, y, w, u], 'k');             ⟶        true
> k;                                        ⟶        3
```

SEE ALSO: **has**

2.1.189   mod, modp, mods – computation over the integers modulo m

Calling Sequence:

e mod m

modp(e, m)

mods(e, m)

`mod`(e, m)

Parameters:

e –   an algebraic expression

m –   a nonzero integer

Synopsis:

- The **mod** operator evaluates the expression e over the integers modulo m. It incorporates facilities for doing finite field arithmetic and polynomial and matrix arithmetic over finite fields, including factorization.

- The operator syntax e **mod** m is equivalent to the function call `mod`(e,m). The global variable `mod` may be assigned either the modp function or the **mods** function. When assigned the value **modp** (the default), the positive representation for integers modulo m is used; i.e. all rational coefficients will be reduced to integers in the range [0,abs(m)-1]. When assigned the value **mods**, the symmetric representation is used; i.e. all rational coefficients will be reduced to integers in the range [-iquo(abs(m)-1,2), iquo(abs(m),2)].

- If the modulus m is a prime integer, then all coefficient arithmetic is done in the finite field of integers modulo m. Elements of finite fields of characteristic m with $q = m\hat{\ }n$ elements are represented as polynomials in *alpha* where *alpha* is a simple algebraic extension over the integers mod m. The extension *alpha* is a **RootOf** a monic univariate irreducible polynomial of degree n over the integers mod m. See **RootOf** and the examples below.

- The following functions for polynomial and matrix arithmetic over finite rings and fields are known to mod. See help for further details.

Content	Det	DistDeg	Divide	Eval
Expand	Factor	Factors	Gcd	Gcdex
Hermite	Interp	Irreduc	Normal	Nullspace
Power	Powmod	Prem	Primitive	Primpart
Quo	Randpoly	Randprime	Rem	Resultant
RootOf	Roots	Smith	Sprem	Sqrfree

- To compute $i\hat{\ }n$ **mod** m where i is an integer, it is undesirable to use this "obvious" syntax because the powering will be performed first over the integers (possibly resulting in a very large integer) before reduction modulo m. Rather, the inert operator &^ should be used: i &^ n mod m. In the latter form, the powering will be performed intelligently by the **mod** operation. Similarly

Powmod(a, n, b, x) mod m computes Rem($a\hat{~}n, b, x$) mod m (where a and b are polynomials in x) without first computing $a\hat{~}n$ mod m.

- Other modular arithmetic operations are stated in their natural form:

$$i+j \bmod m; \quad i\text{-}j \bmod m; \quad i*j \bmod m;$$
$$j\hat{~}(\text{-}1) \bmod m; \quad i/j \bmod m;$$

where the latter case will perform $i*j\hat{~}(\text{-}1)$ modulo m.

- The left precedence of the mod operator is lower than (less binding strength than) the other arithmetic operators. Its right precedence is immediately higher than +, – and lower than *, / .

- There is an interface for user-defined mod functions. For example, if the user has defined the procedure `mod/f` then the operation `f(x,y) mod 23` will generate the function call `mod/f`(x,y,23).

- The mod operator is mapped automatically onto equations, the coefficients of polynomials, and the entries of lists and sets.

Examples:

```
> modp(12,7);
```
$\longrightarrow$ 5

```
> 12 mod 7;
```
$\longrightarrow$ 5

```
> mods(12,7);
```
$\longrightarrow$ -2

```
> 1/3 mod 7;
```
$\longrightarrow$ 5

```
> 5*3 mod 7;
```
$\longrightarrow$ 1

```
> 5 &^ 1000 mod 100;
```
$\longrightarrow$ 25

```
> a := 15*x^2+4*x-3 mod 11;
```
$\longrightarrow$
$$a := 4 x^2 + 4 x + 8$$

```
> `mod` := mods:
  b := 3*x^2+8*x+9 mod 11;
```
$\longrightarrow$
$$b := 3 x^2 - 3 x - 2$$

```
> gcd(a,b);
```
$\longrightarrow$ 1

```
> g := Gcd(a,b) mod 11;
```
$\longrightarrow$ $g := x + 5$

```
> Divide(a,g,'q') mod 11;
```
$\longrightarrow$ true

```
> q;
```
$\longrightarrow$ $4 x - 5$

```
> factor(x^3+2);
```
$\longrightarrow$
$$x^3 + 2$$

```
> Factor(x^3+2) mod 5;
```
$\longrightarrow$
$$(x - 2) (x^2 + 2 x - 1)$$

```
> alias(alpha=RootOf(y^2+2*y-1)):
  Normal(1/alpha) mod 5;
```
$\longrightarrow$ alpha + 2

```
> Factor(x^3+2,alpha) mod 5;        ⟶     (x - alpha) (x - 2) (x + alpha + 2)
> Expand(") mod 5;                  ⟶                  3
                                                      x  + 2
```

SEE ALSO: msolve, modp1, GF, iquo, irem

2.1.190 modp1 – univariate polynomial arithmetic modulo n

Calling Sequence:

   modp1(e, m);

Parameters:

   e – an algebraic expression

   m – a positive integer

Synopsis:

- The purpose of the modp1 function is to make available very efficient arithmetic and other operations in domain of univariate polynomials over the integers modulo n written Zn[x]. The mod function, which supports operations in the domain of multivariate polynomials over the integers modulo n, makes use of modp1 for the univariate case. Hence interactive calculations can be done using mod.

- To achieve a high level of efficiency, modp1 uses a special representation. Explicit conversions are provided for converting Maple's "sum of products" representation to and from this representation. Knowledge of this representation is not required by the user in order to use modp1.

- The actual representation used depends on the size of n, the modulus. If n < prevprime(floor(sqrt MAXINT)) where MAXINT is the largest positive integer representable by the hardware (on 32 bit machines MAXINT = 2^31-1) then a modp1 polynomial is represented internally as an array of machine integers. If the modulus n is greater than this number, the modp1 polynomial is represented as a Maple list of Maple multi-precision integers.

- A typical use of modp1 would be modp1(Eval(a, x), n), meaning evaluate the modp1 polynomial a at the value x (an integer) modulo n. The following functions are known to modp1:

Add	Chrem	Coeff	Constant	ConvertIn
ConvertOut	Degree	Det	Diff	Divide
Embed	Eval	Factors	Gcd	Gcdex
Interp	Irreduc	Lcm	Lcoeff	Ldegree
Monomial	Multiply	One	Power	Powmod
Prem	Quo	Randpoly	Rem	Resultant
Roots	Smith	Shift	Sqrfree	Subtract
Tcoeff	UNormal	Zero		

- The `Add` function is nary. `Subtract` is unary or binary. The nullary functions `Zero` and `One` create the 0 and 1 `modp1` polynomials. The `Constant` function creates a `modp1` polynomial for the given integer constant. The `Multiply` function is binary. Note that scalar multiplication requires explicit conversion using the `Constant` function.

- The function `Degree` computes the degree, `Ldegree` the low degree, `Coeff` the coefficient, `Lcoeff` the leading coefficient, `Tcoeff` the trailing Coefficient, `Diff` the derivative, and `UNormal` the monic part or unit normal part. The `Randpoly` function takes a degree as an argument and generates a polynomial of the given degree with random coefficients modulo n.

- The two functions `ConvertIn` and `ConvertOut` are used to convert to and from the `modp1` representation. `modp1(ConvertIn(b,x),n)` converts from a univariate polynomial b in x over the integers to a `modp1` polynomial modulo n. `modp1(ConvertIn(a),n)` converts from a list of integers to a `modp1` polynomial. The function `ConvertOut` does the reverse of `ConvertIn`.

- Note that unlike the `mod` function, `modp1` arithmetic operations do not take the variable as an argument. For example, the calling sequences of the `Rem` operation are `Rem(a,b)` and `Rem(a,b,'q')`. Calling sequences are similar for `Coeff`, `Degree`, `Diff`, `Discrim`, `Gcdex`, `Interp`, `Lcoeff`, `Ldegree`, `Powmod`, `Prem`, `Quo`, `Resultant`, `Smith`, and `Tcoeff`.

- Separate help exists for most of the remaining functions.

Examples:

```
> p := 11;                              ⟶              p := 11

> a := x^4-1;                           ⟶                 4
                                                     a := x  - 1

> a := modp1(ConvertIn(a,x),p);         ⟶        a := 10000000000000010

> modp1(ConvertOut(a,x),p);             ⟶                 4
                                                       x  + 10

> modp1(ConvertOut(a),p);               ⟶          [10, 0, 0, 0, 1]

> b := modp1(Randpoly(3),p);            ⟶         b := 9000000040008

> Rem(a,b);                             ⟶    Rem(10000000000000010, 9000000040008)

> Roots(a);                             ⟶        Roots(10000000000000010)

> Factors(a);                           ⟶        Factors(10000000000000010)
```

SEE ALSO: `` `mod` ``, `GF`, `evalgf`

2.1.191 msolve – solve equations in Z mod m

Calling Sequence:

    `msolve(`*eqns*`,`*vars*`,`*m*`)` or `msolve(`*eqns*`,`*m*`)`

Parameters:

    eqns   –   a set of equations or a single equation

    vars   –   (optional) set of names or a name

    m   –   an integer

Synopsis:

- The procedure `msolve` solves the equations in *eqns* over the integers (`mod` *m*). It solves for all of the indeterminates occurring in the equations.

- If the solution is indeterminate, and if it is possible, a family of solutions is expressed in terms of the names given in *vars* (or the global names _NN1, _NN2, _NN3 if *vars* is omitted). These names (which should not coincide with the indeterminates) are then allowed to take any integer values.

- It returns `NULL` if there are no solutions over the integers (`mod` *m*).

- The procedure `msolve` has special code for efficiently handling large systems of equations `mod` 2.

- If you want to solve a single univariate polynomial `mod` *m* where *m* is not necessarily prime, see the `Roots` function.

Examples:

```
> msolve({3*x-4*y=1,7*x+y=2},19);              ⟶              {x = 15, y = 11}

> msolve(x^2=3,5);
  msolve(2^i=3,19);                            ⟶              {i = 13 + 18 _NN1}

> msolve(8^j=2,x,17);                          ⟶              {j = 3 + 8 x}
```

SEE ALSO: `` `mod` ``, `Roots`, `solve`

2.1.192   nextprime – determine the next largest prime
             prevprime – determine the next smallest prime

Calling Sequence:

    `nextprime(`*n*`)`

    `prevprime(`*n*`)`

Parameters:

    n   –   an integer

Synopsis:

- The `nextprime` function returns the smallest prime that is larger than *n*. The `prevprime` function returns the largest prime that is less than *n*.

SEE ALSO: `isprime`, `ithprime`

2.1.193   norm – norm of a polynomial

Calling Sequence:

    norm(a, n, v)

Parameters:

    a – an expanded polynomial

    n – a real constant > 1 or the name infinity

    v – (optional) variable specification

Synopsis:

- The `norm` function computes the nth norm of the polynomial a in the indeterminates v. For $n \geq$ 1 the norm is defined:

    ```
    norm(a,n,v) = sum(abs(c)^n for c in [coeffs(a,v)])^(1/n)
    ```

- If v is not specified, the `indets`(a) are used.

- This function must be explicitly loaded via `readlib(norm)`:

Examples:
```
> readlib(norm):
  norm( x-3*y, 1 );
```
$\longrightarrow$ 4
```
> norm( x-3*y, 2 );
```
$\longrightarrow$ $\begin{array}{c} 1/2 \\ 10 \end{array}$
```
> norm( x-3*y, infinity );
```
$\longrightarrow$ 3

SEE ALSO: `linalg[norm]`,`indets`

2.1.194   Normal – inert normal function

Calling Sequence:

    Normal(a)

Parameters:

    a – an expression involving RootOfs

Synopsis:

- The `Normal` function is a placeholder for representing the `normal` function. It is used in conjunction with `mod` and `evala` as described below which defines the coefficient domain.

- The call `Normal`(a) `mod p` applies the `normal` function to a where a is a rational expression over a finite field.

- The call `evala(Normal`(a)) applies the `normal` function to a where a may contain algebraic extensions defined by RootOfs.

Examples:
```
> Normal( (x^3-2*x^2+2*x+1)/(x^4+1) ) mod 5;
                            x + 3
                          ------
                            2
                          x  + 3
> evala(Normal( (x^2-2)/(x-RootOf(_Z^2-2)) ));
                              2
                  x + RootOf(_Z  - 2)
```

SEE ALSO: `normal`, `evala`, `mod`, `RootOf`, `Expand`, `Gcd`

2.1.195   normal – normalize a rational expression

Calling Sequence:
>   normal(f)
>   normal(f, expanded)

Parameters:
> f –   an algebraic expression

Synopsis:

- The **normal** function provides a basic form of simplification. It recognizes those expressions equal to zero which lie in the domain of "rational functions". This includes any expression constructed from sums, products and integer powers of integers and variables. The expression is converted into the "factored normal form". This is the form numerator/denominator, where the numerator and denominator are relatively prime polynomials with integer coefficients.

- If f is a list, set, range, series, equation, relation, or function, **normal** is applied recursively to the components of f. In the case of a series, for example, this means that the coefficients of the series are normalized.

- If f contains subexpressions not in the domain of rational functions such as square roots, functions, and series, **normal** is first applied recursively to these objects, then they are "frozen" (see the frontend function) to unique names so that the "form" of f can be simplified. For such cases, **normal** may not recognize when an expression is equal to zero.

- If normal is called with one argument, the numerator and denominator will be a product of expanded polynomials. If normal is called with the second argument, *expanded*, the numerator and denominator will be expanded polynomials.

Examples:
```
> normal( x^2-(x+1)*(x-1)-1 );            ⟶              0
> normal( (x^2-y^2)/(x-y)^3 );            ⟶            x + y
                                                     --------
                                                         2
                                                     (x - y)
```

```
> normal( (f(x)^2-1)/(f(x)-1) );
```
$\longrightarrow$

$$f(x) + 1$$

```
> normal( {2/x + y/3 = 0} );
```
$\longrightarrow$

$$\left\{ \frac{1}{3} \; \frac{6 + y\,x}{x} = 0 \right\}$$

```
> normal( sin(x*(x+1)-x) );
```
$\longrightarrow$

$$\sin(x^2)$$

```
> normal( 1/x+x/(x+1) );
```
$\longrightarrow$

$$\frac{x + 1 + x^2}{x\,(x + 1)}$$

```
> normal( 1/x+x/(x+1), expanded );
```
$\longrightarrow$

$$\frac{x + 1 + x^2}{x^2 + x}$$

SEE ALSO: denom, expand, factor, map, numer, simplify, Normal

2.1.196 Nullspace – compute the nullspace of a matrix mod p

Calling Sequence:

  Nullspace(A) mod p

Parameters:

  A – a matrix over a finite field

  p – a positive integer

Synopsis:

- Nullspace(A) mod p computes a basis for the null space (Nullspace) of the linear transformation defined by the matrix A. The result is a (possibly empty) set of vectors.

Examples:

```
> A := array( [[1,2,3],[1,2,3],[0,0,0]] )
  ;
```
$\longrightarrow$

$$A := \begin{bmatrix} 1 & 2 & 3 \\ 1 & 2 & 3 \\ 0 & 0 & 0 \end{bmatrix}$$

```
> Nullspace( A ) mod 5;
```
$\longrightarrow$

$$\{[\; 1, \; 4, \; 0 \;], \; [\; 0, \; 0, \; 1 \;]\}$$

SEE ALSO: linalg[nullspace]

2.1.197   numer – numerator of an expression
denom – denominator of an expression

Calling Sequence:

numer(e) and denom(e)

Parameters:

e – any algebraic expression

Synopsis:

• The procedures **numer** and **denom** are typically called after first using the **normal** function. The procedure **normal** is used to put an expression in "normal form" which is the form numerator/denominator where both the numerator and denominator are polynomials. In this case, **numer** simply picks off the numerator of e and **denom** picks off the denominator of e. Note that if e is in normal form, the numerator and denominator will have integer coefficients.

• If e is not in normal form (e contains a subexpression which has one or more terms which are quotients of expressions), it is first converted into a normal form. A common denominator is found so that e can be expressed in the form numerator/denominator.

Examples:

> numer(2/3);	$\longrightarrow$	2
> denom(2/3);	$\longrightarrow$	3
> numer(x^2-(x-1)*(x+1));	$\longrightarrow$	1
> numer((1+x)/x^(1/2)/y);	$\longrightarrow$	x + 1
> numer(2/x + y);	$\longrightarrow$	2 + y x
> numer(x+1/(x+1/x));	$\longrightarrow$	$\dfrac{2}{x\ (x^2 + 2)}$
> denom(x+1/(x+1/x));	$\longrightarrow$	$\dfrac{2}{x^2 + 1}$

SEE ALSO: **normal**

2.1.198   O – the order of function

Synopsis:

• This function is used to indicate the order term in series. See **type/series** and **Order** for more details.

• An example

$$1 + x + x^2 + x^3 + O(x^4)$$

2.1.199   op – extract operands from an expression
nops – the number of operands of an expression

Calling Sequence:

op(i, e)   op($i..j, e$)   op(e)
nops(e)

Parameters:

i,j – non-negative integers marking positions of operands

e – any expression

Synopsis:

- The op function extracts the components of an expression. The **nops** function returns the number of components of an expression.

- If the first argument is a nonnegative integer i then the result is the ith operand of e.

- If the first argument is a range $i..j$ then the result is an expression sequence of the ith to jth operands of e. If there is only one argument, op(e) = op(1..nops(e),e).

- If e is an integer or a string, nops(e) yields 1 and op(1,e) yields e.

- For some data structures, op 0 is defined as a special case. For subscripts and functions, op 0 is the name of the subscript or function and for series, op 0 is the expansion point of the series.

Examples:

> u := [1,4,9];	$\longrightarrow$	u := [1, 4, 9]
> nops(u);	$\longrightarrow$	3
> op(2,u);	$\longrightarrow$	4
> op(2..3,u);	$\longrightarrow$	4, 9
> [op(u),16];	$\longrightarrow$	[1, 4, 9, 16]
> v := f(x,y,z);	$\longrightarrow$	v := f(x, y, z)
> nops(v);	$\longrightarrow$	3
> op(0,v);	$\longrightarrow$	f
> op(v);	$\longrightarrow$	x, y, z

SEE ALSO: type, index[datatypes]

2.1.200   order – determine the truncation order of a series

Calling Sequence:

order($expr$)

Parameters:

    expr   –   an expression of type series

Synopsis:

- The function `order` will return the degree of the indeterminate in the `O(x)` term of a series. If there is no order term in the series then the name `infinity` is returned.

- This function will only work with expressions of type series.

Examples:

```
> series(1/(1-x),x);
```
$$\longrightarrow \qquad 1 + x + x + x + x + x + O(x)$$
$$ \quad 2 \quad\; 3 \quad\; 4 \quad\; 5 \qquad 6$$

```
> order(");
```
$$\longrightarrow \qquad\qquad 6$$

SEE ALSO: `numtheory[order]`, `Order`

2.1.201   plot – create a 2-dimensional plot of functions

Calling Sequence:

    plot(f, h, v)

    plot(f, h, v,...)

Parameters:

    f –   function(s) to be plotted

    h –   horizontal range

    v –   vertical range (optional)

Synopsis:

- A typical call to the `plot` function is `plot(`$f(x)$`,`x`=a..b)`, where f is a real function in x and `a..b` specifies the horizontal real range on which f is plotted.

- The `plot` function provides support for two-dimensional plots of one or more functions specified as expressions, procedures, parametric functions, or lists of points. See `plot[function)]` for more information on plotting functions. For three-dimensional plots, see `plot3d`. A call to plot produces a `PLOT` data structure, which is then printed. For information on the `PLOT` data structure, see `plot[structure]`.

- The horizontal and vertical range arguments h and v define the axis labels and the range over which the function(s) are displayed. They take one of the following forms: `string`, `low..hi`, or `string=low..hi`, where `low` and `hi` are real constants. See `plot[range]` for further information.

- Remaining arguments are interpreted as options which are specified as equations of the form `option = value`. In particular, the `style` option allows one to plot the points as points only, or to interpolate them using cubic spline or line mode. See `plot[options]` for more information.

- Setting the global variable `plotdevice` (via the `interface` command) to one of the values listed under `plot[device]` controls the type of plot which will be produced. On most implementations, the default is a character plot. See `plot[device]` and `plot[setup]` for information on how to set up plots on a particular device (printer or terminal).

Examples:
```
plot(cos(x), x=0..Pi);
plot(tan(x), x=-Pi..Pi, -5..5);
plot([sin(t), cos(t), t=-Pi..Pi]);
plot({sin(x), x-x^3/6}, x=0..2);
plot(sin(x), x=0..infinity);
l := [0,0,1,1,2,1,2,0,1,-1,0,0];
plot(l, x=0..2, style=POINT);
l := [[ n, sin(n)] $n=1..10];
plot(l, x=0..15, style=POINT);
plot([x, tan(x), x=-Pi..Pi], -4..4, -5..5, xtickmarks=8, ytickmarks=10);
```

SEE ALSO: `plot3d`, OR `plot,`*spec* where *spec* is one of `infinity`, `polar`, `parametric`, `spline`, `point`, `line`, `multiple`, `ranges`, `functions`, `options`, `structure`, `setup`, `device`, `replot`

2.1.202   plot[plotdevice] − plotting devices

Calling Sequence:

   `interface(plotdevice = ` *x* `)`

Parameters:

x	−	one of the following:
char	−	character plot
i300	−	imagen 300 laser printer
ln03	−	DEC LN03 laser printer
mac	−	Macintosh plot
pic	−	output file for use with the troff pic preprocessor
postscript	−	PostScript-compatible printer
regis	−	terminals with Regis graphics
tek	−	Tektronix graphics terminals
unix	−	output file for use with the UNIX plot command
vt100	−	VT100 line graphics
x11	−	X11 graphics

Synopsis:
- Currently Maple supports the following plotting devices:

`char`

> A character plot will be produced. This is the default
> (on most implementations) and will work with any ASCII terminal.

`i300`

> Set `plotdevice` to i300, set `plotoutput` to the name
> of a temporary file, and run as many plot commands as needed.
> To print the plots issue the shell command: `lpr -Fl`
> `Pprintername filename`, where `printername` is the
> name of an imagen300 printer. This command is site-dependent.

`ln03`

> Set `plotdevice` to ln03, set `plotoutput` to the name
> of a temporary file, and run as many plot commands as
> needed. To print the plots issue the shell command:
> `lpr -Fl Pprintername filename`, where printername
> is the name of a DEC LN03 printer.
> This command is site-dependent.

`mac`

> For plots on Macintosh computers. The plot will appear
> in a separate window on the Macintosh screen.

`pic`

> Set `plotdevice` to pic, set `plotoutput` to the name
> of a temporary file, and run as many plot commands as
> needed. Now incorporate (insert) the file output by the
> Maple `plot` command into your troff file (the Maple `plot`
> output begins with `.PS` and ends with `.PE`). In the Maple
> plot output file, edit the line containing `scale=` to
> suit your printout (this could be determined by previewing
> your troff file using ximpv or impv). Then use the shell
> command `typeset -p -..... filename`.
> Issue the shell command `man typeset` for more details.
> This command is site-dependent.

postscript

> Set `plotdevice` to postscript, set `plotoutput` to the
> name of a temporary file, and run as many plot commands
> as needed. To print the plots issue the shell command:
> `lpr -Fl Pprintername filename`, where `printername`
> is the name of a PostScript-compatible printer.
> This command is site-dependent.

regis

> For plotting on terminals supporting the Regis graphics
> protocol.

tek

> For plotting on Tektronix 4014 terminals or other
> terminals supporting Tektronix emulation mode.

unix

> Set `plotdevice` to `unix`, set `plotoutput` to the name
> of a temporary file, and run as many plot commands as
> needed. To print the plots issue the shell command:
> `lpr -Fg -Pprintername filename`, where `printername`
> is the name of any printer supported by UNIX plot.
> You can also preview the graph on the terminal; if the
> terminal has Tektronix emulation, put the terminal into
> `tek` mode and issue the shell command `plot -T4014 filename`;
> otherwise you can use `plot -Tdumb filename`.
> Issue the shell command `man plot` for more details.
> This command is site-dependent.

vt100

> Like character plots, but will use the horizontal line
> graphics characters of the VT100. Especially effective
> in 132 column mode.

x11

> For plots on computers using the X11 window system. The plot will

appear in a separate window on the screen.

Examples:
```
interface(plotdevice=regis);
plot(f,h,v,option);
```

SEE ALSO: `plot['options']`, `plot[structure]`, `plot[setup]` `plotsetup`;

2.1.203   plot[function] – acceptable plot functions

Calling Sequence:

  `plot(f, h, v, options)`

Parameters:

  f – function(s) to be plotted

  h – horizontal range

  v – vertical range

Synopsis:

- The procedure `plot` in Maple accepts three function types: a real function in a single variable expressed as an expression, or a Maple procedure, a parametric function, or a list of points.

- A parametric function has the form `[x(t),y(t),t=range of t]`. If the option `coords=polar` is given, the parametric function is interpreted in polar coordinates. Points will be equally spaced along the interval specified by the `range of t` in the parametric plot specification. The points will not be evenly spaced according to the `x(t)` `y(t)` specified. The set of lists must contain one undefined variable in it. If each list contained an undefined variable, then a terminal error would occur. Hence, all parametric plots must be parametric with respect to the same variable.

- A list of points has the form `[x1,y1,x2,y2,...,xn,yn]`. An even number of elements is required.

Examples:
```
plot(sin(x),x=-Pi..Pi,-1..1);
plot([sin(t),t^3,t=7..22]);
plot([5,5,6,6,7,8,9,20]);
```

SEE ALSO: `plot[parametric]`, `plot['options']`, `plot[polar]`

2.1.204   plot[infinity] – infinity plots

Calling Sequence:

  `plot(f, h, v, options)`

Parameters:

f(x)	–	an acceptable function
h	–	horizontal range
v	–	vertical range (optional)

Synopsis:

- If either of the range end points of the horizontal range contains `+-infinity`, an `infinity` plot is generated.

- An infinity plot is obtained by transforming `-infinity` .. `infinity` to `-1` .. `1` by a transformation that approximates `arctan`. This is a nice way of getting the entire `picture` of `f(x)` on the display. Such a graph, although distorted near `x = -infinity` and `infinity`, contains a lot of information about the features of `f(x)`.

Examples:
```
plot( sin(x), x=0..infinity );
plot( {exp,ln}, -infinity..infinity );
```

SEE ALSO: `plot[ranges]`, `plot['options']`, `plot[function]`

2.1.205 plot[line] – style = LINE option for plot

Calling Sequence:
```
plot(f,h,v,style=LINE)
```

Synopsis:

- There are three styles of plots available: `POINT`, `LINE`, `SPLINE`. The default is `SPLINE`.

- If the style is `LINE`, the points computed are connected by straight lines.

Examples:
```
plot([5,6,7,8,9,10],style=LINE);
plot(sin(x),x=-Pi..Pi,style=LINE);
```

SEE ALSO: `plot['options']`, `plot[structure]`

2.1.206 plot[multiple] – multiple plots

Calling Sequence:
```
plot({f_1,f_2,...},h,v,options);
```

Parameters:

h	–	horizontal range
v	–	vertical range (optional)

$f_1, f_2, \ldots$ –   functions to be plotted

Synopsis:

- When **plot** is passed a set of functions, it plots all of these functions on the same graph.

- In the **regis** mode three different colours of lines are used. In the **tektronix** mode eight different line modes are used. In the **char** mode eight different lines are also used, namely the characters **A** through **H**.

- If some of the functions listed in the multiple plot are parametric plots, then the same variable must be used in all parametric plots, as the parameter.

Examples:
```
plot({x-x^3/3,sin(x)},x=0..1);
plot({x,[x^2,x,-1..1]},x=0..1);
```

SEE ALSO: `plot['options']`, `plot[functions]`, `plot[ranges]`

2.1.207   plot[options] – plot options

Calling Sequence:

     plot($f, h, v, opt1, opt2, \ldots$)

Parameters:

     f –   some real function
     h –   horizontal range
     v –   vertical range (optional)
     $opt1, opt2, \ldots$ –   various desired options

Synopsis:

- The options to **plot** are given after the function(s), horizontal range, and vertical range, as equations of the form **option = value**. The following options are supported:

coords=polar Indicates that a parametric plot is in polar coordinates. The first parameter is the radius and the second parameter is the angle.

numpoints=n Specifies the minimum number of points to be generated (the default is n = 49). Note: plot employs an adaptive plotting scheme which automatically does more work where the function values do not lie close to a straight line. Hence plot will often generate more than the minimum number of points.

resolution=n Sets the horizontal display resolution of the device in pixels (the default is n = 200). The value of n is used to determine when the adaptive plotting scheme terminates. A higher value will result in more function evaluations for non-smooth functions.

xtickmarks=n Indicates the number of tickmarks along the horizontal axis; n must be a positive integer. There is also a corresponding option ytickmarks=n to specify the number of tickmarks along the vertical axis.

style=s The interpolation style must be one of the following (default is SPLINE):

POINT	plot points only
LINE	linear interpolation
SPLINE	cubic spline interpolation

title=t The title for the plot. t must be a character string. The default is no title.

Examples:
```
plot(sin(x),x=-2*Pi..2*Pi,title=`Sine Graph`);
```

SEE ALSO: plot[structure], plot[polar],

2.1.208 plot[parametric] – parametric plots

Calling Sequence:
```
plot([x(t),y(t),t=range of t],h,v,options)
```

Parameters:

h –   horizontal range

v –   vertical range

[x,y,range]   –   parametric specifications

Synopsis:

- A parametric plot is specified by a list of three items; the first two are real functions of a parameter, the third is the range for the parameter.

- When used in conjunction with the option `coords=polar`, `parametric plots` produces polar plots.

- If several parametric plots are to be plotted on the same graph, then the same variable or variables must be used for all the parameters of the specified parametric plots.

Examples:
```
plot([sin(t),cos(t),t=0..Pi]);
plot([x^2,x,x=0..3*Pi],-8..8,-10..10,coords=polar);
plot( [(t^2-1)/(t^2+1),2*t/(t^2+1), t=-infinity..infinity] );
```

SEE ALSO: `plot['options']`, `plot[polar]`

2.1.209   plot[point] – style = POINT option for plot

Calling Sequence:

   `plot(f, h, v, style=POINT)`

Synopsis:

- There are three styles of plot available: `POINT`, `LINE`, and `SPLINE`. The default is `SPLINE`.

- If the style is `POINT`, `plot` simply plots the calculated or given points. These should be given as a list of ordered pairs, i.e. $[x_1, y_1, x_2, y_2, \ldots, x_n, y_n]$.

Examples:
```
plot([2,3,4,7,2,9], style=POINT);
plot(sin(x), x=-Pi..Pi, style=POINT);
```

SEE ALSO: `plot[structure]`, `plot['options']`

2.1.210   plot[polar] – polar coordinate plots

Calling Sequence:

   `plot([r(t),theta(t),t=   low..high ],h,v,coords=polar)`

Parameters:

r(t)	–	the distance from the origin as a function of *t*
theta(t)	–	the angle of rotation in terms of *t*
h	–	the horizontal range
v	–	the vertical range

Synopsis:
- If the option `coords=polar` is specified, parametric functions will be interpreted in polar coordinates. The **radius** is the first parameter and the angle is the second parameter.

Examples:
```
plot([1,t,t=0..2*Pi],coords=polar);
```

SEE ALSO: `plot[parametric]`, `plots[polarplot]`

2.1.211   plot[range] – plot ranges

Calling Sequence:
```
plot(f,x=low..hi,y=low..hi,options)
```

Parameters:

f	–	function(s)
x,y	–	x and y are the axis labels (y is optional)
low..hi	–	a real range (optional)
options	–	plot options (optional)

Synopsis:
- The horizontal and vertical range specifications define the axis labels and the range over which the function(s) are displayed.

- If no range is specified, `-10..10` is used for the horizontal range. If one range is specified, this is taken as the horizontal range. If the vertical range is unspecified, a suitable range will be deduced from the values computed.

- The *low* and *hi* elements of an axis range must be real constants. This includes numbers, plus or minus infinity and constants such as `Pi`, `exp(4)`.

- If either `infinity` or `-infinity` is included in an axis range specification then an infinity plot is computed.

Examples:
```
plot(tan,-Pi..Pi,-5..5);
plot(sin(x),x=-Pi..Pi,`y axis`);
```

SEE ALSO: plot[`options`], plot[infinity], type[range], linalg[range], main[range]

2.1.212   plot[setup] – entering graphics modes

Synopsis:

- The Maple interface variables `preplot` and `postplot` are used in defining the strings that a terminal needs to enter and leave graphics mode. These variables should be assigned a list of positive integers in the range 0..127, representing the ASCII codes of the characters to be sent, and negative integers in the range -1..-127 representing delays in seconds. Preplot and postplot are limited to 80 elements. For example, to put a VT240 terminal into Tektronix emulation one would set

        interface(preplot = [27,91,63,51,56,104]);

- which corresponds to the characters

        ESC [? 3 8 h

- To put a KD500G,KD404 into Tektronix mode, one would set

        interface(preplot = [27,49]);

- which corresponds to the characters

        ESC 1

- To put a wy99gt into Tektronix mode, one would set

        interface(preplot = [27,127,63]);

- which corresponds to the characters

        ESC~>

- The corresponding settings to get the terminal back out of graphics mode after plotting has completed would be, respectively:

        interface(postplot = [27,91,63,51,56,108]);
        interface(postplot = [27,50]);
        interface(postplot = [27,127,61]);

- The `preplot` and `postplot` sequences are required for some settings of the `plotdevice` variable. For example, when `plotdevice` is set to `tek` and Tektronix emulation mode is being used, various terminals use different sequences to enter and leave Tektronix emulation mode. When using a real Tektronix 4010/4014 terminal, or when the `plotdevice` variable is set to `regis` or `char`, no such sequences need be defined, although they may be for access to special features such as to clear the screen before a plot. For example, to clear the screen of a VT100 terminal before plotting, and wait ten seconds after plotting, the settings would be

```
interface(preplot = [27,91,72,27,91,50,74]);
interface(postplot = [-10]);
```

- For further information, please consult the documentation accompanying your terminal.

SEE ALSO: `plotsetup`

2.1.213   plot[spline] – style = SPLINE option for plot

Calling Sequence:
   plot(f, h, v, style=SPLINE)

Synopsis:

- There are three styles of plots available: `POINT`, `LINE`, `SPLINE`. The default is `SPLINE`.

- If the style is `SPLINE`, the points are interpolated by a natural cubic spline.

Examples:
```
plot([2,3,4,5,2,9],style=SPLINE);
```

SEE ALSO: `plot['options']`, `plot[functions]`

2.1.214   plot[structure] – The PLOT data structure

Synopsis:

- A word about the way plotting in Maple is organized: The Maple function `plot` generates a list of points using Maple's arbitrary precision model of floating-point arithmetic. A dummy `PLOT` data structure representing the plot is returned as the result of calling the `plot` function. The `PLOT` data structure is understood by Maple's prettyprinter which prints the plot in a "pretty" form just like it does any other Maple object. Thus, note the following:

- 1) `lprint` of a `PLOT` will print out the `PLOT` object;

- 2) a `PLOT` can be saved in a file for later redisplay;

- 3) a `PLOT` can be manipulated like any Maple object.

- The PLOT data structure is quite complicated. As mentioned above, its value can be seen by using lprint. Of particular interest is that it supports three different modes of plots and allows the user to view a particular part of the plot. The modes are POINT (for plotting just the points), LINE (for linear interpolation between points), and SPLINE (for cubic spline interpolation between points — the default). To obtain a LINE plot simply substitute SPLINE = LINE in the PLOT data structure. To change the range on which the plot is displayed, include an argument of the form RANGE(VERTICAL,-2..2) in the plot data structure.

- The plot routine is "adaptive". It does a minimum of 25 function evaluations, then does more work where f is changing the most. Thus it is not necessary to tell plot how many function evaluations to do.

Examples:
```
a := plot(sin,-Pi..Pi):
a;
lprint(a);
subs(SPLINE=POINT,a);
b := plot([tan(t),t,t=0..1],coords=polar):
b;
save a,b,`myplots.m`;
```

SEE ALSO: plot[`options`]

2.1.215   plot3d – three-dimensional plotting

Calling Sequence:
> plot3d($expr1, x = a..b, y = c..d$)
> plot3d($f, a..b, c..d$)
> plot3d([$exprf, exprg, exprh$], $s = a..b, t = c..d$)
> plot3d([f, g, h], $a..b, c..d$)

Parameters:

f, g, h	–	function(s) to be plotted
$expr1$	–	expression in x and y.
$exprf, exprg, exprh$	–	expressions in s and t.
a, b, c, d	–	real constants
x, y, s, t	–	names

Synopsis:
- The four different calling sequences to the plot3d function above all define surfaces. There is no facility yet for describing curves and polyhedra in three dimensions. The first two calling sequences describe surface plots in Cartesian co-ordinates while the second two describe parametric surface plots.

- In the first call, `plot3d(expr1),x=a..b,y=c..d)`, the expression *expr1* must be a Maple expression in the names *x* and *y*. The ranges *a..b* and *c..d* must evaluate to real constants. They specify the range over which *expr1* which will be plotted. In the second call, `plot3d(f,a..b,c..d)`, *f* must be a Maple procedure or operator which takes two arguments.

- A parametric surface can be defined by three expressions *expr1, expr2, expr3* in two variables. In the third call, `plot3d([ expr1,expr2,expr3 ],s=a..b,t=c..d)`, *expr1*, *expr2*, and *expr3* must be Maple expressions in the names *s* and *t*. Finally, in the fourth call, `plot3d([ f,g,h ],a..b,c..d)`, *f*, *g* and *h* must be Maple procedures or operators taking two arguments.

- Any additional arguments are interpreted as options which are specified as equations of the form `option = value`. For example, the option `grid = [m,n]` where `m` and `n` are positive integers specifies that the `plot` is to be constructed on an `m` by `n` grid at equally spaced points in the ranges *a..b* and *c..d* respectively. By default a 25 by 25 grid is used, thus 625 points are generated. See the help page for `plot3d['options']`.

- The result of a call to `plot3d` is a PLOT3D data structure containing enough information to render the plot. The user may assign a PLOT3D value to a variable, save it in a file, then read it back in for redisplay. See the help page for `plot3d[structure]`.

Examples:
```
plot3d(sin(x+y),x=-1..1,y=-1..1);
plot3d(binomial,0..5,0..5,grid=[10,10]);
plot3d((1.3)^x * sin(y),x=-1..2*Pi,y=0..Pi,coords=spherical,style=PATCH);
plot3d(sin(x*y),x=-Pi..Pi,y=-Pi..Pi,style=PATCH);
plot3d([x*sin(x)*cos(y),x*cos(x)*cos(y),x*sin(y)],x=0..2*Pi,y=0..Pi);
plot3d(x*exp(-x^2-y^2),x=-2..2,y=-2..2,grid=[49,49]);
#multiple 3d plots can also be done
plot3d({sin(x*y), x + 2*y},x=-Pi..Pi,y=-Pi..Pi);
c1:= [cos(x)-2*cos(0.4*y),sin(x)-2*sin(0.4*y),y];
c2:= [cos(x)+2*cos(0.4*y),sin(x)+2*sin(0.4*y),y];
c3:= [cos(x)+2*sin(0.4*y),sin(x)-2*cos(0.4*y),y];
c4:= [cos(x)-2*sin(0.4*y),sin(x)+2*cos(0.4*y),y];
plot3d({c1,c2,c3,c4},x=0..2*Pi,y=0..10,grid=[25,15],style=PATCH);
```

SEE ALSO: `plot`, `read`, `save`, `plot3d['options']`, `plot3d[structure]`, `plots`

2.1.216   plot3d[option] − plot3d options

Synopsis:

- Options to the `plot3d` function are given after the first three arguments, as equations of the form `option = value`. The following options are supported.

- The option `numpoints=n` specifies the minimum total number of points to be generated (default $625 = 25^2$). Plot3d will use a rectangular grid of `dimensions = sqrt(n)`.

- The option `grid=[m,n]` specifies the dimensions of a rectangular grid on which the points will be generated (equally spaced).

- The option `title=t` specifies a title for the plot. The value of `t` must be a string. The default is no title.

- The option `labels=[x,y,z]` specifies labels for the axes. The value of `x`, `y`, and `z` must be a string. The default label is no label.

- The option `axes=f` where `f` is one of {BOXED,NORMAL,FRAME,NONE} specifies how the axes are to be drawn. The default axis is NONE.

- The option `style=s` where `s` is one of {POINT,HIDDEN,PATCH,WIREFRAME} specifies how the surface is to be drawn. Some of the options may not be available on some devices. The default style is HIDDEN (for hidden line rendering).

- The option `coords=c` where `c` is one of {cartesian,spherical,cylindrical} specifies the coordinate system to be used. The default is Cartesian.

- The option `scaling=s` where `s` is either UNCONSTRAINED or CONSTRAINED specifies whether the surface should be scaled so that it fits the screen with axes using a relative or absolute scaling. The default is relative (UNCONSTRAINED).

- The option `projection=r` where `r` is a real number between 0 and 1 specifies the perspective from which the surface is viewed. The 1 denotes orthogonal projection, and the 0 denotes wide-angle perspective rendering. The default is `projection = 1`.

- The option `orientation=[theta,phi]` specifies the theta and phi angles of the point in 3-dimensions from which the plot is to be viewed. The default is at a point that is out perpendicular from the screen (negative `z` axis) so that the entire surface can be seen. The point is described in spherical coordinates where theta and phi are angles in degrees, with default 45 degrees in each case.

- The option `view=zmin..zmax` or `[xmin..xmax,ymin..ymax,zmin..zmax]` indicates the minimum and maximum coordinates of the surface to be displayed on the screen. The default is the entire surface.

- The option `shading=s` where `s` is one of XYZ, XY, Z specifies how the surface is coloured. The default is device dependent.

2.1.217   plot3d[structure] – plot3d structure

Synopsis:

- The data structure of the result of a `plot3d` function call is an unevaluated function with name PLOT3D with the following arguments. Note that some options may not be implemented on a particular device.

- The arguments to the PLOT3D function take the form

PLOT3D({*function, object, option*}*)

- where *function* takes the form

```
FUNCTION( { [algebraic,algebraic,algebraic, range, range],
            [algebraic, range, range] },
```

- *range* takes the form { `numeric..numeric, name=numeric..numeric` },
- *object* takes the form {*grid, curvelist, points*}
- *grid* takes the form `GRID( {list(point)*, list(numeric)*} )`,
- *curvelist* takes the form `CURVELIST( list(point)* )`
- *points* takes the form `POINTS( list(point) )`
- *point* is of type `[numeric,numeric,numeric]`.
- The options are specified as additional parameters in any order. The known options (and their defaults) are:
- `TITLE(string)` – default: no title,
- `LABELS(string,string,string)` – default: no axes labels,
- `STYLE({POINT,HIDDEN,PATCH,WIREFRAME})` – default: device dependent,
- `VIEW(numeric..numeric,numeric..numeric,numeric..numeric)` – default: minimum/maximum values in the given points,
- `ORIENTATION(numeric,numeric)` – default: 45,45
- `PROJECTION({numeric})` – default: 1,
- `AXES({BOXED,FRAME,NORMAL,NONE})` – default: no axes
- `SCALING({CONSTRAINED,UNCONSTRAINED})` – default: `UNCONSTRAINED`,
- `SHADING({XYZSHADING,XYSHADING,ZSHADING})` – default: `XYZSHADING`.

2.1.218   plotsetup – set up appropriate parameters for plotting

Calling Sequence:
```
plotsetup(DeviceType)
plotsetup(DeviceType, TerminalType)
```

Parameters:

DeviceType	–	a supported device type
TerminalType	–	a supported terminal type

Synopsis:
- The `plotsetup` function sets up appropriate parameters for plotting. The setup parameters affected are: `pre/postplot`, `plotoutput`, and `plotdevice`.

- Supported device types:

laser printers:	ln03, imagen300 (i300), postscript
terminals:	dumb (char), tek (Tektronix), regis
software interface:	unix (UNIX plot), pic (troff)

- Supported terminal types:

vt series
kd500g, kd404g (graphics terminal)
wy99gt
Mac
xterm

Examples:
```
> plotsetup(i300);
Warning, plotoutput set to i300.out by default
> plotsetup(tek, vt100);
> plotsetup(pic);
Warning, plotoutput set to pic.out by default
```

SEE ALSO: `interface`, `plot[setup]`, `plot[device]`

2.1.219 Power – inert power function

Calling Sequence:
>   Power(a,n)

Parameters:
> a – multivariate polynomial

> n – non-negative integer

Synopsis:
- The `Power` function is a placeholder for representing $a\char94 n$. It is used in conjunction with either `mod` or `modp1`.

- The call `Power(`a,n`) mod p` computes $a\char94 n$ mod p. The multivariate polynomial a must have rational coefficients or coefficients from a finite field specified by `RootOfs`.

- The call `modp1(Power(`a,n`),p)` also computes $a\char94 n$ mod p. The polynomial a must be in the `modp1` representation and p must be a positive integer.

Examples:
```
> Power(x+1,3) mod 2;
```
$$\longrightarrow \qquad x^3 + x^2 + x + 1$$

```
> Power(x+1,4) mod 2;
```
$$\longrightarrow \qquad x^4 + 1$$

SEE ALSO: `power`, `Powmod`, `mod`, `modp1`

2.1.220   Powmod – inert power function with remainder

Calling Sequence:

>   Powmod(a, n, b)
>   Powmod(a, n, b, x)

Parameters:

> a –   polynomial in x
> n –   integer
> b –   polynomial in x
> x –   name

Synopsis:

- The `Powmod` function is a placeholder for representing `Rem(`$a\,\hat{}\,n,b$`)` or `Rem(`$a\,\hat{}\,n,b,x$`)`. `Powmod` is more efficient than computing `Power(`a,n`)` separately. It is used in conjunction with either `mod` or `modp1`.

- The call `Powmod(`a,n,b,x`) mod p` computes `Rem(`$a\,\hat{}\,n,b,x$`) mod p`. The polynomials a and b must have rational coefficients or coefficients over a finite field specified by `RootOfs`.

- The call `modp1(Powmod(`a,n,b`),p)` computes `Rem(`$a\,\hat{}\,n,b$`) mod p`. The polynomials a and b must be in the modp1 representation and p must be a positive integer.

Examples:

```
> Powmod(x,16,x^4+x+1,x) mod 2;
```
$$\longrightarrow \qquad x$$

```
> Powmod(x,-5,x^4+x+1,x) mod 2;
```
$$\longrightarrow \qquad x^2 + x + 1$$

SEE ALSO: `mod`, `modp1`, `Power`, `Rem`, `Gcdex`

2.1.221   Prem – inert pseudo-remainder function
Sprem – inert sparse pseudo-remainder function

Calling Sequence:

>   Prem($a,b,x,\text{´}m\text{´},\text{´}q\text{´}$)
>   Sprem($a,b,x,\text{´}m\text{´},\text{´}q\text{´}$)

Parameters:

> a, b –   multivariate polynomials in the variable x

x —   indeterminate

m, q —   (optional) unevaluated names

Synopsis:

• The `Prem` and `Sprem` functions are placeholders for the pseudo-remainder and sparse pseudo-remainder of a divided by b where a and b are polynomials in the variable x. They are used in conjunction with either `mod` or `evala` which define the coefficient domain, as described below.

• The function Prem returns the pseudo-remainder r such that:

```
m a = b q + r
```

• where `degree(r,`x`)` $<$ `degree(`b,x`)` and m (the multiplier) is:

```
m = lcoeff(b,x) ^ (degree(a,x) - degree(b,x) + 1)
```

• If the fourth argument is present it is assigned the value of the multiplier m defined above. If the fifth argument is present, it is assigned the pseudo-quotient q defined above.

• The function Sprem has the same functionality as Prem except that it uses as a multiplier m the smallest possible power of `lcoeff(`b,x`)` such that the division process does not introduce fractions into q and r. When Sprem can be used it is preferred because it is more efficient.

• The calls `Prem(a,b,x,´m´,´q´) mod p` and `Sprem(a,b,x,´m´,´q´) mod p` compute the pseudo-remainder and sparse pseudo-remainder respectively of `a` divided by `b` modulo p, a prime integer. The coefficients of `a` and `b` must be multivariate polynomials over the rationals or coefficients over a finite field specified by `RootOf` expressions.

• The calls `evala(Prem(a,b,x,´m´,´q´))` and `evala(Sprem(a,b,x,´m´,´q´))` do likewise for `a` and `b` multivariate polynomials over the algebraic constants defined by `RootOf` expressions.

SEE ALSO: `prem`, `sprem`, `mod`, `evala`, `RootOf`, `Rem`

2.1.222   prem – pseudo-remainder of polynomials
sprem – sparse pseudo-remainder of polynomials

Calling Sequence:

   `prem(`$a, b, x, ´m´, ´q´$`)`
   `sprem(`$a, b, x, ´m´, ´q´$`)`

Parameters:

a, b —   multivariate polynomials in the variable x

x —   indeterminate

m, q —   (optional) unevaluated names

Synopsis:
- The function `prem` returns the pseudo-remainder `r` such that

      m a = b q + r

- where `degree(r,x)` < `degree(b,x)` and `m` (the multiplier) is:

      m = lcoeff(b,x) ^ (degree(a,x) - degree(b,x) + 1)

- If the fourth argument is present it is assigned the value of the multiplier `m` defined above. If the fifth argument is present, it is assigned the pseudo-quotient `q` defined above.
- The function `sprem` has the same functionality as `prem` except that it uses as a multiplier `m` the smallest possible power of `lcoeff(b,x)` such that the division process does not introduce fractions into `q` and `r`. When `sprem` can be used it is preferred because it is more efficient.

Examples:
```
> a := x^4+1: b := c*x^2+1:
  r := prem(a,b,x,'m','q'):
  r,m,q;
```
$$\longrightarrow \quad c^2(c^2+1), \ c^3, \ c^2 x^2 - c$$
```
> r := sprem(a,b,x,'m','q'):
  r,m,q;
```
$$\longrightarrow \quad c^2 + 1, \ c^2, \ c x^2 - 1$$

SEE ALSO: `rem`, `quo`, `Prem`, `Sprem`

2.1.223 Primitive – test whether a polynomial is primitive mod p

Calling Sequence:
    `Primitive(`a`) mod` p

Parameters:
    a – univariate polynomial
    p – prime integer

Synopsis:
- `Primitive(`a`) mod` p returns `true` if the univariate polynomial a over the integers mod p is "primitive", and `false` otherwise.
- If a is an irreducible polynomial in Zp[x] then it is primitive if x is a primitive element in the finite field Zp[x]/(a) . Thus x^i for i=1..p^k-1 is the set of all non-zero elements in the field.

Examples:
```
> Primitive( x^4+x+1 ) mod 2;
```
$\longrightarrow$ `true`
```
> Primitive( x^4+x^3+x^2+x+1 ) mod 2;
```
$\longrightarrow$ `false`

SEE ALSO: `Irreduc`, `Powmod`

2.1.224   print – pretty-printing of expressions

Calling Sequence:

> print(*e1*, *e2*, ...)

Parameters:

> *e1, e2, ...*   –   any expressions

Synopsis:

- The function **print** displays the values of the expressions appearing as arguments, and returns NULL as the function value. The expressions printed are separated by a comma and a blank. Note that **print()** with no arguments will print a blank line.

- The following Iris variables control the format. See **help(interface)** for setting Iris variables.

- The Iris variable **prettyprint** is checked to determine the format in which the expressions are to be printed. If set to **true** (the default value) the expressions will be displayed in a two-dimensional format, centered if possible. If set to **false** the expressions will be line printed in one dimension.

- The Iris variable **verboseproc** is checked to determine how the body of Maple procedures (i.e. the code) will be displayed. It can be displayed in full, with indentation, or abbreviated to simply "...". By default user procedures are displayed in full, and Maple library procedures are abbreviated.

- The Iris variable **screenwidth** specifies the number of characters that fit on the output device being used. Since most CRTs are 80 characters wide, the default value is 79.

- The Iris variable **labelling** (or **labeling**) enables the use of % variables to reduce the size of the output. These % labels identify common subexpressions (those appearing more than once) in the output.

- The evaluation of arrays, tables, procedures, and operators is different from other objects in that when assigned to a name, they will normally print with simply the name. For example, if the name *A* has been assigned **array([1,2,3])** then *A*; will display just *A*. To print the array object in full use **print(*A*)**;

- There is a limited facility for user-defined formatting of functions. For example, if the user assigns

    ```
    `print/complex` := proc(real,imag) real + 'j'*imag end
    ```

then during output of the value **complex(2,3)**, the `print/complex` procedure will be called with the arguments 2, 3 resulting in the value 2 + 3*j which will finally be displayed as 2 + 3 j. Note that arguments are formatted first, and alias substitutions take place after formatting is completed.

Examples:
> print(red,rouge,rot); $\longrightarrow$ red, rouge, rot

```
> v := array([1,2,3]):
  v;                                    ⟶                              v
> print(v);                            ⟶                          [ 1, 2, 3 ]
> sin;                                 ⟶                             sin
> print(sin);                          ⟶        proc(x) ... end
```

SEE ALSO: lprint, interface, alias, writeto, save, sort

2.1.225 product – definite and indefinite product
Product – inert form of product

Calling Sequences:

    product(f, k);   product($f, k=m..n$);   product($f, k=alpha$);
    Product(f, k);   Product($f, k=m..n$);   Product($f, k=alpha$);

Parameters:

    f –   an expression
    k –   a name, the product index
    m, n –   integers or arbitrary expressions
    $alpha$ –   a RootOf

Synopsis:

- The call product(f,k) computes the indefinite product of $f(k)$ with respect to k. That is, it computes a formula g such that $g(k+1)/g(k) = f(k)$ for all k.

- The call product(f,k=m..n) computes the definite product of $f(k)$ over the given range $m..n$, i.e. it computes $f(m) f(m+1) ... f(n)$. The definite product is equivalent to $g(n+1)/g(m)$ where g is the indefinite product. For example, product(n,n) = product(k,k=1..n-1) = GAMMA(n).

- If $m = n+1$ then the value returned is 1. If $m > n+1$ then 1/product(f,k=n+1..m-1) is the value returned.

- The call product($f, k=alpha$) computes the definite product of $f(k)$ over the roots of a polynomial where *alpha* must be a RootOf.

- If Maple cannot find a closed form for the product, the function call itself is returned. (The prettyprinter displays the product function using a stylized product sign.)

- The capitalized function name Product is the inert product function, which simply returns unevaluated. The prettyprinter understands Product to be equivalent to product for printing purposes.

Examples:

```
> product( k^2, k=1..4 );            ⟶                          576
> product( k^2, k );                 ⟶                           2
                                                            GAMMA(k)
```

```
> product( a[k], k=1..4 );          ⟶        a[1] a[2] a[3] a[4]

> product( n+k, k=1..4 );           ⟶     (n + 1) (n + 2) (n + 3) (n + 4)

> product( n+k, k );                ⟶           GAMMA(n + k)

> product( n+k, k=1..m );           ⟶         GAMMA(n + m + 1)
                                              -----------------
                                                GAMMA(n + 1)

> Product( n+k, k=1..m );           ⟶             m
                                              --------- ′
                                            ′  |   |
                                               |   |    (n + k)
                                               |   |
                                               |   |
                                              k = 1

> product( k, k=RootOf(x^3-2) );    ⟶              2
```

SEE ALSO: `sum`, `resultant`, `RootOf`

2.1.226   Psi – the Psi and polygamma functions

Synopsis:

- The Psi function is defined by

$$\mathtt{Psi}(x) = \psi(x) = \frac{d}{dx}\ln(\Gamma(x))$$

- Its derivatives are represented by

$$\mathtt{Psi}(n,x) = \psi^{(n)}(x) = (\frac{d}{dx})^n \, \Psi(x)$$

- Automatic simplifications will be applied to these functions.

2.1.227   radsimp – simplification of an expression containing radicals

Calling Sequence:

    radsimp(*expr*)
    radsimp(*expr*, ´*ratden*´)

Parameters:

expr	–	algebraic expression
´*ratden*´	–	(optional) unevaluated name

Synopsis:

- A call to `radsimp` simplifies radicals contained in *expr*.

- If the second argument, ´*ratden*´, is present, it will be assigned the simplified expression with its denominator rationalized.

Examples:

`> radsimp((1 + 2*x + x^2)^(1/2));`	$\longrightarrow$	$1 + x$
`> radsimp((1 + 2^(1/2))^(-1), ´d´);`	$\longrightarrow$	$\dfrac{1}{1 + 2^{1/2}}$
`> d;`	$\longrightarrow$	$\dfrac{1}{-1 + 2^{1/2}}$

SEE ALSO: `simplify`

2.1.228   rand – random number generator

Calling Sequence:

    `rand()`

    `rand(r)`

Parameters:

    r – (optional) an integer range or an integer

Synopsis:

- With no arguments, the call `rand()` returns a 12 digit non-negative random integer.

- With an integer range as an argument, the call `rand(a..b)` returns a procedure which, when called, generates random integers in the range `a..b`.

- With a single integer as an argument, the call `rand(n)` is the abbreviated form of `rand(0..n-1)`.

- Because `rand(a..b)` returns a Maple procedure, more than one random number generator may be used at the same time. However, since all random number generators use the same underlying random number sequence, calls to one random number generator will affect the random numbers returned from another.

- The global variable `_seed` can be used to alter the sequence of random numbers generated, by setting it to any non-zero integer.

Examples:

`> rand();`	$\longrightarrow$	427419669081
`> rand();`	$\longrightarrow$	321110693270
`> die := rand(1..6):` `  die();`	$\longrightarrow$	4
`> die();`	$\longrightarrow$	6

SEE ALSO: `randpoly, randmatrix, stats, combinat`

2.1.229   Randpoly – random polynomial over a finite field
Randprime – random monic prime polynomial over a finite field

Calling Sequence:

Randpoly(n,x) mod p Randpoly($n,x,alpha$) mod p
Randprime(n,x) mod p Randprime($n,x,alpha$) mod p

Parameters:

n –   non-negative integer

x –   name

$alpha$ –   RootOf

p –   integer

Synopsis:

- Randpoly(n,x) mod p returns a polynomial of degree n in the variable x whose coefficients are selected at random from the integers mod p.

- Randprime(n,x) mod p returns a random monic irreducible polynomial of degree $n > 0$ in the variable x over the integers mod p where p must be a prime integer.

- The optional third argument $alpha$ specifies a finite ring which is an algebraic extension of the integers mod p. The extension alpha is specified by a RootOf a monic univariate polynomial of degree k which in the case of Randprime must be irreducible.

- Thus Randprime($n,x,alpha$) mod p creates a random monic irreducible polynomial of degree $n > 0$ in the variable x over a finite field with p^k elements of characteristic p.

Examples:

```
> Randpoly(4,x) mod 2;
```
$\longrightarrow$
$$x^4 + x^3 + x^2$$

```
> Randprime(4,x) mod 2;
```
$\longrightarrow$
$$x^4 + x^3 + x^2 + x + 1$$

```
> alias(alpha=RootOf(y^2+y+1)):
  Randpoly(2,x,alpha) mod 2;
```
$\longrightarrow$
$$(alpha + 1) x^2 + 1$$

```
> Factor(") mod 2;
```
$\longrightarrow$
$$(alpha + 1) (x + alpha + 1)^2$$

```
> Randprime(2,x,alpha) mod 2;
```
$\longrightarrow$
$$x^2 + x + alpha$$

```
> Irreduc(") mod 2;
```
$\longrightarrow$
$$true$$

SEE ALSO: RootOf, Irreduc, Factor, alias

2.1.230   randpoly – random polynomial generator

Calling Sequence:
>   randpoly(*vars*, *eqns*)

Parameters:
>   *vars* –   indeterminate or list or set of indeterminates
>   *eqns* –   (optional) equations specifying properties

Synopsis:

- A call to `randpoly` generates a random polynomial in *vars*. It is useful for generating test problems for debugging, testing and demonstration purposes. Several options can be specified, determining the form of the polynomial. This allows for quite general expressions with certain properties.

- The first argument *vars* specifies the variables in which the polynomial is to be generated. If *vars* is a single variable, a univariate polynomial in that variable will be generated. If *vars* is a list or set of variables, then a multivariate polynomial will be generated.

- The remaining arguments *eqns* are equations of the form `option = value`.

- The possible options (and their default values) are:

Option	Use	Default Value
`coeffs`	generate the coefficients	`rand(-99..99)`
`expons`	generate the exponents	`rand(6)`
`terms`	number of terms generated	6
`degree`	total degree for a dense polynomial	5
`dense`	the polynomial is to be dense	

Examples:
```
> randpoly(x);
```
$$- 85\ x^5 - 55\ x^4 - 37\ x^3 - 35\ x^2 + 97\ x + 50$$
```
> randpoly([x, y], terms = 20);
```
$$56 + 49\ x + 63\ y + 57\ x^4 - 59\ x^5 + 45\ x^3 - 8\ x^2 - 93\ x\ y + 92\ x^2\ y + 43\ x\ y^2$$
$$- 62\ y^3 + 77\ x^3\ y + 66\ x^2\ y^2 + 54\ x\ y^3 - 5\ x^4\ y + 99\ x^3\ y^2 - 61\ x^2\ y^3$$
$$- 50\ y^4 - 12\ y^5 - 18\ x\ y^4$$
```
> randpoly([x, sin(x), cos(x)]);
```
$$- 47\ x\ \cos(x)^2 - 61\ x\ \sin(x)\ \cos(x) + 41\ \sin(x)\ \cos(x)^3 - 58\ x\ \cos(x) - 90\ x$$

$$+ 53 \sin(x)^3 \cos(x)^2$$

```
> randpoly(z, expons = rand(-5..5));
```

$$\frac{82}{z^5} + \frac{23}{z} + 104\ z^5 + 88\ z$$

```
> randpoly([x], coeffs = proc() randpoly(y) end);
```

$$(78\ y^5 + 17\ y^4 + 72\ y^3 - 99\ y^2 - 85\ y - 86)\ x^5$$

$$+ (30\ y^5 + 80\ y^4 + 72\ y^3 + 66\ y^2 - 29\ y - 91)\ x^4$$

$$+ (-53\ y^5 - 19\ y^4 - 47\ y^3 + 68\ y^2 - 72\ y - 87)\ x^3$$

$$+ (79\ y^5 + 43\ y^4 - 66\ y^3 - 53\ y^2 - 61\ y - 23)\ x^2$$

$$+ (-37\ y^5 + 31\ y^4 - 34\ y^3 - 42\ y^2 + 88\ y - 76)\ x - 65\ y^5 + 25\ y^4 + 28\ y^3$$

$$- 61\ y^2 - 60\ y + 9$$

```
> randpoly([x, y], dense);
```

$$29\ x^5 - 66\ x^4\ y - 32\ x^4 + 78\ x^3\ y^2 + 39\ x^3\ y + 94\ x^3 + 68\ x^2\ y^3 - 17\ x^2\ y^2$$

$$- 98\ x^2\ y - 36\ x^2 + 40\ x\ y^4 + 22\ x\ y^3 + 5\ x\ y^2 - 88\ x\ y - 43\ x - 73\ y^5$$

$$+ 25\ y^4 + 4\ y^3 - 59\ y^2 + 62\ y - 55$$

SEE ALSO: **rand**

2.1.231   readlib – read a library file to define a specified name

Calling Sequence:

  readlib(*f*, *file1*, *file2*, ...)

Parameters:

  f   –   the name of a procedure to be read

  file1, ... –   (optional) filenames

Synopsis:

• With only one argument, calling **readlib**(*f*) causes the following statement to be executed:

  read   `` `.libname.`/`.f.`.m` ``;

This causes the .m file for *f* to be loaded from the standard Maple library.

- In the case of two or more arguments, all arguments after the first are taken to be complete filenames of files to be read, and the following statements are executed:

  read file1; read file2; . . . .

- In both cases, the value returned from the **readlib** function is the value assigned to the first argument *f* after executing the read statement(s).

- It is an error if *f* does not become defined by this action.

- The **readlib** function has option **remember**; hence subsequent invocations of **readlib** with the same arguments will not cause files to be re-loaded.

- Many of Maple's library functions are initially "readlib-defined" in the system. For example, the initial definition of the name **gcd** is:

  gcd := 'readlib('gcd')';

  Whenever the name **gcd** is evaluated, it causes the corresponding **readlib** to be executed, thus loading the gcd procedure.

- Other library functions are not "readlib-defined", and must be defined using **readlib** before they are used. This is indicated in the on-line help and manual descriptions for these functions.

- It is often desirable to follow a call to **readlib** with a colon (:) rather than a semicolon (;). Ending a Maple command with a colon causes Maple to suppress printing the output from that command. Thus using a colon when calling **readlib** avoids having the procedure definition output to the screen.

Examples:
```
> readlib(randpoly);                          ⟶   proc(v) ... end
```

SEE ALSO: libname, `read`, files, remember, index[libmisc]

2.1.232   readstat – read one statement from the input stream

Calling Sequence:
   readstat(*string*)

Parameters:
   string – a name

Synopsis:
- The function **readstat** reads the next statement from the input stream (the terminal or a file) and returns the value of that statement.

- The argument *string* is used as a prompt string, which is printed before Maple´s prompt.

Examples:
```
a := readstat(`a will be assigned `);
```

SEE ALSO: `read`

2.1.233   Rem – inert rem function
Quo – inert quo function

Calling Sequence:

   Rem(a,b,x)

   Rem($a,b,x,´q´$)

   Quo(a,b,x)

   Quo($a,b,x,´r´$)

Parameters:

   x –   name (variable)

   a, b –   polynomials in x

   q, r –   unevaluated name

Synopsis:

- The Rem and Quo functions are placeholders for representing the remainder and quotient respectively of a divided by b where a and b are polynomials in the variable x over a field. They are used in conjunction with either mod or evala as described below which define the coefficient domain.

- Functionality: Rem returns the remainder r and if the fourth argument q is present then the quotient is assigned to q. Quo returns the quotient q and if the fourth argument r is present then the remainder is assigned to r. The remainder r and quotient q satisfy: $a = b*q + r$.

- The calls Rem(a,b,x) mod p and Quo(a,b,x) mod p compute the remainder and quotient respectively of a divided by b modulo p, a prime integer. The coefficients of a and b must be rational expressions over the rationals or over a finite field specified by RootOf expressions. In particular, if the coefficients are integers then the computation is done over the field of integers modulo p.

- The calls evala(Rem(a,b,x)) and evala(Quo(a,b,x)) do likewise for a and b multivariate polynomials with algebraic extensions defined by RootOfs.

Examples:
```
> a := x^4+5*x^3+6:
  b := x^2+2*x+7:
  r := Rem(a,b,x,´q´) mod 13;                    ⟶         r := 5 x + 6

> q;                                             ⟶              2
                                                           x  + 3 x

> Expand(a-b*q-r) mod 13;                         ⟶              0
```

```
> c := x^2-x+3:
  d := x-RootOf(_Z^2-3):
  evala(Quo(c,d,x));
```
$\longrightarrow$
$$x - 1 + RootOf(\_Z^2 - 3)$$

```
> evala(Rem(c,d,x));
```
$\longrightarrow$
$$6 - RootOf(\_Z^2 - 3)$$

SEE ALSO: `rem`, `quo`, `mod`, `evala`, `RootOf`, `Divide`, `Powmod`

2.1.234   rem – remainder of polynomials
quo – quotient of polynomials

Calling Sequence:

    `rem(`a`, ` b`, ` x`)`

    `rem(`a`, ` b`, ` x`, ` $'q'$`)`

    `quo(`a`, ` b`, ` x`)`

    `quo(`a`, ` b`, ` x`, ` $'r'$`)`

Parameters:

    a, b –   polynomials in x

    x –   a name

    $'q', 'r'$ –   (optional) unevaluated names

Synopsis:

- The `rem` function returns the remainder of a divided by b. The `quo` function returns the quotient of a divided by b. The remainder r and quotient q satisfy: a = b*`q` + `r` where `degree(`r, x`)` < `degree(`b, x`)`.

- If a fourth argument is present to `rem` or `quo` it will be assigned the quotient q or remainder r respectively.

Examples:

```
> rem(x^3+x+1, x^2+x+1, x, 'q');
```
$\longrightarrow$
$$2 + x$$

```
> q;
```
$\longrightarrow$
$$x - 1$$

```
> quo(x^3+x+1, x^2+x+1, x);
```
$\longrightarrow$
$$x - 1$$

SEE ALSO: `irem`, `iquo`, `prem`, `sprem`, `divide`, `Rem`, `Quo`

2.1.235   Resultant – inert resultant function

Calling Sequence:

    `Resultant(` a, b, x `)`

Parameters:

    a,b – polynomials

    x – name

Synopsis:

- The `Resultant` function is a placeholder for representing the resultant of the polynomials a and b with respect to the main variable x. It is used in conjunction with either `mod` or `evala` which define the coefficient domain as described below.

- The call `Resultant(a,b,x) mod p` computes the resultant of a and b with respect to the main variable x modulo p, a prime integer. The argument a must be a multivariate polynomial over the rationals or over a finite field specified by `RootOfs`.

- The call `evala(Resultant(a,b,x))` does likewise for a and b multivariate polynomials with algebraic extensions defined by `RootOfs`.

- The call `modp1(Resultant(a,b),p)` computes the resultant of a and b modulo p a prime integer where a and b are in the `modp1` representation.

Examples:
```
> Resultant(2*x+1, 3*x+4, x) mod 7;                        ⟶          5

> r := x + RootOf(_Z^2-2):
  s := RootOf(_Z^2-2)*x + 1:
  evala( Resultant(r, s, x) );                             ⟶          -1
```

SEE ALSO: `evala`, `mod`, `modp1`, `resultant`

2.1.236   resultant – compute the resultant of two polynomials

Calling Sequence:

    `resultant(`a`, `b`, `x`)`

Parameters:

    a,b – polynomials in x

    x – a name

Synopsis:

- The function `resultant` computes the resultant of the two polynomials a and b with respect to the indeterminate x.

- The resultant can be computed from the Euclidean algorithm, or computed as the determinant of Sylvester's matrix or Bezout's matrix. For univariate and bivariate resultants over the rationals, modular methods are used for polynomials of high degree and the **subresultant** algorithm is used for polynomials of low degree. Otherwise Bezout's determinant is computed using minor expansion.

- For efficient computation, **resultant** takes advantage of any factorization of a and b that is present, although no explicit factorization is attempted.

- Reference: *Computer Algebra: Symbolic & Algebraic Computation* Edited by B. Buchberger, G. E. Collins, and R. Loos, Springer-Verlag, Wien, 1982, pp. 115-138 .

Examples:

```
> resultant(a*x+b, c*x+d, x);                    ⟶           - c b + d a

> resultant((x+a)^5,(x+b)^5,x);                  ⟶               25
                                                             (- a + b)
```

SEE ALSO: **discrim, gcd, Resultant, linalg[sylvester], linalg[bezout]**

2.1.237   RETURN – explicit return from a procedure

Calling Sequence:

> RETURN($expr_1$, $expr_2$, ...)

Parameters:

> $expr_1$, ...   –   expression sequence (possibly null)

Synopsis:

- One common form of return from a procedure invocation occurs when execution "falls through" the end of the *statement sequence* which makes up the procedure body, in which case the value of the procedure invocation is the value of the last statement executed.

- A call to the **RETURN** function causes an immediate return to the point where the current procedure was invoked.

- If return is via the **RETURN** function then the value of the procedure invocation is the expression sequence specified in the call to **RETURN**.

- It is an error if a call to **RETURN** occurs at a point which is not within a procedure definition.

- A particular form of return is the "fail return", in which it is desired to return the unevaluated function invocation as the result of a computation. This can be done by using the construct **RETURN(´procname(args)´)**. Here, **procname** and **args** are special names which are substituted by, respectively, the name of the current procedure, and the sequence of actual arguments with which the procedure was called.

- The special name **FAIL** is commonly used by Maple library procedures as a return value which indicates that the computation failed or was abandoned, in cases where returning the unevaluated function invocation is not appropriate.

SEE ALSO: **procedures, procname, args, error**

2.1.238   Roots – roots of a polynomial mod n

Calling Sequence:
> Roots(*a*)
> Roots(*a,K*)

Parameters:

    a   –   univariate polynomial

    K   –   RootOf

Synopsis:

- The `Roots` function is a placeholder for representing the roots of the univariate polynomial `a`. The roots are returned as a list of pairs of the form `[[r[1],m[1]], ..., [r[n],m[n]]]` where `r[k]` is a root and `m[k]` its multiplicity, i.e. `(x - r[k])^m[k]` divides *a*.

- The call `Roots(`*a*`) mod n` computes the roots of the polynomial *a* modulo n .

- The call `Roots(`*a,K*`) mod p` computes the roots over the finite field defined by K an algebraic extension of the integers mod p where K is a `RootOf`.

- The call `modp1(Roots(`*a*`),p)` computes the roots of the polynomial *a* in the `modp1` representation modulo the prime integer p.

Examples:

`> Roots(x^3-x) mod 6;`	$\longrightarrow$	`[[0, 1], [1, 1], [2, 1], [3, 1],`
		`[4, 1], [5, 1]]`
`> Roots(x^3-1) mod 2;`	$\longrightarrow$	`[[1, 1]]`
`> alias(alpha=RootOf(x^2+x+1)):`		
`Roots(x^3-1,alpha) mod 2;`	$\longrightarrow$	`[[1, 1], [alpha + 1, 1], [alpha, 1]]`

SEE ALSO: `roots`, `mod`, `modp1`, `msolve`, `Factors`, `RootOf`

2.1.239   RootOf – a representation for roots of equations

Calling Sequence:
> RootOf(*expr*)
> RootOf(*expr*, *x*)

Parameters:

    expr   –   an algebraic expression or equation

    x   –   a variable name

Synopsis:

- The function `RootOf` is a place holder for representing all the roots of an equation in one variable. In particular, it is the standard representation for Maple´s algebraic numbers, algebraic functions, and finite fields `GF(p^k)`, `p prime`, `k > 1`.

- If x is not specified, then *expr* must be either a univariate expression or an expression in _Z. In this case, the RootOf represents the roots of *expr* with respect to its single variable or _Z, respectively. If the first argument is not an equation, then the equation *expr* = 0 is assumed.

- The RootOf function checks the validity of its arguments, and solves it for polynomials of degree one. The RootOf is expressed in a single-argument canonical form, obtained by making the argument primitive and expressing the RootOf in terms of the global variable _Z.

- If *expr* is an irreducible polynomial over a field F then alpha = RootOf(*expr*) represents an algebraic extension field K over F of degree degree(*expr*, x) where elements of K are represented as polynomials in alpha.

- Evaluation in the context of evala uses the RootOf notation for the representation of algebraic numbers and algebraic functions.

- Evaluation in the context of the mod operator uses the RootOf notation for the representation of finite fields. The elements of the finite field GF(p^k) are represented as polynomials in RootOf(*expr*) where *expr* is an irreducible polynomial of degree k mod p (i.e. an algebraic extension over the integers mod p).

- Maple automatically generates RootOfs to express the solutions to polynomial equations and systems of equations, eigenvalues, and rational function integrals. Maple knows how to apply diff, series, evalf, and simplify to RootOf expressions. The alias function is often used with RootOfs to obtain a more compact notation.

- RootOfs are not restricted to algebraic extensions. They can be used to represent the solutions of transcendental equations, for example RootOf(cos(x)=x, x).

Examples:

```
> RootOf(x^2+1=0, x);
```
$\longrightarrow$
$$\text{RootOf}(\_Z^2 + 1)$$

```
> RootOf(x^2-a*x=b, x);
```
$\longrightarrow$
$$\text{RootOf}(\_Z^2 - a\,\_Z - b)$$

```
> RootOf(a*b^2+a/b, b);
```
$\longrightarrow$
$$\text{RootOf}(\_Z^3 + 1)$$

```
> RootOf(a*x+b, x);
```
$\longrightarrow$
$$- b/a$$

```
> RootOf(x^3-1, x) mod 7;
```
$\longrightarrow$
$$\text{RootOf}(\_Z^3 + 6)$$

```
> RootOf(x^2-2*x+1, x) mod 5;
```
$\longrightarrow$
$$1$$

```
> alias(alpha = RootOf(x^2+x+1)):
  Normal(1/alpha) mod 2;
```
$\longrightarrow$
$$\text{alpha} + 1$$

SEE ALSO: convert[RootOf], alias, evala, mod, allvalues, Roots, algnum, solve, type[RootOf]

2.1.240   roots – roots of a univariate polynomial

Calling Sequence:

    roots(a)

    roots(a, K)

Parameters:

    a –   a univariate polynomial

    K –   (optional) algebraic number field extension

Synopsis:

- The `roots` function computes the exact roots of a univariate polynomial over an algebraic number field. The roots are returned as a list of pairs of the form [[r_1,m_1], ..., [r_n,m_n]] where r_i is a root of the polynomial a with multiplicity m_i, that is, $(x - r_i)\hat{\ }m_i$ divides a.

- The call `roots(a)` returns roots over the field implied by the coefficients present. For example, if all the coefficients are rational, then the rational roots are computed. If a has no roots in the implied coefficient field, then an empty list is returned.

- The call `roots(a, K)` computes the roots of a over the algebraic number field defined by K. Here K must be a single `RootOf`, or a list or set of `RootOf`s, or a single radical, or a list or set of radicals. For example, if K is given as `I`, then `roots` looks for the roots of a over the complex rationals.

Examples:

```
> roots(2*x^3+11*x^2+12*x-9);
```
$\longrightarrow$ [[-3, 2], [1/2, 1]]

```
> roots(x^4-4);
```
$\longrightarrow$ []

```
> roots(x^4-4, sqrt(2));
```
$\longrightarrow$ $[[-2^{1/2}, 1], [2^{1/2}, 1]]$

```
> roots(x^4-4, {sqrt(2),I});
```
$\longrightarrow$ $[[-2^{1/2}, 1], [I\,2^{1/2}, 1], [2^{1/2}, 1], [-I\,2^{1/2}, 1]]$

```
> alias(alpha = RootOf(x^2-2)):
  alias(beta = RootOf(x^2+2)):
  roots(x^4-4, alpha);
```
$\longrightarrow$ [[- alpha, 1], [alpha, 1]]

```
> roots(x^4-4, {alpha, beta});
```
$\longrightarrow$ [[- alpha, 1], [alpha, 1], [beta, 1], [- beta, 1]]

SEE ALSO: `Roots`, `RootOf`, `factor`, `solve`

2.1.241   rsolve – recurrence equation solver

Calling Sequence:

    rsolve($eqns$, fcn)

    rsolve($eqns$, fcn, $ztrans$)

Parameters:

  eqns   –   a single equation or a set of equations
  fcn   –   function to solve for
  ztrans –   (optional) solve using Z-transforms

Synopsis:

- The function `rsolve` attempts to solve the recurrence relation specified in *eqns*, returning an expression for the general term of the function.

- The first argument should be a single recurrence relation or a set of recurrence relations and boundary conditions. Any expressions in *eqns* which are not equations will be understood to be equal to zero.

- The second argument *fcn* indicates what `rsolve` should solve for. This expression should be either an unevaluated function call (or calls) of the form `f(n)`, indicating that `rsolve` should return a general solution for `f(n)`, or simply a function `f`, in which case the arguments to `f` are deduced from the variables that occur in the calls to `f` in *eqns*.

- For difference equations of the form `f(n) = a*f(n/b) + g(n)`, a method based on substitutions is used. If no boundary condition is given then `f(1) = 0` is assumed.

- A second method based on characteristic equations is used for linear difference equations with constant coefficients. If no boundary conditions are given, the answer is expressed in terms of `f(0)`, `f(1)`, and so forth.

- First order linear difference equations are handled; in addition, certain classes of first order nonlinear difference equations are recognized.

- The `ztrans` option causes `rsolve` to solve the difference equations using Z-transforms.

- If `rsolve` is unable to compute a solution, it returns the unevaluated function invocation. This unevaluated `rsolve` invocation may be understood by other functions; for example, the `asympt` function is able to compute an asymptotic series expansion for the solution in some cases.

Examples:

```
> rsolve(f(n) = -3*f(n-1) - 2*f(n-2), f);     ⟶
```
$$(2\ f(0) + f(1))\ (-1)^{n}$$
$$+ (-\ f(0) - f(1))\ (-2)^{n}$$

```
> rsolve({f(n) = -3*f(n-1) - 2*f(n-2), f(1)
  =1, f(2)=1}, f);     ⟶
```
$$-\ 3\ (-1)^{n} + (-2)^{n}$$

```
> rsolve(f(n) = 3*f(n/2) + 5*n, f(n));     ⟶
```
$$\frac{\left(\frac{\ln(3)}{\ln(2)}\right)\ \left(-\ 15\ (2/3)^{n} \left(\frac{\ln(n)}{\ln(2)} + 1\right) + 10\right)}{\ }$$

```
> rsolve({y(n) = n*y(n-1), y(0)=1}, y);     ⟶
```
$$GAMMA(n + 1)$$

```
> rsolve({y(n)*y(n-1)+y(n)-y(n-1)=0, y(0)
  =a}, y);
```
$$\longrightarrow \qquad \frac{a}{n\,a + 1}$$

```
> rsolve({y(n+1)+f(n)=2*2^n+n, f(n+1)-y(n)
  =n-2^n+3,
    y(1)=1, f(5)=6}, {y,f}, ztrans);
```
$$\longrightarrow \qquad \{f(n) = n + 1,\ y(n) = -\,1 + 2^{n}\}$$

```
> rsolve(x(n+1) = sin(x(n)), x(n));
```
$$\longrightarrow \qquad \text{rsolve}(x(n + 1) = \sin(x(n)),\ x(n))$$

```
> asympt(", n);
```
$$\longrightarrow \qquad \frac{3^{1/2}}{n^{1/2}} + \frac{\_C + 3/5\ 3^{1/2}\ \ln\!\left(\frac{1}{n^{1/2}}\right)}{n^{3/2}} + O\!\left(\frac{1}{n^{2}}\right)$$

SEE ALSO: dsolve, solve, msolve, ztrans, asympt

2.1.242   select – selection from a list, set, sum, or product

Calling Sequence:
> select(*f*, *expr*)
> select(*f*, *expr*, b_1, ..., b_n)

Parameters:

f	–	a Boolean-valued procedure
expr	–	a list, set, sum, or product
b_1, ..., b_n	–	optional extra arguments

Synopsis:

- The function **select** keeps the operands of *expr* which satisfy the Boolean-valued procedure **f**, creating a new object of the same type as *expr*. Those operands for which *f* does not return **true** are discarded.

- Additional arguments b_1, ..., b_n are passed to *f*.

Examples:
```
> select(isprime, [$10..20]);
```
$$\longrightarrow \qquad [11,\ 13,\ 17,\ 19]$$

```
> select(has,2*x*exp(x)*ln(y),x);
```
$$\longrightarrow \qquad x\,\exp(x)$$

```
> f := indets(2*exp(a*x)*sin(x));
```
$$\longrightarrow \qquad f := \{a,\ x,\ \sin(x),\ \exp(a\,x)\}$$

```
> select(type, f, name);
```
$$\longrightarrow \qquad \{a,\ x\}$$

2.1.243   seq – create a sequence

Calling Sequence:

>   seq(y, i = $m..n$)
>   seq(y, i = x)

Parameters:

y	–	any expression
i	–	name
m, n	–	numerical values
x	–	an expression

Synopsis:

- The `seq` function is related to the `for-loop` construct. It behaves in a manner similar to `for` except that `seq` constructs a sequence of values.

- The most typical call is `seq(f(`i`), `i` = 1..`n`)` which generates the sequence `f(1)`, `f(2)`, ..., `f(`n`)`. More generally, `seq(f(`i`), `i` = `$m..n$`)` generates the sequence `f(`m`)`, `f(`m`+1)`, `f(`m`+2)`, ..., `f(`n`)`. Here m and n do not need to be integers.

- The call `seq(f(i), `i` = `x`)` generates a sequence by applying `f` to each operand of x. Here, x would most commonly be a `set` or `list`, but could be a `sum`, `product`, etc.

- The two versions of the `seq` call can best be explained by defining them in terms of the `for-loop` as follows. Typically, y is a function of i.

>   seq(y, i = m..n)   =   T := NULL; for i from m to n do T := T,y od; T
>   seq(y, i = x)   =   T := NULL; for i in x do T := T,y od; T

- In either form, the `seq` version is more efficient than the `for-loop` version because the `for-loop` version constructs many intermediate sequences. Specifically, the `seq` version is linear in the length of the sequence generated but the `for-loop` version is quadratic.

Examples:

`> seq(sin(Pi*i/6), i = 0..3);`	$\longrightarrow$	$0, 1/2, 1/2 \, 3^{1/2}, 1$
`> L := [seq(i, i = 0..6)];`	$\longrightarrow$	`L := [0, 1, 2, 3, 4, 5, 6]`
`> {seq(i^2 mod 7, i = L)};`	$\longrightarrow$	`{0, 1, 2, 4}`

SEE ALSO: `set`, `list`, `dollar`, `sequence`, `for`, `op`, `map`

2.1.244   series – generalized series expansion

Calling Sequence:

>   series($expr$, eqn)
>   series($expr$, eqn, n)

Parameters:

> *expr* – an expression
>
> *eqn* – an equation (such as $x = a$) or name (such as x)
>
> *n* – (optional) a non-negative integer

Synopsis:

- The `series` function computes a truncated series expansion of *expr*, with respect to the variable x, about the point a, up to order n.

- If *eqn* evaluates to a name x then the equation $x = 0$ is assumed.

- If the third argument n is present then it specifies the "truncation order" of the series. If n is not present, the "truncation order" is determined by the global variable `Order`. The user may assign any non-negative integer to `Order`. The default value of `Order` is 6.

- If the series is not exact then an "order term" (for example `O(x^6)`) is the last term in the series.

- It is possible to invoke `series` on user-defined functions. For example, if the procedure `series/f` is defined then the function call `series(f(x,y),x)` will invoke `series/f`(x,y,x) to compute the series. Note that this user-defined function `series/f` must return a series data structure, not just a polynomial (see `type/series`).

- If `series` is applied to an unevaluated integral then the series expansion of the integral will be computed (if possible).

- The result of the `series` function is a generalized series expansion. This could be a Taylor series or a Laurent series or a more general series. Formally, the coefficients in a "generalized series" are such that

  ```
  k1*(x-a)^eps < |coeff[i]| < k2/(x-a)^eps
  ```

  for some constants k1 and k2, for any eps>0 as x approaches a. In other words, the coefficients may depend on x but their growth must be less than the polynomial in x. O(1) represents such a coefficient, rather than an arbitrary constant.

- Usually, the result of the series function is represented in the form of Maple´s `series` data structure. For an explanation of the data structure, see the help page for `type/series`.

- However, the result of the series function will be represented in ordinary sum-of-products form, rather than in Maple´s series data structure, if it is a generalized series requiring fractional exponents.

Examples:

```
> series(x/(1-x-x^2), x=0);
```
$$\longrightarrow \quad x + x^2 + 2 x^3 + 3 x^4 + 5 x^5 + O(x^6)$$

```
> series(x+1/x, x=1, 3 );
```
$$\longrightarrow \quad 2 + (x - 1)^2 + O((x - 1)^3)$$

```
> series(exp(x), x=0, 4 );
```
$$1 + x + 1/2\ x^2 + 1/6\ x^3 + O(x^4)$$

```
> series(GAMMA(x), x=0, 2 );
```
$$x^{-1} - \text{gamma} + (1/12\ \text{Pi}^2 + 1/2\ \text{gamma}^2)\ x + O(x^2)$$

```
> int(exp(x^3), x );
```
$$\int \exp(x^3)\ dx$$

```
> series(", x=0);
```
$$x + 1/4\ x^4 + O(x^7)$$

```
> series(x^x, x=0, 3);
```
$$1 + \ln(x)\ x + 1/2\ \ln(x)^2\ x^2 + O(x^3)$$

```
> s := series(sqrt(sin(x)), x=0, 4);
```
$$s := x^{1/2} - 1/12\ x^{5/2} + O(x^{7/2})$$

```
> type(s, series);
```
$$\text{false}$$

```
> whattype(s);
```
$$+$$

SEE ALSO: inifcns, powseries, taylor, type[series], type[taylor], type[laurent],
          convert, series[leadterm]

2.1.245   series/leadterm – find the leading term of a series expansion

Calling Sequence:
    series(leadterm(*expr*), *eqn*)
    series(leadterm(*expr*), *eqn*, *n*)

Parameters:

  expr – an expression

  eqn   – an equation (such as $x = a$) or name (such as x)

  n   – (optional) a non-negative integer

Synopsis:

- The **series** function computes a truncated series expansion of *expr*, with respect to the variable x, about the point a, up to order n.

- If the third argument n is present then it specifies the "truncation order" of the series. If n is not present, the "truncation order" is determined by the global variable Order. The user may assign any non-negative integer to Order. The default value of Order is 6.

- If the first argument to **series** consists of a function call to **leadterm**, then only the first (the lowest degree) term of the series of *expr* is computed.

- When `series` is invoked with `leadterm`, it is still subject to the constraint of the global variable `Order`; the series will only be computed up to degree `Order`. Thus, if `series/leadterm` is called on an expression whose leading term is of higher degree than `Order`, an error results. However, `Order` may be set arbitrarily high when using the `leadterm` option, without wasted computation, since only the first term is actually computed.

Examples:

```
> int(exp(x^3), x);
```
$\longrightarrow$
$$\int exp(x^3)\, dx$$

```
> series(leadterm("), x=0);
```
$\longrightarrow$ x

```
> series(leadterm(x^x), x=0, 3);
```
$\longrightarrow$ 1

```
> sqrt(sin(x)):
  series(leadterm("), x=0, 4);
```
$\longrightarrow$ $\dfrac{1/2}{x}$

SEE ALSO: `series`, `Order`

2.1.246   Si – the sine integral

Synopsis:

- The sine integral is defined as follows:

$$Si(x) = \int_0^x \frac{\sin t}{t}\, dt$$

2.1.247   sign – sign of a number or a polynomial

Calling Sequence:

  sign(*expr*, [*x1, x2, ...*], ´*y*´)

Parameters:

  expr   –   a multivariate polynomial
  [x1, ...] –   (optional) list of indeterminates
  y   –   (optional) unevaluated name

Synopsis:

- The `sign` function computes the sign of the leading coefficient of *expr*.

- The leading coefficient of *expr* is determined with respect to the indeterminates given. If none are given, the leading coefficient is taken with respect to all its indeterminates. Note therefore that the leading coefficient is dependent on the order of the indeterminates which may vary from one Maple session to another, but not within a session.

- The unevaluated name specified as the optional third argument is assigned the leading term.

Examples:

```
> sign(0);                                                    1

> sign(-2/3);                                                 -1

> expr := 3*x^2*y^4 - 2*x*y^5 + x;                 2  4       5
                                          expr := 3 x  y  - 2 x y  + x

> indets(expr);                                            {x, y}

> sign(expr);                                                 1

> sign(expr, [x,y]);                                         1

> sign(expr, [y,x]);                                        -1

> sign(expr, [y,x], 'a');                                   -1

> a;                                                          5
                                                           x y
```

SEE ALSO: `indets, lcoeff, signum`

2.1.248   signum – sign function for real and complex expressions

Calling Sequence:

signum(z)

Parameters:

z – any algebraic expression

Synopsis:

- The `signum` function is a mathematical function of a real or complex argument. If the argument is a real or complex constant then it evaluates to the "sign" of the argument in the sense defined below.

- If z is a real constant (of type `rational` or `float`) then the value of `signum`(z) is -1 if the argument is negative, 1 otherwise. Note that `signum(0)` is defined to be 1 .

- If z is a complex number then the value of `signum`(z) is $z/$`abs`(z).

- The function `signum`(*expr*) returns unevaluated if *expr* does not evaluate to a real or complex number.

- The `signum` function maps onto products, uses 10-digit floating-point arithmetic for constants, and performs a few other simplifications.

Examples:

```
> signum(-2/3);                                              -1
```

```
> signum(9.876);                    ⟶              1

> signum(Pi);                       ⟶              1
                                                          1/2
> signum(1+I);                      ⟶        1/2 (1 + I) 2

> evalf(");                         ⟶      .7071067810 + .7071067810 I
                                                 2  4        5
> expr := 3*x^2*y^4 - 2*x*y^5 + x;  ⟶   expr := 3 x  y  - 2 x y  + x

                                                 2  4        5
> signum(expr);                     ⟶     signum(3 x  y  - 2 x y  + x)

> x := 1/3:  y := exp(I):
  signum(expr);                     ⟶                   4            5
                                          signum(1/3 exp(I)  - 2/3 exp(I)  + 1/3)

> evalf(evalc("));                  ⟶      - .1869621696 + .9823671142 I
```

SEE ALSO: evalf, evalc, sign

2.1.249   simplify − apply simplification rules to an expression

Calling Sequence:

    simplify(*expr*)

    simplify(*expr, n1, n2, ...*)

Parameters:

    expr   −   any expression

    n1, n2,... −   (optional) names or sets or lists

Synopsis:

- The `simplify` function is used to apply simplification rules to an expression. If only one argument is present, then `simplify` will search the expression for function calls, square roots, radicals, and powers. Next it will invoke the appropriate simplification procedures, which include: `BesselI`, `BesselJ`, `BesselK`, `BesselY`, `Ei`, `GAMMA`, `RootOf`, `W`, `dilog`, `exp`, `ln`, `sqrt`, `trig` (for trig functions), `hypergeom` (for hypergeometrics), `radical` (occurrence of fractional powers), `power` (occurrence of powers, `exp`, `ln`), and `atsign` (for operators).

- Further information on particular simplification procedures is available for the subtopics `simplify[`*name*`]` where *name* is one of: `atsign`, `GAMMA`, `hypergeom`, `power`, `radical`, `RootOf`, `sqrt`, `trig` .

- In the case of two or more arguments where the additional arguments are names, `simplify` will only invoke the simplification procedures specified by the additional arguments.

- A user can make his own simplifications known to the `simplify` function by defining a Maple procedure. If the procedure `simplify/f` is defined then the function call `simplify(a,f)` will invoke `simplify/f`(a) .

- The case of two or more arguments where the additional arguments are sets or lists is used for simplification with respect to side relations. See the subtopic `simplify[siderels]` for details.

Examples:

```
> simplify(4^(1/2)+3);                    ⟶                          5

> simplify((x^a)^b+4^(1/2), power);       ⟶            (a b)    1/2
                                                      x       + 4

> simplify(exp(a+ln(b*exp(c))));          ⟶            b exp(a + c)

> simplify(sin(x)^2+cos(x)^2, trig);      ⟶                   1

> e := cos(x)^5 + sin(x)^4 + 2*cos(x)^2 -
  2*sin(x)^2 - cos(2*x):
  simplify(e);                            ⟶              5          4
                                                    cos(x)  + cos(x)

> f := -1/3*x^5*y + x^4*y^2 + 1/3*x*y^3 +
  1:
  simplify(f, {x^3 = x*y, y^2 = x+1});    ⟶                2      2
                                                     1 + x  y + x  + x
```

SEE ALSO: `collect`, `combine`, `convert`, `expand`, `factor`, `normal`, `radsimp`, `simplify[name]` where *name* is one of: `atsign`, `GAMMA`, `hypergeom`, `power`, `radical`, `RootOf`, `siderels`, `sqrt`, `trig`

2.1.250   simplify/atsign – simplify expressions involving operators

Calling Sequence:

   simplify(*expr*, atsign)

Parameters:

   expr – any expression

Synopsis:

- The `simplify/atsign` function is used to simplify expressions which contain operators.

- Specifically, it simplifies compositions of inverses.

Examples:

```
> simplify(sin@arcsin, atsign);           ⟶              x -> x

> simplify(exp@ln, atsign);               ⟶              x -> x
```

SEE ALSO: `@`, `operators[functional]`

2.1.251   simplify/GAMMA – simplifications involving the GAMMA function

Calling Sequence:

   simplify(*expr*, GAMMA)

Parameters:

   expr –   any expression

Synopsis:

- The `simplify/GAMMA` function is used to simplify expressions containing the gamma function.

Examples:

```
> simplify(GAMMA(n+1)/GAMMA(n));
```
$$\longrightarrow \qquad n$$

```
> simplify(GAMMA(n+1)*(n^2+3*n+2));
```
$$\longrightarrow \qquad \text{GAMMA}(n + 3)$$

SEE ALSO: `GAMMA`, `simplify`

2.1.252   simplify/hypergeom – simplify hypergeometric expressions

Calling Sequence:

   simplify(*expr*, hypergeom)

Parameters:

   expr –   any expression

Synopsis:

- The `simplify/hypergeom` function is used to simplify expressions which contain hypergeometric functions.

- When `simplify/hypergeom` is executed, the hypergeom routine is read in so any future expressions containing hypergeom will automatically be simplified.

Examples:

```
> simplify(hypergeom([-1],[1],z), hypergeom);
```
$$1 - z$$

SEE ALSO: `simplify`, `hypergeom`

2.1.253   simplify/power – simplify powers

Calling Sequence:

   simplify(*expr*, power)

Parameters:

   expr –   any expression

Synopsis:

- The `simplify/power` procedure is used to simplify powers, exponentials, and logarithms in an expression.

Examples:

```
> simplify((a^b)^c, power);
```
$$\longrightarrow \qquad a^{(b\ c)}$$

```
> simplify(x^a*x^b, power);
```
$$\longrightarrow \qquad x^{(a\ +\ b)}$$

```
> simplify(exp(5*ln(x)+1), power);
```
$$\longrightarrow \qquad x^5\ \exp(1)$$

```
> simplify(ln(x*y), power);
```
$$\longrightarrow \qquad \ln(x)\ +\ \ln(y)$$

SEE ALSO: `simplify`

2.1.254 simplify/radical – simplify expressions with radicals

Calling Sequence:

   `simplify(expr, radical)`

Parameters:

   expr – any expression

Synopsis:

- The `simplify/radical` function is used to simplify expressions which contain radicals.
- This routine just invokes the `radsimp` function.

Examples:

```
> simplify((x-2)^(3/2)/(x^2-4*x+4)^(1/4), radical);
```
$$x\ -\ 2$$

SEE ALSO: `radsimp`, `simplify`

2.1.255 simplify/RootOf – simplify expressions with the RootOf function

Calling Sequence:

   `simplify(expr, RootOf)`

Parameters:

   expr – any expression

Synopsis:

- The `simplify/RootOf` function is used to simplify expressions which contain the `RootOf` function.

- Polynomials and inverses of polynomials in `RootOf` are simplified. Any nth roots involving `RootOf` are resolved.

Examples:

```
> r := RootOf(x^2-2=0, x):
  simplify(r^2, RootOf);
```
$\longrightarrow$ $\qquad\qquad$ 2

```
> simplify(1/r, RootOf);
```
$\longrightarrow$ $\qquad\qquad$ $1/2\ \text{RootOf}(\_Z^2 - 2)$

SEE ALSO: `RootOf`, `simplify`

2.1.256   simplify/siderels – simplify with respect to side relations

Calling Sequence:
>   simplify(*expr*, *eqns*)
>   simplify(*expr*, *eqns*, *vars*)

Parameters:

> *expr* – an expression
>
> *eqns* – a set or list of equations (an expression **e** is understood as the equation **e=0**)
>
> *vars* – (optional) a set or list of variables

Synopsis:

- Simplification of *expr* with respect to the side relations *eqns* is performed. The result is an expression which is mathematically equivalent to *expr* but which is in "normal form" with respect to the specified side relations.

- The specific meaning of "normal form" is determined by Gröbner basis concepts. The Gröbner basis for *eqns* (with respect to *vars*) is computed and then the value returned is computed by applying to *expr* the procedure `grobner[normalf]` — or if *expr* is not a polynomial then `grobner[normalf]` is applied separately to the **numer** and **denom** of `normal(`*expr*`)`.

- If *vars* is not specified then it is determined using indets. There are two reasons for pre-specifying *vars*:

  (i) perhaps some indeterminates are meant to be considered as parameters rather than variables;

  (ii) the precise form of simplification to be performed can be controlled by specifying *vars* as a list (see below).

- The definition of the Gröbner basis varies according to the choice of ordering on the variables:

  (i) if *vars* is not specified, or is specified as a set, then the "total degree" ordering is used;

  (ii) if *vars* is specified as a list then the "pure lexicographic" ordering is used, with respect to the specific order of variables in the list (most main variable first).

Examples:

```
> eqn := {sin(x)^2 + cos(x)^2 = 1}:
> e := sin(x)^3 - 11*sin(x)^2*cos(x) + 3*cos(x)^3 - sin(x)*cos(x) + 2:
> simplify(e, eqn);
```

$$14 \cos(x)^3 - \sin(x) \cos(x) + 2 - \sin(x) \cos(x)^2 + \sin(x) - 11 \cos(x)$$

```
> simplify(e, eqn, [cos(x),sin(x)]);
```

$$\sin(x)^3 - 14 \sin(x)^2 \cos(x) - \sin(x) \cos(x) + 2 + 3 \cos(x)$$

```
> f := sin(x)^7 + sin(x)^5*cos(x)^2 - sin(x)^5*cos(x) - sin(x)^3*cos(x)^3:
> simplify(f, eqn, [cos(x),sin(x)]);
```

$$- \sin(x)^3 \cos(x) + \sin(x)^5$$

```
> g := 8*sin(x)^4*cos(x) + 15*sin(x)^2*cos(x)^3 - 15*sin(x)^2*cos(x)
>                + 7*cos(x)^5 - 14*cos(x)^3 + 7*cos(x):
> simplify(g, eqn);
```

$$0$$

```
> siderels := {z^3 - z^2 - z*y + 2*y^2 = 1,  z^3 + y^2 = 1,
>                         z^2 + z*y - y^2 = 0,  x + y = z}:
> h1 := 36*z^4*y^2+36*z*y^4-36*z*y^2-1/2*z^2+z*y-1/2*y^2+1/2*x*z-1/2*x*y
>                +2/3*z^4+4/3*z^3*y-2/3*z^2*y^2-4/3*z*y^3+2/3*y^4:
> simplify(h1, siderels);
```

$$0$$

```
> h2 := z*y^2+z^3*y^3-3*z^2*y+23/3+5/6*z^2+7/3*z*y-11/6*y^2+1/2*x*z-1/2*x*y
>                +2/3*z^4+4/3*z^3*y-2/3*z^2*y^2-4/3*z*y^3+2/3*y^4:
> simplify(h2, siderels);
```

$$- 34 z^2 y + 14 z y + 16 z^2 - 34/3 + 8 z + 13 y$$

SEE ALSO: grobner

2.1.257   simplify/sqrt – simplify square roots

Calling Sequence:

> simplify(*expr*, sqrt)

Parameters:

> *expr*   –   any expression

Synopsis:

- The `simplify/sqrt` function is used to simplify expressions which contain square roots or powers of square roots.

- It extracts any square integer or polynomial factor from inside the square root.

- The positive square root is used.

Examples:

> `simplify(16^(3/2), sqrt);` $\longrightarrow$ 64

> `simplify((10*x^2+60*x+90)^(1/2), sqrt);` $\longrightarrow$ $10^{1/2}(x+3)$

> `simplify((90/121)^(1/2), sqrt);` $\longrightarrow$ $\dfrac{10^{1/2}}{3/11}$

SEE ALSO: `simplify`

2.1.258   simplify/trig – simplify trigonometric expressions

Calling Sequence:
  `simplify(`*expr*`, trig)`

Parameters:
  expr – any expression

Synopsis:
- The `simplify/trig` function is used to simplify trigonometric expressions by applying the identities `sin(x)^2 + cos(x)^2 = 1` and `cosh(x)^2 - sinh(x)^2 = 1`.
- This function views the input as a polynomial in `sin(x)` and `sinh(x)` and reduces the degree of the input to 1 in `sin(x)` and `sinh(x)`.

Examples:

> `simplify(sin(x)^3, trig);` $\longrightarrow$ $-\sin(x)\cos(x)^2 + \sin(x)$

> `simplify(tan(x)^2, trig);` $\longrightarrow$ $-\dfrac{\cos(x)^2 - 1}{\cos(x)^2}$

> `simplify(1+tan(x)^2, trig);` $\longrightarrow$ $\dfrac{1}{\cos(x)^2}$

> `simplify(cos(2*x)+sin(x)^2);` $\longrightarrow$ $\cos(x)^2$

SEE ALSO: `simplify`

2.1.259   solve – solve equations

Calling Sequence:
  `solve(`*eqns*`,  `*vars*`)`

Parameters:

- *eqns* – an equation or set of equations
- *vars* – (optional) an unknown or set of unknowns

Synopsis:

- The most common application of `solve` is to solve a single equation, or to solve a system of equations in some unknowns. A solution to a single equation *eqns* solved for the unknown *vars* is returned as an expression. To solve a system of equations *eqns* for unknowns *vars*, the system is specified as a set of equations and a set of unknowns. The solution is returned as a set of equations.

- Multiple solutions are returned as an expression sequence. Wherever an equation is expected, if an expression *expr* is specified then the equation *expr* = 0 is understood. If *vars* is not specified, `indets(`*eqns*`,name)` is used in place of *vars*.

- When `solve` is unable to find any solutions, the expression `NULL` is returned. This may mean that there are no solutions or that `solve` was unable to find the solutions.

- To assign the solutions to the variables, use the command `assign`.

- For solving differential equations use `dsolve`; for purely floating-point solutions use `fsolve`; use `isolve` to solve for integer roots; `msolve` to solve modulo a prime; `rsolve` for recurrences, and `linalg[linsolve]` to solve matrix equations.

- Further information is available for the subtopics `solve[`*subtopic*`]` where *subtopic* is one of

floats	functions	identity	ineqs	linear
radical	scalar	series	system	

- For systems of polynomial equations, the function `grobner[gsolve]` which uses a Gröbner-basis approach may be useful.

Examples:

```
> solve(cos(x) + y = 9, x);                    ⟶           Pi - arccos(y - 9)

> solve({cos(x) + y = 9}, {x});                ⟶         {x = Pi - arccos(y - 9)}

> solve(x^3 - 6*x^2 + 11*x - 6, x);            ⟶               1, 2, 3

> solve({x+y=1, 2*x+y=3}, {x,y});              ⟶             {x = 2, y = -1}

> solve({a*x^2*y^2, x-y-1}, {x,y});            ⟶    {x = 1, y = 0}, {x = 0, y = -1}
```

SEE ALSO: `dsolve`, `fsolve`, `isolve`, `msolve`, `rsolve`, `assign`, `isolate`, `match`, `linalg[linsolve]`, `simplex`, `grobner`, `solve[`*subtopic*`]` where *subtopic* is one of: `floats`, `functions`, `identity`, `ineqs`, `linear`, `radical`, `scalar`, `series`, `system`

2.1.260   solve/floats – expressions involving floating-point numbers

Calling Sequence:

solve(*eqns, vars*)

Parameters:

eqns – equations (as for solve), but with floating-point values

vars – variables (as for solve)

Synopsis:

- The solve function with floating-point numbers works by converting the floating-point numbers to approximate rationals, calling solve with these converted arguments, and converting the results back to floating-point numbers.

- This is useful for solving equations with a combination of floating-point numbers and parameters (since fsolve will not solve equations with unassigned parameters).

Examples:
```
> solve(x^2-3*x+0.01, x);
```
$$2.996662955, \ .003337045$$
```
> solve({x^2-y=z, 3.7*y+z=sin(x)});
```
$$\{x = x, \ z = 1.370370370 \ x^2 \ - \ .3703703704 \ \sin(x),$$
$$y = \ - \ .3703703704 \ x^2 \ + \ .3703703704 \ \sin(x)\}$$

SEE ALSO: fsolve, evalf, realroots

2.1.261   solve/functions – for a variable which is used as a function name

Calling Sequence:

solve(*eqn, fcnname*)

Parameters:

eqn   – a single equation (as for solve)

fcnname – a single variable (as for solve)

Synopsis:

- When the unknown is a name which is only used as a function in the equation, the solver constructs a function or procedure which solves the equation.

Examples:
```
> solve(cos(f(x))=sin(x), f);
proc(x) arccos(sin(x)) end
```

```
> solve(f(x)^2-3*f(x)+2*x, f);
      proc(x) 3/2+1/2*(9-8*x)^(1/2) end, proc(x) 3/2-1/2*(9-8*x)^(1/2) end
```

SEE ALSO: unapply, operators[functional]

2.1.262   solve/identity – expressions involving identities

Calling Sequence:

>   solve(identity(*eqn*, *x*), *vars*)

Parameters:

eqn – an equation, or an expression (equated to 0)

x – the identity variable

vars – variables to be determined (as for solve)

Synopsis:

- The expression (or equation) *eqn* is considered an identity in terms of the variable x, and solve attempts to find a solution in terms of *vars* which satisfies the equation *eqn* for any value of x.

- This is a one-dimensional case of a generalized pattern matching.

- x should not be one of the *vars*.

Examples:
```
> solve(identity(sin(x) = cos(a*x+b), x), {a,b});
                        {b = 1/2 Pi, a = -1}
> solve(identity(3/(x^2+4*x+4) = A*(x+B)^P, x), {A,B,P});
                        {P = -2, A = 3, B = 2}
> solve(identity(y/x = a*(x+c)^b, x), {a,b,c});
                        {b = -1, c = 0, a = y}
```

SEE ALSO: match

2.1.263   solve/ineqs – inequalities

Calling Sequence:

>   solve(*ineq*, *var*)

Parameters:

ineq – one inequality ($<$, $\leq$, $>$, or $\geq$)

var – one variable (as in solve)

Synopsis:

- The solve function will solve one inequality in one variable for all the cases where the equality points of *ineq* can be properly ordered.

• Some additional cases are handled by using the signum function.

Examples:

```
> solve(a*x < b, x);
```
$\longrightarrow$
$$\{signum(a)\ x < \frac{signum(a)\ b}{a}\}$$

```
> solve(a^2*x^2 >= 0, x);
```
$\longrightarrow$
$$\{x = x\}$$

```
> solve(x+1/x > 0, x);
```
$\longrightarrow$
$$\{0 < x\}$$

```
> solve((x-1)*(x-2)*(x-3) < 0, x);
```
$\longrightarrow$
$$\{x < 1\},\ \{2 < x,\ x < 3\}$$

```
> solve((x-1+a)*(x-2+a)*(x-3+a) < 0, x);
```
$\longrightarrow$
$$\{x < 1 - a\},\ \{2 - a < x,\ x < 3 - a\}$$

```
> solve(exp(x) > x+1);
```
$\longrightarrow$
$$\{x <> 0\}$$

SEE ALSO: `simplex`, `isolate`

2.1.264 solve/linear – systems of linear equations

Calling Sequence:

   solve(*eqns*, *vars*)

Parameters:

   eqns – a set of linear equations and inequations

   vars – a set of names (unknowns)

Synopsis:
• The linear system defined by *eqns* is solved for the unknowns *vars*. If a solution exists, the solution is returned as a set of equations. If the system is underdetermined, the solver will parameterize the solutions in terms of one or more of the unknowns.

• Currently, the solver uses a number of special algorithms in addition to ordinary Gaussian elimination. All algorithms are geared to large sparse systems though they also perform well on dense systems. The special algorithms (selected automatically) are for:

   (a) rational coefficients, using a "primitive" fraction-free algorithm;

   (b) floating-point coefficients, using Gaussian elimination with partial pivoting for stability;

   (c) rational function coefficients over the rationals, more generally, an algebraic number field, using a "primitive" fraction-free algorithm.

Examples:

```
> solve({a*x+b*y=3, x-y=b}, {x,y});
```
$\longrightarrow$
$$\{x = \frac{b^2 + 3}{a + b},\ y = - \frac{- 3 + b\ a}{a + b}\}$$

```
> solve({I*x+y=1,I*x-y=2}, {x,y});
```
$\longrightarrow$
$$\{x = - 3/2\ I,\ y = -1/2\}$$

SEE ALSO: `solve[system]`, `linalg[linsolve]`

2.1.265 solve/radical – expressions involving radicals of unknowns

Calling Sequence:

solve(*eqns*, *vars*)

Parameters:

eqns – equations (as for solve), but with radicals

vars – variables (as for solve)

Synopsis:

- The `solve` function will solve equations with radicals by using new auxiliary variables and adding new equations. The resulting problem is always a system of equations, and hence it is solved like a system and not as a scalar.

- Radicals will be considered to be multivalued functions, and hence verification of the solution requires finding the right combination of branches of the radicals.

Examples:

```
> solve(sqrt(x+1)-sqrt(x-1)=a, x);
```

$$1/4 \ \frac{(4+a^4)}{a^2}$$

```
> solve(x+sqrt(x)-2, x);
```

$$1, \ 4$$

```
> solve(x+sqrt(x)+x^(1/3)=3,x);
```

$$1, \ \mathrm{RootOf}(\_Z^5 + \_Z^4 + \_Z^3 + 2\ \_Z^2 + 3\ \_Z + 3)^6$$

```
> solve(sqrt(x)+sqrt(x+1)=3);
```

$$16/9$$

```
> solve({sqrt(x)+sqrt(y)-2, x-y+1},{x,y});
```

$$\{y = \frac{25}{16}, \ x = 9/16\}$$

SEE ALSO: solve[system], solve[scalar]

2.1.266 solve/scalar – scalar case (a single variable and equation)

Calling Sequence:

solve(*eqn*, *var*)

Parameters:

eqn – an algebraic expression (to be equated to 0), or an equation, or a set containing a single expression

var – (optional) a name, or a set containing a single name

Synopsis:

- The `solve` function with a single equation and a single variable uses the scalar solver. When called with a single argument, it is equivalent to `solve(eqn, indets(eqn,name))`.

- The scalar solver will report all solutions found, even if some roots are multiple roots. In particular, for polynomials, `solve` will find all solutions.

- For polynomials of degree 2, 3 and 4, the scalar solver will compute the explicit solutions. For irreducible and non-composable polynomials of degree higher than 4, the scalar solver will return the solutions using `RootOf` notation.

- The `solve` function normally returns an expression sequence of sets. However, if the input arguments are expressed without using sets (for instance, as an algebraic expression and a variable name), then it returns an expression sequence of the solution values for that variable.

Examples:
```
> solve(cos(x) + y = 9, x);                    ⟶          Pi - arccos(y - 9)
> solve(x^3 - 6*x^2 + 11*x - 6, x);            ⟶               1, 2, 3
> solve({x^3 - 6*x^2 + 11*x - 6}, {x});        ⟶       {x = 1}, {x = 2}, {x = 3}
```

SEE ALSO: `solve[floats]`, `solve[identity]`, `solve[radical]`

2.1.267   solve/series – expressions involving general series

Calling Sequence:

> `solve(eqn, var)`

Parameters:

eqn – an equation involving a series in *var*

var – the variable to be solved for

Synopsis:

- For an equation which contains a series, solving for the series variable achieves a generalized inversion of series.

- The result, usually a series, will be in one of the remaining indeterminates. The indeterminate which produces the simplest answer is usually chosen.

- The global variable `Order` will be used to determine the order of the series result.

- Formally, if `t = solve(f(series_in_x, y), x)` is a series in y with sufficient terms, then substituting `t` for x in `f` will result in 0 or `O(y^Order)`.

Examples:
```
> Order := 3;
                    Order := 3
```

```
> solve(series(x*exp(x),x) = y, x);
```
$$y - y^2 + O(y^3)$$
```
> solve(series(exp(x),x) = a+b, x);
```
$$- 1 + b + a - 1/2 (- 1 + b + a)^2 + O((- 1 + b + a)^3)$$

SEE ALSO: `series`, `RootOf`

2.1.268 solve/system – systems of equations, multiple equations or unknowns

Calling Sequence:
> solve(*eqns*, *vars*)

Parameters:

eqns – a set of equations and inequations

vars – a set of names (unknowns)

Synopsis:

- The system solver behaves slightly differently than the scalar solver. For linear systems of equations see `solve[linear]`.

- Multiple solutions are reported only once. Solutions of irreducible polynomials of degree 2, 3, or 4 are represented implicitly by `RootOf`s.

- Inequalities (*expr*$_1$ <> *expr*$_2$) are either part of the input or are generated from denominators, singularities, and so on. That the solutions will satisfy the equations and inequalities is guaranteed.

- If there are more unknowns than equations, the solver will select the variables for which the system is easiest to solve.

- If the global variable _MaxSols is set to an integer value, the solver will stop after _MaxSols solutions are found. By default this value is set to 100.

Examples:
```
> solve({x+2*y=3, y+1/x=1}, {x,y});
```
$$\{x = -1, y = 2\}, \{x = 2, y = 1/2\}$$
```
> solve(a*x^2+b*x+c, {a,b,c,x});
```
$$\{c = - a x^2 - b x, x = x, a = a, b = b\}$$
```
> solve({x^2-y^2-y, x+y^2, x<>0});
```
$$\{y = RootOf(\_Z^3 - \_Z - 1), x = - RootOf(\_Z^3 - \_Z - 1)^2 \}$$

SEE ALSO: `solve[linear]`

2.1.269   sort – sort a list of values or a polynomial

Calling Sequence:
    sort(L)
    sort(L, F)
    sort(A)
    sort(A, V)

Parameters:

L – a list of values to be sorted

F – (optional) a Boolean procedure of two arguments

A – an algebraic expression

V – (optional) variables

Synopsis:

- The **sort** function sorts the elements of a list L into ascending order and the terms of polynomials in an algebraic expression A into descending order.

- If F is given, it is used to define the ordering for sorting a list. If F is the string $<$ or **numeric** then L is sorted into numerical order. If F is the string **lexorder** or **string** then strings are sorted into lexicographic order. If F is the string **address** then the elements are sorted by address. Otherwise, F must be a Boolean-valued function of two arguments. Specifically, $F(a,b)$ should return **true** if a precedes b in the desired ordering, false otherwise.

- If no ordering function F is specified, a list of numbers is sorted into numerical order, and a list of strings are sorted into lexicographical order. Otherwise, elements of L are sorted by machine address. Note that sorting by machine address is session dependent.

- In Maple, polynomials are not automatically stored in sorted order. They are stored in the order they were first created and printed in the order they are stored. The sort function can be used to sort polynomials. But please note that sorting polynomials is a destructive operation: the input polynomial will be sorted "in-place".

- If V is given it specifies the variable ordering to be used when sorting polynomials. It can be a name or function, or a list or set of names (for the multivariate case). All polynomials in the expression A are sorted into decreasing order in V. If V is not specified, the indets appearing in A will be used.

- An additional third argument, either the string **plex** or **tdeg**, can be given to define the ordering for the multivariate case. If **tdeg** is specified (default) then polynomials in V are sorted in total degree with ties broken by lexicographical order. If **plex** is specified, polynomials in V are sorted in pure lexicographical order.

- The sorting algorithm used for sorting lists is a recursive implementation of Merge sort with early detection of sorted sequences. The sorting algorithm used for sorting polynomials is an in-place Shell sort.

Examples:

```
> sort([3,2,1]);
```
$\longrightarrow$ [1, 2, 3]

```
> sort(1+x+x^2);
```
$\longrightarrow$
$$x^2 + x + 1$$

```
> sort([c,a,d],lexorder);
```
$\longrightarrow$ [a, c, d]

```
> p := y^3+y^2*x^2+x^3:
  sort(p,[x,y]);
```
$\longrightarrow$
$$x^2 y^2 + x^3 + y^3$$

```
> sort(p,[x,y],plex);
```
$\longrightarrow$
$$x^3 + x^2 y^2 + y^3$$

```
> sort((y+x)/(y-x),x);
```
$\longrightarrow$
$$\frac{x + y}{-x + y}$$

SEE ALSO: `list`, `indets`, `polynom`, `degree`

2.1.270   Sqrfree – inert square free factorization function

Calling Sequence:

   Sqrfree(a)

Parameters:

   a –   multivariate polynomial

Synopsis:

- The `Sqrfree` function is a placeholder for representing the square free factorization of the multivariate polynomial a over a unique factorization domain. It is used in conjunction with either `mod`, `evala` or `evalgf` which define the coefficient domain as described below.

- The Sqrfree function returns a data structure of the form `[u,[[f[1],e[1]],...,[f[n],e[n]]]` such that `a = u * f[1]^e[1] * ... * f[n]^e[n]` and `f[i]` is primitive where `Gcd(f[i],f[j])` `= 1` for `i <> j` where `u` is the leading coefficient of a.

- The call `Sqrfree(`a`) mod p` computes the square free factorization of a modulo `p` a prime integer. The multivariate polynomial a must have rational coefficients or coefficients from an algebraic extension of the integers modulo p.

- The call `modp1(Sqrfree(`a`),p)` computes the square free factorization of the polynomial a in the `modp1` representation modulo p a prime integer.

Examples:

```
> Sqrfree(2*x^2+6*x+6) mod 7;
```
$\longrightarrow$
$$[2, [[x^2 + 3 x + 3, 1]]]$$

```
> Sqrfree(4*x^2+4*x+1) mod 7;
```
$\longrightarrow$
$$[4, [[x + 4, 2]]]$$

> `alias(alpha=RootOf(x^2+x+1));` $\longrightarrow$ `I, alpha`

> `Sqrfree(alpha*x^3+(alpha+1)*x^2+x+alpha)` $\longrightarrow$ `[alpha, [[x + alpha, 3]]]`
  `mod 2;`

SEE ALSO: `sqrfree, isqrfree, mod, modp1, Factors, RootOf`

2.1.271   sqrt – square root

Calling Sequence:
   `sqrt(x)`

Parameters:
   x – any algebraic expression

Synopsis:
- The function `sqrt` forms the square root of its argument x. It returns the square root of integers and polynomials which are perfect squares. It is mapped onto products.

Examples:
> `sqrt(3);` $\longrightarrow$ $3^{1/2}$

> `sqrt(4);` $\longrightarrow$ 2

> `sqrt(12);` $\longrightarrow$ $2\ 3^{1/2}$

> `sqrt(3+4*I);` $\longrightarrow$ $2 + I$

> `sqrt(x^2/9);` $\longrightarrow$ $1/3\ x$

> `sqrt(a);` $\longrightarrow$ $a^{1/2}$

SEE ALSO: `simplify[sqrt]`

2.1.272   subs – substitute subexpressions into an expression

Calling Sequence:
   $subs(s_1, s_2, \ldots, s_n, expr)$

Parameters:
   $s_1,\ldots$ – equations or sets or lists of equations
   $expr$ – any expression

Synopsis:
- The function `subs` returns an expression resulting from applying the substitutions specified by the first arguments to the last argument $expr$.

- The substitutions are performed sequentially starting with *s1*. The substitutions within a set or list are performed simultaneously.

- Every occurrence of the left hand side of a substitution equation that appears in *expr* is replaced by the right hand side of the equation.

- Note that only sub-expressions in *expr* that correspond to operands (**ops**) of a Maple object are matched. This is termed "syntactic substitution".

- The action of substitution is not followed by evaluation. In cases where full evaluation is desired, it is necessary to use the **eval** function to force an evaluation. For example, **subs(y=ln(x), exp(y))**, as shown below.

Examples:

```
> subs( x=r^(1/3), 3*x*ln(x^3) );
```
$$\longrightarrow \qquad 3\ r^{1/3}\ \ln(r)$$

```
> subs( a+b=y, (a+b+c)^2 );
```
$$\longrightarrow \qquad (a + b + c)^2$$

```
> subs( x=y, y=x, [x,y]);
```
$$\longrightarrow \qquad [x,\ x]$$

```
> subs( {x=y, y=x}, [x,y]);
```
$$\longrightarrow \qquad [y,\ x]$$

```
> subs( sin(x)+cos(x) );
```
$$\longrightarrow \qquad \sin(x) + \cos(x)$$

```
> subs( y=ln(x), exp(y) );
```
$$\longrightarrow \qquad \exp(\ln(x))$$

```
> eval(subs( y=ln(x), exp(y) ));
```
$$\longrightarrow \qquad x$$

SEE ALSO: op, subsop, eval

2.1.273   subsop – substitute for specified operands in an expression

Calling Sequence:

   subsop(*eq1*, *eq2*, ..., *eqN*, *expr*)

Parameters:

   eqI   –   an (optional) equation of the form: *numI = exprI*

   numI   –   a non-negative integer

   exprI   –   an expression

   expr   –   an expression

Synopsis:

- The function **subsop** will do the simultaneous substitutions specified by the equation arguments in the last argument *expr*.

- This function will return the expression resulting from replacing op(num1, *expr*) by $expr_1$, op(num2, *expr*) by $expr_2$, ..., and op(numN, *expr*) by $expr_n$ in *expr*.

Examples:

> `subsop( 3=y^4, x^7+8*x^6+x-9 );` $\longrightarrow$

$$x^7 + 8x^6 + y^4 - 9$$

> `subsop( 1=NULL, 2=b, [x,y,x] );` $\longrightarrow$ $[b, x]$

> `subsop( x^2+5 );` $\longrightarrow$

$$x^2 + 5$$

> `subsop( 0=g, 3=a, f(x,f(x,y,z),x) );` $\longrightarrow$ $g(x, f(x, y, z), a)$

SEE ALSO: `subs`, `op`

2.1.274   substring – extract a substring from a string

Calling Sequence:

   `substring( `*string*`, `*m..n*` )`

Parameters:

   string – a string

   m..n – a range

Synopsis:

- If m and n evaluate to integers then *substring* will extract a substring from *string* starting with the mth character and ending with the nth character. If m is less than one then the substring will start at the first character of *string*. If n is greater than the length of *string* then the substring will return a string from the mth character to the end of *string*. If m is greater than n then substring will return the null string.

- If either m or n fails to evaluate to an integer, then *substring* will remain unevaluated.

Examples:

> `substring( abcdefgh, 3..7 );` $\longrightarrow$ `cdefg`

> `substring( wxyz, 2..2 );` $\longrightarrow$ `x`

SEE ALSO: `length`, `cat`

2.1.275   sum – definite and indefinite summation
Sum – inert form of summation

Calling Sequences:

   `sum(`*f*`, `*k*`);`   `sum(`*f*`, `*k=m..n*`);`   `sum(`*f*`, `*k=alpha*`);`

   `Sum(`*f*`, `*k*`);`   `Sum(`*f*`, `*k=m..n*`);`   `Sum(`*f*`, `*k=alpha*`);`

Parameters:

f	–	an expression
k	–	a name, the summation index
m, n	–	integers or arbitrary expressions
$alpha$	–	a `RootOf` expression

Synopsis:

- The call `sum(f, k)` computes the indefinite sum of $f(k)$ with respect to k. Thus it computes a formula g such that $g(k+1)-g(k)=f(k)$ for all k.

- The call `sum(f, k=m..n)` computes the definite sum of $f(k)$ over the given range $m..n$, so it computes $f(m) + f(m+1) + ... + f(n)$. The definite sum is equivalent to $g(n+1)-g(m)$ where g is the indefinite sum. For example, `sum(n, n) = sum(k, k=0..n-1) = (n^2-n)/2`.

- If $m = n+1$ then the value returned is 0. If $m > n+1$ then the value returned is `-sum(f, k=n+1..m-1)`.

- The call `sum(f, k=alpha)` computes the definite sum of $f(k)$ summed over the roots of a polynomial $alpha$ where $alpha$ must be a RootOf.

- Note: It is recommended (and often necessary) that both f and k be enclosed in single quotes to prevent premature evaluation. (For example, k may have a previous value.) Thus the common format is `sum('f', 'k'=m..n)`.

- For indefinite summation, Maple uses the following methods: Polynomials are summed using a formula based on Bernoulli polynomials. Moenck's method is used for summing rational functions of k. The result is a rational function plus a sum of terms involving the Polygamma function $Psi(k)$ and its derivatives. Gosper's decision procedure is used to sum expressions containing factorials (including binomial coefficients) and powers.

- For definite sums, if $n-m$ is a small integer, the sum is computed directly. Otherwise it is computed via indefinite summation and taking limits, and/or using various hypergeometric summation identities.

- If Maple cannot find a closed form for the summation, the function call itself is returned. (The prettyprinter displays the sum function using a stylized summation sign.)

- The capitalized function name `Sum` is the inert `sum` function, which simply returns unevaluated. The prettyprinter understands `Sum` to be equivalent to `sum` for printing purposes.

Examples:

```
> sum(k^2, k=1..4);
```
$\longrightarrow$
$$30$$

```
> sum(k^2, k);
```
$\longrightarrow$
$$\tfrac{1}{3} k^3 - \tfrac{1}{2} k^2 + \tfrac{1}{6} k$$

```
> sum(a[k], k=1..4);
```
$\longrightarrow$
$$a[1] + a[2] + a[3] + a[4]$$

```
> sum(k/(k+1), k=1..4);
```
$\longrightarrow$

$$\frac{163}{60}$$

```
> sum(k/(k+1), k=0..n);
```
$\longrightarrow$

$$n + 1 - Psi(n + 2) - gamma$$

```
> Sum(k/(k+1), k=0..n);
```
$\longrightarrow$

$$\sum_{k = 0}^{n} \frac{k}{k + 1}$$

```
> sum(k*a^k, k);
```
$\longrightarrow$

$$\frac{a^k (k a - k - a)}{(a - 1)^2}$$

```
> sum(binomial(n+k,k), k);
```
$\longrightarrow$

$$\frac{k \, binomial(n + k, k)}{n + 1}$$

```
> sum(- exp(-k), k);
```
$\longrightarrow$

$$\frac{exp(1)}{(- 1 + exp(1)) \, exp(k)}$$

```
> sum(k*binomial(n,k), k=0..n);
```
$\longrightarrow$

$$1/2 \, 2^n \, n$$

```
> sum(1/k!, k=0..infinity);
```
$\longrightarrow$

$$exp(1)$$

```
> sum(1/k^2, k=1..infinity);
```
$\longrightarrow$

$$1/6 \, Pi^2$$

```
> sum(k/(k+1), k=RootOf(x^3-2));
```
$\longrightarrow$

$$2$$

2.1.276   Svd – compute the singular values/vectors of a numeric matrix

Calling Sequence:

   Svd(X)

   Svd($X, U,$ `left`)

   Svd($X, V,$ `right`)

   Svd(X, U, V)

Parameters:

   X – an n x p matrix

   U – (optional) the left singular vectors are to be returned in U

   V – (optional) the right singular vectors are to be returned in V

Synopsis:

- Svd(X) returns a 1 by `min(n,p)` array of the singular values of X.

- The entries of X must be all numerical.

- $\text{Svd}(X, U, \text{`left`})$ returns the singular values and the left singular vectors in U.

- $\text{Svd}(X, V, \text{`right`})$ returns the singular values and the right singular vectors in V.

- $\text{Svd}(X, U, V)$ returns the singular values and the left and right singular vectors in U and V respectively. The singular vectors together with the singular values satisfy $U'XV = \text{D}$ where U' is the transpose of U and U is n by n, V is p by p, X is n by p, and D is n by p where $D[i,i]$ is/are the singular value/values of X.

- This procedure Svd is compatible with the Fortran library `linpack`.

- Note that nothing happens when the user invokes $\text{Svd}(X)$ (same for other calling sequences); the user must use $\text{evalf}(\text{Svd}(X))$ to actually compute the singular values and singular vectors.

Examples:
```
> A := linalg[matrix](2,2,[1,2,3,4]);
```
$$\longrightarrow \qquad A := \begin{bmatrix} 1 & 2 \\ 3 & 4 \end{bmatrix}$$

```
> evalf(Svd(A));
```
$$\longrightarrow \qquad [\ 5.464985704,\ .3659661906\ ]$$

SEE ALSO: `linalg[singularvals]`

2.1.277   system – invoke a command in the host operating system

Calling Sequence:

    system(*command*)

Parameters:

    command   –   a Maple string

Synopsis:

- The function `system` passes *command* to the host operating system (for example UNIX, VMS, or CMS) which performs the appropriate function. The result of a system command is the "return status" of the command performed.

- Note: On some operating systems, some operating system commands may not be accessible via this mechanism.

Examples:
```
> system( date );
```
$$\longrightarrow \qquad \text{Thu Jun 6 17:26:11 EDT 1991}$$
$$0$$

SEE ALSO: escape, `option` for ``option system''

2.1.278   table – create a table

Calling Sequence:
    `table( F, L )`

Parameters:
    F – (optional) the indexing function
    L – (optional) list of initial table entries

Synopsis:

- A table is created either explicitly by a call to the **table** function or implicitly by assigning to an indexed name. Unlike arrays, where indices must be integers, the indices (or keys) of a table can be any value.

- The **table** function creates a table with initial values specified by L. If L is a list of equations then the left-hand side is taken to be the table index (key), and the right-hand side the entry (value). Otherwise, L is taken to be a list of entries with indices 1, 2, ... If L is not specified, then an empty table is created.

- New entries can be added to a table using the subscript notation. Thus T := `table([4])` is equivalent to: T := `table()` and `T[1] := 4`, or equivalently: $T[1]$:= 4 if T was unassigned (implicit creation). Entries can be removed from a table by assigning a table entry to its own name. Thus $T[1]$:= ´$T[1]$´ removes the entry 4 from T.

- The indexing function F can be a procedure or a name specifying how indexing is to be performed; if null then ordinary indexing is implied. The built-in indexing functions are: **symmetric**, **antisymmetric**, **sparse**, **diagonal**, and **identity**.

- Tables have special evaluation rules (like procedures) so that if the name T has been assigned a table then T evaluates to T. The call **op(** T **)** yields the actual table structure; and **op(op(** T **))** yields the components of the table, which are the indexing function (if there is one) and a list of equations for the tables values.

- The **indices** function can be used to obtain a sequence of a table´s indices and likewise, the **entries** function returns a sequence of the table´s entries.

Examples:
```
> table();                      ⟶        table([])

> table([22,42]);               ⟶        table([
                                             1 = 22
                                             2 = 42
                                          ])

> S := table([(2)=45,(4)=61]);  ⟶        S := table([
                                                   2 = 45
                                                   4 = 61
                                             ])

> S[1], S[2];                   ⟶        S[1], 45
```

```
> T := table(symmetric,[(c,b)=x]);        ⟶      T := table(symmetric,[
                                                        (c, b) = x
                                                   ])

> T[c,b];                                  ⟶                x

> T[b,b];                                  ⟶              T[b, b]

> op(T);                                   ⟶         table(symmetric,[
                                                        (c, b) = x
                                                   ])

> op(op(T));                               ⟶      symmetric, [(c, b) = x]

> F := table([sin=cos,cos=-sin]):
  op(op(F));                               ⟶      [cos = - sin, sin = cos]

> F[cos](Pi/2);                            ⟶                -1
```

SEE ALSO: `indexed`, `array`, `symmetric`, `antisymmetric`, `sparse`, `diagonal`, `identity`, `indexfcn`, `indices`, `entries`, `copy`, `selection`, `op`, `evaln`, `index[tables]`

2.1.279   taylor – Taylor series expansion

Calling Sequence:

   taylor(*expr*, *eq/nm*, *n*)

Parameters:

   expr   – an expression

   eq/nm   – an equation (such as $x = a$) or name (such as x)

   n   – (optional) non-negative integer

Synopsis:

- The function **taylor** computes the Taylor series expansion of *expr*, with respect to the variable x, about the point a, up to order n.

- The **taylor** function is a restriction of the more general **series** function. See **series** for a complete explanation of the parameters.

- If the result of the **series** function applied to the specified arguments is a Taylor series then this result is returned; otherwise, an error-return occurs.

Examples:

```
> taylor( exp(x), x=0, 4 );       ⟶                      2       3       4
                                               1 + x + 1/2 x  + 1/6 x  + O(x )

> taylor( 1/x, x=1, 3 );          ⟶                          2            3
                                               1 - (x - 1) + (x - 1)  + O((x - 1) )
```

```
> taylor( 1/x + y + x^3, x=0 );          ⟶    Error, does not have a taylor expansion\
                                                , try series()
> int( exp(x^3), x );                    ⟶        /
                                                  |        3
                                                  |  exp(x ) dx
                                                  |
                                                  /

> taylor(", x=0);                        ⟶                4      7
                                                 x + 1/4 x  + O(x )
```

SEE ALSO: `series`, `mtaylor`, `type[taylor]`

2.1.280   `testeq` – random polynomial-time equivalence tester

Calling Sequence:
> `testeq( a = b )`
> `testeq( a, b )`
> `testeq( a )`

Parameters:
> `a, b` – arbitrary expressions

Synopsis:
- The function `testeq` tests for equivalence probabilistically. It returns `false` if the expressions are not equal (or not equal to 0) and true otherwise for the class of expressions that `testeq` understands. The result `false` is always correct; the result `true` may be incorrect with very low probability.

- This function will succeed over expressions formed with rational constants, independent variables, and `I`, combined by arithmetic operations, exponentials, trigonometrics and a few others. It may also succeed with some expressions involving radicals, `Pi` as an argument of trigonometrics, and algebraic constants and functions. If the expressions do not fall in this class, `testeq` returns `FAIL`. `Testeq` may also return `FAIL` if it cannot find an appropriate modulus that works after seven trials.

- Based on: *Determining Equivalence of Expressions in Random Polynomial Time* by Gaston Gonnet, Proceedings of the 16th ACM Symposium on the Theory of Computing, Washington DC, April 1984, pp. 334-341.

Examples:
```
> a := (sin(x)^2 - cos(x)*tan(x)) * (sin(x)^2 + cos(x)*tan(x))^2:
> b := 1/4*sin(2*x)^2 - 1/2*sin(2*x)*cos(x) - 2*cos(x)^2
>         + 1/2*sin(2*x)*cos(x)^3 + 3*cos(x)^4 - cos(x)^6:
> evalb( a = b );
                              false
```

```
> evalb( expand(a) = expand(b) );
                                        false

> testeq( a = b );
                                        true
```

2.1.281   time – total CPU time used for the session

Calling Sequence:
> time()

Synopsis:
- The function **time** returns the total CPU time used since the start of the Maple session. The units are in seconds and the value returned is a floating-point number.

- To time particular statements or groups of statements, one method would be the following:

```
st := time():
. . .   statements to be timed   . . .
time() - st;
```

- Note the use of a colon to prevent the display of an uninteresting value; it is the value **time()** − st which gives the desired information.

SEE ALSO: **profile**, **showtime**, ´**status**´, **words**

2.1.282   trace, untrace – trace procedures in order to debug them

Calling Sequence:
> trace(f)
> trace($f, g, h, \ldots$)
> untrace(f)
> untrace($f, g, h, \ldots$)

Parameters:
> $f, g, h, \ldots$ –   name(s) of procedure(s) to be traced

Synopsis:
- The trace function is a debugging tool for tracing the execution of the procedure(s) f, g, h, and so forth.

- During execution, the entry points, the statements that are executed, and the exit points of the traced procedure(s) are printed. At entry points, the actual parameters are displayed. At exit points, the returned function value is displayed.

- The **untrace** function turns off the tracing of its arguments.

- The `trace` (`untrace`) function returns an expression sequence of the names of the procedures for which tracing was turned on (off).

- Note that it is not possible to trace any function which has special evaluation rules, namely any of: `assigned`, `eval`, `evalhf`, `evaln`, `traperror`.

- The `trace` function will silently ignore any argument which is not the name of a traceable procedure. In particular, it is possible to invoke `trace(anames())` to cause tracing of all procedures which correspond to currently-assigned names even though many of the names in `anames()` may not be assigned procedures.

SEE ALSO: `profile`, `linalg[trace]`

2.1.283   traperror (and variable lasterror) – trap an error condition

Calling Sequence:

   `traperror(` $expr_1$, $expr_2$, ... `)`

Parameters:

   $expr_1$, $expr_2$, ...   –   any expressions

Synopsis:

- The function `traperror` sets an error trap prior to evaluating its arguments. If an error occurs either during evaluation or during simplification, `traperror` returns the string corresponding to the error message of the first error occurring.

- Otherwise, it simply returns its simplified evaluated arguments.

- The most recently occurring error is stored in the variable `lasterror`. This variable can be manipulated as any other variable. `lasterror` is useful for checking the results of `traperror` as in the following:

      `if  traperror(...) = lasterror  then ...`

   Note: `traperror` and `lasterror` must appear in this order.

Examples:

```
> f := u -> (u^2-1)/(u-1);                    ⟶
```

$$f := u \rightarrow \frac{u^2 - 1}{u - 1}$$

```
> printlevel := 3:

  for x in [0, 1, 2] do
     r := traperror( f(x) );
     if r = lasterror then
        if r = `division by zero` then
           r := limit(f(u), u=x)
        else
           ERROR(lasterror)
        fi
     fi;
     lprint(`Result:  x =`, x, `f(x) =`, r)

  od;
```

```
                                      ⟶                    r := 1
                                         Result:  x =   0    f(x) =    1
                                                  r := division by zero

                                                     r := 2
                                         Result:  x =   1    f(x) =    2
                                                     r := 3
                                         Result:  x =   2    f(x) =    3
```

SEE ALSO: ERROR

2.1.284   **trunc – truncate a number to an integer**
               round – round a number to an integer
               frac – fractional part of a number

Calling Sequence:

```
trunc(x)
round(x)
frac(x)
```

Parameters:

x – any expression

Synopsis:

- The `trunc` function truncates a numeric argument to an integer.

- The `round` function rounds a numeric argument to an integer.

- The `frac` function computes the fractional part of a number. It is defined as $x-\text{trunc}(x)$.

- All functions return unevaluated if their arguments are not numbers.

Examples:

```
> trunc(7);
```
⟶   7

`> trunc(8/3);`	$\longrightarrow$	2
`> round(8/3);`	$\longrightarrow$	3
`> frac(8/3);`	$\longrightarrow$	2/3
`> trunc(-2.4);`	$\longrightarrow$	-2
`> trunc(x);`	$\longrightarrow$	`trunc(x)`

2.1.285   type – type-checking function

Calling Sequence:

   type(x, t)

Parameters:

   x –   any expression

   t –   type name or set of type names

Synopsis:

- The type function is a Boolean function which returns **true** if x is of type t, **false** otherwise.

- In the case where t is specified as a set of type names, the value returned is **true** if x is any of the types in the set, **false** otherwise.

- The following type names are known to Maple:

*	**	+	.	..	<
$\leq$	<>	=	PLOT	PLOT3D	RootOf
^	algebraic	algext	algfun	algnum	algnumext
and	anything	array	biconnect	bipartite	boolean
colourabl	connected	constant	cubic	digraph	equation
even	evenfunc	expanded	facint	float	fraction
function	graph	indexed	integer	intersect	laurent
linear	list	listlist	logical	mathfunc	matrix
minus	monomial	name	negative	negint	nonneg
nonnegint	not	numeric	odd	oddfunc	operator
or	planar	point	polynom	posint	positive
primeint	procedure	quadratic	quartic	radext	radfun
radfunext	radical	radnum	radnumext	range	rational
ratpoly	realcons	relation	scalar	series	set
sqrt	square	string	subgraph	symmfunc	taylor
tree	trig	type	undigraph	uneval	union
vector					

- Note: If a type name is an operator, it must be back-quoted to prevent a syntax error. See the examples below.

- For further information, see `type[datatype]` where *datatype* is one of the names in the above list.

- A user can make a datatype known to the type function in the following way. If the user has defined the procedure `type/mytype` then a call to type of the form: `type(a, mytype(b, c, ...))` will generate the function call `type/mytype`(*a, b, c, ...*).

- See `type[definition]` and `type[structured]` for information on defining and using type expressions.

- See `type[surface]` for information on the distinction between a "surface type" and a "nested type".

- See `type[argcheck]` for information on the use of structured data types for argument-checking in procedures.

Examples:

`> type( a + b, polynom );`	$\longrightarrow$	true
`> type( a + b, `+` );`	$\longrightarrow$	true
`> type( a * b, `+` );`	$\longrightarrow$	false
`> type( a and b, `and` );`	$\longrightarrow$	true

SEE ALSO: `index[datatypes]`, `hastype`, `whattype`, `convert`, `define`, `type[topic]` where *topic* is one of: `definition`, `structured`, `surface`, `argcheck`

2.1.286 type/algebraic – check for an algebraic expression

Calling Sequence:

   `type(expr, algebraic)`

Parameters:

   expr – any expression

Synopsis:

- This function returns `true` if *expr* is of type `algebraic`, and `false` otherwise.

- An expression is of type `algebraic` if it is one of the following types:

integer	fraction	float	string	indexed
`+`	`*`	`^`	`**`	series
function	`!`	`.`	uneval	

Examples:

`> type(1/2*sin(1), algebraic);` $\longrightarrow$ true

> type(a[0] + a[1]*x, algebraic);	$\longrightarrow$	true
> type([1,2,3], algebraic);	$\longrightarrow$	false
> type(2 < 3, algebraic);	$\longrightarrow$	false

SEE ALSO: integer, fraction, float, string, indexed, arithop, series, function, factorial, concat, uneval, type[algnum], type[radnum], type

2.1.287   type/algext – check for an algebraic extension

Calling Sequence:
>   type(*expr*, algext))
>   type(*expr*, algext(K))

Parameters:
>   *expr*   –   any expression
>   *K*   –   (optional) type name for the coefficient domain

Synopsis:

- type(*expr*, algext) checks if *expr* is a RootOf expression. It is equivalent to type(*expr*, RootOf).

- type(*expr*, algext(K)) checks whether *expr* is a RootOf expression over a given coefficient domain K. For example, K could be integer or rational.

- The function returns true if *expr* is such an expression, and false if it is not.

Examples:

> type(RootOf(x^2+5), algext);	$\longrightarrow$	true
> type(RootOf(x^2+5), algext(integer));	$\longrightarrow$	true
> type(RootOf(x^2-y,x), algext(rational)) ;	$\longrightarrow$	false
> type(RootOf(x^2-y,x), algext(algebraic)) ;	$\longrightarrow$	true

SEE ALSO: type[RootOf], type[algnum], type[algnumext], type[radext]

2.1.288   type/algfun – check for an algebraic function

Calling Sequence:
>   type(*expr*, algfun)
>   type(*expr*, algfun(K))
>   type(*expr*, algfun(K, V))

Parameters:

expr – an expression

K – a type – for coefficient domain

V – name or list or set of names – the variable(s)

Synopsis:

- An expression *expr* is of type algfun (algebraic function) if it is an expression in the variable(s) V over the domain K extended by `RootOf`s.

- The optional argument V is an indeterminate or a list or set of indeterminates.

- If V is not specified, then all the indeterminates of *expr* which are names are used. That is, *expr* must be an algebraic function in all of its variables.

Examples:

```
> type(x/(1-x),algfun(rational,x));                          true
> f := 1+2*RootOf(x^3-y,x)+y*z;
```
$$f := 1 + 2\ \mathrm{RootOf}(\_Z^3 - y) + y\ z$$
```
> type(f,algfun(anything));                                  true
> type(f,algfun(rational));                                  true
> type(f,algfun(rational,y));                                false
> type(f,algfun(rational,[y,z]));                            true
```

SEE ALSO: `RootOf`, `type[algnum]`, `type[algext]`, `type[radfun]`

2.1.289   type/algnum – check for an algebraic number

Calling Sequence:

type(*expr*, `algnum`)

Parameters:

expr – any expression

Synopsis:

- The call type(*expr*, `algnum`) checks to see if *expr* is an algebraic number.

- An algebraic number is defined as either a rational number, or a root of a univariate polynomial with algebraic number coefficients, specified by a `RootOf`. A sum, product, or quotient of these is also an algebraic number.

- For example, the algebraic number $2^{(1/2)}$ could be specified as `RootOf(z^2-2, z)`.

Examples:
```
> type(2/3, algnum);                                         true
```

> type(ln(2), algnum);	$\longrightarrow$	false
> type(RootOf(z^2+1, z), algnum);	$\longrightarrow$	true
> type(5 / RootOf(z^2-2, z), algnum);	$\longrightarrow$	true
> type(RootOf(z^2+y, z), algnum);	$\longrightarrow$	false

SEE ALSO: RootOf, convert[RootOf]

2.1.290   type/algnumext – check for an algebraic number extension

Calling Sequence:

   type(*expr*, algnumext)

Parameters:

   expr – any expression

Synopsis:

- The call type(*expr*, algnumext) checks to see if *expr* is an algebraic number extension.

- An algebraic number extension is a root of a univariate polynomial with algebraic number coefficients, specified by a RootOf.

- For example, the algebraic number 2^(1/2) could be specified as RootOf(z^2-2, z).

- The expression type(*expr*, algnumext) is equivalent to the expression: type(*expr*, RootOf) and type(*expr*, algnum) .

Examples:

> type(2/3, algnumext);	$\longrightarrow$	false
> type(RootOf(z^2+1, z), algnumext);	$\longrightarrow$	true
> type(RootOf(z^2+y, z), algnumext);	$\longrightarrow$	false
> type (RootOf(z^2-RootOf(x^2-5), z), algnumext);	$\longrightarrow$	true

SEE ALSO: RootOf, convert[RootOf], type[algnum], type[RootOf]

2.1.291   type/anything – check for any type

Calling Sequence:

   type(*expr*,anything)

Parameters:

   expr – any expression

Synopsis:

- Any valid Maple expression, except an expression sequence, will satisfy this call.

- Expression sequences cannot be allowed, because each element would be interpreted as a separate argument to `type`.

Examples:

```
> type(5^(1/2),anything);                    ⟶                true
> type(sin(3*x)/Pi,anything);                ⟶                true
```

SEE ALSO: `type`

2.1.292 Function-arguments type checking

Synopsis:

- This is a convention for doing argument checking in procedure definitions in an efficient and functional way. Part of the usefulness of this scheme is based on how much the main library follows it, so that the rest of the system can dynamically use type definitions of other functions. To this end some standards are being suggested here, though they are not enforced throughout the Maple library at the time of this writing.

- Each function which is user-accessible (which a user may call directly) should have a type definition. The type information is defined together with the procedure and will follow the format:

```
# Definition of the function xxx
`type/xxxargs` := <argument-type-definition for xxx>;
xxx := proc(...) .....
   . . . . .
   if not type([args], xxxargs) then ERROR(`invalid arguments`) fi;
   . . . . .
end:
```

- The following are some of the advantages of this proposed type-checking scheme.

  (1) With one `type` testing, we count the arguments and check the validity of one or many alternatives.

  (2) The types accepted by a function can be used or inspected by other functions.

  (3) This scheme is more readable than the equivalent nested "if" statements.

  (4) A `type` expression executes an order of magnitude faster than an equivalent "if" statement.

Examples:

```
`type/intargs`  := [algebraic, {name, name=algebraic..algebraic}];
`type/gcdargs`  := {[polynom, polynom], [polynom, polynom, name, name]};
```

2.1.293   type/array – check for an array

Calling Sequence:
    type(A, array)
    type(A, ´array´(K))

Parameters:
    A –   any expression
    K –   any type

Synopsis:

• The call **type**(A, array) checks to see if A is an array. It will return **true** if A is an array, and **false** otherwise. It does not check the entries of A. See the information under **array** for a description of the array data structure and how to create arrays.

• The call **type**(A, ´array´(K)) checks to see if A is an array whose entries are of type K. It will return **true** if A is an array, and if **type**(x, K) is true for each entry x of A. It will return **false** otherwise. See the information under **type** for a description of available types in Maple.

• If any entries of A are undefined, then their type will not be checked. Thus if A is an array which has no defined elements, then **type**(A, ´array´(K)) will return true for any domain K.

• It is necessary to surround the word **array** with quotes (´) when using this function in the second form. This prevents invocation of the **array** function, which is used to create arrays.

Examples:

```
> A := array(1..2, 1..2, [[1, 3], [1/2, 5]]
);
```
$\longrightarrow$
$$A := \begin{bmatrix} 1 & 3 \\ 1/2 & 5 \end{bmatrix}$$

```
> type(A, ´array´(rational));
```
$\longrightarrow$ true

```
> type(A, ´array´(integer));
```
$\longrightarrow$ false

```
> B := array(1..3, [x^2+5, x^3-x, x+1]);
```
$\longrightarrow$
$$B := [\, x^2 + 5, \ x^3 - x, \ x + 1\,]$$

```
> type(B, ´array´(polynom(integer, x)));
```
$\longrightarrow$ true

```
> C := array (0..2, 0..2, symmetric);
```
$\longrightarrow$ C :=

array(symmetric, 0 .. 2, 0 .. 2, [])

```
> type(C, array);
```
$\longrightarrow$ true

SEE ALSO: **array**, **type[matrix]**, **type**

2.1.294 type/boolean – check for type Boolean
type/relation – check for type relation
type/logical – check for type logical

Calling Sequence:
type(*expr*,boolean)
type(*expr*,relation)
type(*expr*,logical)

Parameters:
expr – any expression

Synopsis:
- These functions return **true** if *expr* is of the specified type, and **false** otherwise.

- A Boolean expression is an expression of type **relation** or of type **logical** or one of the Boolean constants **true**, **false**, or **FAIL**.

- A relational expression is an expression of type `=`, `<>`, `<` or `≤`.

- A logical expression is an expression of type `and`, `or`, or `not`.

Examples:

> type (x < 5, relation);	$\longrightarrow$	true
> type (a or not b, logical);	$\longrightarrow$	true
> type (1 < 2 and x > 5, boolean);	$\longrightarrow$	true

SEE ALSO: type, evalb

2.1.295 type/constant – check for a constant

Calling Sequence:
type(*x*,constant);

Parameters:
x – any expression

Synopsis:
- The call type(*x*,constant) checks to see if *x* is a complex constant.

Examples:

> type(5, constant);	$\longrightarrow$	true
> type(0.05, constant);	$\longrightarrow$	true
> type(ln(-Pi),constant);	$\longrightarrow$	true
> type(infinity,constant);	$\longrightarrow$	true

```
> type(x^2,constant);                    ⟶                    false
```

SEE ALSO: `type/realcons`, `type/numeric`

2.1.296   Definition of a type in Maple
Synopsis:
- By definition, a type in Maple is any expression or sequence of expressions which is recognized by the `type` function and causes it to return true or false for some set of expressions. For example, in `type( expr, typexpr )`, *expr* is of type *typexpr* if and only if the above expression evaluates to `true`.

- A *typexpr* may be a single expression or an expression sequence (such as `integer` or `polynom`, `x`, `integer`). In the case that the type is an expression sequence, the first component of this sequence will be called the `type` and the remainder expressions are the `type-modifiers`.

- Any particular type belongs to one of the following four categories:

  (1) A system type: a type which is a name that is defined in the kernel of the Maple system. System types usually correspond to primitive algebraic or data structure concepts, such as integer, float, list, and relation.

  (2) A procedural type where the type is a name, for example **xxx**, and there is a procedure `` `type/xxx` `` which will perform the type analysis of the argument and will return `true` or `false` accordingly. This procedure may be available from the global Maple environment or from the library (for the latter case it is automatically loaded into the environment). This is one of the mechanisms to define new types in Maple. `monomial`, `algnum`, and `odd` are presently implemented as external Maple functions.

  (3) An assigned type where the type is a name, such as **xxx**, and the global name `` `type/xxx` `` is assigned a type expression. The type evaluation proceeds as if the type checking were done with the expression assigned to `` `type/xxx` ``. Thus

```
        `type/intargs` := [algebraic, {name,name=algebraic..algebraic}];
             . . . .
        if not type([args],intargs) then ERROR(...)
```

  (4) A type expression which is a Maple general expression as described in type[structured].

SEE ALSO: `type[structured]`, `type`, `index[datatypes]`

2.1.297   type/. – check for an expression of type `` `.` ``
Calling Sequence:
    type (*expr*, `` `.` ``)

Parameters:

  expr – any expression

Synopsis:

- This function returns **true** if *expr* is an unevaluated concatenated object, and **false** otherwise.

- A concatenated object remains unevaluated if its right-hand argument does not evaluate to a string, integer, or range.

Examples:

`> type (p.i, `.`);`	$\longrightarrow$	false
`> type (a.(4/3), `.`);`	$\longrightarrow$	true
`> type (n.(45.67), `.`);`	$\longrightarrow$	true
`> r := poly.(x^3-2*x);`	$\longrightarrow$	$r := poly.(x^3 - 2 x)$
`> type (r, `.`);`	$\longrightarrow$	true

SEE ALSO: dot, cat, type

2.1.298   type/even – check for an even integer
           type/odd – check for an odd integer

Calling Sequence:

  type(*n*, even)
  type(*n*, odd)

Parameters:

  n – an integer

Synopsis:

- These procedures test the parity of the given integer *n*.

Examples:

`> type(100,even);`	$\longrightarrow$	true
`> type(100,odd);`	$\longrightarrow$	false
`> type(x^2,even);`	$\longrightarrow$	false

SEE ALSO: type

2.1.299   type/expanded – check for an expanded polynomial

Calling Sequence:

  type(*expr*, expanded)

Parameters:

  expr – a polynomial

Synopsis:
- Assuming that type(*expr*,polynom) is true, then type(*expr*,expanded) will return true if *expr* is an expanded polynomial, false otherwise.

Examples:
```
> type((4*x*y-x)*(x+y), expanded);                    false
> type(4*x^2*y + 4*x*y^2 - x^2 - x*y,
    expanded);                                         true
```

SEE ALSO: type, type[polynom], expand

2.1.300   type/facint – test for factored integer form

Calling Sequence:

  type (*expr*, facint)

Parameters:

  expr – any expression

Synopsis:
- This function will return true if *expr* is an expression of the form returned by the function ifactor, and false otherwise.

- The numbers 0, 1, and –1 are considered to be of type facint, but all other integers and rationals must be passed through ifactor before type/facint will return true when applied to them.

Examples:
```
> a := ifactor (2520);
```
$$a := (2)^3 \ (3)^2 \ (5) \ (7)$$
```
> type(a, facint);                                     true
> b := ifactor (81/8);
```
$$b := \frac{(3)^4}{(2)^3}$$
```
> type(b, facint);                                     true
> type(1, facint);                                     true
> type(5, facint);                                     false
> type(ifactor(5), facint);                            true
```

SEE ALSO: ifactor, type[integer]

2.1.301   type/float – check for an object of type float

Calling Sequence:

    type(x, float)

Parameters:

   x – any expression

Synopsis:

- The call type(x, float) checks to see if x is of type float. It returns true if x is of type float, and false otherwise.

- An object is of type float if it is a floating point number, that is, a sequence of digits and a decimal point. An alternative format for floating-point numbers is the notation Float(mantissa, exponent) which corresponds to Maple's internal representation of floating point numbers.

Examples:

> type(5, float);	$\longrightarrow$	false
> type(5., float);	$\longrightarrow$	true
> type(0.05, float);	$\longrightarrow$	true
> type(.2, float);	$\longrightarrow$	true
> type(Float(1234, -20), float);	$\longrightarrow$	true

SEE ALSO: type[numeric], type[realcons], type[integer], float, Digits, evalf, evalhf

2.1.302   type/fraction – check for an object of type fraction

Calling Sequence:

    type(x, fraction)

Parameters:

   x – any expression

Synopsis:

- The call type(x, fraction) checks to see if x is of type fraction. It returns true if x is of type fraction, and false otherwise.

- An object is of type fraction if it is represented by a signed integer divided by a positive integer.

- Note that Maple will automatically remove common factors from the numerator and denominator of a fraction, so that, for example, 4/2 is not of type fraction since it is automatically simplified to 2. Maple will also automatically attach the sign of a fraction to the numerator rather than the denominator.

Examples:

> type(5/3, fraction);	$\longrightarrow$	true

`> type(-1/2, fraction);`	$\longrightarrow$	true
`> type(4/2, fraction);`	$\longrightarrow$	false
`> type(35/(-14), fraction);`	$\longrightarrow$	true
`> type(a/b, fraction);`	$\longrightarrow$	false

SEE ALSO: `type[integer]`, `type[rational]`, `type[numeric]`

2.1.303   type/integer – check for an integer

Calling Sequence:

   `type (n, integer)`

Parameters:

   n –   any expression

Synopsis:

- The call `type(n, integer)` checks to see if n is an `integer`. It returns `true` if n is an `integer`, and `false` otherwise.

- An expression is of type `integer` if it is an (optionally signed) sequence of one or more digits.

SEE ALSO: `integer`, `numeric`, `constant`, `type/posint`

2.1.304   type/laurent – check for Laurent series

Calling Sequence:

   `type(expr, laurent)`

Synopsis:

- The call `type(expr, laurent)` returns `true` if $expr$ is of type `series` and only contains power terms of the variable in which it was expanded.

- In mathematical terminology it returns `true` if $expr$ is a Laurent series with finite principal part, since type `series` in Maple represents series with only a finite number of negative powers and with an order-term representing the truncation of a potentially infinite number of positive power terms.

Examples:

`> series(sin(x), x, 5);`	$\longrightarrow$	$x - 1/6\, x^3 + O(x^5)$
`> type(", laurent);`	$\longrightarrow$	true
`> series(1/sin(x), x, 5);`	$\longrightarrow$	$x^{-1} + 1/6\, x + O(x^3)$

```
> type(", laurent);                    ⟶            true

> series(ln(x+x^2), x, 3);            ⟶
```

$$\ln(x) + x - 1/2\ x^2 + O(x^3)$$

```
> type(", laurent);                    ⟶            false
```

SEE ALSO: series, type, type[series], type[taylor]

2.1.305   type/linear – check for linear functions
         type/quadratic – check for quadratic functions
         type/cubic – check for cubic functions
         type/quartic – check for quartic functions

Calling Sequence:

     type(a,linear(v))
     type(a,quadratic(v))
     type(a,cubic(v))
     type(a,quartic(v))

Parameters:

     a –   an expression

     v –   an indeterminate or a list or set of indeterminates

Synopsis:

- Check if the expression a is linear (quadratic, cubic or quartic) in the indeterminates v. If v is not specified, this is equivalent to the call type(a,linear(indets(a))) That is, a must be linear (quadratic, cubic, quartic) in all of its indeterminates.

- The definition of linear in the indeterminates v is type(a, polynom(v)) and (degree(a, v) = 1) where **degree** means "total degree" in the case of several variables. The definitions for quadratic, cubic and quartic are analogous with degree(a, v) = 2, 3, and 4 respectively.

Examples:

```
> type(x*y+z, linear(x));             ⟶            true

> type(x*y+z, linear);                ⟶            false

> type(x*y+z, quadratic);             ⟶            true

> type(x^2+y, cubic);                 ⟶            false

> type(f(1)*x+2^(1/2), linear);       ⟶            true

> type(x/z+y/g(2), linear([x,y]));    ⟶            true
```

SEE ALSO: indets, degree, type[polynom]

2.1.306   type/listlist – check for a list of lists

Calling Sequence:

    type(*e*, listlist)

Synopsis:

- The function **type/listlist** returns **true** if *e* is a list of lists, with each inner list having the same number of elements. Otherwise, it returns **false**.

Examples:

> type([[1,2],[3,4]], listlist);	$\longrightarrow$	true
> type([8,7,6,12], listlist);	$\longrightarrow$	false
> type([[1,2],[3,4,5]], listlist);	$\longrightarrow$	false
> type([[]], listlist);	$\longrightarrow$	true
> type([[[a],[b]], [[c],[d]]], listlist);	$\longrightarrow$	true

---

SEE ALSO: convert[listlist]

2.1.307   type/mathfunc – check for mathematical functions

Calling Sequence:

    type(*f*,mathfunc)

Parameters:

    f – a name

Synopsis:

- This procedure checks to see if the given name, *f*, is the name of a mathematical function known to Maple.

- The following mathematical functions are known to **type/mathfunc**:

Ai	BesselI	BesselJ	BesselK	BesselY	Beta
Bi	Ci	Diff	Ei	FresnelC	FresnelS
GAMMA	Int	Limit	MeijerG	MeijerG	Product
Psi	Si	Sum	Zeta	abs	arccos
arccosh	arccot	arccoth	arccsc	arccsch	arcsec
arcsech	arcsin	arcsinh	arctan	arctanh	binomial
cos	cosh	cot	coth	csc	csch
dilog	erf	exp	hypergeom	int	limit
ln	product	sec	sech	signum	sin
sinh	sqrt	sum	tan	tanh	

- Further information is available for each of the names in the above list.

SEE ALSO: type, inifcns

2.1.308 type/matrix – check for a matrix

Calling Sequences:

```
type(A, matrix)
type(A, 'matrix'(R))
type(A, 'matrix'(R, square))
```

Parameters:

A – expression

R – a type – type of the coefficient ring

Synopsis:

- This function checks whether its first argument is a Maple matrix. A matrix is defined to be a two-dimensional array with indices starting at 1.

- If the argument R is given, the function checks whether A is a matrix with entries from the domain R.

- If the argument **square** is given, then the function also checks if A is a square matrix.

- When the second or third syntax is used, it is essential to quote the word matrix in order to avoid confusion with the **matrix** function in the **linalg** package.

Examples:

```
> type(array(1..2,1..2),matrix);        ⟶              true
> type(array(0..2,0..2),matrix);        ⟶              false
> A := linalg[matrix](2,3,[1,2,3,4,5/2,6])
  ;                                       ⟶
                                                        [ 1   2   3 ]
                                              A :=  [            ]
                                                        [ 4  5/2  6 ]
> type(A, 'matrix'(integer));           ⟶              false
> type(A, 'matrix'(rational));          ⟶              true
> type(A, 'matrix'(rational, square));  ⟶              false
```

SEE ALSO: matrix, array, linalg[matrix], type[vector]

2.1.309 type/monomial – check for a monomial

Calling Sequence:

```
type(m, monomial)
type(m, monomial(K))
type(m, monomial(K, v))
```

Parameters:

m – any expression

K – (optional) type name for the coefficient domain

v – (optional) variable(s)

Synopsis:

- The call **type**(m,**monomial**(K, v)) checks to see if m is a monomial in the variable(s) v over the coefficient domain K, where v is either an indeterminate or a list or set of indeterminates.

- A monomial is defined to be a polynomial in v which does not contain a sum. This function returns **true** if m is such a monomial, and **false** otherwise.

- If v is omitted, it is taken to be the set of all indeterminates appearing in m, that is, it checks if m is a monomial in all of its variables.

- The domain specification K should be a type name, such as **rational** or **algebraic**. If K is specified, then this function will check that the coefficients of m come from the domain K. If the coefficient domain K is omitted, then only coefficients of type **constant** are allowed.

Examples:

> `type(sin(1)*x^2, monomial);`	$\longrightarrow$	true
> `type(sin(1)/x^2, monomial);`	$\longrightarrow$	false
> `type((1+y)*x^2, monomial(anything, y));`	$\longrightarrow$	false
> `type(sin(x)*y, monomial(anything, y));`	$\longrightarrow$	true

SEE ALSO: `indets`, `type`, `type[polynom]`

2.1.310   type/numeric – check for an object of type numeric

Calling Sequence:

`type($x$, numeric)`

Parameters:

x – any expression

Synopsis:

- The call **type**(x, **numeric**) checks to see if x is of type **numeric**. It returns true if x is of type **numeric**, and **false** otherwise.

- An object is of type **numeric** if it is either an **integer**, a **fraction**, or a **float**.

Examples:

> `type(5, numeric);`	$\longrightarrow$	true
> `type(0.05, numeric);`	$\longrightarrow$	true

> type(Pi, numeric); $\longrightarrow$ false

> type(4/3, numeric); $\longrightarrow$ true

> type(x^2, numeric); $\longrightarrow$ false

SEE ALSO: type[integer], type[fraction], type[rational], type[constant], type[realcons]

2.1.311   type/operator – check for a functional operator

Calling Sequence:

type(*f*, operator)

Parameters:

f – any expression

Synopsis:

- A functional operator in Maple is a special form of a function. See operators[functional] or Chapter 7 of the Maple Language Reference Manual for a full description of functional operators.

- The call type(*f*, operator) will return true if *f* is a functional operator, and false otherwise.

Examples:

> f := x -> 3*x +5; $\longrightarrow$ f := x -> 3 x + 5

> type(f, operator); $\longrightarrow$ true

> g := (x,y) -> sin(x)*cos(y) + x*y; $\longrightarrow$ g := (x,y) -> sin(x) cos(y) + x y

> type(g, operator); $\longrightarrow$ true

> h := <y^2|y>; $\longrightarrow$
$$h := <y^2|y>$$

> type(h, operator); $\longrightarrow$ true

> p := proc(x) option operator; x^3-5
  end; $\longrightarrow$
$$p := <x^3 - 5|x>$$

> type(p, operator); $\longrightarrow$ true

> type(q, operator); $\longrightarrow$ false

SEE ALSO: operators[functional], unapply, @, @@, D, operators[D], operators[examples]

2.1.312   type/PLOT – check for a PLOT data structure

Calling Sequence:

type(*P*, PLOT)

Parameters:

P –   any algebraic expression

Synopsis:

- This function will return **true** if P is a PLOT data structure, and **false** otherwise.

- A PLOT data structure is the result of a call to the **plot** command.

Examples:
```
> a := plot(sin(x), x=0..Pi):
  type(a, PLOT);                          ⟶              true
> b := plot([0,0,1,1,2,1,2,0,1,-1,0,0],
  style=POINT):
  type(b, PLOT);                          ⟶              true
```

SEE ALSO: plot, plot[structure], type[PLOT3D]

2.1.313   type/PLOT3D – check for a PLOT3D data structure

Calling Sequence:

type(P, PLOT3D)

Parameters:

P –   any algebraic expression

Synopsis:

- This function will return **true** if P is a PLOT3D data structure, and **false** otherwise.

- A PLOT3D data structure is the result of a call to the **plot3d** command.

Examples:
```
> a := plot3d(sin(x) * cos(y), x=0..Pi, y=0..Pi):
> type(a, PLOT3D);
                              true
> b := plots[pointplot]({[0,0,1],[1,2,1],[2,0,1],[-1,0,0]}):
> type(b, PLOT3D);
                              true
```

SEE ALSO: plot3d, plot3d[structure], type[PLOT]

2.1.314   type/point – check for a point

Calling Sequence:

type(P, point)

Parameters:

P – any algebraic expression

Synopsis:

- This function will return **true** if and only if P is an equation with only a name on the left-hand side and an algebraic expression on the right-hand side.

- If P is a set, it will return **true** if all members of the set are of type point and **false** otherwise.

- The left-hand side names are intended to define the coordinate space, and the right-hand side values are the corresponding coordinate values.

- Objects of type **point** differ from objects of type **Point** (in the **student** package) in that the coordinates are defined by names rather than by position in a list.

Examples:

`> type( a=4, point );`	$\longrightarrow$	`true`
`> type( {a=4,b=2}, point );`	$\longrightarrow$	`true`
`> type( {a+b=c,d=3} , point );`	$\longrightarrow$	`false`

SEE ALSO: `type`, `limit`, `student[Point]`

2.1.315   type/polynom – check for a polynomial

Calling Sequence:

type(a, polynom)
type(a, polynom(d))
type(a, polynom(d, v))

Parameters:

a – any expression

d – (optional) type name for the coefficient domain

v – (optional) variable(s)

Synopsis:

- The call **type(a, polynom(d, v))** checks to see if a is a polynomial in the variables v with coefficients in the domain d. A typical calling sequence would be

        type(a, polynom(integer, x))

which tests to see if a is a polynomial in x over the integers.

- The variable(s) v can be a single indeterminate, or can be a list or set of indeterminates. In the latter case, one would be testing for a multivariate polynomial.

- If v is omitted, then it is taken to be a set of all the indeterminates appearing in a. Thus the function will check that a is a polynomial in all of its variables.

- The domain specification d should be a type name, such as `rational` or `algnum` (algebraic number). If the domain specification is given as `anything` then no restriction is placed on the coefficients. If d is omitted, then it defaults to type `constant`.

Examples:

```
> type(x^2+y-z, polynom);                  ⟶        true
> type(sin(x)+y, polynom(anything, x));    ⟶        false
> type(sin(x)+y, polynom(anything, y));    ⟶        true
> type(f(1)*x+2^(1/2), polynom);           ⟶        true
> type(x^2+y^3/3, polynom(anything, [x, y])
  );                                       ⟶        true
> type(x+1/2, polynom(integer));           ⟶        false
> type(x+1/2, polynom(rational));          ⟶        true
```

SEE ALSO: `indets`, `type`, `polynom`, `type[ratpoly]`, `type[monomial]`, `type[constant]`

2.1.316   type/positive – check for a positive number
            type/negative – check for a negative number
            type/nonneg – check for a non-negative number

Calling Sequence:
    `type(n, positive)`
    `type(n, negative)`
    `type(n, nonneg)`

Parameters:
    n – a numeric value

Synopsis:
- These procedures test the sign of the numeric value n. If n is not of type numeric, then **false** is returned.

- The functions **type/positive**, **type/negative**, and **type/nonneg** return **true** if n is positive, negative, or non-negative, respectively.

Examples:

```
> type(0,positive);       ⟶        false
> type(-1.79,negative);   ⟶        true
> type(0,nonneg);         ⟶        true
```

```
> type(x^2,nonneg);
```
$\longrightarrow$ false
```
> type(sqrt(8),positive);
```
$\longrightarrow$ false

SEE ALSO: type, type[numeric]

2.1.317   type/posint – check for a positive integer
type/negint – check for a negative integer
type/nonnegint – check for a non-negative integer

Calling Sequence:

type(n,posint)
type(n,negint)
type(n,nonnegint)

Parameters:

n – any value

Synopsis:

- These procedures test the value of n. If n is not of type integer, then false is returned.

- The functions type/posint, type/negint, and type/nonnegint return true if the integer n is positive, negative, or non-negative, respectively, and false otherwise.

Examples:
```
> type(3, posint);
```
$\longrightarrow$ true
```
> type(-5, negint);
```
$\longrightarrow$ true
```
> type(0, nonnegint);
```
$\longrightarrow$ true
```
> type(x, negint);
```
$\longrightarrow$ false
```
> type(3.5, posint);
```
$\longrightarrow$ false

SEE ALSO: type, type/integer, type/positive, type/negative, type/nonneg

2.1.318   type/primeint – check for a prime integer

Calling Sequence:

type(*expr*, primeint)

Parameters:

expr – any expression

Synopsis:

- This function returns true if *expr* is a prime integer, and false otherwise.

- The function `isprime` is used to check the primality of *expr*, once *expr* has been determined to be an integer.

SEE ALSO: `type/integer`, `isprime`

2.1.319   type/radext – check for an algebraic extension in terms of radicals

Calling Sequence:

    type (*expr*, radext))
    type (*expr*, radext(*K*))

Parameters:

    expr   –   any expression

    K   –   (optional) type name for the coefficient domain

Synopsis:

- The function `type(`*expr*`, radext)` checks if *expr* is a `radical` expression. It is equivalent to `type(`*expr*`, radical)`.

- The function `type(`*expr*`, radext(`*K*`))` checks whether *expr* is a `radical` expression where the expression under the root sign belongs to the domain *K*. For example, *K* could be `integer` or `rational`.

- The function returns `true` if *expr* is such an expression, and `false` if it is not.

Examples:

`> type (sqrt (x^2+5), radext);`	$\longrightarrow$	`true`
`> type ((x^2+5)^(4/3), radext(polynom));`	$\longrightarrow$	`true`
`> type ((5-sqrt(3)^(5/6)), radext(rational) );`	$\longrightarrow$	`false`
`> type ((5-sqrt(3))^(5/6), radext(radnum)) ;`	$\longrightarrow$	`true`
`> type (2/3, radext);`	$\longrightarrow$	`false`

SEE ALSO: `type/RootOf`, `type/radnum`, `type/algext`, `type/radnumext`, `type/radext`

2.1.320   type/radfun – check for a radical function

Calling Sequence:

    type(*expr*,radfun)
    type(*expr*,radfun(*K*))
    type(*expr*,radfun(*K*, *V*))

Parameters:

> *expr*   –   an expression
>
> *K*   –   type name for the coefficient domain
>
> *V*   –   a variable or a list or set of variables

Synopsis:

- An expression *expr* is of type `radfun` (radical function) if it is a rational function in the variable(s) *V* over the domain *K* extended by radicals.

- If no variables were specified, all the indeterminates of *expr* which are names are used, so *expr* must be an algebraic function in all of its variables.

- If no domain is specified, the default domain `'constant'` is used.

Examples:

`> type(x/(1-x), radfun(integer));`	$\longrightarrow$	true
`> type(sqrt(x-y), radfun(rational,[x,y]))` `;`	$\longrightarrow$	true
`> type(sqrt(x-sin(y)), radfun(anything,x))` `;`	$\longrightarrow$	true
`> type(sqrt(x-sin(y)), radfun(anything,y))` `;`	$\longrightarrow$	false
`> type(sqrt(x+1),radfun);`	$\longrightarrow$	true
`> type(sqrt(x+sqrt(2)),radfun(rational,x))` `;`	$\longrightarrow$	true
`> type(sqrt(x+sqrt(y)),radfun);`	$\longrightarrow$	true
`> type(sqrt(x+exp(x)),radfun);`	$\longrightarrow$	false

SEE ALSO: `type`, `type/algfun`, `type,radext`, `type/radnum`, `radical`

2.1.321   type/radfunext – check for a radical function extension

Calling Sequence:

> type(*expr*,radfunext)

Parameters:

> *expr*   –   any expression

Synopsis:

- The function `type(`*expr*`,radfunext)` checks to see if *expr* is a radical function extension.

- A radical function extension is an algebraic function extension specified in terms of radicals. This is a root of a combination of rational functions and roots of rational functions specified in terms of radicals.

- The function type(*expr*, radfunext) is equivalent to type(*expr*, radical) and type(*expr*, radfun)

Examples:

> type((x+y)^(1/2), radfunext);	$\longrightarrow$	true
> type((sqrt(z) - 3/2*x + y)^(4/3), radfunext);	$\longrightarrow$	true
> type((4*x - 5^(1/3)*y^2)^(1/4), radfunext);	$\longrightarrow$	true
> type(x^(1/4), radfunext);	$\longrightarrow$	true

SEE ALSO: sqrt, type/algfunext, type/radfun, type/radical, type/radext, type/radnumext

2.1.322   type/radical – check for fractional powers

Calling Sequence:

    type(*expr*, radical)

Parameters:

    expr – an expression

Synopsis:

- The definition of type radical is that *expr* is of type `^` and op(2, *expr*) (the exponent) is of type fraction.

Examples:

> type(2^(1/2), radical);	$\longrightarrow$	true
> type(y^(2/5), radical);	$\longrightarrow$	true
> type(y^3, radical);	$\longrightarrow$	false
> type (sqrt(x^2+y^2), radical);	$\longrightarrow$	true

SEE ALSO: `^`, sqrt, type/sqrt, fraction, type/radnum, type/radext

2.1.323   type/radnum – check for an algebraic number in terms of radicals

Calling Sequence:

    type(*expr*, radnum)

Parameters:

    expr – any expression

Synopsis:

- The call type(*expr*, radnum) checks to see if *expr* is a radical number.

- A radical number is defined as either a rational number, or a combination of roots of rational numbers specified in terms of radicals. A sum, product or quotient of these is also a radical number.

Examples:

> type(2/3, radnum);	$\longrightarrow$	true
> type(ln(2), radnum);	$\longrightarrow$	false
> type((-1)^(1/2), radnum);	$\longrightarrow$	true
> type(5/sqrt(2), radnum);	$\longrightarrow$	true
> type((3/2)^(4/3), radnum);	$\longrightarrow$	true
> type(x^(1/4), radnum);	$\longrightarrow$	false

SEE ALSO: sqrt, type/algnum, type/radfun, type/radical

2.1.324   type/radnumext – check for a radical number extension

Calling Sequence:

   type(*expr*,radnumext)

Parameters:

   expr   –   any expression

Synopsis:

- The call type(*expr*,radnumext) checks to see if *expr* is a radical number extension.

- A radical number extension is an algebraic number extension specified in terms of radicals. This is a root of a combination of rational numbers and roots of rational numbers specified in terms of radicals.

- type(*expr*, radnumext) is equivalent to type(*expr*, radical) and type(*expr*, radnum) .

Examples:

> type((-1)^(1/2), radnumext);	$\longrightarrow$	true
> type((sqrt(5) - 3/2)^(4/3), radnumext);	$\longrightarrow$	true
> type((4-5^(1/3))^(1/4), radnumext);	$\longrightarrow$	true
> type(x^(1/4), radnumext);	$\longrightarrow$	false

SEE ALSO: sqrt, type/algnumext, type/radnum, type/radical, type/radext,
           type/radfunext

2.1.325   type/range – check for a range

Calling Sequence:

> type (*expr*, range)
>
> type (*expr*, `` `..` ``)

Parameters:

     expr – any expression

Synopsis:

- This function returns **true** if *expr* is of type **range**, and **false** otherwise.

- An expression of type **range** (also called type `` `..` ``) has two operands, the expression on the left-hand side and the expression on the right-hand side. These two operands are separated by an ellipsis (..) .

Examples:

`> type (a..b, `..`);`	$\longrightarrow$	true
`> type (1..4, `..`);`	$\longrightarrow$	true
`> type (i..j, range);`	$\longrightarrow$	true
`> type (1..n, range);`	$\longrightarrow$	true

SEE ALSO: `main[range]`, `linalg[range]`, `plot[range]`

2.1.326   type/rational – check for an object of type rational

Calling Sequence:

> type(*x*, rational)

Parameters:

     x – any expression

Synopsis:

- The call **type(*x*, rational)** checks to see if *x* is of type **rational**. It returns true if *x* is of type **rational**, and **false** otherwise.

- An object is of type **rational** if it is either of type **fraction** or of type **integer**.

Examples:

`> type(-17, rational);`	$\longrightarrow$	true
`> type(5/3, rational);`	$\longrightarrow$	true
`> type(-1/2, rational);`	$\longrightarrow$	true
`> type(4/2, rational);`	$\longrightarrow$	true
`> type(35/(-14), rational);`	$\longrightarrow$	true
`> type(a/b, rational);`	$\longrightarrow$	false

SEE ALSO: `type[integer]`, `type[fraction]`, `type[numeric]`

2.1.327   type/ratpoly – check for a rational polynomial

Calling Sequence:

   type(*expr*, ratpoly)
   type(*expr*, ratpoly(*K*))
   type(*expr*, ratpoly(*K*, *v*))

Parameters:

   expr –   any expression

   K   –   type name for the coefficient domain

   v   –   variable(s)

Synopsis:

- The call **type(*expr*,ratpoly(*K*, *v*))** checks to see if *expr* is a rational function in the variables *v* with coefficients in the domain *K*.

- A typical calling sequence would be **type(*a*,ratpoly(integer, *x*))** which tests to see if *a* is a rational polynomial in *x* over the integers.

- The variable(s) *v* can be a single indeterminate or a list or set of indeterminates. If *v* is omitted, then it defaults to a list of all indeterminates in *expr*.

- The domain specification *K* is a type name such as **integer** or **algnum** (algebraic number). If the domain specification is omitted, then it defaults to type **constant**.

Examples:

```
> type((1+x)/(1-y),ratpoly);                    ⟶           true
> type((1+sin(x))/y,ratpoly);                   ⟶           false
> type((1+sin(x))/y,ratpoly(anything,y));       ⟶           true
> type(1/f(1)+1/x,ratpoly);                     ⟶           true
> type(sin(z)-cos(z)/x,ratpoly(anything,x))
  ;                                             ⟶           true
> type(x+sqrt(2)*x^2, ratpoly(radnum));         ⟶           true
```

SEE ALSO: `indets`, `ratpoly`, `type[polynom]`

2.1.328   type/realcons – check for a real constant

Calling Sequence:

   type(*x*,realcons)

Parameters:

   x   –   any expression

Synopsis:
- The call `type(`*x*`,realcons)` checks if *x* is a real constant. A real constant is `infinity`, `-infinity`, or an expression for which `evalf` will return a floating point number.

Examples:

> `type(4, realcons);`	$\longrightarrow$	true
> `type(2+I, realcons);`	$\longrightarrow$	false
> `type(3.5, realcons);`	$\longrightarrow$	true
> `type(ln(Pi),realcons);`	$\longrightarrow$	true
> `type(sin(2),realcons);`	$\longrightarrow$	true
> `type(infinity,realcons);`	$\longrightarrow$	true
> `type(x^2,realcons);`	$\longrightarrow$	false

SEE ALSO: `type/constant`

2.1.329   type/RootOf – check for a `RootOf` expression

Calling Sequence:
    `type(`*expr*`,RootOf)`

Parameters:
    expr   –   any expression

Synopsis:
- An expression is of type `RootOf` if it is of the form `RootOf(x)`, where x is an algebraic expression.

Examples:

> `type(RootOf(x^2-3,x),RootOf);`	$\longrightarrow$	true
> `type(sin(3*x),RootOf);`	$\longrightarrow$	false
> `type(RootOf(x-3,x),RootOf);`	$\longrightarrow$	false
> `type(RootOf(Z^7-2*Z^5+3),RootOf);`	$\longrightarrow$	true

SEE ALSO: `RootOf`, `type[algnum]`, `type[algext]`

2.1.330   type/scalar – check for scalar (in the matrix sense)

Calling Sequence:
    `type(`*expr*`, scalar)`

Parameters:

   expr – expression

Synopsis:

- Checks if *expr* is a scalar. A scalar is anything which is not a vector or matrix.

Examples:

```
> a := array([[1,2],[3,4]]):
  type(a,scalar);                    ⟶             false
> type(det(a),scalar);               ⟶             true
> type(x,scalar);                    ⟶             true
> type(42,scalar);                   ⟶             true
```

2.1.331 type/series - the series data structure

Calling Sequence:

   type(*expr*, series)

Synopsis:

- The function **type/series** returns **true** if the value of *expr* is Maple's series data structure, explained below.

- The **series** data structure represents an expression as a truncated series in one specified indeterminate, expanded about a particular point. It is created by a call to the **series** function.

- The function op(0,*series*) is *x-a* where *x* denotes the "series variable" and *a* denotes the particular point of expansion. op(2*i-1,*series*) is the *i*th coefficient (a general expression) and op(2*i,*series*) is the corresponding integer exponent.

- The exponents are "word-size" integers, in increasing order.

- The representation is sparse; zero coefficients are not represented.

- Usually, the final pair of operands in this data type are the special **order** symbol O(1) and the integer *n* which indicates the order of truncation. However, if the series is exact then there will be no **order** term, for example, the series expansion of a low-degree polynomial.

- Formally, the coefficients of the series are such that

$$k1*(x-a)^{eps} < |coeff[i]| < k2/(x-a)^{eps}$$

for some constants k1 and k2, for any eps > 0, and as x approaches a. In other words, the coefficients may depend on x, but their growth must be less than polynomial in x. O(1) represents such a coefficient, rather than an arbitrary constant.

- A zero series is immediately simplified to the integer zero.

Examples:

```
> a := series(sin(x), x=0, 5);
```
$\longrightarrow$
$$a := x - 1/6\ x^3 + O(x^5)$$

```
> type(a,series);
```
$\longrightarrow$ true

```
> type(a,taylor);
```
$\longrightarrow$ true

```
> op(0,a);
```
$\longrightarrow$ x

```
> op(a);
```
$\longrightarrow$ 1, 1, -1/6, 3, O(1), 5

```
> b := series(1/sin(x), x=0, 5);
```
$\longrightarrow$
$$b := x^{-1} + 1/6\ x + O(x^3)$$

```
> type(b,series);
```
$\longrightarrow$ true

```
> type(b,taylor);
```
$\longrightarrow$ false

```
> op(0,b);
```
$\longrightarrow$ x

```
> op(b);
```
$\longrightarrow$ 1, -1, 1/6, 1, O(1), 3

```
> type(x^3,series);
```
$\longrightarrow$ false

```
> series(sqrt(sin(x)), x=0, 4);
```
$\longrightarrow$
$$x^{1/2} - 1/12\ x^{5/2} + O(x^{7/2})$$

```
> type(",series);
```
$\longrightarrow$ false

```
> whattype("");
```
$\longrightarrow$ +

```
> c := series(x^x, x=0, 3);
```
$\longrightarrow$
$$c := 1 + \ln(x)\ x + 1/2\ \ln(x)^2\ x^2 + O(x^3)$$

```
> type(c,series);
```
$\longrightarrow$ true

```
> type(c,taylor);
```
$\longrightarrow$ false

```
> op(0,c);
```
$\longrightarrow$ x

```
> op(c);
```
$\longrightarrow$ 1, 0, $\ln(x)$, 1, 1/2 $\ln(x)^2$, 2, O(1), 3

```
> d := series(sin(x+y), x=y, 2);
```
$\longrightarrow$
$$d := \sin(2\ y) + \cos(2\ y)\ (x - y) + O((x - y)^2)$$

```
> type(d,series);
```
$\longrightarrow$ true

```
> op(0,d);
```
$\longrightarrow$ x - y

```
> op(d);
```
$\longrightarrow$ $\sin(2\ y)$, 0, $\cos(2\ y)$, 1, O(1), 2

SEE ALSO: series, taylor, op, type, type[laurent], type[taylor]

2.1.332   type/list – check for a list
type/set – check for a set

Calling Sequence:

type(*expr*, list)
type(*expr*, set)
type(*expr*, list(*K*))
type(*expr*, set(*K*))

Parameters:

expr   –   any expression

K   –   a type name

Synopsis:

- In the first form (where *K* is not specified) these functions check if *expr* is a valid Maple list or set, respectively. They return true if *expr* is a list or set, and false otherwise.

- See the information under list and set for a description of the list and set datatypes.

- In the second form, where *K* is a specified type name, these functions check if *expr* is a list or set whose entries are of type *K*. That is, type(*expr*, list(*K*)) will return true if type(*expr*, list) is true and type(x, *K*) is true for each entry x of *expr*. Sets are checked in a similar manner.

- See the information under type for a description of valid types in Maple.

Examples:

> S := {1, 3/2, 2};	$\longrightarrow$	S := {1, 2, 3/2}
> type(S, set);	$\longrightarrow$	true
> type(S, set(rational));	$\longrightarrow$	true
> type(S, set(integer));	$\longrightarrow$	false
> L := [x^4-1, x^2, x+3];	$\longrightarrow$	$L := [x^4 - 1, x^2, x + 3]$
> type(L, list);	$\longrightarrow$	true
> type(L, list(polynom(integer, x)));	$\longrightarrow$	true

SEE ALSO: set, list, type

2.1.333   type/sqrt – check for a square root

Calling Sequence:

type(*expr*, sqrt)
type(*expr*, ´sqrt´(domain))

Parameters:

    expr   –   any expression

    domain –   any valid type domain

Synopsis:

- An expression is of type `sqrt` if it is a radical and the exponent has a denominator of 2.

- An expression is of type `sqrt(`*domain*`)` if it is of type `radext(`*domain*`)` and the exponent has a denominator of 2.

- Note that a square root of a product or quotient is not of type `sqrt` since it gets reduced to a product or quotient of square roots.

- When used in the second form, it is necessary to enclose `sqrt` in forward (unevaluation) quotes to prevent the `sqrt` function from being invoked.

Examples:

`> type(5^(1/2),'sqrt');`	$\longrightarrow$	`true`
`> type(5^(1/2), 'sqrt'(integer));`	$\longrightarrow$	`true`
`> type (y^(1/2), 'sqrt'(name));`	$\longrightarrow$	`true`
`> type(5^(1/4),sqrt);`	$\longrightarrow$	`false`
`> type((x+7)^(3/2), 'sqrt'(integer));`	$\longrightarrow$	`false`
`> type((x+7)^(3/2), 'sqrt'(polynom));`	$\longrightarrow$	`true`

SEE ALSO: `sqrt`, `type`, `type[radical]`, `type[radext]`

2.1.334   type/square – check for a perfect square

Calling Sequence:

    `type(`*expr*`,square)`

Parameters:

    expr   –   any expression

Synopsis:

- This procedure checks to see if an expression is a (simple) perfect square. By "simple", it means that the square root of *expr* should be an expression free of radicals.

- The power of `type/square` is bounded by the power of `sqrt()`.

Examples:

`> type(100,square);`	$\longrightarrow$	`true`
`> type(10,square);`	$\longrightarrow$	`false`
`> type(x^2+2*x+1,square);`	$\longrightarrow$	`true`

```
> type(x^2+y^2,square);
```
$\longrightarrow$ false

SEE ALSO: `sqrt`, `type`

2.1.335   Definition of a structured type in Maple
Synopsis:

- A structured type is a Maple expression other than a string which can be interpreted as a type. A typical example would be `set(name=algebraic)`. This specifies a set of equations whose left-hand sides are (variable) names, and whose right-hand sides are of type `algebraic`.

- This file first gives a formal grammatical description of the valid structured types, then notes on some of the special types, and lastly, examples.

- In the formal definition below, read ":=" to mean "is defined to be", "|" to mean "or", and "*" to mean "zero or more occurrences of".

	Syntax	Matches
type ::=	{ type* }	alternation; any of the types
	| [type*]	a list of the given types
	| numeric	match a numerical constant exactly
	| string	a system, procedural, or assigned type
	| type = type	an equation of the corresponding types
	| type <> type	an inequality of the corresponding types
	| type < type	a relation of the corresponding types
	| type ≤ type	a relation of the corresponding types
	| type > type	a relation of the corresponding types
	| type ≥ type	a relation of the corresponding types
	| type .. type	a range of the corresponding types
	| type and type	an and of the corresponding types
	| type or type	an or of the corresponding types
	| not type	a not of the corresponding type
	| type &+ type ...	a sum of the corresponding types
	| type &* type ...	a product of the corresponding types
	| type ^ type	a power of the corresponding types
	| type . type	a concatenation of the corresponding types
	| ´type´	an unevaluated expression of the given type
	| fcntype	a function or special type
	| name[type*]	an indexed reference of the given types
fcntype ::=	set(type)	a set of elements of the given type
	| list(type)	a list of elements of the given type
	| `+`(type)	a sum of terms of the given type
	| `*`(type)	a product of factors of the given type
	| identical(expr)	an expression identical to expr
	| specfunc(type,foo)	the function foo with type arguments
	| anyfunc(type*)	any function of the given types
	| foo(type*)	type defined by a procedure `type/foo`
	| foo(type*)	the function foo of the given types

- The square brackets [and] are used to check for a fixed argument sequence. The type `[name,set]` matches a list with exactly 2 arguments, a name followed by a set.

- The set brackets { and } are used for alternation. The type `{set(name),list(integer)}` matches either a set of names, or a list of integers.

- The type `anything` matches any expression except a sequence.

- The type `identical(expr)` matches the expression `expr` identically.

- The type `anyfunc(t1,...,tn)` matches any function with n arguments of type t1, ..., tn .

Thus `type(f,anyfunc(t1,...,tn)` is equivalent to both `type(f,function)` and `type([op(f)]`, `[t1,...,tn])` .

- The type `specfunc(t,n)` matches the function `n` with 0 or more arguments of type `t`. Thus `type(f,specfunc(t,n))` is equivalent to all three of the forms, `type(f,function)`, `op(0,f) = n`, and `type( [op(f)], list(t) )`.

Examples:

`> type(x^(-2),name^integer);`	$\longrightarrow$	true
`> type(x^(-2),name^posint);`	$\longrightarrow$	false
`> type(x^(-2),algebraic^integer);`	$\longrightarrow$	true
`> type(exp(x),exp(name));`	$\longrightarrow$	true
`> T := TEXT(`line 1`,`line 2`):` `type(T,TEXT(string));`	$\longrightarrow$	false
`> type(T,TEXT(string,string));`	$\longrightarrow$	true
`> type(T,specfunc(string,TEXT));`	$\longrightarrow$	true
`> type(T,anyfunc(string));`	$\longrightarrow$	false
`> type(T,anyfunc(string,string));`	$\longrightarrow$	true
`> type([x,1],[name,integer]);`	$\longrightarrow$	true
`> type([x,1],list(integer));`	$\longrightarrow$	false
`> type([x,1],list(name));`	$\longrightarrow$	false
`> type([x,1],list({name,integer}));`	$\longrightarrow$	true

SEE ALSO: `type[surface]`

2.1.336 Definition of surface and nested types

Synopsis:

- The type checks that require information only about the top level of the expression tree will be called "surface types".

- Types that check a complete expression tree (probably recursively) will be called "nested types".

- Most of the system types are surface types since these are encoded in the top node of the expression tree. Thus

```
type({ .. }, set);
type([a] + [b,c], algebraic);
```

Both return **true** regardless of the types of the components of the set in the first case, and regardless of the types of the terms of the sum in the second.

- The following are surface types:

`` `=` ``	`` `<>` ``	`` `<` ``	`` `≤` ``	`` `.` ``	`` `..` ``
`` `!` ``	`` `+` ``	`` `*` ``	`` `^` ``	`` `**` ``	`` `and` ``
`` `or` ``	`` `not` ``	`` `union` ``	`` `intersect` ``	`` `minus` ``	algebraic
anything	array	boolean	equation	even	float
fraction	function	indexed	integer	laurent	linear
list	listlist	logical	mathfunc	matrix	monomial
name	negative	nonneg	numeric	odd	point
positive	procedure	radical	range	rational	relation
RootOf	scalar	series	set	sqrt	square
string	table	taylor	trig	type	uneval
vector					

- The type `` `constant` ``, on the other hand, will completely scan an expression to determine whether or not it is composed of any non-constant parts. Hence it is a nested type. The following types are nested types:

algfun	algnum	constant	cubic	expanded	polynom
linear	quadratic	quartic	radnum	radfun	ratpoly

2.1.337 type/taylor – check for Taylor series

Calling Sequence:

    type(*expr*, taylor)

Synopsis:

- The function **type/taylor** returns **true** if *expr* is a series which is a polynomial in the variable in which it was expanded. An order term may be present.

Examples:

```
> series(sin(x),x,5);
```
$$\longrightarrow \quad x - \tfrac{1}{6} x^3 + O(x^5)$$

```
> type(",taylor);
```
$$\longrightarrow \quad \text{true}$$

```
> series(1/sin(x),x,5);
```
$$\longrightarrow \quad x^{-1} + \tfrac{1}{6} x^3 + O(x^3)$$

```
> type(",taylor);
```
$$\longrightarrow \quad \text{false}$$

```
> type(x^3,taylor);
```
$$\longrightarrow \quad \text{false}$$

```
> series(x/y,x);
```
$$\longrightarrow \quad 1/y\ x$$

```
> type(",taylor);
```
$$\longrightarrow \quad \text{true}$$

```
> series(x/y,y);
```
$\longrightarrow$

$$x\,y^{-1}$$

```
> type(",taylor);
```
$\longrightarrow$
false

SEE ALSO: `taylor`, `series`, `type`, `type[laurent]`, `type[series]`

2.1.338   type/trig – check for trigonometric functions

Calling Sequence:

    type(*expr*,trig(*x*))
    type(*expr*,trig)

Parameters:

   expr – any expression

   x   – a variable name

Synopsis:

- The call `type(`*expr*`,trig)` returns true if *expr* is a function and the function name is one of the trigonometric functions:

    sin, cos, tan, sec, csc, cot, sinh, cosh, tanh, sech, csch, coth

- The call `type(`*expr*`,trig(`*x*`))` checks, in addition, that some trigonemetric function in *expr* contains the variable name *x*.

Examples:

```
> type(sin(x), trig);
```
$\longrightarrow$
true
```
> type(exp(x), trig);
```
$\longrightarrow$
false
```
> type(sin(x) + cos(x), trig);
```
$\longrightarrow$
false
```
> type(sin(1), trig(x));
```
$\longrightarrow$
false
```
> type(tanh(3*x-1), trig(x));
```
$\longrightarrow$
true

2.1.339   type/type – check for type expressions

Calling Sequence:

    type(*a*, type)

Parameters:

   a – any expression

Synopsis:

- The call `type(`*a*`, type)` checks to see if *a* is a valid type expression. By definition, a type expression is any expression *a* for which the command `type(`*expr*, *a*`)` could succeed.

- A type defined by the system (such as **integer**, **numeric**, and **name**), a type defined by a procedure (`` `type/a` ``), a type defined by an assignment, or a combination of types are valid type expressions.

Examples:

> type(integer, type);	$\longrightarrow$	true
> type(a, type);	$\longrightarrow$	false
> type (name=numeric..numeric, type);	$\longrightarrow$	true
> mytype := {integer,name,list};	$\longrightarrow$	mytype := {name, integer, list}
> type(mytype,type);	$\longrightarrow$	true

SEE ALSO: **indets**, **type**, **type[structured]**

2.1.340   type/vector – check for vector (one-dimensional array)

Calling Sequences:

    type(v, vector)

    type(A, ´vector´(K))

Parameters:

    v – expression

    F – type of the coefficient field

Synopsis:

- This function checks whether its first argument is a Maple vector. A vector is defined to be an one-dimensional array indexed from 1.

- If the argument K is given, the function checks whether A is a vector with entries of type K.

- Also, if K is given, it is **essential** to quote the word vector in order to avoid confusion with the vector function in the **linalg** package.

Examples:

> type(array(1..3,[1,2,3]),vector);	$\longrightarrow$	true
> type(array(0..2,[1,2,3]),vector);	$\longrightarrow$	false
> v := array([2,2/3,1]);	$\longrightarrow$	v := [2, 2/3, 1]
> type(v,´vector´(integer));	$\longrightarrow$	false
> type(v,´vector´(rational));	$\longrightarrow$	true

SEE ALSO: **array**, **vector**, **linalg[vector]**, **type[matrix]**

2.1.341 unames – sequence of unassigned names

Calling Sequence:

    unames()

Synopsis:

- The function **unames** returns an expression sequence consisting of all the active names in the current Maple session which are "unassigned names".

- An "unassigned name" is a name (or string) that has been used in the current Maple session but has no value other than its own name.

- Note, that since file names and error messages are strings, these are also unassigned names.

SEE ALSO: **anames**

2.1.342 unapply – returns an operator from an expression and arguments

Calling Sequence:

    unapply(expr, x, y, ..)

Parameters:

expr	–	any expression
x,y,..	–	variable names

Synopsis:

- The result of **unapply**(*expr*, *x*) is a functional operator. If we apply this operator to **x** we get our original expression.

      unapply(expr,x)(x) ==> expr

- In particular, for a function **f**(*x*),

      unapply(f(x),x) ==> f

- Whenever it is desired to construct an operator using contents of variables or evaluated expressions, **unapply** should be used.

- This function implements the lambda-expressions of lambda calculus.

- For reference see *An Implementation of Operators for Symbolic Algebra Systems* by G.H. Gonnet, SYMSAC July 1986 .

Examples:

```
> p:= randpoly(x);
                      5        4        3        2
           p := - 85 x  - 55 x  - 37 x  - 35 x  + 97 x + 50
```

```
> unapply(p,x);
                       5        4        3        2
            x -> - 85 x  - 55 x  - 37 x  - 35 x  + 97 x + 50
> unapply(p,x)(x);
                  5        4        3        2
            - 85 x  - 55 x  - 37 x  - 35 x  + 97 x + 50
```

SEE ALSO: `operators[functional]`, `operators[example]`, `operators[D]`.

2.1.343   userinfo – print useful information to the user

Calling Sequence:

userinfo(*lev, fn, e1, e2* ...)

Parameters:

lev – a non-negative integer

fn – a procedure name or a set of procedure names

e1 – any expression

e2 – (optional) any expression

Synopsis:

- The procedure `userinfo` is used to print useful information to the user.

- The first argument *lev* is a non-negative integer which determines the level at which information will be printed.

- The second argument *fn* is a procedure name or set of procedure names for which this information is printed. The information will be printed if the global assignment `infolevel[`*fn*`]` `:= ` *lev*; or infolevel[all] := *lev*; was entered before invoking a procedure that contains `userinfo`.

- First the name of the invoked procedure is printed using `lprint`, followed by a colon and 3 spaces. Then the third and optional other arguments are evaluated and lprinted together, but separated by 3 spaces. If one of the arguments is of the form `print(...)` then that argument is prettyprinted. If one of the arguments is of the form `lprint(...)` then that argument is printed on a separate line.

- The user must assign a non-negative integer to some of the entries in the global table `infolevel` before invoking the procedure. If the entry `infolevel[all]` is a non-negative integer then every `userinfo` call will print if its level is less than or equal to `userinfo[all]` .

Examples:

```
> a1 := proc(x,y) userinfo(1,myname,`entered with`,x,y); x+y end:
> infolevel[myname] := 1:
> a1(3,4);
a1:   entered with   3   4
```

2.1.344   W – the omega function

Synopsis:
- W is the function that satisfies the property

$$\omega(x)\, e^{\omega(x)} = x$$

2.1.345   whattype – query the basic data type of an expression

Calling Sequence:

whattype(*expr*)

Parameters:

expr   –   any expression

Synopsis:
- The function **whattype** returns the data type name of *expr*, which may be any of the following basic data types:

`` `+` ``	`` `*` ``	`` `^` ``	`` `=` ``	`` `<>` ``	`` `<` ``
`` `≤` ``	`` `.` ``	`` `..` ``	`` `and` ``	`` `or` ``	`` `not` ``
exprseq	float	fraction	function	indexed	integer
list	procedure	series	set	string	table
uneval					

- Although **exprseq** is not a type name known to the **type** function, it is the name of the internal data structure for expression sequences.

- For a general expression, **whattype** returns the "top level" data type as determined by the order of precedence of the operators.

Examples:
```
> whattype(x + y);                    ⟶              +
> whattype(x - y);                    ⟶              +
> whattype(-x);                       ⟶              *
> whattype(x^2*f(y));                 ⟶              *
> whattype(x/y);                      ⟶              *
> whattype(x^y);                      ⟶              ^
> whattype(1/x);                      ⟶              ^
> whattype(x, y);                     ⟶          exprseq
```

SEE ALSO: `type`, `precedence`, `index[datatypes]`

2.1.346   with – define the names of functions from a library package

Calling Sequence:

   with(*package*) or with(*package*, *f1*, *f2*, ...)

Parameters:

   package   –   name of a Maple package

   f1, ...   –   (optional) names of functions

Synopsis:

- The `with` function defines a function name f (or a group of function names) to point to the same procedure as the corresponding package name `package[f]`.

- If only the package name is supplied to `with` then all the functions in the package will be defined. Additional arguments (if present) restrict the defining process to the specified functions.

- The value returned is a list of the functions defined. A warning message is printed for each name that had a previous value.

- For example, the result of invoking `with(linalg)` is to define a group of names (such as `add` or `inverse`) to point to the same procedures as their corresponding package names (`linalg[add]` or `linalg[inverse]`).

- The name `_liblist` may be used to provide `with` with alternative places (possibly your own directory) to look for packages. For example, the command `_liblist := ['libname', mylibrary]:` would allow `with` to search for the package `dequeue` first in the main library, and then in the directory named `mylibrary`;

Examples:

> A := array([7]);	$\longrightarrow$	A := [7]
> with(linalg, add);	$\longrightarrow$	[add]
> add(A, A);	$\longrightarrow$	[14]

SEE ALSO: `index[packages]`

2.1.347   words – query memory usage (words used)

Calling Sequence:

   words(*n*)

Parameters:

   n   –   an optional integer

Synopsis:

- The function `words` returns the total words used. On most implementations, the size of a word is 32 bits which corresponds to 4 bytes. This measure is a cumulative count of all memory requests made up to that point to the Maple storage manager, which typically includes some reused memory.

- During a Maple session, a message of the form:

```
bytes used=xxxx, alloc=yyyy, time=zzzz
```

- is printed out at the first "safe point" after every 100,000 words used (approximately 400K bytes) or some other system-specific default value. Here **xxxx** is the total bytes used, **yyyy** is the total bytes actually allocated, and **zzzz** is the total CPU time used (in seconds) since the start of the session. This same message is printed out after each garbage collection occurs.

- The integer parameter n (if present) controls the frequency of the "bytes used ..." messages, measured in units of 4-byte words. If n is less than or equal to zero then the "bytes used ..." messages will not be printed (except when a garbage collection occurs). The current value of this frequency is available as **status[4]**, the fourth element of the **status** sequence.

SEE ALSO: gc, ´status´

2.1.348   writeto – write output to a file
appendto – write output to a file in append mode

Calling Sequence:

writeto(*filename*) or appendto(*filename*)

Parameters:

filename – a name

Synopsis:

- When the command writeto(*filename*); is typed all future commands will have their results immediately stored in *filename* and will not be displayed on the screen.

- While in this mode, no prompt will appear at the terminal. The prompt and the input statements will be echoed into *filename* (unless echoing is turned off).

- The special command writeto(terminal) will restore the standard mode of printing the results on the terminal.

- If *filename* exists the contents of the file are overwritten.

- An alternative mode of writing to a file is available via the appendto function. In the case of appendto(*filename*); rather than overwriting the contents (if any) of *filename*, the new output is appended to *filename*. In other respects, **appendto** and **writeto** have the same functionality.

SEE ALSO: `save`, echo, write

2.1.349   Zeta – the Riemann zeta function

Synopsis:
- The Riemann zeta function has syntax `Zeta(s)` and is defined by

$$\texttt{Zeta}(s) = \zeta(s) = \sum_{i=1}^{\infty} \frac{1}{i^s} , \quad \text{for } s > 1$$

- Its derivatives are represented by

$$\texttt{Zeta}(n, s) = \zeta^{(n)}(s) = (\frac{d}{ds})^n \, \zeta(s) , \quad \text{for } n >= 1$$

- Automatic simplifications will be applied to these functions.

2.1.350   zip – zip together two lists or vectors

Calling Sequence:
   `zip(f, u, v)` or `zip(f, u, v, d)`

Parameters:
   f –   binary function
   u, v –   lists or vectors
   d –   value (optional)

Synopsis:
- The function `zip` applies the binary function f to the components of two lists/vectors u and v creating a new list/vector r defined as follows. If m is the length of u and n is the length of v then r is a list/vector of length `min(m,n)` with `r[i]` = `f(u[i],v[i])` for i in `1..min(m,n)`.

- If the optional fourth argument d is given, it is used as a default value for f when one of the lists/vectors is shorter than the other. Thus r is a list/vector of length $max(m,n)$ with `r[i]` = $f(u[i],v[i])$ for i in `1..min(m,n)` and `r[i]` = $f(t[i],d)$ for i in `1+min(m,n)..max(m,n)` where t is the longer of u and v.

Examples:
```
> zip((x,y)->x+y,[1,2,3],[4,5,6]);
```
$\longrightarrow$ [5, 7, 9]
```
> zip(gcd,[0,14,8],[2,6,12]);
```
$\longrightarrow$ [2, 2, 4]
```
> zip((x,y)->x+y,[1,2,3],[4,5],0);
```
$\longrightarrow$ [5, 7, 3]

2.1.351   ztrans – Z transform

Calling Sequence:

> ztrans(*f*, *n*, *z*)

Parameters:

f – expression

n – name

z – name

Synopsis:

- The function **ztrans** finds the Z transformation of $f(n)$ with respect to z. Formally, **ztrans**$(f(n),n,z)$ = sum$(f(n)/z\hat{\ }n,n$=0..infinity$)$.

- **ztrans** recognizes and specially handles a large class of expressions, and only resorts to using the definition to calculate the transformation if the given expression has an unknown form. If the **Z** transform of the given expression cannot be found in a closed form, then the left-hand side of the formal definition is returned, rather than the right-hand side.

- **ztrans** recognizes the delta function as **Delta(...)** and the step function as **Step(...)**.

- This function has to be defined using **readlib(ztrans)** before it can be used.

Examples:

```
> ztrans(f(n+1),n,z);                  ⟶        z ztrans(f(n), n, z) - f(0) z

> ztrans(sin(Pi/2*t),t,z);             ⟶                  z
                                                        ------
                                                            2
                                                        1 + z

> ztrans(3^n/n!,n,w);                  ⟶              exp(3/w)

> ztrans(beta*5^n*(n^2+3*n+1),n,z);    ⟶                            2
                                                  beta z (10 z - 25 + z )
                                                  -----------------------
                                                               3
                                                          (z - 5)

> ztrans(Delta(t-5)*Psi(t),t,w);       ⟶                - 25 + 12 gamma
                                                  - 1/12 ---------------
                                                                5
                                                                w

> ztrans(n*Step(n-3),n,z);             ⟶              3 z - 2
                                                     -----------
                                                      2        2
                                                     z  (z - 1)

> ztrans(invztrans(f(z),z,n),n,z);     ⟶                f(z)
```

SEE ALSO: **invztrans** , **laplace** , **rsolve**

2.2 Miscellaneous Library Functions

2.2.1   bernstein – Bernstein polynomial approximating a function

Calling Sequence:

    bernstein(n, f, x)

Parameters:

    n – an integer

    f – a function (specified as a procedure or operator)

    x – an algebraic expression

Synopsis:

- This procedure returns the nth degree Bernstein polynomial in x approximating the function $f(x)$ on the interval [0,1]. Note that f must be a function of one variable specified as a procedure or operator.

- This function should be defined by the command `readlib(bernstein)` before it is used.

Examples:

```
> readlib(bernstein):
  bernstein(3,<1/(x+1)|x>,z);
```
$\longrightarrow$
$$1 - 3/4\ z + 3/10\ z^2 - 1/20\ z^3$$

```
> f := proc(t) if t < 1/2 then 4*t^2 else
  2 - 4*t^2 fi end:
  bernstein(2,f,x);
```
$\longrightarrow$
$$2\ x - 4\ x^2$$

2.2.2   bianchi – find the Bianchi type of any 3-dimensional Lie algebra

Calling Sequence:

    classi();

Synopsis:

- This procedure must first be defined by the command `readlib(bianchi):` before it is invoked.

- One must define the array `structure` as follows:

    structure:=array([C123,C231,C312,C112,C113,C221,C223,C331,C332]);

where the quantities C.i.j.k denote nine independent structure constants of the 3-dimensional Lie algebra; they are assumed to be antisymmetric in the last pair of indices. They are defined by

```
[e.i,e.j] = e.1*C.1.i.j + e.2*C.2.i.j + e.3*C.3.i.j
```

or explicitly by

```
[e2,e3] =   e1*C123 + e2*C223 - e3*C332
[e3,e1] = -e1*C113 + e2*C231 + e3*C331
[e1,e2] =   e1*C112 - e2*C221 + e3*C312
```

where [e1,e2,e3] is a basis for the Lie algebra and the square brackets on the left-hand side denote the commutator (the Lie algebra product).

- The procedure is started by entering `classi();`. If the variable `structure` has not been defined, the procedure prompts you for the above structure constants.

- The procedure first checks that the structure constants actually define a Lie algebra by checking the Jacobi identities. If these identities are not satisfied an error message is printed.

- If the structure constants indeed define a Lie algebra the procedure prints the Bianchi type.

- It may prompt you for input about certain expressions being `ZERO` or `NONZERO`.

Examples:
```
> readlib(bianchi):
# define the structure constants
> structure := array(1..9,[1,0,0,0,0,0,0,0,0]):
# determine Bianchi type
> classi();
```
$$\text{Jacobi identities are satisfied}$$
$$\text{Bianchi-type is II}$$

2.2.3   bspline – compute the B-spline segment polynomials

Calling Sequence:

$\qquad$ bspline(d, v)

$\qquad$ bspline(d, v, k)

Parameters:

$\qquad d$ – non-negative integer

$\qquad v$ – name

$\qquad k$ – (optional) list of $d + 2$ numbers or symbols

Synopsis:

- This function computes the segment polynomials for the B-spline of degree d in the symbol v on the knot sequence k. If k is not specified, the uniform knot sequence [0, 1, ..., $d + 1$] is used.

- If the knot sequence contains symbols, the symbolic representation of the B-spline segment polynomials is returned. Additionally, the knot sequence can have multiple knots. Multiple knots result in a loss of continuity at the knot. If the multiplicity of a knot is $1 \leq m \leq d$, then the continuity at that knot is $d - m$.

- For numerical knots, the B-spline can be plotted by

$$b := \text{convert(bspline(d,v),procedure);} \quad \text{plot(b,v=0..d+1);}$$

- Note that this results in a temporary file called `temp` being created.

- This function should be defined by the command `readlib(bspline)` before it is used.

Examples:
```
> readlib(bspline):
> bspline(1,u);
           [[u < 0, 0], [u < 1, u], [u < 2, - u + 2], [2 <= u, 0]]

> bspline(1,u,[0,a,2]);
                                          - u + a
         [[u < 0, 0], [u < a, u/a], [u < 2, ------- + 1], [2 <= u, 0]]
                                           2 - a
```

2.2.4   C – generate C code

Calling Sequence:

C(*s*)

Parameters:

s – an expression, array of expressions, list of equations

Synopsis:

- The C function generates C code for evaluating the input. The input *s* must be either a single algebraic expression, a named array of algebraic expressions, or a list of equations of the form `name = algebraic` where the latter is understood to mean a sequence of assignment statements. Currently, C cannot generate output for Maple procedures.

- By default the output is sent to standard output. An additional argument of the form `filename = foo` can be used to direct the output to the file `foo`.

- If the keyword `optimized` is specified as an additional argument, common subexpression optimization is performed. The result is a sequence of assignment statements in which temporary values are stored in local variables beginning with the letter `t`. The global names `t0`, `t1`, `t2`, ... are reserved for use by C for this purpose.

- The global variable precision can be assigned either single or double (single by default) for single or double precision respectively.

- NOTE: if the input is a named table or procedure, no error will be generated. It will be treated just as a variable name. NOTE: array subscripts are translated as is; they are not shifted.

- This function should be defined by the command `readlib(C)` before it is used.

Examples:

```
> readlib(C):
  s := ln(x)+2*ln(x)^2-ln(x)^3:
  C(s);
```
$\longrightarrow$
```
              t0 = log(x)+2.0*pow(log(x),2.0)-p\
ow(log(x),3.0);
```
```
> C(s,optimized);
```
$\longrightarrow$
```
              t1 = log(x);
              t2 = t1*t1;
              t6 = t1+2.0*t2-t2*t1;
```

SEE ALSO: `optimize`, `fortran`, `eqn`, `latex`

2.2.5   cartan — A collection of procedures for the computation of the connection coefficients and curvature components using Cartan´s structure equations

Calling Sequence:

    `cartan(`*h, hinv, coords, g*`)`
    `simp1(`*temp, normal, gamma*`)`
    `riemann(`*hinv, gamma, coords, g*`)`
    `simp2(`*temp, normal, R*`)`
    `printcartan(`*R*`)`

Parameters:

h	—	an array indexed from 0 to 15
hinv	—	the inverse of h
coords	—	a list of four variable names
g	—	a constant metric with indices ranging from 0 to 3
temp	—	a variable name
normal	—	a procedure name
gamma	—	the connection coefficient array computed by `cartan`
R	—	the Riemann tensor array computed by `riemann`

Synopsis:

- The parameter *coords* should give a list of the coordinate names. For example, we could define `c` to be

    `c := [t,x,y,z];`

- The tetrad covariant components are entered as a one-dimensional array with subscripts from 0 to 15. The name assigned to the array is up to you. For example, we would define **f** as:

```
f := array(0..15);
```

and enter the components in the indicated slots:

```
n[0] to n[3]              in   f[0]  to  f[3]
l[0] to l[3]              in   f[4]  to  f[7]
-mbar[0] to -mbar[3]      in   f[8]  to  f[11]
-m[0] to -m[3]            in   f[12] to  f[15]
```

- The contravariant components of the tetrad are defined in a similar manner. The name of the array is again up to you. Note the change in the ordering.

```
h := array(0..15);
l[0], l[1], l[2], l[3]                      in   h[0], h[4], h[8],  h[12]
n[0], n[1], n[2], n[3]                      in   h[1], h[5], h[9],  h[13]
m[0], m[1], m[2], m[3]                      in   h[2], h[6], h[10], h[14]
mbar[0], mbar[1], mbar[2], mbar[3]          in   h[3], h[7], h[11], h[15]
```

- Another array which must be entered is the **g** array which is a two-dimensional array containing the constant metric applicable to the dimension and signature of the space being considered. Note that if you are in a four-dimensional space-time, the call at the start automatically sets up the **g** array for a complex null tetrad as follows:

```
g := array(0..3,0..3,[[0,1,0,0], [1,0,0,0], [0,0,0,-1], [0,0,-1,0]]);
```

Note that the above concentrates on a four-dimensional space-time. It is possible to work in higher dimensional spaces with any signature by changing the dimensions and definition of the above arrays accordingly.

- With the above set-up, it is now possible to calculate the connection coefficients with a call to the **cartan** procedure:

```
cartan(f, h, c, g);
```

Each of the arguments should be defined as described above. The procedure returns the connection coefficients in an array with name **gamma** . It is defined in the procedure as

```
gamma := array(0..3,0..3,0..3);
```

Since it is used in the Riemann tensor components calculation, it should be simplified with calls to the procedure `simp1` described below.

- The procedure `simp1` is designed to ease the task of simplifying the **gamma** array. It applies the given procedure to each element of **gamma** and assigns the result to the unassigned variable. It is called

```
simp1(unassigned_variable, procedure, parameters);
```

For example,

```
simp1(temp, normal, gamma);
```

will normalize all components of **gamma** by assigning the result to **temp** .

```
simp1(newgamma, subs, f(r)=23*r^2 - 2, temp);
```

will then substitute for the arbitrary function f(r) in all components of **temp** returning the results in **newgamma** .

- The Riemann tensor components can now be calculated with a call to the procedure **riemann** as follows:

```
riemann(h, gamma, c, g);
```

where **h** was the above contravariant tetrad components, **gamma** is the array of simplified connection components, **c** is the above list of coordinates and **gamma** is the above constant metric. The Riemann tensor components are returned in an array called **R** which is defined in the procedure as

```
R := array(0..3,0..3,0..3,0..3);
```

Note that here again the components will be unsimplified. It is advised that they be simplified with the procedure `simp2` before any displaying or calculating is done with them.

- The procedure `simp2` is similar to `simp1` except that it is used to simplify the 4-dimensional array containing the Riemann tensor components. The arguments to `simp2` are the same as to `simp1` . For example,

```
simp2(temp, normal, R);
```

will normalize all components of **R** by assigning the result to **temp** .

```
simp2(newR, subs, f(r)=23*r^2 - 2, temp);
```

will then substitute for the arbitrary function f(r) in all components of R by assigning the result to newR .

- The procedure **printcartan** takes any array as argument and prints the non-zero entries . If all entries are zero it prints a message.

Examples:
```
> readlib(cartan):
# Plane wave metric
> coord := [u,x,y,v]:
> f := array(0..15,[a(u)*x^2+2*b(u)*y*x+c(u)*y^2,0,0,1,
> 1,0,0,0, 0,0,-1,0, 0,-1,0,0]):
> h := array(0..15,[0,1,0,0, 0,0,0,-1, 0,0,-1,0,
> 1, -(a(u)*x^2+2*b(u)*x*y+c(u)*y^2),0,0]):
> cartan(f, h, coord, g);
elapsed time:    .600   sec
```
$$\text{gamma}$$
```
> riemann(h, gamma, coord, g);
elapsed time:    .416   sec
```
$$R$$
```
> printcartan(gamma);
```
$$\text{gamma}[1, 2, 1] = 2\ b(u)\ x + 2\ c(u)\ y$$
$$\text{gamma}[1, 3, 1] = 2\ a(u)\ x + 2\ b(u)\ y$$
$$\text{gamma}[2, 1, 1] = -\ 2\ b(u)\ x - 2\ c(u)\ y$$
$$\text{gamma}[3, 1, 1] = -\ 2\ a(u)\ x - 2\ b(u)\ y$$
```
> printcartan(R);
```
$$R[1, 2, 1, 2] = -\ 2\ c(u)$$
$$R[1, 2, 1, 3] = -\ 2\ b(u)$$
$$R[1, 3, 1, 3] = -\ 2\ a(u)$$

SEE ALSO: **debever**

2.2.6   coeftayl – coefficient of (multivariate) expression

Calling Sequence:
> coeftayl(*expr*, *eqn*, *k*)

Parameters:

   expr   –   an arbitrary expression

   eqn   –   an equation of the form **x = alpha** where **x** is a name (univariate case) or list (multivariate case)

   k   –   a non-negative integer (univariate case) or a list of non-negative integers (multivariate case)

Synopsis:

- This function computes a coefficient in the (multivariate) Taylor series representation of *expr* without forming the series (it uses differentiation and substitution). Often, *expr* is a polynomial.

- The one-variable and several-variable cases are distinguished by the types of the input parameters.

- UNIVARIATE CASE: x is a name and k a non-negative integer.

- In this case, the value returned is the coefficient of `(x-alpha)^k` in the Taylor series expansion of *expr* about x = alpha. This is equivalent to executing `coeff(taylor(`*expr*`, x = alpha, k + 1), x - alpha, k)` but it is more efficient (because only a single term is computed).

- MULTIVARIATE CASE: x is a nonempty list [x1, ..., xv] of indeterminates appearing in *expr* and `alpha` is a list [alpha1, ..., alphav] specifying the point of expansion with respect to the given indeterminates; k is a list [k1, ..., kv] of non-negative integers corresponding to elements in x and `alpha`.

- In this case, the value returned is the coefficient of the term specified by the monomial

      (x[1]-alpha[1])^k[1] * . . . * (x[v]-alpha[v])^k[v]

- In the multivariate Taylor series expansion of *expr* about the **point** x = alpha, if k is the list of zeros then the value returned is the value resulting from substituting x = alpha into *expr*.

- This function should be defined by the command `readlib(coeftayl)` before it is used.

2.2.7   commutat − commutator routines

      c, expand/c, convert/c, simplify/c, &*, expand/&*, convert/&*

Synopsis:
- Definition : the commutator c(x,y) = x y - y x

- where multiplication here is non-commutative. This group of routines provides for the manipulation and simplification of commutators, expanding commutators in terms of &* (Maple´s non-commutative multiplication operator), and converting an expression in terms of &* to commutator form. The following identities are applied automatically:

      c(y,x) = - c(x,y)
      c(x,x) = 0   for all x
      c(k,x) = 0   for any constant k
      c(x+y,z) = c(x,y) + c(y,z)
      c(k*x,y) = k*c(x,y)   for any constant k

- A canonical form for commutators is obtained by applying the following two rules given the ordering: order by number of nested commutators, and break ties by address of object.

```
c(y,x)      ==> -c(x,y) iff x < y
c(x,c(y,z)) ==> c(y,c(x,z)) - c(z,c(x,y)) iff x < y
```

- In particular, this form recognizes the Jacobi identity:

```
c(x,c(y,z)) + c(y,c(z,x)) + c(z,c(x,y)) = 0
```

- Commutators can be expressed in terms of &* as in

```
convert( c(x,(y,z)), `&*` );
    - &*(y, z, x)   &*(x, z, y) + &*(z, y, x) + &*(x, y, z)
```

- Or converted back in terms of c

```
convert(",c);
    c(y, c(x, z)) - c(z, c(x, y))
```

- Note: If the expression being converted to a commutator cannot be expressed as a commutator, the result returned will be pure nonsense. The utility routine **commutat** can be used to display commutators as a list of lists syntax.

- These routines must first be loaded via **readlib(commutat)**

2.2.8   convergs – print convergents of continued fraction

Calling Sequence:

  convergs(*listA*, *listB*, *n*)

Parameters:

listA	–	list of numbers, or a function
listB	–	(optional) list of numbers
n	–	(optional) integer

Synopsis:
- The **convergs** function takes the continued fraction

```
a1 + b2/(a2 + b3/(a3 + b4/(a4 + . . .
```

and prints the successive convergents in the form n, nth convergent. The function convergs returns NULL.

- *listA* is of the form [a1, a2, ...] and similarly *listB* is of the form [1, b2, b3,...] where aI and bI correspond to the values in the displayed continued fraction.

- If *listB* is of the form [1, 1, 1, ...] convergs may be omitted.

- If a third argument, n, is present, convergs indicates the number of convergents to compute.

- This function should be defined by the command readlib(convergs) before it is used.

SEE ALSO: convert, convert[confrac]

2.2.9 cost – operation evaluation count

Calling Sequence:

cost($x1$, $x2$,...,xn)

Parameters:

xk – of type algebraic, name = algebraic, array(algebraic)

Synopsis:

- Cost is used to compute an operation count for the numerical evaluation of the given arguments. The operation count is expressed as a polynomial in the names additions, multiplications, divisions, functions, subscripts and assignments with non-negative integer coefficients.

- Assignment of positive real values to these global names yields a weighted cost.

- Note that the cost used for computing powers is as follows. For an integral power, repeated multiplication is assumed. For a general power it is assumed to be computed using exp and ln.

- This function should be defined by the command readlib(cost) before it is used.

Examples:

```
> readlib(cost):
  a := x+x^2+x^3+x^4:
  cost(a);                          ⟶        3 additions + 6 multiplications
> a := convert(a,horner):
  cost(a);                          ⟶        3 multiplications + 3 additions
```

SEE ALSO: optimize

2.2.10 debever – A collection of procedures for the computation of the Newman-Penrose spin coefficients and curvature components using Debever's formalism

Calling Sequence:

```
npspin(h, info, coords, flag)
curvature(spincf, info, flag)
simp(temp, normal, spincf)
printspin(spincf)
printcurve(curve)
```

Parameters:

h	–	an array indexed from 0 to 15
info	–	an unassigned name
coords	–	a list of four variable names
flag	–	(optional) parameter which will suppress printing
spincf	–	an array indexed from 0 to 11
temp	–	an unassigned name
normal	–	a procedure name
curve	–	an array indexed from 0 to 11

Synopsis:

- In order to identify the coordinates, the parameter *coords* must be a list of four coordinate names, such as `[t,x,y,z]` .

- The tetrad covariant components are entered as a one-dimensional array with subscript from 0 to 15. The name given to the array is up to you. For example, we would define *h* as

```
h := array(0..15);
```

and enter the components in the indicated slots:

```
n[0] to n[3]            in    h[0]  to  h[3]
l[0] to l[3]            in    h[4]  to  h[7]
-mbar[0] to -mbar[3]    in    h[8]  to  h[11]
-m[0] to -m[3]          in    h[12] to  h[15]
```

Now the spin coefficients can be calculated. This is done with a call to the **npspin** procedure:

```
npspin( h, info, coords, flag );
```

```
h       This refers to the array containing the tetrad as defined above.
info    This variable should be any undefined variable.  It is used to hold
        certain values that are needed in the curvature components calculation.
coords  This refers to the coordinates you are using as defined previously.
flag    This parameter is optional.  If any value is entered, printing of the
        spin coefficients after computation is suppressed.  Otherwise the
        coefficients are displayed in the same manner as printspin.
```

When the procedure is completed, the spin coefficients are returned in an array subscripted from 0 to 11 named *spincf* as follows:

```
spincf  spin coefficient
     0  Nu
     1  Lambda
     2  Mu
     3  Pi
     4  Tau
     5  Rho
     6  Sigma
     7  Kappa
     8  Gamma
     9  Alpha
    10  Beta
    11  Epsilon
```

- Now the curvature components may be calculated using the `curvature` procedure. It is advisable to simplify the spin coefficients using the `simp` procedure before calculating the curvature components. The following call is used to calculate the curvature components:

```
curvature( spincf, info, flag );
```

```
spincf    This is the name of the array containing the spin coefficients as
          calculated previously in npspin.
info      This is the name of the temporary variable returned from npspin.
          It contains internal data used in the calculation of the curvature
          components.
flag      This has a similar purpose as the flag in npspin.  If any value
          is entered here, the printing of the curvature components after
          calculation is suppressed.
```

The curvature components are returned in an array named *curve*, subscripted 0 to 11. The following chart shows what each of the values is:

```
curve  curvature component
    0  Psi[0]
    1  Psi[1]
    2  Psi[2]
    3  Psi[3]
    4  Psi[4]
    5  R
```

```
 6   Phi[0,0]
 7   Phi[0,1]
 8   Phi[0,2]
 9   Phi[1,1]
10   Phi[1,2]
11   Phi[2,2]
```

- The `simp` procedure allows you to simplify the components of an array using other procedures. When accessed, `simp` invokes the second argument on each component of the last argument returning the new value in the first argument. For example, the following procedure performs the normal function on each spin coefficient, returning the new values in the array `temp` .

```
simp( temp, normal, spincf );
```

Extending upon this, the following procedure substitutes A=1 into each curvature component, returning the new values in the array **newcurve** .

```
simp( newcurve, subs, A=1, curve );
```

- The procedure `printspin` will print out all of the spin coefficients. Its argument should be the simplified form of the spin coefficient array.

```
printspin(spincf);
```

- The procedure `printcurve` will print out all of the curvature components. Its argument should be the simplified form of the curvature component array.

```
printcurve(curve);
```

- Note that any element of either array may be accessed directly. For example, the following statements will access Psi[2] and Tau respectively:

```
curve[2];
spincf[4];
```

Also, since some of these values can become quite long, it is advisable that simplification of the values be done before printing them out. Some procedures you may wish to use are: `evalc`, `normal`, `factor`, `expand`, and `subs`.

Examples:
```
> readlib(debever):
# Null tetrad for the Plane wave metric
> p := array(0..15,[a(u)*x^2+2*b(u)*x*y+c(u)*y^2,0,0,1,
> 1,0,0,0,  0,0,-1,0,  0,-1,0,0]):
> npspin(p,info,[u,x,y,v]);
                      -i * SQRT(-det(g)) = 1
```

```
                    Newman-Penrose Spin Coefficients
                    Nu = - 2 a(u) x - 2 b(u) y
                              Lambda = 0
                                  Mu = 0
                                  Pi = 0
                                 Tau = 0
                                 Rho = 0
                               Sigma = 0
                               Kappa = 0
                               Gamma = 0
                               Alpha = 0
                                Beta = 0
                             Epsilon = 0
elapsed time:    .184    sec
> curvature(spincf,info);

                        Curvature Components
                           psi[0] = 0
                           psi[1] = 0
                           psi[2] = 0
                           psi[3] = 0
                           psi[4] = 2 a(u)
                                R = 0
                         phi[0,0] = 0
                         phi[0,1] = 0
                         phi[0,2] = 0
                         phi[1,1] = 0
                         phi[1,2] = 0
                         phi[2,2] = 2 b(u)
elapsed time:    .116    sec
```

SEE ALSO: cartan

2.2.11   dinterp – probabilistic degree interpolation

Calling Sequence:
>   dinterp(f, n, k, d, p)

Parameters:

f	–	a Maple procedure
n, k, d	–	integers
p	–	prime modulus

Synopsis:
- Given a function f that evaluates a polynomial in n variables modulo p, and a degree bound d on the kth variable, determine probabilistically the degree of the kth variable.

- The `dinterp` function may return `FAIL` if it encounters a division by zero when evaluating f. It may also return a result for the degree of the kth variable which is too low. The probability that this happens can be decreased by using a larger modulus. A 12 to 20 digit modulus is considered ideal.

- This function should be defined by the command `readlib(dinterp)` before it is used.

Examples:
```
> readlib(dinterp):
  f := proc(x,y,z,p) x^2+y^3+z^4 mod p
  end:
  dinterp(f,3,1,6,997);                    ⟶            2
> dinterp(f,3,2,6,997);                    ⟶            3
> dinterp(f,3,3,6,997);                    ⟶            4
```

2.2.12   Dirac – the Dirac delta function
Heaviside – the Heaviside step function

Calling Sequence:
>   Dirac(t)
>   Dirac(n, t)
>   Heaviside(t)

Parameters:

t	–	algebraic expression
n	–	nonnegative integer

Synopsis:
- The `Dirac($t$)` delta function is defined as zero everywhere except at $t = 0$ where it has a singularity. It has an additional property, specifically:

$$\texttt{Int(Dirac(t),t = -infinity..infinity) = 1}$$

- Derivatives of the `Dirac` function are denoted by the two-argument `Dirac` function. The first argument denotes the order of the derivative. For example, `diff(Dirac(t), t$n)` will be automatically simplified to `Dirac(n, t)` for any integer n.

- The `Heaviside(t)` unit step function is defined as zero for $t < 0$, 1 for $t \geq 0$. It is related to the Dirac function by `diff(Heaviside(t),t) = Dirac(t)`.

- These functions are typically used in the context of integral transforms such as `laplace()` or `mellin()`.

- This function should be defined by the command `readlib(Dirac)` before it is used.

- This function should be defined by the command `readlib(Heaviside)` before it is used.

SEE ALSO: `laplace`, `mellin`

2.2.13   edit − expression editor

Calling Sequence:

   edit(*expr*)

Parameters:

   expr   −   any expression

Synopsis:

- This expression editor is invoked as: `edit( `*expr*` )`; The *expr* is then displayed with a **BOX** around it. The **BOX** is moved around within *expr* by commands. The name _ is assigned the contents of the **BOX**.

- COMMANDS :

up;	move **BOX** to enclosing expression
down;	move **BOX** to first operand of expression
top;	move **BOX** to include the whole expression
succ;	move **BOX** to successive subexpr at current level
prev;	move **BOX** to previous subexpr at current level
first;	move **BOX** to first subexpr at current level
last;	move **BOX** to last subexpr at current level
child(4,1,2);	move **BOX** to `op(2,op(1,op(4,the expression)))`
find(XXX);	move **BOX** to the next occurrence of **XXX**
use(YYY);	**YYY** replaces the **BOX**´s contents.
	(**YYY** is usually some function of _)
show;	display the expression

- This function should be defined by the command `readlib(edit)` before it is used.

Examples:
```
> readlib(edit):
  edit(x^2+x^3);
```
$$\longrightarrow \qquad BOX(x^2 + x^3)$$

```
> down;
```
$$\longrightarrow \qquad BOX(x^2) + x^3$$

```
> down;
```
$$\longrightarrow \qquad BOX(x)^2 + x^3$$

```
> use(sin(_));
```
$$\longrightarrow \qquad BOX(sin(x))^2 + x^3$$

```
> top;
```
$$\longrightarrow \qquad BOX(sin(x)^2 + x^3)$$

```
> _;
```
$$\longrightarrow \qquad sin(x)^2 + x^3$$

2.2.14   ellipsoid – surface area of an ellipsoid

Calling Sequence:

> ellipsoid(a,b,c)

Parameters:

> a,b,c –   the three principal semi-axes of the ellipsoid

Synopsis:
- The procedure `ellipsoid` returns the surface area of an ellipsoid whose semi-axes are a, b, and c where $a \geq b$, $b \geq c$, $c \geq 0$. For a general ellipsoid the result is expressed as an unevaluated integral in terms of `Int` , which may be numerically evaluated by `evalf`.

- When two or more of the arguments are the same, then a solvable integral can be created. Otherwise, elliptical integrals are needed to evaluate the general case.

- This procedure must be defined by calling `readlib(ellipsoid):` before it is invoked.

Examples:
```
> readlib(ellipsoid):
  evalf(ellipsoid(5,3,2));
```
$$\longrightarrow \qquad 134.7751767$$

2.2.15   eulermac – Euler-Maclaurin summation

Calling Sequence:

> eulermac($expr$, x)
> eulermac($expr$, x, n)

Parameters:

  $expr$ — an expression in x

  x — independent variable

  n — (optional) integer (degree of summation)

Synopsis:

- The `eulermac` function computes an nth degree Euler-Maclaurin summation formula of $expr$ (thus n terms of the expansion are given). If n is not specified, it is assumed to be `Order - 1`.

- The `eulermac` function is an asymptotic approximation of `sum(`$expr$`, x )`. If `F(`x`) = eulermac(` `f(`x`), x)`, then `F(`$x + 1$`) - F(`x`)` is asymptotically equivalent to `f(`x`)`.

- This function should be defined by the command `readlib(eulermac)` before it is used.

Examples:
```
> readlib(eulermac):
  eulermac( 1/x , x , 4);
```

$$\longrightarrow \quad \ln(x) - \frac{1}{2\,x} - \frac{1}{12\,x^2} + \frac{1}{120\,x^4} + O\left(\frac{1}{x^6}\right)$$

2.2.16   evalgf – evaluate in an algebraic extension of a finite field

Calling Sequence:

  `evalgf(`$expr$`, `p`)`

Parameters:

  $expr$ — an expression or unevaluated function call

  p — a prime number

Synopsis:

- If $expr$ is an unevaluated function call (such as `evalgf(Gcd(u,v)),p)`, then the function is performed in the smallest algebraic extension of `Z mod` p possible. See `help(`*function*`)` for more information, where *function* is one of the following:

Content	Divide	Expand	Gcd	Gcdex	Normal
Prem	Primpart	Quo	Rem	Resultant	Sprem

- Otherwise, $expr$ is returned unchanged, after first checking for dependencies between the `RootOf`s in the expression.

- If a dependency is noticed between `RootOf`s during the computation, then an error occurs, and the dependency is indicated in the error message (this is accessible through the variable `lasterror`).

- This function should be defined by the command `readlib(evalgf)` before it is used.

Examples:
```
> readlib(evalgf): `mod` := mods:
  evalgf(Gcd(x^2-1, x^2+x*RootOf(_Z^2-1)+2)
  , 3);                                          ⟶                                    1

> evalgf(Quo(x^2+x*RootOf(_Z^2-1),
  x-RootOf(_Z^2+1), x, 'r'), 3);                 ⟶              2                 2
                                                      x + RootOf(_Z  - 1) + RootOf(_Z  + 1)

> r;                                             ⟶              2                 2
                                                      - 1 + RootOf(_Z  + 1) RootOf(_Z  - 1)
```

SEE ALSO: `mod`, RootOf, GF

2.2.17   evalr – evaluate an *expr* using range arithmetic   shake – compute a bounding interval

Calling Sequence:

> evalr(*expr*)
> shake(*expr*, *ampl*)

Parameters:

> *expr*   –   any *expr*
> *ampl*   –   a positive integer

Synopsis:

- The **evalr** function evaluates an *expr* containing ranges written [a..b] with a≤b, sequences of ranges written [a..b, c..d,...], or bounded variables written [x,a..b]. By default a variable x is converted to [x, -infinity ..infinity]. The result of **evalr** in this case can be a sequence of ranges, or an *expr* when both sides of the resulting range are equal, the unprecised variables [x, -infinity..infinity] are converted back to x when the result is the same for all values of x.

- The function **evalr** can be called with an *expr* without ranges in it. In this case the decision functions min, max, abs and Signum are evaluated using range arithmetic. Thus **evalr(Signum(*expr*))** may return **FAIL**.

- The power of **evalr** is limited by the power of **solve**.

- The **shake** function replaces the constants at the leaves of the *expr* tree by an interval of width Float(10, −*ampl*), and then uses **evalr** to propagate these intervals bottom-up. If *ampl* is not precised, then the current value of Digits is used.

- This function should be defined by the command **readlib(evalr)** before it is used.

- This function should be defined by the command **readlib(shake)** before it is used.

Examples:

```
> readlib(evalr):
  readlib(shake):
  evalr(min(2,sqrt(3)));
```
$\longrightarrow$
$$\frac{1/2}{3}$$

```
> evalr(sin([2..7]));
```
$\longrightarrow$ [-1 .. sin(2)]

```
> evalr(abs(x));
```
$\longrightarrow$ [[x, 0 .. infinity],

                    - [x, - infinity .. 0]]

```
> shake(Pi,3);
```
$\longrightarrow$ [3.1102 .. 3.1730]

SEE ALSO: abs, min, max, signum.

2.2.18   extrema – find relative extrema of an expression

Calling Sequence:

extrema(*expr*, *constraints*)

extrema(*expr*, *constraints*, *vars*)

extrema(*expr*, *constraints*, *vars*, *'s'*)

Parameters:

expr	–	expression whose extrema are to be found
constraints	–	a constraint or set of constraints
vars	–	a variable or set of variables
s	–	unevaluated name

Synopsis:

- The **extrema** function can be used to find extreme values of a multivariate expression with zero or more constraints. Candidates for extreme value points can also be returned. The **extrema** are returned as a set, and the candidates are returned as a set of sets of equations in the appropriate variables.

- *expr* must be an algebraic expression. The constraints may be specified as either expressions or equations. When a constraint is given as an expression, it is understood that *constraint* = 0. If no constraints are to be given, then the empty set {} is used in the parameter list. If *vars* is not given then all name indeterminates in the *expr* and constraints are used. *vars* must be specified if the fourth parameter **s** is given. The candidates for the extreme value points are returned in *s*.

- When the candidates cannot be expressed in closed form, *s* will contain the system of equations which when solved will produce these candidates.

- This function employs the method of Lagrange multipliers.

- This function should be defined by the command **readlib(extrema)** before it is used.

Examples:
```
> readlib(extrema):
> extrema( a*x^2+b*x+c,{},x );
```
$$\{1/4 \ \frac{- b^2 + 4 c a}{a}\}$$

```
> extrema( a*x*y*z, x^2+y^2+z^2=1, {x,y,z} );
```
$$\{\min(0, \ - 1/9 \ a \ 3^{1/2} \ , \ 1/9 \ a \ 3^{1/2} \ ), \ \max(0, \ - 1/9 \ a \ 3^{1/2} \ , \ 1/9 \ a \ 3^{1/2} \ )\}$$

```
> f := (x^2+y^2)^(1/2)-z; g1 := x^2+y^2-16; g2 := x+y+z = 10;
```
$$f := (x^2 + y^2)^{1/2} - z$$

$$g1 := x^2 + y^2 - 16$$

$$g2 := x + z + y = 10$$

```
> extrema(f,{g1,g2},{x,y,z},'s');
```
$$\{\min(- \ 6 \ - \ 4 \ 2^{1/2} \ , \ - \ 6 + 4 \ 2^{1/2} \ ), \ \max(- \ 6 \ - \ 4 \ 2^{1/2} \ , \ - \ 6 + 4 \ 2^{1/2} \ )\}$$

```
> s;
```
$$\{\{z = - \ 4 \ 2^{1/2} \ + 10, \ x = 2 \ 2^{1/2} \ , \ y = 2 \ 2^{1/2} \ \},$$

$$\{y = - \ 8^{1/2} \ , \ z = 2 \ 8^{1/2} \ + 10, \ x = - \ 8^{1/2} \ \}\}$$

SEE ALSO: `minimize`, `maximize`

2.2.19   factors – factor a multivariate polynomial

Calling Sequence:

    `factors(a);` or `factors(a, K);`

Parameters:

    a – a multivariate polynomial

    K – (optional) algebraic number field extension

Synopsis:

- The `factors` function computes the factorization of a polynomial over an algebraic number field. Note the multivariate case is only implemented over the rationals.

- Unlike the `factor` function where the input is any expression and the output is a Maple sum of products, the input to the `factors` function must be a polynomial, and the output is a data-structure more suitable for programming purposes.

- The factorization is returned in the form [u, [[f[1], m[1]], .., [f[n], m[n]]]] where a = u*f[1]^m[1] ... f[n]^m[n] where each f[k] (the factor) is a unit normal irreducible polynomial and each m[k] (its multiplicity) is a positive integer.

- The call factors(a) factors over the field implied by the coefficients present: thus, if all the coefficients are rational, then the polynomial is factored over the rationals.

- The call factors(a, K) factors the polynomial a over the algebraic number field defined by K. K must be a single RootOf, a list or set of RootOf's, a single radical, or a list or set of radicals.

- This function should be defined by the command readlib(factors) before it is used.

Examples:

```
> readlib(factors):
> factors( 3*x^2+6*x+3 );
```
$$[3, [[x + 1, 2]]]$$
```
> factors( x^4-4 );
```
$$[1, [[x^2 - 2, 1], [x^2 + 2, 1]]]$$
```
> alias(alpha=RootOf(x^2-2)):
> alias(beta=RootOf(x^2+2)):
> factors( x^4-4, alpha );
```
$$[1, [[x - alpha, 1], [x^2 + 2, 1], [x + alpha, 1]]]$$
```
> factors( x^4-4, beta );
```
$$[1, [[x^2 - 2, 1], [x + beta, 1], [x - beta, 1]]]$$
```
> factors( x^4-4, {alpha,beta} );
```
$$[1, [[x - alpha, 1], [x + beta, 1], [x - beta, 1], [x + alpha, 1]]]$$
```
> factors( x^4-4, {I,sqrt(2)} );
```
$$[1, [[I\ 2^{1/2} + x, 1], [x - I\ 2^{1/2}, 1], [x - 2^{1/2}, 1], [x + 2^{1/2}, 1]]]$$

SEE ALSO: Factors, ifactors, sqrfree, factor, roots

2.2.20   FFT – fast Fourier transform
iFFT – inverse fast Fourier transform

Calling Sequence:

FFT(m, x, y)

evalhf(FFT(m, var(x), var(y)))

iFFT(m, x, y)

evalhf(iFFT(m, var(x), var(y)))

Parameters:

m –   a non-negative integer

x, y –   arrays of floats indexed from 1 to 2^m

Synopsis:

- The procedures FFT and iFFT are for computing in place the fast Fourier transform and the inverse fast Fourier transform of a complex sequence of length 2^m.

- The first argument m should be a non-negative integer and the second and third arguments x and y should be arrays of floats indexed from 1 to 2^m . The array x contains the real part of the complex sequence on input and contains the real part of the fast Fourier transform on output. The array y contains the imaginary part of the complex sequence on input and contains the imaginary part of the fast Fourier transform on output. Both procedures return 2^m, the number of points in the complex sequence.

- These procedures may be invoked with **evalhf**, which uses the hardware floating point number system.

- You must use **readlib(FFT)** before invoking FFT or iFFT .

- Reference: *Digital Signal Processing*, by Alan V. Oppenheim and Ronald W. Schafer, Prentice-Hall, Englewood Cliffs, N.J., 1975, Fig. P6.5, page 332 .

Examples:

```
> readlib(FFT):
  x := array([7.,5.,6.,9.]):
  y := array([0,0,0,0]):
  FFT(2,x,y);
```
$\longrightarrow$ 4

```
> print(x);
```
$\longrightarrow$ [27., 1., -1., 1.]

```
> print(y);
```
$\longrightarrow$ [0, 4., 0, -4.]

```
> iFFT(2,x,y);
```
$\longrightarrow$ 4

```
> print(x);
```
$\longrightarrow$ [7.000000000, 5.000000000,

6.000000000, 9.000000000]

```
> print(y);
```
$\longrightarrow$ [0, 0, 0, 0]

SEE ALSO: laplace, mellin, int

2.2.21   **finance – amount, interest, payment, or periods**
amortization – amortization schedule
blackscholes – present value of a call option

Calling Sequence:

```
finance(amount=a,interest=i,payment=p); or
finance(amount=a,payment=p,periods=n); etc.
amortization(a,i,p)
blackscholes(EX,t,P,s,rf); or blackscholes(EX,t,P,s,rf,´hr´);
```

Parameters:

a, i, p, n	–	any float or rational
Ex, t, P, s, rf	–	any float or rational
hr	–	(optional) unevaluated name

Synopsis:

- These procedures must be defined by the statement `readlib(finance):` before invoking any of them.

- The procedure `finance` expects three of the following four equations as arguments: {`amount=a`, `interest=i`, `payment=p`, `periods=n`} and it solves for the fourth equation. The variables [a,i,p,n] represent the amount of a loan, the interest rate per period, the payment per period, and the number of periods, respectively.

- The procedure `amortization` computes an amortization schedule for a loan of amount **a** , interest rate per period **i**, and periodic payment **p** . The schedule is returned as an array indexed from 0 to the number of periods, where each entry in the array is a list containing the period number, the periodic payment, the interest, the principal reduction, and the new balance. The last payment may be less than the other payments.

- The procedure `blackscoles` uses the formula derived by Black and Scholes to find the present value of a call option and the option delta or hedge ratio. The arguments are: **Ex** the exercise price of the option, **t** the time to the exercise date, **P** the price of the stock now, **s** the standard deviation per period of the continuously compounded rate of return on the stock, **rf** the continuously compounded risk-free rate of interest, and **hr** an optional name, which will be assigned the hedge ratio or option delta if it is present.

- Reference: Richard A. Brealey and Stewart C. Myers, *Principles of Corporate Finance (Third Edition)* McGraw-Hill, New York, 1988.

Examples:

```
> readlib(finance):
> finance(amount=1000.00,interest=.1375,periods=6);
                    payment = 255.3993222
> amortization(1000.00,.1375,255.40);
                array(0 .. 6,, [
                    0 = [0, 0, 0, 0, 1000.00]
                    1 = [1, 255.40, 137.50, 117.90, 882.10]
                    2 = [2, 255.40, 121.29, 134.11, 747.99]
                    3 = [3, 255.40, 102.85, 152.55, 595.44]
                    4 = [4, 255.40, 81.87, 173.53, 421.91]
                    5 = [5, 255.40, 58.01, 197.39, 224.52]
                    6 = [6, 255.39, 30.87, 224.52, 0]
                ])
> blackscholes(160.00,4,140.00,.4,ln(1+.1247),´hr´);
                    60.35018257
```

```
> hr;
```
                                        .7940821351
_____

2.2.22   fixdiv – compute the fixed divisor of a polynomial

Calling Sequence:

> `fixdiv(a, x)`

Parameters:

> a, x – a is polynomial in x over the integers

Synopsis:

- Compute the fixed divisor of $a(x)$ in $\mathtt{Z}[x]$. The fixed divisor is the largest integer that divides $a(\mathtt{k})$ for all $\mathtt{k}$ in $\mathtt{Z}$. For primitive polynomials, the fixed divisor is no greater than $\mathtt{n}!$ where $\mathtt{n} =$ `degree(a, x)`.

- The probability that the fixed divisor is greater than one for random polynomials of degree $\mathtt{n}$ is approximately 0.278 as $\mathtt{n}$ goes to infinity.

- This function should be defined by the command `readlib(fixdiv)` before it is used.

_____

Examples:
```
> readlib(fixdiv):
  fixdiv(x*(x+1)*(x+2)*(x+3),x);                    ⟶            24
```
_____

2.2.23   forget – remove an entry or entries from a remember table

Calling Sequence:

> `forget(f, a, b, c, ...)`
> `forget(f)`

Parameters:

> f –   any name assigned to a Maple procedure
> $a, b, c, \dots$ –   specific argument sequence for the function f

Synopsis:

- The statement `forget(f)` will remove all entries from the remember table of f, while `forget(f, a, b, c)` will remove only the entry corresponding to $f(a, b, c)$.

- Maple procedures with option remember automatically remember every value computed by inserting an entry in the remember table.

- Users may assign $f(a, b, c)$ a specific value as in $f(a, b, c)$:= 3;. This assigned value is kept in the remember table of the procedure f. Once assigned a value in this manner any future reference to $f(a, b, c)$ will return 3, without recomputing the function value. To allow f to recompute its function value, assigned values must be removed from the remember table.

- You need not use forget if your goal is simply to reassign $f(a, b, c)$ a different value. For example, just use $f(a, b, c)$:= **5**;.

- This function must be loaded via `readlib(forget)`:

Examples:

```
> readlib(forget):
  f(x),f(y);                    ⟶           f(x), f(y)
> f(x) := 456:
  f(x),f(y);                    ⟶           456, f(y)
> forget(f,x);
  f(x),f(y);                    ⟶           f(x), f(y)
```

SEE ALSO: `options`, remember, procedures, totorder

2.2.24   freeze, thaw – replace an expression by a name

Calling Sequence:

> freeze(*expr*)
> thaw(*var*)

Parameters:

expr – the expression to be "frozen"
var   – the "frozen" variable

Synopsis:

- freeze replaces its argument, *expr* (not a name and not a rational number), with a name of the form _R0, _R1, ...

- To return the original expression the thaw function must be used.

- These functions should be defined by the command `readlib(freeze)` before they are used.

Examples:

```
> readlib(freeze):
  freeze(x+y);                  ⟶           _R0
> thaw(_R0);                    ⟶           x + y
```

2.2.25   GF – Galois Field Package

Calling Sequence:

> GF(p, k)
> GF(p, k, a)

Parameters:

  p – prime integer

  k – positive integer

  a – irreducible polynomial of degree k over the integers mod p

Synopsis:

- The GF function returns a table G of functions and constants for doing arithmetic in the finite field $GF(p^k)$, a Galois Field with p^k elements. The field $GF(p^k)$ is defined by the field extension $GF(p)[x]/(a)$ where a is an irreducible polynomial of degree k over the integers mod p.

- If a is not specified, an irreducible polynomial of degree k over the integers mod p is chosen at random. It can be accessed as the constant G[extension]. The elements of $GF(p^k)$ are represented using the modp1 representation.

- The functions G[input] and G[output] convert from an integer in the range $0..p^k-1$ to the corresponding polynomial and back. Alternatively, G[ConvertIn] and G[ConvertOut] convert an element from $GF(p^k)$ to a Maple sum of products, a univariate polynomial where the variable used is that given in the argument a. Otherwise the name `?` is used.

- Arithmetic in the field is defined by the following functions:

$$G[`+`], \; G[`-`], \; G[`*`], \; G[`\hat{\ }`], \; G[inverse], \; G[`/`]$$

- The additive and multiplicative identities are given by G[0] and G[1].

- The operations G[trace], G[norm], and G[order] compute the trace, norm and multiplicative order of an element from $GF(p^k)$ respectively.

- The operation G[random] returns a random element from $GF(p^k)$.

- The operation G[PrimitiveElement] generates a primitive element at random.

- The operation G[isPrimitiveElement] tests whether an element from $GF(p^k)$ is a primitive element, being a generator for the multiplicative group $GF(p^k) - \{0\}$.

- This function should be defined by the command readlib(GF) before it is used.

Examples:

```
> readlib(GF):
  G16 := GF(2,4,alpha^4+alpha+1):
  a := G16[ConvertIn](alpha);
```
$\longrightarrow$ $a := 10000$

```
> G16[`*`](a,a);
```
$\longrightarrow$ 100000000

```
> G16[`^`](a,4);
```
$\longrightarrow$ 10001

```
> x := G16[`^`](a,8);
```
$\longrightarrow$ $x := 100000001$

```
> G16[output](x);
```
$\longrightarrow$ 5

```
> G16[ConvertOut](x);
```
$\longrightarrow$ 2
          $alpha + 1$

```
> G16[isPrimitiveElement](a);                  ⟶                true
> x := G16[`^`](a,-1);                          ⟶          x := 1000000000001
> G16[`*`](a,x);                                 ⟶                 1
```

SEE ALSO: `modp1`, `mod`

2.2.26   heap – Priority Queue Data Structure

Calling Sequences:

> heap[new] (f)
> heap[new] (f, x_1, ..., x_n)
> heap[insert] (x, h)
> heap[extract] (h)
> heap[empty] (h)
> heap[max] (h)
> heap[size] (h)

Parameters:

f –   a Boolean-valued function

h –   a heap

x, x_i –   values to be inserted into the heap

Synopsis:

- The call **heap[new]** (f) returns an empty heap where f is a Boolean-valued function which specifies a total ordering for the elements to be inserted into the heap.

- The call **heap[new]** (f, x_1, ..., x_n) returns a heap with the values x_1, ..., x_n initially inserted into the heap.

- The call **heap[insert]** (x, h) inserts the element x into the heap h while **heap[extract]** (h) returns (and removes) the maximum element from the heap (according to the ordering defined by f).

- The call **heap[empty]** (h) returns **true** if the heap h is empty, and **false** if it is not empty.

- Additionally, **heap[max]** (h) returns the maximum element in the heap (but does not remove it) and **heap[size]** (h) returns the number of elements in the heap.

- These functions should be defined by the command **readlib(heap)** before they are used.

Examples:

```
> readlib(heap):
  h := heap[new](lexorder, greg, tony,
  bruno, michael):
  heap[insert](stefan, h);                      ⟶              stefan
```

```
> heap[size](h);                                    5

> heap[max](h);                                    tony

> while not heap[empty](h) do heap[extract]
  (h) od;                                          tony

                                                   stefan

                                                   michael

                                                   greg

                                                   bruno
```

2.2.27 history – maintain a history of all values computed

Calling Sequence:

> history()

Synopsis:

- The **history** function is used to maintain a history of all results computed within a Maple session, beyond what is available with the Maple quotes ", "", and """. A separate timing facility is also available.

- After loading the **history** function, type **history();** to initiate the history mechanism. The user will be prompted for input by the prompt O1 := and for successive statements by the prompts O2 :=, O3 :=, and so on. Any Maple statement entered is evaluated and displayed normally. The result is assigned to the global variables O1, O2, ... which may be referred to later by the user thus providing a history mechanism.

- The previous value may be referred to by using ". However "" and """ are no longer available under history.

- By typing **off;** the history session is terminated and control is passed back to the normal session.

- By typing **timing(f);** the time to execute the expression **f** will be displayed after the normal output.

- By typing **clear;** the values of O1, O2, ... are cleared.

- Caveats: history is an ordinary Maple function, programmed in Maple. As such it cannot trap interrupts. Hence, if you interrupt a calculation you will be thrown back to the main Maple session. The values of O1, O2, ... are still available.

- If the history function does not do exactly what you want, you may want to consider writing your own version. For example, you may want to use your own print routine. The code for history may be a good starting point. See **help(print);** for how to print the code for Maple library routines.

- This function should be defined by the command **readlib(history)** before it is used.

SEE ALSO: time, profile, showtime, print

2.2.28   hypergeom – generalized hypergeometric function

Calling Sequence:

hypergeom([n1, n2, ...], [d1, d2, ...], z)

Parameters:

[n1, n2, ...] – list of numerator coefficients

[d1, d2, ...] – list of denominator coefficients

Synopsis:

- The function hypergeom(n, d, z) is the generalized hypergeometric function F(n, d, z). This function is also known as Barnes's extended hypergeometric function. If there are j coefficients in n, and k coefficients in d, this function is also known as $_jFk$.

- The definition of F(n, d, z) is

```
sum ( (product( GAMMA( n[i]+k ) / GAMMA( n[i] ), i=1..j)*z^k )
  / (product( GAMMA( d[i]+k ) / GAMMA (d[i] ), i=1..m)*k! ), k=0..infinity)
```

where j is the number of terms in the list [n1, n2, ...] and m is the number of terms in the list [d1, d2, ...].

- If n[i] = -m, a non-positive integer, the series stops after m terms.

- This function should be defined by the command readlib(hypergeom) before it is used.

Examples:

```
> readlib(hypergeom):
  hypergeom( [],[],z );                      ⟶              exp(z)

> hypergeom( [a],[],z );                      ⟶                     (- a)
                                                            (1 - z)

> hypergeom( [1,2],[2,3],z );                 ⟶              exp(z) - 1 - z
                                                          2 -------------
                                                                 2
                                                                 z
```

SEE ALSO: convert

2.2.29   ifactors – integer factorization

Calling Sequence:

ifactors(n)

Parameters:

n – any integer

Synopsis:

- The `ifactors` function returns the complete integer factorization of the integer n.

- The result is returned as in the form `[u, [[p[1], e[1]], ... ,[p[m], e[m]]]]` where n = `u*p[1]^e[1]* ... *p[m]^e[m]`, `p[i]` is a prime integer, `e[i]` is its exponent (multiplicity) and `u` is the sign of n.

- This function should be defined by the command `readlib(ifactors)` before it is used.

Examples:

```
> readlib(ifactors):
  ifactors( 61 );                    ⟶          [1, [[61, 1]]]
> ifactors( -120 );                  ⟶      [-1, [[2, 3], [3, 1], [5, 1]]]
> ifactors( 0 );                     ⟶            [0, []]
> ifactors( 1 );                     ⟶            [1, []]
```

SEE ALSO: `ifactor`, `factor`, `factors`, `isqrfree`

2.2.30   invlaplace – inverse Laplace transform

Calling Sequence:

invlaplace(*expr*, *s*, *t*)

Parameters:

expr	–	expression to be transformed
s	–	variable *expr* is transformed with respect to
t	–	variable to replace *s* in transformed expression

Synopsis:

- The function `invlaplace` applies the inverse Laplace transformation to *expr* with respect to *s*.

- Expressions which are sums of rational functions of polynomials can be transformed. Expressions which are transforms of sums of terms involving exponentials, polynomials and Bessel functions with linear arguments can also be transformed.

- Both `invlaplace` and `laplace` recognize the Dirac-delta (or unit-impulse) function as `Dirac(t)`.

- The global variable `_U` is used as an integration variable if the answer contains convolution integrals.

- This function should be defined by the command `readlib(laplace)` before it is used.

Examples:

```
> readlib(laplace):
> invlaplace(1/(s-a)+1/(s^2-b^2)+1, s, t);
                              sinh(b t)
              exp(a t) + --------- + Dirac(t)
                                b
```

```
> invlaplace(1/(s^3-8)+laplace(y(t), t, s), s, t);
```
$$1/12 \ \exp(2\ t) - 1/12 \ \exp(-\ t) \ \sin(3^{1/2}\ t)\ 3^{1/2} - 1/12 \ \exp(-\ t) \ \cos(3^{1/2}\ t)$$
$$+\ y(t)$$

```
> invlaplace(s^2/(s^2+a^2)^(3/2), s, t);
```
$$-\ t \ \text{BesselJ}(1,\ a\ t)\ a + \text{BesselJ}(0,\ a\ t)$$

```
> invlaplace(s/(s-1)*laplace(F(t), t, s), s, t);
```
$$\int_0^t (\exp(\_U) + \text{Dirac}(\_U))\ F(t - \_U)\ d\_U$$

SEE ALSO: `laplace`, `ztrans`, `dsolve`

2.2.31 invztrans – inverse Z transform

Calling Sequence:

   `invztrans( f, z, n);`

Parameters:

   f – expression

   z – name

   n – name

Synopsis:

- `invztrans` finds the inverse Z transformation of $f(z)$ with respect to n.

- `invztrans` can find the inverse Z transform of rational polynomials and a few special functions.

- This function must be loaded by entering **`readlib(ztrans)`**.

Examples:

```
> readlib(ztrans):
  invztrans(z/(z-2),z,n);                    ⟶              2^n
```

```
> invztrans((6*z^2-13*z)/(z^2-5*z+6),z,t)
  ;                                          ⟶           2^t + 5 3^t
```

```
> invztrans(a*exp(3/z),z,n);                 ⟶            a 3^n
                                                          ----
                                                           n!
```

```
> invztrans(ztrans(f(n),n,w),w,n);           ⟶            f(n)
```

SEE ALSO: `ztrans`, `laplace`, `mellin`, `int`

2.2.32   iratrecon – rational reconstruction

Calling Sequence:
    `iratrecon(u, m, N, D, 'n', 'd')`

Parameters:
  u, m –   integers
  N, D –   positive integers
  n, d –   variables

Synopsis:
- The purpose of this routine is to reconstruct a signed rational number from its image u mod m. Given u **mod** m and positive integers N and D, if **iratrecon** returns true then n and d will be assigned integers such that

      `n/d == u mod m   where   abs(n) <= N, abs(d) <= D`

  Otherwise **iratrecon** returns false. The reconstruction will be unique if the following condition holds:

      `2 N D < m`

- This function should be defined by the command `readlib(iratrecon)` before it is used.

Examples:
```
> readlib(iratrecon):
> m := 11;
                              m := 11

> u := 1/2 mod m;
                               u := 6

> iratrecon(u,m,2,2,'n','d');
                               true

> n/d;
                               1/2

> U := 0,1,2,3,4,5,6,7,8,9,10:
> f := proc(x,N,D) local n,d;
>     if iratrecon(x,m,N,D,'n','d') then n/d else false fi
> end:
> seq( f(u,2,2), u=U );
            0, 1, 2, false, false, -1/2, 1/2, false, false, -2, -1

> seq( f(u,3,2), u=U );
              0, 1, 2, 3, -3/2, -1/2, 1/2, 3/2, -3, -2, -1

> seq( f(u,2,3), u=U );
              0, 1, 2, -2/3, 1/3, -1/2, 1/2, -1/3, 2/3, -2, -1
```

SEE ALSO: `mod`

2.2.33   iscont – test continuity on an interval

Calling Sequence:
>   iscont(*expr*, *x* = *a* .. *b*)
>   iscont(*expr*, *x* = *a* .. *b*, ´*closed*´)

Parameters:

expr	–	an algebraic expression
x	–	variable name
a..b	–	a real interval
´*closed*´	–	(optional) indicates that endpoints should be checked

Synopsis:

- The `iscont` function returns true if the expression is continuous on the interval (thus having no poles), or **false** if the expression is not continuous. If `iscont` cannot determine the result it returns **FAIL**.

- The `iscont` function assumes that any symbols in the expression are real.

- By default, the interval is considered to be an open interval. If the optional third argument ´*closed*´ is specified then continuity at the endpoints is checked; specifically under this option, `limit(`*expr*`,`*x=a*`,right)` and `limit(`*expr*`,`*x=b*`,left)` must exist and be finite.

- The endpoints of the interval must be either real constants or **infinity** or **-infinity**.

- The endpoints of the interval must be ordered with the first value less than the second; otherwise `iscont` will return **FAIL**.

- This function should be defined by the command `readlib(iscont)` before it is used.

Examples:
```
> readlib(iscont):
  iscont( 1/x, x=1..2 );                    ⟶              true
> iscont( 1/x, x=-1..1 );                   ⟶              false
> iscont( 1/x, x=0..1 );                    ⟶              true
> iscont( 1/x, x=0..1, ´closed´ );          ⟶              false
> iscont( 1/(x+a), x=0..1);                 ⟶              FAIL
```

2.2.34   isolate – isolate a subexpression to left side of an equation

Calling Sequence:
>   isolate(*eqn*, *expr*)
>   isolate(*eqn*, *expr*, *iter*)

Parameters:

    eqn   –   an equation or an algebraic expression

    expr –   any algebraic expression

    iter   –   (optional) a positive integer

Synopsis:

- The procedure `isolate` attempts to isolate the second argument *expr* in the first argument *eqn* and solves *eqn* for *expr*.

- If the first argument is not an equation, then the equation *eqn* = 0 is assumed.

- The optional third argument *iter* controls the maximum number of transformation steps that `isolate` will perform.

- This function should be defined by the command `readlib(isolate)` before it is used.

Examples:
```
> readlib(isolate):
  isolate(4*x*sin(x)=3,sin(x));
```
$\longrightarrow$
$$\sin(x) = \frac{3}{4\,x}$$

```
> isolate(x^2-3*x-5,x^2);
```
$\longrightarrow$
$$x^2 = 3\,x + 5$$

SEE ALSO: `student[isolate]`, `solve`

2.2.35   isqrfree – integer square free factorization

Calling Sequence:

    `isqrfree(`*n*`)`

Parameters:

    n   –   any integer

Synopsis:

- The `isqrfree` function returns the square free integer factorization of the integer *n*.

- The function returns the result in the form `[u,[[p[1],e[1]],...,[p[m],e[m]]]]` where `p[i]` is a positive factor of *n*, `e[i]` is its multiplicity, `gcd(p[i], p[j]) = 1` for all i $<>$ j, and u is the sign of *n*. The square free factorization of *n* is: `n = u * p[1]^e[1] * ... * p[m]^e[m]`.

- This function should be defined by the command `readlib(isqrfree)` before it is used.

Examples:
```
> readlib(isqrfree):
  isqrfree(61);
```
$\longrightarrow$
    `[1, [[61, 1]]]`

```
> isqrfree(180);
```
$\longrightarrow$
    `[1, [[5, 1], [6, 2]]]`

| > `ifactors(-120);` | $\longrightarrow$ | `[-1, [[2, 3], [3, 1], [5, 1]]]` |

SEE ALSO: `ifactor`, `ifactors`, `factor`, `factors`, `sqrfree`, `Sqrfree`, `Factor`

2.2.36   isqrt – integer square root
iroot – integer nth root

Calling Sequence:
>   `isqrt(x)`
>   `iroot(x, n)`

Parameters:
> x – an integer
> n – an integer

Synopsis:

- The `isqrt` function computes an integer approximation to the square root of x. The approximation is exact for perfect squares, and the error is less than 1 otherwise. Note: if $x < 0$, `isqrt(x)` returns 0.

- The `iroot` function computes an integer approximation to the nth root of x. The approximation is exact for perfect powers, and the error is less than 1 otherwise. Note: if $x < 0$, and n is even, `iroot(x, n)` returns 0.

- The `iroot` function should be defined by the command `readlib(iroot)` before it is used.

Examples:

| > `isqrt(10);` | $\longrightarrow$ | 3 |
| > `readlib(iroot):`
  `iroot(100, 3);` | $\longrightarrow$ | 5 |

SEE ALSO: `sqrt`, `issqr`, `psqrt`, `numtheory[msqrt]`, `numtheory[mroot]`

2.2.37   issqr – test if an integer is a perfect square

Calling Sequence:
>   `issqr(n)`

Parameters:
> n – a positive integer

Synopsis:

- The function `issqr` returns `true` if n is the square of an integer.

- This function should be defined by the command `readlib(issqr)` before it is used.

Examples:
```
> readlib(issqr):
  issqr(9);                          ⟶            true
> issqr(10);                         ⟶            false
```

SEE ALSO: `isqrt, sqrt, psqrt, numtheory[msqrt]`

2.2.38   lattice – find a reduced basis of a lattice

Calling Sequence:

> `lattice(`*lvect*`)`

Parameters:

> *lvect*   –   an expression sequence of vectors or lists

Synopsis:

- The `lattice` function finds a reduced basis (in the sense of Lovasz) of the lattice specified by the vectors of *lvect* using the LLL algorithm.

- This function requires that the dimension of the subspace generated by the vectors equals the number of vectors.

- This function should be defined by the command `readlib(lattice)` before it is used.

Examples:
```
> readlib(lattice):
> lattice([1,2,3], [2,1,6]);
                       [0, -3, 0], [1, -1, 3]
> lattice([1/5,2,8],[-2,3/8,4],[1/4,-3,9]);
             [-2, 3/8, 4], [21/5, 5/4, 0], [1/20, -5, 1]
```

SEE ALSO: `minpoly`

2.2.39   minimize – compute the minimum
maximize – compute the maximum

Calling Sequence:

> `minimize(`*expr*`)`
> `minimize(`*expr*`, `*vars*`)`
> `maximize(`*expr*`)`
> `maximize(`*expr*`, `*vars*`)`

Parameters:

> *expr*   –   any algebraic expression

    vars   –   (optional) a set of variables

Synopsis:

- The function `minimize` returns the minimum value of *expr* or an expression sequence with all the values that are candidates for minima.

- If *vars* is specified, then *expr* is minimized with respect to the variables given; otherwise *expr* is minimized with respect to all of its indeterminates.

- When the solution is in terms of algebraic numbers of degree larger than two, `minimize` will return a floating point approximation to the minima.

- This function finds unconstrained minima over the real numbers, i.e. for any values of *vars* which are real (including the limits at infinity).

- The function `maximize` is defined to compute the negative of the result of `minimize` applied to the negated *expr*.

- These functions should be defined by the command `readlib(minimize)` before they are used.

Examples:
```
> readlib(minimize):
  minimize(x^2+y^2+3);                          ⟶              3
> minimize(sin(x));                             ⟶             -1
> minimize((x^2-y^2)/(x^2+y^2));                ⟶             -1
> maximize((x^2-y^2)/(x^2+y^2));                ⟶              1
> minimize(x^2 + y^2, {x});                     ⟶              2
                                                               y
> minimize(x^2 + y^2, {x, y});                  ⟶              0
```

SEE ALSO: `simplex`, `student`

2.2.40   minpoly – find minimum polynomial with an approximate root

Calling Sequence:
```
    minpoly( r, n );
    minpoly( r, n, acc );
```

Parameters:

    r   –   the approximate root

    n   –   the degree of the polynomial

    acc   –   desired accuracy of the approximation

Synopsis:

- The `minpoly` function uses the lattice algorithm to find a polynomial of degree **n** (or less) with small integer coefficients which has the given approximation **r** of an algebraic number as one of its roots.

- If a third argument is specified, then the value `acc*abs(f(r))` is given the same weight as the coefficients in determining the polynomial. The default value for `acc` is `10^(Digits-2)`.

- This function must be loaded via `readlib(lattice)`:

Examples:
```
> readlib(lattice):
  minpoly( 1.234, 3 );
```
$$\longrightarrow \quad 109 - 61\ \_X + 5\ \_X^2 - 22\ \_X^3$$

SEE ALSO: `lattice`, `linalg[minpoly]`

2.2.41   modpol – expression evaluation in a quotient field

Calling Sequence:
     `modpol(a, b, x, p)`

Parameters:
   a – a rational expression over the rational numbers

   b – a polynomial over the rational numbers

   x – a name

   p – a prime number

Synopsis:
- This procedure evaluates $a(x)$ over $Zp[x]/(b(x))$.

- This function should be defined by the command `readlib(modpol)` before it is used.

Examples:
```
> readlib(modpol):
  modpol(x^2-1,x,x,7);
> modpol(x^8+2*x,x^3-9*x+2,x,3);
```
$\longrightarrow$ `Error, (in modpol) invalid arguments`
$\longrightarrow$ `Error, (in modpol) invalid arguments`

SEE ALSO: `evala`, `mod`, `Powmod`

2.2.42   MOLS – mutually orthogonal Latin squares

Calling Sequence:
     `MOLS(p, m, n)`

Parameters:
   p – a prime number

   m – a positive integer

   n – a positive integer less than p^m

Synopsis:
- This function returns a list of n mutually orthogonal Latin squares of size $p\hat{\ }m$.

- Two Latin squares, **A** and **B**, are mutually orthogonal if the set of ordered pairs $\{$ (A[i,j], B[i,j]), $0{\le}$i, j${\le}p\hat{\ }m\}$ has no repeated elements.

- This function should be defined by the command **readlib(MOLS)** before it is used.

Examples:

```
> readlib(MOLS):
  MOLS(3, 1, 2);
```

$\longrightarrow$

$$
\begin{bmatrix} 0 & 1 & 2 \\ 1 & 2 & 0 \\ 2 & 0 & 1 \end{bmatrix}, \begin{bmatrix} 0 & 1 & 2 \\ 2 & 0 & 1 \\ 1 & 2 & 0 \end{bmatrix}
$$

2.2.43   mtaylor – multivariate Taylor series expansion

Calling Sequence:

mtaylor(f, v)

mtaylor(f, v, n)

mtaylor(f, v, n, w)

Parameters:

 f – an algebraic expression

 v – a list or set of names or equations

 n – (optional) non-negative integer

 w – (optional) list of positive integers

Synopsis:
- The **mtaylor** function computes a multivariate Taylor series expansion of the input expression f, with respect to the variables v, to order n, using the variable weights w.

- The result type has changed in Maple V to be simply a polynomial in the variables with no order term.

- The variables v can be a list or set of names or equations. If v[i] is an equation, then the left-hand side of v[i] is the variable, and the right-hand side is the point of expansion. If v[i] is a name, then v[i] = 0 is assumed as the point of expansion.

- If the third argument n is present then it specifies the "truncation order" of the series. The concept of "truncation order" used is "total degree" in the variables. If n is not present, the truncation order used is the value of the global variable **Order**, which is 6 by default.

- If the fourth argument w is present it specifies the variable weights to be used (by default all 1). A weight of 2 will halve the order in the corresponding variable to which the series is computed.

- Note: `mtaylor` restricts its domain to "pure" Taylor series, those series with non-negative powers in the variables.

- This function should be defined by the command `readlib(mtaylor)` before it is used.

Examples:
```
> readlib(mtaylor):
> mtaylor(sin(x^2+y^2), [x,y]);
```
$$x^2 + y^2$$
```
> mtaylor(sin(x^2+y^2), [x,y], 8);
```
$$x^2 + y^2 - 1/6\ x^6 - 1/2\ y^2\ x^4 - 1/2\ y^4\ x^2 - 1/6\ y^6$$
```
> mtaylor(sin(x^2+y^2), [x,y], 8, [2,1]);
```
$$y^2 + x^2 - 1/6\ y^6$$
```
> mtaylor(sin(x^2+y^2), [x=1,y], 3);
```
$$\sin(1) + 2\cos(1)\ (x - 1) + (- 2\sin(1) + \cos(1))\ (x - 1)^2 + \cos(1)\ y^2$$
```
> mtaylor(f(x,y), [x,y], 3);
```
$$f(0,\ 0) + D[1](f)(0,\ 0)\ x + D[2](f)(0,\ 0)\ y + 1/2\ D[1,\ 1](f)(0,\ 0)\ x^2$$
$$+ x\ D[1,\ 2](f)(0,\ 0)\ y + 1/2\ D[2,\ 2](f)(0,\ 0)\ y^2$$

SEE ALSO: `taylor`, `series`

2.2.44   oframe – orthonormal tetrads in the Ellis-MacCallum formalism

Calling Sequence:
> const(*c*)
> deriv(*a*, *lett*)
> e0(*a*) or e1(*a*) or e2(*a*) or e3(*a*)
> com01(*a*) or com02(*a*) or com03(*a*)
> com12(*a*) or com23(*a*) or com31(*a*)
> id(*eq*)
> clear(*a*)
> transfo()

Parameters:
> *c*   –   the name of a constant
> *a*   –   any algebraic expression
> *lett* –   one of e0,e1,e2,e3
> *eq*   –   an equation

Synopsis:

- This collection of procedures allows formal manipulation of the variables in the Ellis-MacCallum formalism. The notation follows MacCallum (1973, Cargese) as closely as possible. The lower and upper case omegas are presented as w and W respectively, whereas thetas have become H′s; also the shear components s11,s22,s33 have been replaced systematically by H1-H/3, H2-H/3, and H3-H/3. The identity H=H1+H2+H3 has not been inserted.

- The procedure const is used to define constants which are maintained in the global variable Constants. It adds its arguments to the global set Constants.

- The derivative operator deriv is the building block for the directional derivatives e1, e2, e3, e4.

- The commutator operators are com01, com02, com03, com12, com23, com31. They give the commutation relations; see the example below.

- The procedure id tests if its argument equation eq is an identity.

- The procedure clear will remove its argument from the remember tables of the directional derivative operators e0,e1,e2,e3.

- When using the orthonormal frame package it is customary to fix e0 in some invariant way and play with the spatial axes until one hits the ideal frame. The procedure transfo requires that the array rot:=array(1..3,1..3) define a spatial rotation after which the command transfo() will calculate new spin coefficients xyz_ as functions of old coefficients xyz.

- The global variables j1,j2,j3,j4,j5,j6,j7,j8,j9,j10,j11,j12,j13,j14,j15,j16 contain the Jacobi identities.

- The global variables ein00,ein01,ein02,ein03,ein11,ein12,ein13,ein22,ein23,ein33 contain the Einstein field equations.

- In order to use the oframe functions, you must first use readlib(oframe).

Examples:
```
> readlib(oframe):
> com01(a);
    e0(e1(a)) - e1(e0(a)) =

        u1 e0(a) - H1 e1(a) - (s12 - w3 - W3) e2(a) - (s13 + w2 + W2) e3(a)
```

SEE ALSO: np

2.2.45 optimize – common subexpression optimization

Calling Sequence:
optimize(*expr*)

Parameters:

> $expr$ – an expression, array, or list of equations

Synopsis:

- The input to **optimize** must be a single algebraic expression, an array of algebraic expressions (for example a matrix), an equation of type **name = algebraic**, or a list of equations of the form **name = algebraic** corresponding to a computation sequence.

- The output from the **optimize** function is a sequence of equations of the form **name = algebraic** representing an "optimized computation sequence" for the input expression. Each equation in the sequence corresponds to an assignment. The global variables **t.k** are used for temporary (local) registers in the computation sequence.

- The **optimize** function makes use of Maple´s **option remember** facility to identify common subexpressions in linear time and space. This means, however, that only those subexpressions which Maple has simplified to be identical are found. For example, the expression **x+y** is not recognized as being common to **x+y+z**. That is, **optimize** performs only "syntactic" optimizations.

- The routine `optimize/makeproc` can be used to generate a Maple procedure from a computation sequence. See **help(optimize,makeproc)**.

- This function should be defined by the command **readlib(optimize)** before it is used.

Examples:
```
> readlib(optimize):
  optimize(x^4);
```
$\longrightarrow$
$$t1 = x^2 ,\ t2 = t1^2$$

```
> optimize(2*x^3-y*x^4);
```
$\longrightarrow$
$$t1 = x^2 ,\ t4 = t1^2 ,\ t7 = 2\,t1\,x - y\,t4$$

SEE ALSO: `cost`, `fortran`, `optimize[makeproc]`

2.2.46   optimize/makeproc – procedure construction

Calling Sequence:

> `optimize/makeproc`$(e,\ o_1,\ o_2,\ \ldots)$

Parameters:

> e – a list of equations of type **name = algebraic**
>
> $o_1,\ o_2,\ \ldots$ – (optional) a sequence of options

Synopsis:

- The input to `optimize/makeproc` is a list of equations of the type **name = algebraic** corresponding to a computation sequence. The result returned is a Maple procedure that, when invoked, will execute the assignments represented by the equations.

- Unless otherwise specified, all left-hand sides of the equations will be declared local and other names appearing in the input will be assumed global. Which names are to be considered parameters, locals, and globals can be indicated by specifying optional arguments, given in any order, of the form `parameters=list(string), locals=list(string), globals=list(string)`.

- This function should be defined by the command `readlib(optimize)` before it is used.

Examples:
```
> readlib(optimize):
> s := [t1 = x^2, t2 = t1^2];
```
$$s := [t1 = x^2, \; t2 = t1^2]$$
```
> `optimize/makeproc`(s);
proc() local t1,t2; t1 := x^2; t2 := t1^2 end

> `optimize/makeproc`(s, parameters=[x]);
proc(x) local t1,t2; t1 := x^2; t2 := t1^2 end
```

SEE ALSO: `optimize, cost, fortran`

2.2.47   petrov – finds the Petrov classification of the Weyl tensor

Calling Sequence:

   petrov(*Psi*) or petrov(*Psi*, *flag*)
   nonzero()

Parameters:

   Psi – an array indexed from 0 to 4

   flag – an unassigned name

Synopsis:

- The procedure **petrov** is for determining the Petrov type of a given Weyl tensor in Newman-Penrose form. The Petrov classification of the Weyl tensor is a result of determining the multiplicity of the roots of the following quartic equation (Q) in **z**:

```
Q := Psi[0]*z^4 + 4*Psi[1]*z^3 + 6*Psi[2]*z^2 + 4*Psi[3]*z + Psi[4] = 0
```

where the `Psi` values are the Newman-Penrose Weyl tensor components. Given the coefficients `Psi[i]`, the **petrov** algorithm returns the Petrov type by using quartic theory. The different cases are outlined in the table below.

```
Petrov Type    Solutions of (Q)
-------------------------------------------------
    type I     four distinct roots
    type II    one root of multiplicity 2; other two roots distinct
```

type D	two distinct roots of multiplicity 2
type III	one root of multiplicity 3; other root distinct
type N	one root of multiplicity 4
type O	degenerate, all Psi[i] = 0

The algorithm uses the zero/non-zero nature of the coefficients as well as the zero/non-zero nature of some expressions in the coefficients to determine the Petrov type. Note that, in view of the structure of equation (Q), any common factor amongst the Psi[i] can be removed before the algorithm is called without changing the result.

- The Psi[i] coefficients for a given null tetrad may be calculated using the **debever** procedure. It is important that the coefficients be in a simplified form. This is especially true if one or more of the coefficients is identically zero. For added efficiency it is suggested that any common factor of all the coefficients be removed before calling **petrov** .

- The Petrov classification algorithm may be accessed with the command

petrov(Psi, flag);

where Psi is an array indexed from 0 to 4 which contains the Newman-Penrose Weyl tensor components, and where flag is optional. If flag is present then the Psi[i] will be simplified using an internal routine. This should be avoided by making sure the coefficients are in their most simplified form before invoking **petrov** . The procedure **petrov** will return the Petrov type that it determines.

- The procedure **petrov** calculates expressions and tests when some of them are identically zero . The zero test may fail if some identities are not applied. For example, the expression cos(x)^2+sin(x)^2-1 is not known to be zero by **petrov** . Therefore it is important to verify that any determined non-zero expressions are in fact non-zero. These values will be displayed with the following call:

nonzero();

If you would like to add a simplification to **petrov**, then define the procedure **simpl2** which accepts one argument and returns a simplified form of the argument. For example, the procedure

simpl2 := proc(a) simplify(a,trig) end;

could be used to simplify the expression cos(x)^2+sin(x)^2-1 to zero.

- Reference: F.W. Letniowski and R. G. McLenaghan, *An Improved Algorithm for Quartic Equation Classification and Petrov Classification*, General Relativity and Gravitation, Plenum Publishing Corporation, 1988, Volume 20, pages 463-483.

Examples:
```
> readlib(petrov):
> Psi := array(0..4, [1-x^2, 1, 0, x, 1-x]):
> petrov(Psi);
elapsed time:   Float(33,-3)   sec
```

The Petrov type is   type_I

$$\text{type\_I}$$

SEE ALSO: debever

2.2.48   poisson – Poisson series expansion

Calling Sequence:
>   poisson(f, v)
>   poisson(f, v, n)
>   poisson(f, v, n, w)

Parameters:
> f – an algebraic expression
> v – a list or set of names or equations
> n – (optional) a non-negative integer
> w – (optional) a list of positive integers

Synopsis:
- The `poisson` function generates a multivariate Taylor series expansion of the input expression f, with respect to the variables v, to order n, using the variable weights w. Trigonometric terms in the coefficients are combined.

- The parameters and result are the same type as for the `mtaylor` function. The only difference is that sines and cosines in the coefficients are combined into the Fourier canonical form.

- This function should be defined by the command `readlib(poisson)` before it is used.

Examples:
```
> readlib(poisson):
> f := sin(3*w+x)*cos(2*w-y);
                    f := sin(3 w + x) cos(2 w - y)

> poisson(f, [x,y], 3);
   1/2 sin(5 w) + 1/2 sin(w) + (1/2 cos(w) + 1/2 cos(5 w)) x

                                                            2
     + (1/2 cos(w) - 1/2 cos(5 w)) y + (- 1/4 sin(5 w) - 1/4 sin(w)) x

                        2
     + (- 1/4 sin(5 w) - 1/4 sin(w)) y  + (1/2 sin(5 w) - 1/2 sin(w)) y x
```

SEE ALSO: `combine/trig`, `mtaylor`

2.2.49   priqueue – Priority Queue Functions

Calling Sequence:
>   initialize(*pq*)
>   insert(*NewRecord*, *pq*)
>   extract(*pq*)

Parameters:

pq	–	the name of the priority queue
NewRecord	–	the structure to be inserted in the queue

Synopsis:

* The `initialize` function makes the necessary assignments (side effects) so that the queue *pq* is initialized with an empty queue. The queue *pq* is treated as an unbounded array of the inserted records. Additionally, *pq*[0] maintains the number of entries in the queue.

* The `insert` function adds an element to the queue. The queue is a maximum queue, that is, the maximum element is kept on the head of the queue. The criteria for ordering the queue is giving by *NewRecord*[1] and hence this value should exist and be comparable.

* The `extract` function returns the stored record with the highest value (in its first position) and deletes this record from the queue.

* Multiple priority queues are allowed.

* If inspection alone is desired, the maximum element of the queue is always at *pq*[1].

* These functions should be defined by the command `readlib(´priqueue´)` before they are used.

Examples:
```
> readlib(´priqueue´):
  initialize(DailyTemps);                    ⟶              0
> insert([14,Apr30], DailyTemps);
  insert([24,May4], DailyTemps);
  extract(DailyTemps);                       ⟶         [24, May4]
> extract(DailyTemps);                       ⟶         [14, Apr30]
```

2.2.50   procbody – create a "neutralized form" of a procedure

Calling Sequence:
>   procbody(*procedure*)

Parameters:

procedure	–	any Maple procedure

Synopsis:

* This routine creates a "neutralized form" of a Maple procedure. This "neutralized form" can be examined, or modified in ways the original procedure cannot be. This routine is probably more

useful in writing other routines to manipulate procedures than for interactive use. The function `procmake` creates a procedure from a "neutralized form".

- The neutralized form replaces statements, local variables, parameters, and some functions with "neutral" equivalents. These neutral forms are described in the help file for `procmake`.

- Do not evaluate the "neutral form" since this will almost certainly corrupt it due to such functions as `indets(&args[-1])` being invoked.

- This function should be defined by the command `readlib(procbody)` before it is used.

Examples:
```
> readlib(procbody):
> a := proc(x) if x<0 then -x^2 else x^2 fi end:
> procbody(a);
```

$$\&proc([x], [], [], \&if(\&args[1] < 0, - \&args[1]^2, \&args[1]^2))$$

```
> b := proc(x) local a; for a in x do print(a) od; RETURN(a); end:
> procbody(b);
    &proc([x], [a], [],
        &for(&local[1], &args[1], true, print(&expseq &local[1])) &statseq
            (&RETURN (&expseq &local[1]))                                    )
```

SEE ALSO: `procmake`

2.2.51 procmake – create a Maple procedure

Calling Sequence:
> procmake(*neutralform*)

Parameters:
> *neutralform* – "neutral form" of a procedure

Synopsis:
- This routine takes the "neutral form" of a procedure (such as that generated by `procbody`) and creates an executable procedure.

- There are "neutral forms" for statements, local variables, parameters, and several functions. The function `procmake` does not require a "neutral form" exactly equivalent to the internal representation. For example, `f(a,b)` is accepted as well as `f(`&expseq`(a,b))`. Similar examples exist for table references, sets, and lists.

- Do not fully evaluate the "neutral form" as this will probably invoke functions that are part of procedures, such as `indets(`&args`[-1])`. When entering such function calls, it is best to quote them, as with `´gcd(a,b)´`.

- The specific neutral forms are:

The expression `` `&proc` ``(A, B, C, D) is a procedure, where:

A is a list of arguments	B is a list of local variables
C is a list of options	D is the statement sequence

- Inside a procedure, arguments are represented by `` `&args` ``[n] where n is the position in A and local variables by `` `&local` ``[n] where n is the position by B. The three special names, **nargs**, **args**, and **procname**, are represented by `` `&args` ``[0], `` `&args` ``[-1], and `` `&args` ``[-2]. If there are no statements in the procedure, use `` `&expseq` ``().

- The expression `` `&statseq` ``(A, B, C, ...) is a statement sequence, where A, B, and C are statements.

- The expression `` `&expseq` ``(A, B, C) is an expression sequence, where A, B, and C are expressions.

- The empty expression sequence (NULL) is represented by `` `&expseq` ``().

- The expression `` `&:=` ``(A, B) is an assignment statement, where A is assigned the value of B.

- The expression `` `&if` ``(A[1], B[1], A[2], B[2], ... E) is an **if** statement, where A[i] are the conditions, B[i] are the matching statements, E is the **else** part (if there is an odd number of arguments).

- The expression `` `&for` ``(VAR, INIT, INCR, FIN, COND, STAT) is a **for-from** statement, where:

VAR is the variable	INIT is the initial value
INCR is the increment	FIN is the final value
COND is the looping condition	STAT is the statement sequence

- If any of VAR, FIN, or STAT is missing, use `` `&expseq` ``() instead. If there is no looping condition, use **true**. Note that **break** and **next** are represented by `` `&break` `` and `` `&next` ``.

- The expression `` `&for` ``(VAR, EXPR, COND, STAT) is a **for-in** statement, where:

VAR is the variable	EXPR is the expression
COND is the looping condition	STAT is the statement sequence

- If there is no statement sequence, use `` `&expseq` ``() instead. If there is no looping condition, use **true**. Note that break and next are represented by `` `&break` `` and `` `&next` ``.

- The strings `` `&done` ``, `` `&quit` ``, and `` `&stop` `` are all accepted for the **quit** statement.

- The strings `` `&"` ``, `` `&""` ``, and `` `&"""` `` represent ", "", and """.

- The expressions `` `&read` ``(A, B) and `` `&save` ``(B) are the **read** and **save** statements, where A is a name being saved and B is the file being saved to. The functions `` `&read` `` and `` `&save` `` accept any arrangement of arguments that **read** and **save** would accept.

- The functions `` `&ERROR` ``() and `` `&RETURN` ``() represent `ERROR()` and `RETURN()`.
- This function should be defined by the command `readlib(procmake)` before it is used.

Examples:
```
> readlib(procmake):
> b := `&proc`([x,y], [a], [remember],
>     `&statseq`(`&:=`(`&local`[1], `&args`[1] + `&args`[2]),
>         `&RETURN`(''gcd(`&args`[1],`&local`[1])'' ))):
> procmake(b);
proc(x,y) local a; options remember; a := x+y; RETURN(gcd(x,a)) end
```

SEE ALSO: `procbody`

2.2.52   profile – space & time profile of a procedure
        unprofile – space & time profile of a procedure
        showprofile – space & time profile of a procedure

Calling Sequence:
     profile(*f1*, *f2*, ...)
     unprofile(*f1*, *f2*, ...)
     showprofile()

Parameters:
     f1, f2, ...   –   the names of the procedures to be profiled

Synopsis:
- The functions `profile`, `unprofile` and `showprofile` are for profiling Maple procedures. To find out how much space and time some procedures are using, issue the command `profile`(*f1*, *f2*, ...), then perform whatever calculations you wish to have profiled. To view the results, use the command `showprofile()`. To return the procedures to their normal unprofiled state, use the command `unprofile`(*f1*, *f2*, ...).

- Both `profile` and `unprofile` return the names of the procedures that they were invoked with. The function `showprofile` returns `NULL` and prints the time and space results.

- These functions should be defined by the command `readlib(profile)` before they are used.

SEE ALSO: `time, showtime, trace`

2.2.53   psqrt – square root of a polynomial
        proot – nth root of a polynomial

Calling Sequence:
     psqrt(*p*)
     proot(*p*, *n*)

Parameters:

p – multivariate rational coefficients polynomial

n – integer

Synopsis:

- If p is a perfect square, `psqrt` returns a square root of p. Otherwise it returns the name _NOSQRT.

- If p is an nth power, `proot`(p, n) returns a nth root of p. Otherwise `proot` returns the name _NOROOT.

- These functions should be defined by the commands `readlib(psqrt)` and `readlib(proot)` before they are used.

Examples:

```
> readlib(psqrt):
  readlib(proot):
  psqrt(9);                        ⟶        3
> proot(9, 2);                     ⟶        3
> psqrt(x^2+2*x*y+y^2);            ⟶        y + x
> proot(x^3+3*x^2+3*x+1, 3);       ⟶        x + 1
> psqrt(x+y);                      ⟶        _NOSQRT
> proot(x+y, 2);                   ⟶        _NOROOT
```

SEE ALSO: `sqrt`, `isqrt`, `iroot`

2.2.54   realroot – isolating intervals for real roots of a polynomial

Calling Sequence:

realroot(*poly*, *widthgoal*)

Parameters:

poly   – a univariate polynomial with integer coefficients

widthgoal – (optional) maximal size of each isolating interval

Synopsis:

- The command `realroot`(*poly*, *widthgoal*) returns a list of isolating intervals for all real roots of the univariate polynomial *poly*. The width of the interval is less than or equal to the optional parameter *widthgoal*, a positive number. If *widthgoal* is omitted, the most convenient width is used for each interval returned.

- The intervals are specified as a list of two rational numbers. The list [a,a] represents the single point a. The list [a,b] with a < b represents the open real interval defined by a and b. Multiplicity information is not included.

- The function `realroot` uses Descartes´ rule of signs (see Loos and Collins in *Computer Algebra*, B. Buchberger, ed.).

- This function should be defined by the command `readlib(realroot)` before it is used.

Examples:
```
> readlib(realroot):
> realroot(x^8+5*x^7-4*x^6-20*x^5+4*x^4+20*x^3, 1/1000);
                  181  1449        1449    181
        [[0, 0], [---, ----], [- ----, - ---], [-5, -5]]
                  128  1024        1024    128
```

SEE ALSO: `roots`, `Roots`, `solve`, `fsolve`

2.2.55   recipoly – determine whether a polynomial is self-reciprocal

Calling Sequence:
> `recipoly(a, x)`
> `recipoly(a, x, ´p´)`

Parameters:
> a – an expression
>
> x – an indeterminate
>
> p – a name (optional)

Synopsis:
- Determine whether a is a "self-reciprocal" polynomial in x. This property holds if and only if `coeff(a,x,k) = coeff(a,x,d-k)` for all `k = 0..d`, where `d = degree(a,x)`.

- If `d` is even and if the optional second argument p is specified, p is assigned the polynomial P of degree `d/2` such that `x^(d/2)*P(x+1/x) = a`.

- Note that if `d` is odd, a being self-reciprocal implies a is divisible by x+1. In this case, if p is specified then the result computed is for `a/(x+1)`.

- This function should be defined by the command `readlib(recipoly)` before it is used.

Examples:
```
> readlib(recipoly):
  recipoly(x^4+x^3+x+1, x, ´p´);              ⟶              true

> p;                                          ⟶                    2
                                                         - 2 + x + x
```

2.2.56   relativity - General relativity procedures and packages

Synopsis:
- The following procedures may be used for general relativity calculations:

bianchi   A procedure for the determination of the Bianchi type of any 3-dimensional Lie algebra given its structure constants.

cartan   A collection of procedures for the computation of the connection coefficients and curvature components in any rigid moving frame using Cartan´s structure equations.

debever   A collection of procedures for the computation of the Newman-Penrose spin coefficients and curvature components using Debever´s formalism.

difforms   A package for performing calculations with differential forms.

np   A package for performing abstract calculations with the Newman-Penrose formalism.

oframe   A collection of procedures for performing abstract calculations with the Ellis-MacCallum orthonormal frame formalism.

petrov   A procedure for determining the Petrov type of any Weyl tensor.

tensor   A collection of procedures for the calculation of curvature tensors of any given metric in a coordinate basis.

SEE ALSO: `bianchi`, `cartan`, `debever`, `difforms`, `liesymm`, `np`, `oframe`, `petrov`, `tensor`

2.2.57   residue – compute the algebraic residue of an expression

Calling Sequence:
   residue(f, $x=a$)

Parameters:
   f – an arbitrary algebraic expression

   x – a variable

   a – an algebraic value at which the residue is evaluated

Synopsis:
- Computes the algebraic residue of the expression f for the variable x around the point a. The residue is defined as the coefficient of $(x-a)\hat{\ }(-1)$ in the Laurent series expansion of f.

• This function should be defined by the command **readlib(residue)** before it is used.

Examples:
```
> readlib(residue):
  residue(Zeta(s), s=1);                    ⟶                         1

> residue(Psi(x)*GAMMA(x)/x, x=0);          ⟶                    2              2
                                                          1/12 Pi   + 1/2 gamma

> residue(exp(x), x=1);                     ⟶                         0
```

SEE ALSO: **series, singular**

2.2.58   search – substring search

Calling Sequence:
> search(*s*, *t*)
> search(*s*, *t*, ´*p*´)

Parameters:
> *s*, *t* – strings
> *p* – (optional) a name

Synopsis:
• The **search** function is a Boolean-valued function that searches the subject string *s* for the first occurrence of the target string *t*.

• The third argument if present will be assigned the index where the target string starts, if the search was successful.

• This function should be defined by the command **readlib(search)** before it is used.

Examples:
```
> readlib(search):
  search(abcxxefg, xx, ´p´);               ⟶                     true

> p;                                        ⟶                       4
```

SEE ALSO: **substring**

2.2.59   showtime – display time and space statistics for commands

Calling Sequence:
> showtime()

Synopsis:
• The **showtime** function is used to maintain a history of results computed within a Maple session and to display the time and space used by Maple for the execution of each statement.

- This function should be defined by the command `readlib(showtime)` before it is used.

- After loading in `showtime`, the command `on` initiates the history mechanism, and the command `off` terminates it. After entering the command `on`, the user is prompted initially with the prompt `O1 :=` and for successive statements by the prompts numbered `O2 :=`, `O3 :=`, and so on. Any Maple statement entered is evaluated normally, its result returned followed by a line numbered `O1`, `O2`, .. with the time taken and the amount of memory used being displayed. The result is assigned to the global variable `O1`, `O2`, .. which may be referred to later by the user, thus providing a complete history mechanism.

- By entering the command `off` Maple returns to normal mode using the standard prompt.

- The previous value may be referred to by using `"` but `""` and `"""` are no longer available under `showtime`.

- Caveats: `showtime` is an ordinary Maple function, programmed in Maple. As such it cannot trap interrupts. Hence, if you interrupt a calculation you will be thrown back to the main Maple session. The values of `O1`, `O2`, .. are still available.

SEE ALSO: `time`, `profile`, `history`

2.2.60   singular – find singularities of an expression

Calling Sequence:
> singular(*expr*)
> singular(*expr*, *vars*)

Parameters:
> *expr*   –   an algebraic expression
> *vars*   –   (optional) a variable or a set of variables

Synopsis:
- The function `singular` outputs an expression sequence representing the singularities of *expr*.

- If two arguments are given the expression, *expr* is considered to be a function in *vars*. If `singular` is called with only one argument, then *expr* is considered as a function in the variables returned by the command `indets(`*expr*`,name)`.

- Each singular point is represented by a set of equations, the left-hand side of the equations being the variables.

- The `singular` function will return non-removable as well as removable singularities. For instance, `(x-2)/sin(x-2)` will report a singularity at `x=2`.

- The power of `singular` to find singularities is basically that of `solve`. For example, some zeros that `solve` cannot find may result in singularities that `singular` will not find.

- The `singular` function may return expressions in _N or in _NN. The name _N represents any integer. The name _NN represents any natural number.

- This function should be defined by the command `readlib(singular)` before it is used.

Examples:
```
> readlib(singular):
  singular(x*y + 1/(x*y), x);
```
$\longrightarrow$ {x = 0}, {x = infinity}

```
> singular(ln(x)/(x^2-1));
```
$\longrightarrow$ {x = 0}, {x = 1}, {x = -1}

```
> singular(x/(x-y));
```
$\longrightarrow$ {x = y, y = y}

```
> singular(tan(x));
```
$\longrightarrow$ {x = _N Pi + 1/2 Pi}

```
> singular(Psi(1/x));
```
$\longrightarrow$
$$\{x = 0\}, \{x = \text{infinity}\}, \{x = -\frac{1}{\_NN - 1}\}$$

SEE ALSO: `solve`, `residue`, `roots`, `Roots`

2.2.61   sinterp – sparse multivariate modular polynomial interpolation

Calling Sequence:
   sinterp(*f*, *x*, *t*, *p*)

Parameters:
   f – a Maple procedure
   x – a list of variables to be interpolated
   t – a positive integer
   p – a prime modulus

Synopsis:
- Assume that *f* is a function which, given **n** integers and a prime *p*, evaluates a polynomial in **n** variables modulo *p*. Then `sinterp` returns a polynomial in *x* whose values modulo *p* equal those of *f*. The list *x* must contain **n** variables. The function `sinterp` uses sparse interpolation.

- For the interpolation to succeed, the modulus **p** should be sufficiently large. If it is not, **FAIL** is returned.

- The degree bound **t** is a bound on the total degree of the polynomial being interpolated.

- This function should be defined by the command `readlib(sinterp)` before it is used.

Examples:
```
> readlib(sinterp):
  f := proc(x,y,p) local t1, t2;
          t1 := x^2+y^2;
          t2 := x^2-y^2;
          t1 * t2 mod p;
      end:
  sinterp(f, [x,y], 10, 101);
```
$$\longrightarrow \qquad 100\ y^4 + x^4$$

SEE ALSO: `sdegree`, `interp`

2.2.62   sturm – number of real roots of a polynomial in an interval
sturmseq – Sturm sequence of a polynomial

Calling Sequence:
> sturmseq(p, x)
> sturm(s, x, a, b)

Parameters:

p – a polynomial in x with rational or float coefficients

x – a variable in polynomial p

a, b – rationals or floats such that $a \leq b$

s – a Sturm sequence for polynomial p

Synopsis:

- The procedure `sturmseq` computes a Sturm sequence for the polynomial p in x. It returns the Sturm sequence as a list of polynomials and replaces multiple roots with single roots. It uses the procedures `sturmrem` and `sturmquo` instead of `quo` and `rem` to avoid the fuzzy zero. This can be reinstated by defining

$$\text{sturmrem:=rem: sturmquo:=quo:}$$

- The procedure `sturm` uses Sturm´s theorem to return the number of real roots in the interval $(a,b]$ of polynomial p in x. The first argument to `sturm` should be a Sturm sequence for p. This may be computed by `sturmseq`. Note that the interval includes the upper endpoint b but it does not include the lower endpoint a.

- These functions should be defined by the command `readlib(sturm)` before they are used.

Examples:
```
> readlib(sturm):
> s := sturmseq(expand((x-1)*(x-2)*(x-3)), x);
                3     2              2
         s := [x  - 6 x  + 11 x - 6, x  - 4 x + 11/3, x - 2, 1]
```

```
> sturm(s,x,3/2,4);
```
$$2$$
```
> sturm(s,x,1,2);
```
$$1$$

SEE ALSO: `realroot`, `roots`, `solve`

2.2.63   tensor – compute curvature tensors in a coordinate basis

Calling Sequence:
    `invmetric()`
    `d1metric()`
    `d2metric()`
    `Christoffel1()`
    `Christoffel2()`
    `Riemann()`
    `Ricci()`
    `Ricciscalar()`
    `Einstein()`
    `Weyl()`
    `tensor()`
    `display()` or `display(`n`)`

Parameters:
    n – (optional) name that display understands

Synopsis:

- The dimension `Ndim` of the space(time) has the initial value **4** and is a global variable. The dimension `Ndim` may be changed by an assignment statement such as `Ndim := 3` which would change the dimension to **3**.

- The coordinates of the space(time) have been initialized to `x`, `y`, `z`, and `t` by the following assignment statements:

      `x1:=x;   x2:=y;   x3:=z;   x4:=t;`

    The coordinates may be changed by assignment statements which use the global variables `x1`, `x2`, `x3`, ..., `x.Ndim` on the left-hand side of the assignment statements.

- The metric tensor has been initialized to the Minkowski metric by the following assignment statements:

$$g11:=1; \quad g12:=0; \quad g13:=0; \quad g14:=0;$$
$$g21:=0; \quad g22:=1; \quad g23:=0; \quad g24:=0;$$
$$g31:=0; \quad g32:=0; \quad g33:=1; \quad g34:=0;$$
$$g41:=0; \quad g42:=0; \quad g43:=0; \quad g44:=-1;$$

The covariant metric tensor may be changed by assignment statements which use the global variables `g11`, `g12`, ..., `g.Ndim.Ndim` to define the metric tensor components. The metric tensor should be symmetric as a matrix, so `g21:=g12`; `g31:=g13`; .... The above variables are the input to the `tensor` procedures.

- The procedure `invmetric` computes the contravariant metric tensor which is the inverse of the covariant metric tensor matrix. This procedure assigns the elements of the contravariant metric tensor to the global variables `h11`, `h12`, ... , `h.Ndim.Ndim`. This procedure also computes the determinant of the covariant metric tensor matrix and assigns it to the global variable `detg`.

- The procedure `d1metric` computes all of the first partial derivatives of all the covariant metric tensor components with respect to all of the coordinates and assigns them to the global variables `g111`, `g112`, ... `g.Ndim.Ndim.Ndim`. For example, `g231` is the partial derivative of `g23` with respect to `x1`.

- The procedure `d2metric` computes all the second partial derivatives of all the covariant metric tensor components with respect to all the coordinates and assigns them to the global variables `g1111`, `g1112`, ... `g.Ndim.Ndim.Ndim.Ndim`. For example, `g2314` is the second partial derivative of `g23` with respect to `x1` and `x4`. The procedure `d1metric` must be invoked before `d2metric` is invoked.

- The procedure `Christoffel1` computes all the Christoffel symbols of the first kind and assigns them to the global variables `c111`, `c112`, ... `c.Ndim.Ndim.Ndim`. Christoffel symbols of the first kind are defined by

$$c.i.j.k := (g.k.i.j + g.j.k.i - g.i.j.k)/2;$$

The procedure `d1metric` must be invoked before `Christofel1` is invoked.

- The procedure `Christoffel2` computes all of the Christoffel symbols of the second kind and assigns them to the global variables `C111`, `C112`, ... `C.Ndim.Ndim.Ndim`. Christoffel symbols of the second kind are defined by

$$C.i.j.k := h.k.m*c.i.j.m;$$

where the Einstein summation convention was used to imply that the repeated dummy index `m` is summed from 1 to `Ndim`. The procedures `d1metric`, `invmetric`, and `Christofel1` must be invoked before `Christofel2` is invoked.

- The procedure `Riemann` computes all of the covariant components of the Riemann tensor and assigns them to the global variables `R1111`, `R1112`, ... `R.Ndim.Ndim.Ndim.Ndim`. These components of the Riemann tensor are defined by

```
R.i.j.k.l := (g.i.l.j.k+g.j.k.i.l-g.i.k.j.l-g.j.l.i.k)/2
             + h.m.n*(c.i.l.m*c.j.k.n - c.i.k.m*c.j.l.n);
```

The procedures `invmetric`, `d1metric`, `Christofel1`, and `d2metric` must be invoked before `Riemann` is invoked.

- The procedure `Ricci` computes all of the covariant components of the Ricci tensor and assigns them to the global variables `R11`, `R12`, ... `R.Ndim.Ndim`. Covariant components of the Ricci tensor are defined by

```
R.i.j := h.m.n*R.m.i.j.n;
```

The procedures `invmetric`, `d1metric`, `Christofel1`, `d2metric` and `Riemann` must be invoked before `Ricci` is invoked.

- The procedure `Ricciscalar` computes the Ricci scalar and assigns it to the global variable `R`. The Ricci scalar is defined by

```
R := h.m.n*R.m.n;
```

The procedures `invmetric`, `d1metric`, `Christofel1`, `d2metric`, `Riemann` and `Ricci` must be invoked before `Ricciscalar` is invoked.

- The procedure `Einstein` computes all of the covariant components of the Einstein tensor and assigns them to the global variables `G11`, `G12`, ... `G.Ndim.Ndim`. These components are defined by

```
G.i.j := R.i.j - 1/2*g.i.j*R;
```

The procedures `invmetric`, `d1metric`, `Christofel1`, `d2metric` and `Riemann`, `Ricci` and `Ricciscalar` must be invoked before `Einstein` is invoked.

- The procedure `Weyl` computes all of the covariant components of the Weyl tensor (which is also called the conformal curvature tensor) and assigns them to the global variables `C1111`, `C1112`, ... `C.Ndim.Ndim.Ndim.Ndim`. Components of the Weyl tensor are defined by

```
C.i.j.k.l := R.i.j.k.l
+ 1/(Ndim-2)*(g.i.k*R.j.l - g.i.l*R.j.k + g.j.l*R.i.k - g.j.k*R.i.l)
+ 1/(Ndim-1)/(Ndim-2)*R*(g.i.l*g.j.k - g.i.k*g.j.l);
```

The procedures `invmetric`, `d1metric`, `Christofel1`, `d2metric`, `Riemann`, `Ricci`, and `Ricciscalar` must be invoked before `Weyl` is invoked.

- The procedure `tensor` calls all of the above procedures. The above procedures all return **NULL**.

- The procedure `display` accepts one of the following arguments:

dimension	coordinates	metric	detmetric	invmetric	Christoffel1
Christoffel2	Riemann	Ricci	Ricciscalar	Einstein	Weyl

It prints all the non-zero components of the given geometric object and uses some of the symmetries of the object to avoid duplication. It also returns NULL. If the command `display()` is invoked without any arguments then it prints everything.

- Reference: L. P. Eisenhart, *Riemannian Geometry*, Princeton University Press, Princeton, 1966.

- These functions should be defined by the command `readlib(tensor)` before they are used.

Examples:

```
# Schwarzschild metric
> readlib(tensor):
# define the coordinates
> x1:=t: x2:=r: x3:=th: x4:=ph:
# define the covariant metric
> g11:=(1-2*m/r): g22:=-1/g11: g33:=-r^2: g44:=-r^2*sin(th)^2:
# compute curvature
> tensor();
# Show its a vacuum solution of the Einstein field equations
> display(Einstein);
                    covariant Einstein tensor components

                    Einstein = All components are zero

# Show that it's not flat
> C1212;
```

$$2\,\frac{m}{r^3}$$

SEE ALSO: `debever`, `cartan`

2.2.64   thiele – Thiele´s continued fraction interpolation formula

Calling Sequence:

thiele $(x,\ y,\ v)$

Parameters:

x – a list of independent values, $[x[1],..,x[n]]$

y – a list of dependent values, $[y[1],..,y[n]]$

v – a variable or value to be used in a rational function

Synopsis:

- The function `thiele` computes the rational function of variable v (or evaluated at numerical value v) in continued fraction form which interpolates the points $(x[1],y[1])$, $(x[2],y[2])$, ...,

$(x[n],y[n])$. If `n` is odd then the numerator and denominator polynomials will have degree $(n-1)/2$. Otherwise, `n` is even and the degree of the numerator is $n/2$ and the degree of the denominator is $n/2-1$.

- If the same x-value is entered twice, it is an error, whether or not the same y-value is entered. All independent values must be distinct.

- Reference: The function `thiele` uses Thiele's interpolation formula involving reciprocal differences. See the *Handbook of Mathematical Functions* , by Abramowitz and Stegun, Dover Publications, Inc, New York, 1965. Chapter 25, page 881, formula 25.2.50.

- This function should be defined by the command `readlib(thiele)` before it is used.

Examples:
```
> readlib(thiele):
  thiele([1,2,a],[3,4,5],z);
```
$\longrightarrow$

```
                               z - 1
                     3 + -------------
                                 z - 2
                         1 + ---------
                             1 - a
                             ----- + 1
                             3 - a
```

SEE ALSO: `interp`, `sinterp`

2.2.65   translate – linear translation of a polynomial

Calling Sequence:
> translate(a, x, x_0)

Parameters:

a –   a polynomial

x –   an indeterminate

x_0 –   a constant

Synopsis:
- Translate the polynomial $a(x)$ by $x = x + x_0$ efficiently. The method used requires `O(n)` multiplications and divisions and `O(n^2)` additions where `n = degree`(a,x).

- This function should be defined by the command `readlib(translate)` before it is used.

Examples:
```
> readlib(translate):
  translate(x^2,x,1);
```
$\longrightarrow$

```
                          2
             1 + 2 x + x
```

SEE ALSO: `subs`, `expand`

2.2.66   trigsubs – handling trigonometric identities

Calling Sequence:

   trigsubs(*expr*)

   trigsubs(*s*)

   trigsubs(*s*, *expr*)

Parameters:

   expr   –   an expression

   s   –   an equation

Synopsis:

- The function **trigsubs** manages a table of valid trigonometric identities.

- If **trigsubs** is called with the single argument 0, it returns a set of functions known to the procedure.

- If **trigsubs** is called with a single trigonometric expression *expr*, it returns a list of trigonometric expressions equal to *expr*.

- If **trigsubs** is called with a single equation *s* which represents a trigonometric identity, it returns the string `found` if this identity belongs to the table, and `not found` otherwise.

- If **trigsubs** is called with two arguments, it checks whether the identity *s* belongs to the table or not. In the former case, the function applies this identity to *expr* and returns the result. In the latter case, the function returns an error message.

- For substitution of identities not known to this procedure use **subs**.

- This function should be defined by the command **readlib(trigsubs)** before it is used.

Examples:

```
> readlib(trigsubs):
> trigsubs(0);
                        {cos, cot, csc, sec, sin, tan}

> trigsubs(cos(a+b*w));
```

$$
[\cos(a + b\ w),\ \cos(a + b\ w),\ 2\cos(\%1)^2 - 1,\ 1 - 2\sin(\%1)^2,
$$

$$
\cos(\%1)^2 - \sin(\%1)^2,\ \frac{1}{\sec(a + b\ w)},\ \frac{1}{\sec(a + b\ w)},\ \frac{1 - \tan(\%1)^2}{1 + \tan(\%1)^2},
$$

$$
1/2\ \exp(I\ (a + b\ w)) + 1/2\ \exp(-\ I\ (a + b\ w))]
$$

```
%1 :=                           1/2 a + 1/2 b w

> trigsubs(cos(w) = sin(w));
                              `not found`
```

```
> trigsubs(cos(w) = sin(w), 1);
Error, (in trigsubs) not found in table - use subs to over ride
> trigsubs(sin(2*z) = 2*cos(z)*sin(z), sin(2*z)*cos(z));
                          2
                   2 cos(z)  sin(z)
```

SEE ALSO: subs

2.2.67   unassign – unassign names

Calling Sequence:
>   unassign(*name1*, *name2*, ...)

Parameters:
>   *name1*, *name2*, ...   –   names

Synopsis:
- This procedure unassigns all the unevaluated names given as input. The value returned by unassign is NULL.

- This function should be defined by the command **readlib(unassign)** before it is used.

Examples:
```
> readlib(unassign):
  a := 1:
  a;                              ⟶                1
> unassign(a);                    ⟶    Error, (in assign) invalid arguments
> unassign(´a´);
  a;                              ⟶                a
```

SEE ALSO: assign

2.2.68   write, writeln, open, close

Calling Sequence:
>   write(*r*)
>   writeln(*r*)
>   open(*f*)
>   close()

Parameters:
>   *r*   –   sequence of expressions, possibly empty
>   *f*   –   name of a file

Synopsis:

- The user would typically open a particular file for output by **open**(f), write some expressions to that file using **write** and **writeln**, then close the file by **close**. Expressions are line printed to the file. A buffer is maintained so that only after the output exceeds the line length (screenwidth) or a **writeln** (for **write** followed by new line) or a **close**, is the buffer flushed.

- Caveats: Maple does not allow more than one file to be open for input or output at one time.

- These functions should be defined by the command **readlib(write)** before they are used.

SEE ALSO: **writeto, save**

3
Packages

3.1 The Student Calculus Package

3.1.1 Introduction to the student package

Synopsis:

- This package may be loaded by the command `with(student)`. The package must be loaded before any functions in this package may be used.

- The `student` package is a collection of routines designed to carry out step-by-step solutions to problems.

- Integrals, sums, and limits are handled in an unevaluated form. Routines for expanding or combining such expressions or for performing change of variables, integration by parts, completing the square or replacing factors of expressions by names are included. Other routines compute distances, slopes, and find midpoints of line segments, or provide information about graphs of functions.

- The routines contained in the package are the following:

changevar	combine	completesquare	D
distance	Int	intercept	intparts
isolate	leftbox	leftsum	Limit
makeproc	maximize	middlebox	middlesum
midpoint	minimize	powsubs	rightbox
rightsum	showtangent	simpson	slope
Sum	trapezoid	value	

- The routines in this package have been used to prepare the document *Teaching First Year Calculus through the Use of Symbolic Algebra* by J. S. Devitt, classroom notes, University of Waterloo, Faculty of Mathematics, February 1982.

SEE ALSO: `with`, `student`[*function*] (where *function* is from the above list)

3.1.2   student[changevar] – perform a change of variables

Calling Sequence:

    changevar(s, f)

    changevar(s, f, u)

Parameters:

    s –   an expression of the form $h(x) = g(u)$, defining x as a function of u

    f –   an expression such as Int(F(x)), x = a...b)

    u –   the name of the new integration (summation) variable.

Synopsis:

- The changevar function performs a "change of variables" for integrals, sums, or limits.

- The first argument is an equation defining the new variable in terms of the old variable. If more than two variables are involved, the new variable must be given as the third argument. The second argument is the expression to be rewritten and usually contains Int, Sum, or Limit.

- The changevar command acts like **powsubs** if none of Int, Sum, or Limit appears in f.

- The change of variables may be implicitly defined (e.g. x^2+2 = 2*u^2).

- The unevaluated forms Int, Limit, and Sum should be used, rather than int, limit, and sum. They can be evaluated later by using the **value** command.

- This function is part of the **student** package, and so can be used in the form changevar(..) only after performing the command with(student) or with(student, changevar).

Examples:
```
> with(student):
> changevar(cos(x)+1=u, Int((cos(x)+1)^3*sin(x), x), u);
                        /
                        |    3
                        | - u   du
                        |
                        /

> changevar(x=sin(u), Int(sqrt(1-x^2), x=a...b), u);
                arcsin(b)
                    /
                    |                2 1/2
                    |       (1 - sin(u) )    cos(u) du
                    |
                    /
                arcsin(a)
```

SEE ALSO: **student, powsubs, subs, Int, Limit, Sum, value**

3.1.3 student[completesquare] – complete the square

Calling Sequence:
  completesquare(f)
  completesquare(f, x)

Parameters:
  f – an algebraic expression
  x – one of the indeterminates occurring in f

Synopsis:
- This function completes the square of polynomials of degree 2 in x by re-writing such polynomials as perfect squares plus a remainder. If more than one variable appears in f, then x must be specified.

- The polynomial may appear as a subexpression of f, so that equations, reciprocals, and in general, parts of expressions, may be simplified. Since the polynomial may occur as an argument to **Int** this function can be used to help simplify unevaluated integrals.

- Terms not involving the indicated variable are treated as constants.

- This function is part of the **student** package, and so can be used in the form completesquare(..) only after performing the command with(student) or with(student, completesquare).

Examples:
```
> with(student):
  completesquare(9*x^2 + 24*x + 16);
```
$$\longrightarrow \qquad 9\ (x + 4/3)^2$$

```
> completesquare(x^2 - 2*x*a + a^2 + y^2
  -2*y*b + b^2 = 23, x);
```
$$\longrightarrow \qquad (x - a)^2 + y^2 - 2\ y\ b + b^2 = 23$$

```
> completesquare(", y);
```
$$\longrightarrow \qquad (y - b)^2 + (x - a)^2 = 23$$

SEE ALSO: student, Int

3.1.4 student[combine] – combine terms into a single term

Calling Sequence:
  combine(f)
  combine(f, n)

Parameters:
  f – any expression
  n – a name

Synopsis:

- The `combine` function applies transformations which combine terms in sums, products, and powers into a single term. This function is applied recursively to the components of lists, sets, and relations; f and n may be lists/sets of expressions and names, respectively.

- For many functions, the transformations applied by `combine` are the inverse of the transformations that are applied by `expand`. For example, consider the well-known identity

$$sin(a+b) = sin(a)*cos(b) + cos(a)*sin(b)$$

- The `expand` function applies the identity from left to right whereas the `combine` function does the reverse.

- Subexpressions involving `Int`, `Sum`, and `Limit` are combined into one expression where possible using linearity; thus, `c1*f(a,range) + c2*f(b,range)` $==>$ `f(c1*a+c2*b,range)`.

- A specific set of transformations is obtained by specifying a second (optional) argument n (a name) which is one of the following: `exp`, `ln`, `power`, `trig`, `Psi`. For additional information and examples about the transformations applied by each of these, see `combine[n]`.

- This function is identical to the main library `combine`.

Examples:

```
> combine(Int(x,x=a..b)-Int(x^2,x=a..b));
```
$\longrightarrow$

```
        b
       /
      |        2
      |   x - x   dx
      |
      /
       a
```

```
> combine(Limit(x,x=a)*Limit(x^2,x=a)+c);
```
$\longrightarrow$

```
                3
      Limit   x   + c
      x -> a
```

```
> combine(4*sin(x)^3,trig);
```
$\longrightarrow$ `- sin(3 x) + 3 sin(x)`

```
> combine(exp(x)^2*exp(y),exp);
```
$\longrightarrow$ `exp(2 x + y)`

```
> combine(2*ln(x)-ln(y),ln);
```
$\longrightarrow$
```
            2
           x
      ln(----)
           y
```

```
> combine((x^a)^2,power);
```
$\longrightarrow$
```
       (2 a)
      x
```

```
> combine(Psi(-x)+Psi(x),Psi);
```
$\longrightarrow$ `2 Psi(x) + Pi cot(Pi x) + 1/x`

SEE ALSO: `expand`, `factor`, `student`

3.1.5   student[distance] − compute the distance between Points

Calling Sequence:

   distance(a, b)

Parameters:

   a, b − expressions or objects of type Point

Synopsis:

- The distance function computes the distance between Points in one or higher dimensions. If the Points are of dimension 2 or higher (such as [x, y] or [1, 2, 3]) then the standard Euclidean distance is used. If one or both Points are algebraic expressions other than names, then the distance is computed in one dimension. If the Points are both names then distance returns unevaluated.

- A Point in two or more dimensions is defined as a list of coordinates (such as [x, y], or [x, y, z]). Points in one dimension are just algebraic expressions, though names may be treated as either depending on the context.

- This function is part of the student package, and so can be used in the form distance(..) only after performing the command with(student) or with(student, distance).

Examples:
```
> with(student):
  distance(a, b);
```
$$\longrightarrow \qquad distance(a, b)$$
```
> distance(-3, a+5);
```
$$\longrightarrow \qquad abs(a + 8)$$
```
> distance([a, b], [c, d]);
```
$$\longrightarrow \qquad (a^2 - 2\,a\,c + c^2 + b^2 - 2\,b\,d + d^2)^{1/2}$$

SEE ALSO: student, midpoint, slope, completesquare, Point

3.1.6   student[Int] − inert form of int (integration function)

Calling Sequence:

   Int(f, x)

   Int(f, $x = a..b$)

Parameters:

   f − the expression to be integrated

   x − the variable of integration

   a, b − (optional) lower and upper bounds defining the range of integration

Synopsis:

- This function corresponds to the integral of f with respect to x. It is an "unevaluated" form of Maple´s int function, so only minor simplifications are performed. Such unevaluated expressions can be manipulated by routines in the **student** package such as `changevar` (change of variables), `intparts` (integration by parts), `combine`, and `expand`. Partial fraction decomposition of integrals is available through `convert(..., parfrac)`.

- For definite integration, the range of integration $(x = a..b)$ must be indicated. For indefinite integration, omit the third parameter to `Int`.

- Use `value` to force `Int` to evaluate like `int`.

- This function is part of the **student** package, and so can be used in the form `Int(..)` only after performing the command `with(student)` or `with(student, Int)`.

Examples:
```
> with (student):
  Int(f(x),x);
```
$$\longrightarrow \qquad \int f(x)\ dx$$

```
> Int(f*g,x=1..n);
```
$$\longrightarrow \qquad \int_{1}^{n} f\ g\ dx$$

SEE ALSO: `int`, `student`, `value`, `changevar`, `intparts`, `convert[parfrac]`

3.1.7   student[intercept] – compute the points of intersection of two curves

Calling Sequence:

> `intercept(eqn`$_1$`)`
> `intercept(eqn`$_1$`, eqn`$_2$`, {x, y})`

Parameters:

> eqn_1, eqn_2 — bivariate equations (e.g. `y = x^2 + 3`)
> x, y — (optional) the coordinate variables

Synopsis:

- Compute the points corresponding to the intersection of the two curves defined by eqn_1 and eqn_2.

- If only one equation is given, the y-intercept (the intercept with `x = 0`) is returned. In this case, the dependent variable should appear alone on the left-hand side of the given equation as in `y = 3*x + 5`.

- An optional third argument may be used to indicate the set of coordinate variables.

- This function is part of the **student** package, and so can be used in the form intercept(..) only after performing the command with(student) or with(student, intercept).

Examples:
```
> with(student):
  intercept(y = x+5);                    ⟶        {y = 5, x = 0}
> intercept(y = x+5, y = 0);             ⟶        {y = 0, x = -5}
> intercept(y+4 = x+5, x = 0);           ⟶        {x = 0, y = 1}
> intercept(y = a*x+b, x = 0, {x, y});   ⟶        {x = 0, y = b}
```

SEE ALSO: student, intercept, midpoint, distance

3.1.8   student[intparts] – perform integration by parts

Calling Sequence:
  intparts(*f*, *u*)

Parameters:
  f – an expression of the form Int(u*dv, x)
  u – the factor of the integrand to be differentiated

Synopsis:

- Carry out integration by parts on an unevaluated integral (expressed in terms of Int).

- The function returns u*v - Int(du*v, x) as its value.

- This function is part of the **student** package, and so can be used in the form intparts(..) only after performing the command with(student) or with(student, intparts).

Examples:
```
> with(student):
> intparts(Int(x^k*ln(x), x), ln(x));
                      (k + 1)     /   (k + 1)
              ln(x) x             |  x
              -------------  -  |  --------- dx
                  k + 1          |  x (k + 1)
                                 /
> intparts(Int(sin(x)*x+sin(x), x), sin(x));
                                /
                      2         |            2
          sin(x) (1/2 x  + x) - |  cos(x) (1/2 x  + x) dx
                                |
                                /
```

SEE ALSO: student, powsubs, changevar

3.1.9   student[isolate] – isolate a subexpression to left side of an equation

Calling Sequence:
    isolate(*eqn*, *expr*)
    isolate(*eqn*, *expr*, *iter*)

Parameters:

eqn	–	an equation or an algebraic expression
expr	–	any algebraic expression
iter	–	(optional) a positive integer

Synopsis:
- The procedure `isolate` attempts to isolate the second argument *expr* from the first argument *eqn* and solves *eqn* for *expr*.

- If the first argument is not an equation, then the equation *eqn* = 0 is assumed.

- The optional third argument *iter* controls the maximum number of transformation steps that `isolate` will perform.

- This function is part of the **student** package, and so can be used in the form `isolate(..)` only after performing the command `with(student)` or `with(student, isolate)`.

Examples:
```
> with(student):
  isolate(4*x*sin(x)=3, sin(x));
```
$$\longrightarrow \qquad \sin(x) = \frac{3}{4\,x}$$

```
> isolate(x^2-3*x-5, x^2);
```
$$\longrightarrow \qquad x^2 = 3\,x + 5$$

SEE ALSO: `isolate, solve`

3.1.10   student[leftbox] – graph an approximation to an integral

Calling Sequence:
    leftbox(*f(x)*, *x=a..b*)
    leftbox(*f(x)*, *x=a..b*, *n*)

Parameters:

f(x)	–	an algebraic expression in *x*
x	–	variable of integration
a	–	left bound of integration
b	–	right bound of integration
n	–	(optional) indicates the number of rectangles to use

Synopsis:

- The function `leftbox` will generate a plot of rectangular boxes used to approximate a definite integral. The height of each rectangle (box) is determined by the value of the function at the left side of each interval.

- Four intervals are used by default.

- The value of the corresponding numerical approximation can be obtained by the Maple procedure `leftsum`.

- This function is part of the **student** package, and so can be used in the form `leftbox(..)` only after performing the command `with(student)` or `with(student, leftbox)`.

Examples:
```
with(student):
leftbox(x^4*ln(x), x=2..4);
leftbox(sin(x)*x+sin(x), x=0..2*Pi, 5);
```

SEE ALSO: student, leftsum, rightbox, rightsum, simpson, trapezoid, middlebox, middlesum

3.1.11   student[leftsum] – numerical approximation to an integral

Calling Sequence:

leftsum($f(x)$, $x=a..b$)

leftsum($f(x)$, $x=a..b$, n)

Parameters:

$f(x)$ –   an algebraic expression in x

x –   variable of integration

a –   left end of the interval

b –   right end of the interval

n –   (optional) indicates the number of rectangles to use

Synopsis:

- The function `leftsum` computes a numerical approximation to a definite integral using rectangles. The height of each rectangle (box) is determined by the value of the function at the left side of each interval.

- Four equal-sized intervals are used by default.

- A graph of the approximation can be obtained by the Maple procedure `leftbox`.

- This function is part of the **student** package, and so can be used in the form `leftsum(..)` only after performing the command `with(student)` or `with(student, leftsum)`.

Examples:
```
> with(student):
> leftsum(x^k*ln(x), x=1..3);
```
$$\frac{1}{2} \left(\sum_{i=0}^{3} \; (1 + 1/2\ i)^k \; \ln(1 + 1/2\ i) \right)$$

```
> leftsum(sin(x)*x+sin(x), x=1..3, 12);
```
$$\frac{1}{6} \left(\sum_{i=0}^{11} \; (\sin(1 + 1/6\ i) \; (1 + 1/6\ i) + \sin(1 + 1/6\ i)) \right)$$

SEE ALSO: `student`, `value`, `leftbox`, `rightbox`, `rightsum`, `simpson`, `trapezoid`, `middlebox`, `middlesum`

3.1.12   student[Limit] − inert form of limit

Calling Sequence:
>   Limit(f, $x = a$)
>   Limit(f, $x = a$, *direction*)

Parameters:

f	−	an expression in x
x	−	variable name
a	−	the limiting value for x to approach
direction	−	(optional) one of **right**, **left**, **real**, or **complex**

Synopsis:
- This function corresponds to the limit of f as x approaches a. It is an "unevaluated" form of Maple's `limit` function so only minor simplifications are performed. It can be manipulated by **changevar** and **expand**. The **combine** function can be used to rewrite sums and products as a single Limit.

- Use **value** to force evaluation by the **limit** function.

- This function is part of the **student** package, and so can be used in the form `Limit(..)` only after performing the command `with(student)` or `with(student, Limit)`.

Examples:

```
> with(student):
  Limit(x^3 + 3*x^2 + x+1, x=4);
```
$\longrightarrow$
$$\operatorname{Limit}_{x \to 4} x^3 + 3 x^2 + x + 1$$

```
> Limit(1/x, x=0, right);
```
$\longrightarrow$
$$\operatorname{Limit}_{x \to 0+} 1/x$$

```
> Limit(1/x, x=0, real);
```
$\longrightarrow$
$$\operatorname{Limit}_{x \to 0, real} 1/x$$

```
> expand(Limit(x^3 + 3*x^2 + x + 1, x=4));
```
$\longrightarrow$
$$(\operatorname{Limit}_{x \to 4} x)^3 + 3 (\operatorname{Limit}_{x \to 4} x)^2 + (\operatorname{Limit}_{x \to 4} x) + 1$$

```
>
  value(");
```
$\longrightarrow$ 117

SEE ALSO: limit, student, Int, Sum, value, changevar, combine

3.1.13 student[makeproc] – convert an expression into a Maple procedure

Calling Sequence:

    makeproc(*expr*, *x*)

Parameters:

    expr – any expression
    x   – variable name

Synopsis:

- The result of calling `makeproc(`*expr*, *x*`)` is a procedure which when evaluated at *x* returns *expr*.

- It is useful when evaluating an expression at several values of *x*.

- If *expr* is an unevaluated function call such as `sin(x)`, then the name of the function is returned.

- This function is part of the **student** package, and so can be used in the form `makeproc(..)` only after performing the command `with(student)` or `with(student, makeproc)`.

Examples:

```
> with (student):
  makeproc(sin(x), x);
```
$\longrightarrow$ sin

```
> p:= x^2 + 2*x + 3;
```
$\longrightarrow$
$$p := x^2 + 2 x + 3$$

```
> f := makeproc(p, x);
```
$\longrightarrow$
$$f := x \to x^2 + 2 x + 3$$

```
> f(x);
```
$$\longrightarrow \qquad x^2 + 2x + 3$$

```
> f(u);
```
$$\longrightarrow \qquad u^2 + 2u + 3$$

SEE ALSO: unapply, student, operators, operators[functional]

3.1.14 student[middlebox] – graph an approximation to an integral

Calling Sequence:
> middlebox($f(x)$, $x=a..b$)
> middlebox($f(x)$, $x=a..b$, n)

Parameters:

$f(x)$	–	an algebraic expression in x
x	–	variable of integration
a	–	left bound of integration
b	–	right bound of integration
n	–	(optional) indicates the number of rectangles to use

Synopsis:

- The function `middlebox` will generate a plot of rectangular boxes used to approximate a definite integral. The height of each rectangle (box) is determined by the value of the function at the centre of each interval.

- Four intervals are used by default.

- The value of the corresponding numerical approximation can be obtained by the Maple procedure `middlesum`.

- This function is part of the **student** package, and so can be used in the form `middlebox(..)` only after performing the command `with(student)` or `with(student, middlebox)`.

Examples:
```
with(student):
middlebox(x^4*ln(x), x=2..4);
middlebox(sin(x)*x+sin(x), x=0..2*Pi, 5);
```

SEE ALSO: student, leftsum, leftbox, rightbox, rightsum, simpson, trapezoid, middlesum

3.1.15 student[middlesum] – numerical approximation to an integral

Calling Sequence:
> middlesum($f(x)$, $x=a..b$)
> middlesum($f(x)$, $x=a..b$, n)

Parameters:

$f(x)$ –   an algebraic expression in x

x –   variable of integration

a –   lower bound for the range of integration

b –   upper bound for the range of integration

n –   (optional) indicates the number of rectangles to use

Synopsis:

- The function `middlesum` computes a numerical approximation to a definite integral using rectangles. The height of each rectangle (box) is determined by the value of the function at the midpoint of each interval.

- Four equal-sized intervals are used by default.

- This function is part of the **student** package, and so can be used in the form `middlesum(..)` only after performing the command `with(student)` or `with(student, middlesum)`.

Examples:
```
> with(student):
> middlesum(x^k*ln(x), x=1..3);
                    /   3                                  \
                    |-----                                 |
                    |  \                k                  |
              1/2   |   )    (5/4 + 1/2 i)  ln(5/4 + 1/2 i)|
                    |  /                                   |
                    |-----                                 |
                    \i = 0                                 /
> middlesum(sin(x)*x+sin(x), x=1..3, 12);
                 /   11                                              \
                 |-----                                              |
                 |  \    /      13   \  / 13        \      /  13      \|
           1/6   |   )  |sin(---- + 1/6 i) |---- + 1/6 i| + sin(---- + 1/6 i)||
                 |  /    \      12   /  \ 12        /      \  12      /|
                 |-----                                              |
                 \i = 0                                              /
```

SEE ALSO: `student`, `value`, `leftsum`, `rightsum`, `leftbox`, `rightbox`, `middlebox`, `simpson`, `trapezoid`

3.1.16   student[midpoint] – compute the midpoint of a line segment

Calling Sequence:

  `midpoint(p1, p2)`

Parameters:

  $p1, p2$ –   names, expressions, or points

Synopsis:

- The function `midpoint` will compute the midpoint of the line segment defined by the two points *p1* and *p2*. The points should be either lists of length 2, or expressions. In the latter case, the midpoint of the two expressions is just their average.

- If one argument is a point and the other is a name, then the name is treated as an unevaluated point.

- This function is part of the **student** package, and so can be used in the form `midpoint(..)` only after performing the command `with(student)` or `with(student, midpoint)`.

Examples:
```
> with(student):
  midpoint([a, b], [c, d]);
```
$\longrightarrow$ `[1/2 a + 1/2 c, 1/2 b + 1/2 d]`
```
> midpoint(-5, 8);
```
$\longrightarrow$ `3/2`
```
> midpoint(c, [a, b]);
```
$\longrightarrow$ `[1/2 c[1] + 1/2 a, 1/2 c[2] + 1/2 b]`

SEE ALSO: **student, slope, distance**

3.1.17   student[minimize] – compute the minimum
student[maximize] – compute the maximum

Calling Sequence:

>   minimize(*expr*)
>   minimize(*expr*, *vars*)
>   maximize(*expr*)
>   maximize(*expr*, *vars*)

Parameters:

>   *expr*   –   any algebraic expression
>   *vars*   –   (optional) a set of variables

Synopsis:

- The function `minimize` returns the minimum value of *expr* or an expression sequence with all the values that are candidates for minima.

- If *vars* is specified, then *expr* is minimized with respect to the variables given; otherwise *expr* is minimized with respect to all of its indeterminates.

- When the solution is in terms of algebraic numbers of degree larger than 2, `minimize` will return a floating point approximation to the minima.

- This function finds unconstrained minima over the real numbers, i.e. for any values of *vars* which are real (including the limits at infinity).

- The function `maximize` is defined to compute the negative of the result of `minimize` applied to the negated *expr*.

- These functions are part of the student package, and so can be used only after performing the command `with(student)` or `with(student, minimize)`.

Examples:
```
> with(student):
  minimize(x^2+y^2+3);
```
$\longrightarrow$ 3

```
> minimize(sin(x));
```
$\longrightarrow$ −1

```
> minimize((x^2-y^2)/(x^2+y^2));
```
$\longrightarrow$ −1

```
> maximize((x^2-y^2)/(x^2+y^2));
```
$\longrightarrow$ 1

```
> minimize(x^2 + y^2, {x});
```
$\longrightarrow$ 2
                 y

```
> minimize(x^2 + y^2, {x, y});
```
$\longrightarrow$ 0

SEE ALSO: `simplex`, `minimize`, `student`

3.1.18 student[powsubs] – substitute for factors of an expression

Calling Sequence:
>   powsubs(*eqn*, *f*)

Parameters:
>   *eqn* – an equation **s1** = **s2** specifying the substitution to be made
>   *f*   – an expression with **s1** as an algebraic factor of a subexpression

Synopsis:
- The function `powsubs` differs from subs in that it is defined in terms of algebraic factors rather than in terms of an underlying data structure. Each sub-expression is examined and if **s1** occurs as a factor of that sub-expression, the substitution is carried out. However, `powsubs` cannot find expressions which only occur as part of a sum.

- If more than one substitution is specified they are processed in order from left to right. `powsubs` does not carry out simultaneous substitutions.

- This function is part of the **student** package, and so can be used in the form `powsubs(..)` only after performing the command `with(student)` or `with(student, powsubs)`.

Examples:
```
> with(student):
> p := (x+2)^3 + 1/(x+2) + sin(x+2);
```
$$p := (x + 2)^3 + \frac{1}{x + 2} + \sin(x + 2)$$

```
> powsubs(x+2 = z, p);
```

$$z^3 + 1/z + \sin(z)$$

```
> q := sqrt(x^2+y^2) + f(x^2+y^2) + Int(x^2+y^2, x);
```

$$q := (x^2 + y^2)^{1/2} + f(x^2 + y^2) + \int x^2 + y^2 \ dx$$

```
> powsubs(x^2+y^2 = z, q);
```

$$z^{1/2} + f(z) + \int z \ dx$$

SEE ALSO: `subs`, `student`

3.1.19   student[rightbox] – graph an approximation to an integral

Calling Sequence:
>   rightbox($f(x)$, $x=a..b$)
>   rightbox($f(x)$, $x=a..b$, n)

Parameters:
> $f(x)$ – an algebraic expression in x
> x – variable of integration
> a – lower bound of range of integration
> b – upper bound of range of integration
> n – (optional) indicates the number of rectangles to use

Synopsis:
- The function `rightbox` will generate a plot of rectangular boxes used to approximate a definite integral. The height of each rectangle (box) is determined by the value of the function at the right side of each interval.

- Four equal-sized intervals are used by default.

- The value of the corresponding numerical approximation can be obtained by the Maple procedure `rightsum`.

- This function is part of the `student` package, and so can be used in the form `rightbox(..)` only after performing the command `with(student)` or `with(student, rightbox)`.

Examples:
```
with (student):
```

```
rightbox(x^4*ln(x), x=2..4);
rightbox(sin(x)*x+sin(x), x=0..2*Pi, 5);
```

SEE ALSO: student, rightsum, leftbox, leftsum, simpson, trapezoid, middlebox, middlesum

3.1.20   student[rightsum] – numerical approximation to an integral

Calling Sequence:

    rightsum($f(x)$, $x=a..b$)

    rightsum($f(x)$, $x=a..b$, n)

Parameters:

    $f(x)$ –   an algebraic expression in x

    x –   variable of integration

    a –   lower bound on the range of integration

    b –   upper bound on the range of integration

    n –   (optional) indicates the number of rectangles to use

Synopsis:

- The function **rightsum** computes a numerical approximation to an integral using rectangles. The height of each rectangle (box) is determined by the value of the function at the right side of each interval.

- Four equal-sized intervals are used by default.

- A graph of the approximation can be obtained by the Maple procedure **rightbox**.

- This function is part of the **student** package, and so can be used in the form **rightsum(..)** only after performing the command **with(student)** or **with(student, rightsum)**.

Examples:

```
> with(student):
> rightsum(x^k*ln(x), x=1..3);
                    /  4                                \
                    |-----                              |
                    | \                       k         |
              1/2 | )   (1 + 1/2 i)  ln(1 + 1/2 i)|
                    | /                                 |
                    |-----                              |
                    \i = 1                              /
```

```
> rightsum(sin(x)*x+sin(x), x=1..3, 12);
               /   12                                                    \
               |-----                                                    |
               |  \                                                      |
         1/6   |   )     (sin(1 + 1/6 i) (1 + 1/6 i) + sin(1 + 1/6 i))|
               |  /                                                      |
               |-----                                                    |
               \i = 1                                                    /
```

SEE ALSO: student, value, rightbox, leftsum, leftbox, simpson, trapezoid,
          middlebox, middlesum

3.1.21   student[showtangent] – plot a function and its tangent line

Calling Sequence:

    showtangent $(f(x), \ x = a)$

Parameters:

    $f(x)$ – an expression in x

    x – the independent variable

    a – the value of x at the desired point of tangency

Synopsis:

- This routine generates a graph of the function defined by the expression $f(x)$ and its tangent line at the point $[a, \ f(a)]$.

- This function is part of the **student** package, and so can be used in the form **showtangent(..)** only after performing the command **with(student)** or **with(student, showtangent)**.

Examples:
```
with(student):
showtangent(x^2+5, x = 2);
```

SEE ALSO: student

3.1.22   student[simpson] – numerical approximation to an integral

Calling Sequence:

    simpson $(f(x), x=a..b)$

    simpson $(f(x), \ x=a..b, \ n)$

Parameters:

    $f(x)$ – an algebraic expression in x

    x – variable of integration

    a – lower bound on the range of integration

b –   upper bound on the range of integration

n –   (optional) indicates the number of rectangles to use

Synopsis:

- The function **simpson** approximates a definite integral using Simpson's rule. If the parameters are symbolic, then the formula is returned.

- Four equal-sized intervals are used by default.

- This function is part of the **student** package, and so can be used in the form **simpson(..)** only after performing the command **with(student)** or **with(student, simpson)**.

Examples:
```
> with(student):
> simpson(x^k*ln(x), x=1..3);
```

$$\frac{1}{6}\,3^{k}\,\ln(3) + \frac{2}{3}\left(\sum_{i=1}^{2} (1/2 + i)^{k}\,\ln(1/2 + i)\right)$$

$$+ \frac{1}{3}\left(\sum_{i=1}^{1} (1 + i)^{k}\,\ln(1 + i)\right)$$

```
> simpson(sin(x)*x+sin(x), x=1..3, 12);
```

$$\frac{1}{9}\sin(1) + \frac{2}{9}\sin(3)$$

$$+ \frac{2}{9}\left(\sum_{i=1}^{6} (\sin(5/6 + 1/3\,i)\,(5/6 + 1/3\,i) + \sin(5/6 + 1/3\,i))\right)$$

$$+ \frac{1}{9}\left(\sum_{i=1}^{5} (\sin(1 + 1/3\,i)\,(1 + 1/3\,i) + \sin(1 + 1/3\,i))\right)$$

SEE ALSO: student, rightbox, rightsum, leftsum, leftsum, simpson, trapezoid, middlesum

3.1.23   student[slope] – compute the slope of a line

Calling Sequence:
>   slope(*equation*)
>   slope(*equation, y, x*)
>   slope(*equation, y(x)*)
>   slope(*p1, p2*)

Parameters:

equation	–	an equation of a line
y, x	–	(optional in most cases) the dependent and independent variables
p1, p2	–	two 2-dimensional points (such as [a, b] and [c, d])

Synopsis:
- The function `slope` will compute the slope of a line determined by two points, or by an equation.

- If *equation* is of the form `y = f(x)`, then the dependent variable is taken to be `y`, and the independent variable to be `x`. Ambiguity caused by three or more unknowns or by the use of a different form of *equation* is resolved by providing two additional arguments, *y* and *x*, the names of the dependent and independent variables, respectively.

- This function is part of the **student** package, and so can be used in the form `slope(..)` only after performing the command `with(student)` or `with(student, slope)`.

Examples:
```
> with(student):
  slope(y = 3*x+5);
```
$\longrightarrow$ 3
```
> slope(y = a*x+5, y, x);
```
$\longrightarrow$ a
```
> slope(2*y+4 = x-3, x, y);
```
$\longrightarrow$ 2
```
> slope(2*y+4 = x-3, x(y));
```
$\longrightarrow$ 2

SEE ALSO: student, intersect, midpoint, showtangent

3.1.24   student[Sum] – inert form of sum

Calling Sequence:
>   Sum(*f, i = a..b*)

Parameters:

f	–	the expression to be summed (an expression in *i*)
i	–	the summation variable
a, b	–	(optional) upper and lower bounds defining the range of summation for *i*

Synopsis:
- This function corresponds to the sum of f for i in the given range. It is an unevaluated form of the Maple **sum** function, so only some minor simplifications are performed. Expressions involving Sum can be manipulated by routines such as **expand**, **combine**, and **changevar**.

- Use **value** to force Sum to evaluate like **sum**.

- This function is part of the **student** package, and so can be used in the form Sum(..) only after performing the command **with(student)** or **with(student, Sum)**.

Examples:
```
> with(student):
  Sum(f(i), i=a..b);
```

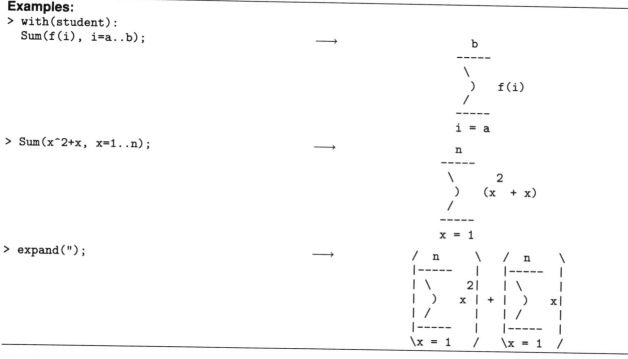

```
> Sum(x^2+x, x=1..n);
```

```
> expand(");
```

SEE ALSO: sum, student, combine, expand, value, changevar

3.1.25   student[trapezoid] – numerical approximation to an integral

Calling Sequence:
    trapezoid($f(x)$, $x=a..b$)
    trapezoid($f(x)$, $x=a..b$, n)

Parameters:
    $f(x)$ –   an algebraic expression in x
    x –   variable of integration
    a –   lower bound on the range of integration

b – upper bound on the range of integration

n – (optional) indicates the number of rectangles to use

Synopsis:

- The function `trapezoid` computes a numerical approximation to an integral using the Trape-zoidal rule. If symbolic arguments are given, an appropriate formula is generated.

- Four equal-sized intervals are used by default.

- This function is part of the **student** package, and so can be used in the form `trapezoid(..)` only after performing the command `with(student)` or `with(student, trapezoid)`.

Examples:
```
> with(student):
> trapezoid(x^k*ln(x), x=1..3);
              /  3                               \
              |-----                             |
              | \                  k             |       k
        1/2 | )   (1 + 1/2 i)  ln(1 + 1/2 i)| + 1/4 3   ln(3)
              | /                               |
              |-----                             |
              \i = 1                            /
> trapezoid(sin(x)*x+sin(x), x=1..3, 12);
                    /  11                                                \
                    |-----                                               |
                    | \                                                  |
      1/6 sin(1) + 1/6 | )   (sin(1 + 1/6 i) (1 + 1/6 i) + sin(1 + 1/6 i))|
                    | /                                                  |
                    |-----                                               |
                    \i = 1                                               /
          + 1/3 sin(3)
```

SEE ALSO: `student`, `rightsum`, `rightbox`, `leftsum`, `leftbox`, `simpson`, `middlesum`

3.1.26   student[value] – evaluate inert functions (formerly student[Eval])

Calling Sequence:
   `value(f)`

Parameters:
   f – any algebraic expression

Synopsis:

- In general, when Maple has two functions of the same name, where one begins with a capital letter and one with lower case, the capitalized function is an "inert" form of the lower case function. These unevaluated functions may be manipulated by various Maple routines (especially in the student package), and are printed in a special way.

- The `value` function forces these "unevaluated" functions such as `Int`, `Limit`, `Sum`, and `Diff` to be evaluated by their corresponding Maple (lower case) function, such as `int`, `limit`, `sum`, or `diff`. This is easily done by substituting for upper case names their lower case equivalents and invoking `eval`.

- Only `Int`, `Limit`, `Sum`, `Diff`, and `Product` are known to `value` at this time.

- This function is part of the `student` package, and so can be used in the form `value(..)` only after performing the command `with(student)` or `with(student, value)`.

Examples:
```
> with(student):
  value(Int(x, x));
```
$$\longrightarrow \qquad 1/2\ x^2$$

```
> value(Limit(sin(x)/x, x=0));
```
$$\longrightarrow \qquad 1$$

```
> value(Sum(i, i=1..n) + 5);
```
$$\longrightarrow \qquad 1/2\ (n + 1)^2 - 1/2\ n + 9/2$$

SEE ALSO: `int`, `limit`, `sum`, `student`, `eval`

3.1.27 type/Point – check for type Point

Calling Sequence:

   `type(P,Point)`

Parameters:

   P – any algebraic expression

Synopsis:

- A Point is defined as a list of coordinate values (such as `[a,b,c]`). The length of the list determines the dimension of the space.

- They differ from objects of Maple type `point` (such as `{x=a,y=b,z=c}`) in that the coordinates are defined by position in the list rather than by names. `subs({x=a,y=b,z=c},[x,y,z])` converts a Point to a `point`.

- This function is part of the student package and can only be used after performing the command `with(student)`.

Examples:
```
> with (student):
  type(a , Point);
```
$$\longrightarrow \qquad false$$

```
> type([a], Point);
```
$$\longrightarrow \qquad true$$

```
> type([a+4,b] , Point);
```
$$\longrightarrow \qquad true$$

SEE ALSO: `student`, `distance`, `midpoint`, `slope`

3.2 The Linear Algebra Package

3.2.1 Introduction to the `linalg` package

Calling Sequence:

    function(`args`)

    `linalg`[*function*]`(args)`

Synopsis:

- To use a linalg function, either define that function alone using the command `with(linalg, `*function*`)`, or define all linalg functions using the command `with(linalg)`. Alternatively, invoke the function using the long form `linalg`[*function*]. This long form notation is necessary whenever there is a conflict between a package function name and another function used in the same session.

- The functions available are:

add	addcol	addrow	adj	adjoint	angle
augment	backsub	band	basis	bezout	BlockDiagonal
charmat	charpoly	col	coldim	colspace	colspan
companion	concat	cond	copyinto	crossprod	curl
definite	delcols	delrows	det	diag	diverge
dotprod	eigenvals	eigenvects	equal	exponential	extend
ffgausselim	fibonacci	frobenius	gausselim	gaussjord	genmatrix
grad	GramSchmidt	hadamard	hermite	hessian	hilbert
htranspose	ihermite	indexfunc	innerprod	intbasis	inverse
ismith	iszero	jacobian	jordan	JordanBlock	kernel
laplacian	leastsqrs	linsolve	matrix	minor	minpoly
mulcol	mulrow	multiply	norm	nullspace	orthog
permanent	pivot	potential	randmatrix	range	rank
row	rowdim	rowspace	rowspan	rref	scalarmul
singularvals	smith	stack	submatrix	subvector	sumbasis
swapcol	swaprow	sylvester	toeplitz	trace	transpose
vandermonde	vecpotent	vectdim	vector		

- For more information on a particular function see `linalg`[*function*].

- As an example, to multiply two matrices **A** and **B**, use

    `with(linalg,multiply);  multiply(A,B);`

or alternatively, use Maple´s `evalm` evaluator as follows

```
evalm( A &* B );
```

- Note that the `linalg` functions for manipulating vectors and matrices expect as input the Maple types `vector` and `matrix`, rather than more general arrays. In particular, subscripts are indexed from 1.

SEE ALSO: `evalm`, `with`, `vector`, `matrix`, `mod`

3.2.2   linalg[add] – matrix or vector addition

Calling Sequence:
> add(A, B)
> add(A, B, $c1$, $c2$)

Parameters:
> A, B –   matrices or vectors (with consistent dimensions)
> $c1$, $c2$ –   (optional) scalar parameters

Synopsis:
- The function add(A, B) computes the matrix (vector) sum of A and B.

- If the optional scalar parameters $c1$ and $c2$ are given, then **add** computes the matrix (vector) sum $c1*A + c2*B$.

- The **evalm** command is an easy method of accessing **add**.

- This function is part of the `linalg` package, and so can be used in the form `add(..)` only after performing the command `with(linalg)` or `with(linalg, add)`. The function can always be accessed in the long form `linalg[add](..)`.

Examples:
```
> with (linalg):
  A := matrix(3,3,[1,2,3,2,3,4,3,4,5]);
```
$\longrightarrow$
$$A := \begin{bmatrix} 1 & 2 & 3 \\ 2 & 3 & 4 \\ 3 & 4 & 5 \end{bmatrix}$$

```
> B := array( 1..3, 1..3, identity );
```
$\longrightarrow$ `B := array(identity, 1 .. 3, 1 .. 3, [])`

```
> add( A, B, 1, 10 );
```
$\longrightarrow$
$$\begin{bmatrix} 11 & 2 & 3 \\ 2 & 13 & 4 \\ 3 & 4 & 15 \end{bmatrix}$$

```
> v := vector(3,[2,3,4]);
```
$\longrightarrow$ $v := [2, 3, 4]$

```
> u := vector(3,[-3,1,-5]);
```
$\longrightarrow$ $u := [-3, 1, -5]$

```
> add(u,v,-2,1);                          ⟶              [ 8, 1, 14 ]
> add(transpose(u),transpose(v));         ⟶         transpose([ -1, 4, -1 ])
```

SEE ALSO: evalm

3.2.3   linalg[addrow] – form linear combinations of matrix rows
linalg[addcol] – form linear combinations of matrix columns

Calling Sequence:

   addrow(A, $r1$, $r2$, m)
   addcol(A, $c1$, $c2$, m)

Parameters:

A	–	matrix
$r1,r2,c1,c2$	–	integers; row (column) indices
m	–	a scalar expression

Synopsis:

- The call addrow($A,r1,r2,m$) returns a copy of the matrix A in which row $r2$ is replaced by m * row($A,r1$) + row($A,r2$). Similarly addcol($A,c1,c2,m$) returns a copy of the matrix A in which column $c2$ is replaced by m * col($A,c1$) + col($A,c2$).

Examples:
```
> with (linalg):
  a := array( 1..3,1..3, [ [1,2,3], [2,3,4]                        [ 1   2   3 ]
  , [3,4,5] ] );                       ⟶                          [           ]
                                                             a := [ 2   3   4 ]
                                                                  [           ]
                                                                  [ 3   4   5 ]

> addrow(a,1,2,10);                    ⟶                          [  1    2    3 ]
                                                                  [              ]
                                                                  [ 12   23   34 ]
                                                                  [              ]
                                                                  [  3    4    5 ]

> addcol(a,1,2,-x);                    ⟶                     [ 1   - x + 2    3 ]
                                                            [                  ]
                                                            [ 2   - 2 x + 3   4 ]
                                                            [                  ]
                                                            [ 3   - 3 x + 4   5 ]
```

SEE ALSO: linalg[row], linalg[col]

3.2.4   linalg[adjoint] – compute the adjoint of a matrix

Calling Sequence:

    adjoint(A)
    adj(A)

Parameters:

   A – square matrix (with all elements assigned)

Synopsis:

- The function `adjoint` computes the matrix such that the matrix product of A and `adjoint($A$)` is the product of `det($A$)` and the matrix identity. This is done through the computation of minors of A.

- This function is part of the `linalg` package, and so can be used in the form `adjoint(..)` only after performing the command `with(linalg)` or `with(linalg, adjoint)`. The function can always be accessed in the long form `linalg[adjoint](..)`.

Examples:

```
> with(linalg):
  A := matrix(3,3,[1,2,3,4,5,6,7,8,9]);
```

$$A := \begin{bmatrix} 1 & 2 & 3 \\ 4 & 5 & 6 \\ 7 & 8 & 9 \end{bmatrix}$$

```
> adj(A);
```

$$\begin{bmatrix} -3 & 6 & -3 \\ 6 & -12 & 6 \\ -3 & 6 & -3 \end{bmatrix}$$

```
> B := array (1..2,1..2,[[1,4],[0,2]]);
```

$$B := \begin{bmatrix} 1 & 4 \\ 0 & 2 \end{bmatrix}$$

```
> adjoint(B);
```

$$\begin{bmatrix} 2 & -4 \\ 0 & 1 \end{bmatrix}$$

SEE ALSO: `linalg[det]`, `linalg[minor]`

3.2.5   linalg[angle] – compute the angle between vectors

Calling Sequence:

    angle(u, v)

Parameters:

   u,v – n-dimensional vectors

Synopsis:
- The call `angle(u,v)` computes the angle theta between the n-dimensional vectors u and v via the formula $\cos(\text{theta}) = \text{sqrt}(u.v \,/\, |u| \,|v|)$, using the Euclidean inner product and norm.

- This function is part of the `linalg` package, and so can be used in the form `angle(..)` only after performing the command `with(linalg)` or `with(linalg, angle)`. The function can always be accessed in the long form `linalg[angle](..)`.

Examples:
```
> with (linalg):
  angle( vector([1,0]), vector([0,1]) );
```
$\longrightarrow$ 1/2 Pi

```
> angle( array([1,1]), array([0,1]) );
```
$\longrightarrow$ 1/4 Pi

```
> angle( vector([1,0]), vector([1,0]) );
```
$\longrightarrow$ 0

SEE ALSO: `linalg[norm]`, `linalg[dotprod]`, `linalg[innerprod]`

3.2.6   linalg[augment] – join two or more matrices together horizontally
linalg[concat] – a synonym for augment

Calling Sequence:
```
augment(A, B, ...)
concat(A, B, ...)
```

Parameters:
$A, B, ...$ –   matrices or vectors

Synopsis:
- The function `augment` joins two or more matrices or vectors together horizontally. A vector is interpreted as a column vector. The matrices and vectors must have the same number of rows.

- This function is part of the `linalg` package, and so can be used in the form `augment(..)` only after performing the command `with(linalg)` or `with(linalg, augment)`. The function can always be accessed in the long form `linalg[augment](..)`.

Examples:
```
> with(linalg):
  a := matrix([[1,2],[2,3]]);
```
$\longrightarrow$
$$a := \begin{bmatrix} 1 & 2 \\ 2 & 3 \end{bmatrix}$$

```
> b := matrix(2,3,[3,4,5,6,7,8]);
```
$\longrightarrow$
$$b := \begin{bmatrix} 3 & 4 & 5 \\ 6 & 7 & 8 \end{bmatrix}$$

```
> augment(a,b);
```
$\longrightarrow$
$$\begin{bmatrix} 1 & 2 & 3 & 4 & 5 \\ 2 & 3 & 6 & 7 & 8 \end{bmatrix}$$

```
> v := vector(2,[1,2]);
```
$\longrightarrow$
$$v := [\ 1,\ 2\ ]$$

```
> concat(v,b,v);
```
$\longrightarrow$
$$\begin{bmatrix} 1 & 3 & 4 & 5 & 1 \\ 2 & 6 & 7 & 8 & 2 \end{bmatrix}$$

SEE ALSO: linalg[stack]

3.2.7   linalg[backsub] – back substitution on a matrix

Calling Sequence:

backsub(*F*)

Parameters:

F   –   a rectangular matrix

Synopsis:

- If *F* is the result of applying forward Gaussian elimination to the augmented matrix of a system of linear equations, as might be obtained from **gausselim** or **gaussjord**, **backsub** completes the solution by back substitution. If a solution exists, it is returned as a vector. If the solution is not unique, it will be parameterized in terms of the global symbols t1, t2, ... etc. If no solution exists, an error will be generated.

- The input matrix must be in upper triangular form with all zero rows grouped at bottom. Such a matrix is produced by applying **gausselim** or **gaussjord** to the augmented matrix of a system of linear equations. The matrix entries in general must be rational expressions over the rational numbers.

- This function is part of the **linalg** package, and so can be used in the form **backsub(..)** only after performing the command **with(linalg)** or **with(linalg, backsub)**. The function can always be accessed in the long form **linalg[backsub](..)**.

Examples:

```
> with(linalg):
  A := randmatrix(3,4):
  F := gausselim(A);
```
$\longrightarrow$

$$F := \begin{bmatrix} -85 & -55 & -37 & -35 \\[4pt] 0 & -\dfrac{217}{17} & \dfrac{3126}{85} & \dfrac{273}{17} \\[8pt] 0 & 0 & \dfrac{19504}{155} & -\dfrac{1234}{31} \end{bmatrix}$$

```
> backsub(F);
```
$\longrightarrow$

$$\left[\ \dfrac{133337}{68264},\ -\dfrac{74049}{34132},\ -\dfrac{3085}{9752}\ \right]$$

```
> H := matrix([[1,2,3],[2,1,3],[1,-1,0]])
  :
  v := vector([1,2,1]):
  A := augment(H,v);
```
$$\longrightarrow \qquad A := \begin{bmatrix} 1 & 2 & 3 & 1 \\ 2 & 1 & 3 & 2 \\ 1 & -1 & 0 & 1 \end{bmatrix}$$

```
> F := gaussjord(A);
```
$$\longrightarrow \qquad F := \begin{bmatrix} 1 & 0 & 1 & 1 \\ 0 & 1 & 1 & 0 \\ 0 & 0 & 0 & 0 \end{bmatrix}$$

```
> backsub(F);
```
$$\longrightarrow \qquad [\ -\ t3 + 1,\ -\ t3,\ t3\ ]$$

SEE ALSO: `linalg[gausselim]`, `linalg[gaussjord]`, `linalg[linsolve]`

3.2.8   linalg[band] – create a band matrix

Calling Sequence:
    band(b, n)

Parameters:
    b – list or vector
    n – positive integer

Synopsis:
- The call `band`(b,n) creates an n x n banded matrix using the elements of b. The number of elements of b must be odd and not greater than n.
- All the elements of the ith subdiagonal of the result are initialized to $b[i]$, for the first half of b. The elements of the main diagonal of the matrix result are initialized to the middle element of b. The last half of b is used to initialize the superdiagonals.
- If n is larger than the length of b, then the remaining sub- and super- diagonals are set to zero.
- The matrix result uses a sparse indexing function.
- This function is part of the `linalg` package, and so can be used in the form `band(..)` only after performing the command `with(linalg)` or `with(linalg, band)`. The function can always be accessed in the long form `linalg[band](..)`.

Examples:
```
> with (linalg):
  band([1],3);
```
$$\longrightarrow \qquad \begin{bmatrix} 1 & 0 & 0 \\ 0 & 1 & 0 \\ 0 & 0 & 1 \end{bmatrix}$$

```
> band([1,2,-1],4);
```
                                                    ⟶
$$\begin{bmatrix} 2 & -1 & 0 & 0 \\ 1 & 2 & -1 & 0 \\ 0 & 1 & 2 & -1 \\ 0 & 0 & 1 & 2 \end{bmatrix}$$

SEE ALSO: linalg[diag]

3.2.9   linalg[basis] – find a basis for a vector space

Calling Sequence:

   basis(*V*)

Parameters:

   V – a vector, set of vectors or list of vectors

Synopsis:

- If the argument *V* is a single vector, then this vector is returned as a set { *V* }.

- If *V* is a set of vectors, then **basis** will return a basis for the vector space spanned by those vectors in terms of the original vectors.

- For an ordered basis, use a list of vectors rather than a set.

- A basis for the zero-dimensional space is an empty set or list.

- This function is part of the **linalg** package, and so can be used in the form **basis(..)** only after performing the command **with(linalg)** or **with(linalg, basis)**. The function can always be accessed in the long form **linalg[basis](..)**.

Examples:

```
> with(linalg):
  v1 := vector([1,0,0]):
  v2 := vector([0,1,0]):
  v3 := vector([0,0,1]):
  v4 := vector([1,1,1]):
  basis({v1,v2,v3});
```
                                    ⟶   {v1, v2, v3}
```
> basis([v3,v2,v1]);
```
                                    ⟶   [v3, v2, v1]
```
> basis({v1,v2,v3,v4});
```
                                    ⟶   {v1, v2, v3}
```
> basis({vector([1,1,1]),vector([2,2,2]),
  vector([1,-1,1]),
  vector([2,-2,2]), vector ([1,0,1]),
  vector([0,1,1])});
```
                    ⟶   {[1, 1, 1], [1, -1, 1], [0, 1, 1]}

3.2.10   linalg[bezout] – create the Bezout matrix of two polynomials

Calling Sequence:
   bezout(p, q, x)

Parameters:
   p, q –   polynomials in the variable x

   x –   a variable (name)

Synopsis:
- The call `bezout`(p,q,x) returns the Bezout matrix of the polynomials p and q with respect to x. Note that the determinant of this matrix is equal to `resultant`(p,q,x).

- If p is of degree m in x and q is of degree n in x then the output is a square matrix of dimension `max`(m,n).

- The polynomials p and q must be in expanded form because the **coeff** function is used on p and q to determine the entries of the matrix.

- This function is part of the `linalg` package, and so can be used in the form `bezout(..)` only after performing the command `with(linalg)` or `with(linalg, bezout)`. The function can always be accessed in the long form `linalg[bezout](..)`.

Examples:
```
> with (linalg):
  p := a+b*x+c*x^2;
```
$\longrightarrow$
$$p := a + b\ x + c\ x^2$$

```
> q := d+e*x+f*x^2;
```
$\longrightarrow$
$$q := d + e\ x + f\ x^2$$

```
> bezout(p,q,x);
```
$\longrightarrow$
$$\begin{bmatrix} c\ d - f\ a & d\ b - a\ e \\ c\ e - f\ b & c\ d - f\ a \end{bmatrix}$$

SEE ALSO: `linalg[det]`, `linalg[sylvester]`, `collect`, `resultant`

3.2.11   linalg[charmat] – construct the characteristic matrix

Calling Sequence:
   charmat($A, lambda$)

Parameters:
   A –   a square matrix

   $lambda$ –   a name or algebraic expression

Synopsis:
- The function **charmat** constructs the characteristic matrix M = $lambda$*I - A, where I is the identity matrix. If $lambda$ is a name, then det(M) is the characteristic polynomial.

- This function is part of the `linalg` package, and so can be used in the form `charmat(..)` only after performing the command `with(linalg)` or `with(linalg, charmat)`. The function can always be accessed in the long form `linalg[charmat](..)`.

Examples:
```
> with(linalg):
  A := matrix(3,3,[1,2,3,1,2,3,1,5,6]);
```
$\longrightarrow$

$$A := \begin{bmatrix} 1 & 2 & 3 \\ 1 & 2 & 3 \\ 1 & 5 & 6 \end{bmatrix}$$

```
> charmat(A,lambda);
```
$\longrightarrow$

$$\begin{bmatrix} \lambda - 1 & -2 & -3 \\ -1 & \lambda - 2 & -3 \\ -1 & -5 & \lambda - 6 \end{bmatrix}$$

SEE ALSO: `linalg[charpoly]`, `linalg[eigenvals]`, `linalg[nullspace]`

3.2.12   linalg[charpoly] – compute the characteristic polynomial of a matrix

Calling Sequence:

charpoly(*A*, *lambda*)

Parameters:

A   – a square matrix

lambda – a name or algebraic expression

Synopsis:

- The function `charpoly` computes the characteristic polynomial of the matrix $A = \det(lambda*\text{I} - A)$, where I is the identity matrix and n is the dimension of *A*.

- This function is part of the `linalg` package, and so can be used in the form `charpoly(..)` only after performing the command `with(linalg)` or `with(linalg, charpoly)`. The function can always be accessed in the long form `linalg[charpoly](..)`.

Examples:
```
> with(linalg):
  A := matrix(3,3,[1,2,3,1,2,3,1,5,6]);
```
$\longrightarrow$

$$A := \begin{bmatrix} 1 & 2 & 3 \\ 1 & 2 & 3 \\ 1 & 5 & 6 \end{bmatrix}$$

```
> charpoly(A,x);
```
$\longrightarrow$

$$x^3 - 9 x^2$$

SEE ALSO: `linalg[charmat]`, `linalg[minpoly]`, `linalg[eigenvals]`

3.2.13   linalg[companion] – return a companion matrix associated with a polynomial

Calling Sequence:

   companion(p, x)

Parameters:

   p –   a monic univariate polynomial in x

   x –   a name (the variable)

Synopsis:

- The call companion(p, x) will return a companion matrix associated with the polynomial $p(x)$.

- If C := companion(p, x) and p is a0 + a1*x +...+x^n (a univariate monic polynomial), then C[i,n] = -coeff(p,x,i-1) (i = 1..n), C[i,i-1] = 1 (i = 2..n) and C[i,j] = 0 for other values of i and j.

- The polynomial p must be in expanded form so that the coefficients of powers of x may be computed.

- This function is part of the linalg package, and so can be used in the form companion(..) only after performing the command with(linalg) or with(linalg, companion). The function can always be accessed in the long form linalg[companion](..).

Examples:

```
> with(linalg,companion);
```
$\longrightarrow$ [companion]

```
> p := x^4 + 9*x^3+2*x^2+17*x+5;
```
$\longrightarrow$
$$p := x^4 + 9 x^3 + 2 x^2 + 17 x + 5$$

```
> companion(p,x);
```
$\longrightarrow$
$$\begin{bmatrix} 0 & 0 & 0 & -5 \\ 1 & 0 & 0 & -17 \\ 0 & 1 & 0 & -2 \\ 0 & 0 & 1 & -9 \end{bmatrix}$$

```
> q := expand((z+2)*(z-5)*(z+3));
```
$\longrightarrow$
$$q := z^3 - 19 z - 30$$

```
> companion (q,z);
```
$\longrightarrow$
$$\begin{bmatrix} 0 & 0 & 30 \\ 1 & 0 & 19 \\ 0 & 1 & 0 \end{bmatrix}$$

SEE ALSO: linalg[diag], linalg[JordanBlock]

3.2.14 linalg[cond] – condition number of a matrix

Calling Sequence:

>   cond(A)

>   cond(A, *normname*)

Parameters:

> A – a square matrix

>   *normname*   – (optional) matrix norm, must be one of: 1, 2, ´infinity´, or ´frobenius´.

Synopsis:

- The function **cond** computes the "standard" matrix condition number, defined as **norm**(A) $*$ **norm**(**inverse**(A)). The matrix norm is the default employed by the **linalg[norm]** function, namely the infinity norm (maximum row sum).

- More generally, **cond**(A, *normname*) computes **norm**(A, *normname*) $*$ **norm**(**inverse**(A), *normname*). This is the same measure, but using the specified norm instead of the infinity norm.

- This function is part of the **linalg** package, and so can be used in the form **cond**(..) only after performing the command **with(linalg)** or **with(linalg, cond)**. The function can always be accessed in the long form **linalg[cond]**(..).

Examples:

```
> with (linalg):
  B := array(1..2,1..2, identity);        ⟶   B := array(identity, 1 .. 2, 1 .. 2, [])
> cond(B);                                ⟶                  1
> A := matrix (3,3,[1,0,3,-4,2,0,0,3,-2])
  ;
                                                          [  1   0   3 ]
                                                          [            ]
                                          ⟶     A :=  [ -4   2   0 ]
                                                          [            ]
                                                          [  0   3  -2 ]
> cond(A,1);                              ⟶                  3
```

SEE ALSO: linalg[norm]

3.2.15 linalg[copyinto] – move entries from one matrix into another

Calling Sequence:

>   copyinto(A, B, m, n)

Parameters:

> A, B – matrices

> m, n – non-negative integers

Synopsis:
- The function `copyinto(A, B, m, n)` copies the entries of the matrix A into the matrix B beginning at index position $[m,n]$, so $B[m,n]$ is assigned $A[1,1]$. This function and `linalg[extend]` are used to create new matrices from old ones.

- This function is part of the `linalg` package, and so can be used in the form `copyinto(..)` only after performing the command `with(linalg)` or `with(linalg, copyinto)`. The function can always be accessed in the long form `linalg[copyinto](..)`.

Examples:
```
> with(linalg):
  A := matrix(2,2,[1,2,3,4]);
```
$$A := \begin{bmatrix} 1 & 2 \\ 3 & 4 \end{bmatrix}$$

```
> B := extend(A,2,2,0);
```
$$B := \begin{bmatrix} 1 & 2 & 0 & 0 \\ 3 & 4 & 0 & 0 \\ 0 & 0 & 0 & 0 \\ 0 & 0 & 0 & 0 \end{bmatrix}$$

```
> copyinto(A,B,3,3);
```
$$\begin{bmatrix} 1 & 2 & 0 & 0 \\ 3 & 4 & 0 & 0 \\ 0 & 0 & 1 & 2 \\ 0 & 0 & 3 & 4 \end{bmatrix}$$

SEE ALSO: `linalg[extend]`, `linalg[submatrix]`, `linalg[concat]`, `linalg[concat]`

3.2.16   linalg[crossprod] – vector cross product

Calling Sequence:
   `crossprod(u, v)`

Parameters:
   u, v – lists or vectors, each with three elements

Synopsis:
- This function computes the vector cross-product of u and v, defined to be vector($[u[2]*v[3]-u[3]*v[2],\ u[3]*v[1]\ u[1]*v[3],\ u[1]*v[2]\ u[2]*v[1]]$).

- This function is part of the `linalg` package, and so can be used in the form `crossprod(..)` only after performing the command `with(linalg)` or `with(linalg, crossprod)`. The function can always be accessed in the long form `linalg[crossprod](..)`.

Examples:

```
> with(linalg):
  v1 := vector([1,2,3]);                  ⟶        v1 := [ 1, 2, 3 ]
> v2 := vector([2,3,4]);                  ⟶        v2 := [ 2, 3, 4 ]
> crossprod(v1,v2);                       ⟶          [ -1, 2, -1 ]
```

SEE ALSO: linalg[dotprod], linalg[innerprod]

3.2.17   linalg[curl] – curl of a vector

Calling Sequence:

curl(f, v)

Parameters:

f – a list or vector of three expressions

v – a list or vector of three variables

Synopsis:

- The curl of f with respect to v is computed, where f is a three-dimensional function of the three variables v.

- The curl is defined to be a vector of length 3: [diff($f[3]$,$v[2]$) - diff($f[2]$,$v[3]$), diff($f[1]$,$v[3]$) - diff($f[3]$,$v[1]$), diff($f[2]$,$v[1]$) - diff($f[1]$,$v[2]$)];

- This function is part of the `linalg` package, and so can be used in the form `curl(..)` only after performing the command `with(linalg)` or `with(linalg, curl)`. The function can always be accessed in the long form `linalg[curl](..)`.

Examples:

```
> with (linalg):
  f := [ x^2, x*z, y^2*z ];               ⟶                    2          2
                                                         f := [x , x z, y  z]
> v := [ x, y, z ];                       ⟶               v := [x, y, z]
> curl(f, v);                             ⟶             [ 2 y z - x, 0, z ]
> g := vector ([2*x + z, y^3*x, 3*x*z^2 +
  y]);                                    ⟶                      3          2
                                                   g := [ 2 x + z, y  x, 3 x z  + y ]
> curl (g, v);                            ⟶                    2    3
                                                       [ 1, 1 - 3 z , y  ]
```

SEE ALSO: linalg[grad], linalg[diverge]

3.2.18   linalg[definite] – test for positive (negative) definite matrices

Calling Sequence:

   definite(A, *kind*)

Parameters:

   A –   a square symmetric matrix

   kind   –   one of the following symbols: ´positive_def´, ´positive_semidef´,
              ´negative_def´, or ´negative_semidef´

Synopsis:

- For numerical matrices, **definite** returns true if the matrix has the property specified by the parameter *kind*. The properties are positive definite, positive semidefinite, negative definite, or negative semidefinite.

- For matrices with non-numerical entries, **definite** returns a conjunction of Boolean expressions, all of which must be true if the matrix is to have the property specified by the parameter *kind*.

- This function is part of the **linalg** package, and so can be used in the form **definite(..)** only after performing the command **with(linalg)** or **with(linalg, definite)**. The function can always be accessed in the long form **linalg[definite](..)**.

Examples:
```
> with (linalg):
  A := matrix (2,2,[2,1,1,3]);
```
                                       $\longrightarrow$ $A := \begin{bmatrix} 2 & 1 \\ 1 & 3 \end{bmatrix}$

```
> definite( A,´positive_def´ );
```
                                       $\longrightarrow$ true

```
> B := array( 1..2, 1..2, symmetric ):
  definite( B,´negative_semidef´ );
```
                                       $\longrightarrow$ true

---

3.2.19   linalg[delrows] – delete rows of a matrix
linalg[delcols] – delete columns of a matrix

Calling Sequence:

   delrows(A, $r..s$)

   delcols(A, $r..s$)

Parameters:

   A –   a matrix

   $r..s$ –   integer range

Synopsis:

- The function **delrows** returns the submatrix of the matrix A obtained by deleting rows r through s. Similarly **delcols**($A,r..s$) returns the submatrix of A obtained by deleting columns r through s.

Examples:
```
> with (linalg):
  a := matrix(3,3,[1,2,3,4,5,6,7,8,9]);
```
$$a := \begin{bmatrix} 1 & 2 & 3 \\ 4 & 5 & 6 \\ 7 & 8 & 9 \end{bmatrix}$$

```
> delrows(a,2..3);
```
$$\begin{bmatrix} 1 & 2 & 3 \end{bmatrix}$$

```
> delcols(a,1..1);
```
$$\begin{bmatrix} 2 & 3 \\ 5 & 6 \\ 8 & 9 \end{bmatrix}$$

SEE ALSO: `linalg[submatrix]`, `linalg[row]`, `linalg[col]`

3.2.20 linalg[det] – determinant of a matrix

Calling Sequence:

det(A)

det(A, sparse)

Parameters:

A –   a two-dimensional square matrix

sparse   –   optional directive

Synopsis:

- The function **det** computes the determinant of the given matrix.

- The second argument (optional) specifies that the computational method of minor expansion should be used. This works well for sparse matrices.

- With no extra argument, a decision is made automatically whether to use minor expansion or Gaussian elimination (or a combination of both). Specifying the extra argument eliminates the overhead of the decision process if you happen to know ahead of time that minor expansion will work best.

- This function is part of the **linalg** package, and so can be used in the form `det(..)` only after performing the command `with(linalg)` or `with(linalg, det)`. The function can always be accessed in the long form `linalg[det](..)`.

Examples:
```
> with(linalg):
  A := hilbert(3);
```
$\longrightarrow$
$$A := \begin{bmatrix} 1 & 1/2 & 1/3 \\ 1/2 & 1/3 & 1/4 \\ 1/3 & 1/4 & 1/5 \end{bmatrix}$$

```
> det(A);
```
$\longrightarrow$ $1/2160$

```
> B := matrix(2,2);
```
$\longrightarrow$ `B := array(1 .. 2, 1 .. 2, [])`

```
> det(B);
```
$\longrightarrow$ `B[1, 1] B[2, 2] - B[1, 2] B[2, 1]`

3.2.21   linalg[diag] – create a block diagonal matrix
linalg[BlockDiagonal] – a synonym for diag

Calling Sequence:
  diag(B_1, B_2, .., B_n)
  BlockDiagonal(B_1, B_2, ..., B_n)

Parameters:
  B_1, B_2, ..., B_n –   square matrices or scalar matrix entries

Synopsis:
- The call `diag`(B_1, B_2, .., B_n) returns a matrix on whose diagonal are the matrix blocks B_1, B_2, .., B_n.

- If used in conjunction with `JordanBlock`, `diag` can be used to easily create a Jordan form matrix.

- This function is part of the `linalg` package, and so can be used in the form `diag(..)` only after performing the command `with(linalg)` or `with(linalg, diag)`. The function can always be accessed in the long form `linalg[diag](..)`.

Examples:
```
> with(linalg):
  B1 := array([[1,3,5], [9,a,z], [x^2,
  sin(q),10]]);
```
$\longrightarrow$
$$B1 := \begin{bmatrix} 1 & 3 & 5 \\ 9 & a & z \\ 2 & & \\ x & \sin(q) & 10 \end{bmatrix}$$

```
> B2 := matrix(2, 2, [1,3,2,4]);
```
$\longrightarrow$
$$B2 := \begin{bmatrix} 1 & 3 \\ 2 & 4 \end{bmatrix}$$

```
> diag(B1, B2);
```
$\longrightarrow$

$$
\begin{bmatrix}
1 & 3 & 5 & 0 & 0 \\
9 & a & z & 0 & 0 \\
2 & & & & \\
x & \sin(q) & 10 & 0 & 0 \\
0 & 0 & 0 & 1 & 3 \\
0 & 0 & 0 & 2 & 4
\end{bmatrix}
$$

```
> diag(lambda1, lambda2);
```
$\longrightarrow$

$$
\begin{bmatrix}
\text{lambda1} & 0 \\
0 & \text{lambda2}
\end{bmatrix}
$$

SEE ALSO: `linalg[JordanBlock]`, `linalg[companion]`

3.2.22 linalg[diverge] – compute the divergence of a vector function

Calling Sequence:

diverge(f, v)

Parameters:

f – a vector or list of expressions

v – a vector or list (same length as f) of the variables of f

Synopsis:

- The function **diverge** computes the divergence of f with respect to v, where f is a vector function of the variables given by v.

- The divergence of f with respect to v is defined as the sum of `diff(f[i],v[i])` as i ranges over the length of f.

- This function is part of the **linalg** package, and so can be used in the form `diverge(..)` only after performing the command `with(linalg)` or `with(linalg, diverge)`. The function can always be accessed in the long form `linalg[diverge](..)`.

Examples:
```
> with (linalg):
  f := vector( [x,y^2,z] );
```
$\longrightarrow$
$$f := [x, y^2, z]$$

```
> v := vector( [x,y,z] );
```
$\longrightarrow$
$$v := [x, y, z]$$

```
> diverge( f,v );
```
$\longrightarrow$
$$2 + 2y$$

```
> g := [x^2*y, 2*y*z, 3*x*z^2];
```
$\longrightarrow$
$$g := [x^2 y, 2 y z, 3 x z^2]$$

```
> diverge (g, [x,y,z]);
```
$\longrightarrow$
$$2 x y + 2 z + 6 x z$$

SEE ALSO: `linalg[grad]`, `linalg[curl]`

3.2.23   linalg[dotprod] – vector dot (scalar) product

Calling Sequence:

    dotprod(u, v)
    dotprod(u, v, ´orthogonal´)

Parameters:

  u,v – lists or vectors of the same length

  orthogonal – (optional) assume an orthogonal vector space

Synopsis:

- If called with only two arguments, **dotprod** computes the vector dot product using the standard definition for a vector space over the complex field: sum $u[i]*conjugate(v[i])$, as i ranges over the length of u and v.

- If the third argument ´orthogonal´ is specified, **dotprod** will compute the vector dot product using the definition: sum $u[i]*v[i]$, as i ranges over the length of u and v.

- If u and v are real vectors, then the two forms of **dotprod** are equivalent.

- This function is part of the **linalg** package, and so can be used in the form **dotprod(..)** only after performing the command **with(linalg)** or **with(linalg, dotprod)**. The function can always be accessed in the long form **linalg[dotprod](..)**.

Examples:
```
> with (linalg):
  u := vector( [1,x,y] );                 ⟶              u := [ 1, x, y ]

> v := vector( [1,0,0] );                 ⟶              v := [ 1, 0, 0 ]

> dotprod(u,v);                           ⟶                    1

> a := [1,I];  b := [I,1];                ⟶              a := [1, I]

                                                          b := [I, 1]

> dotprod(a,b);                           ⟶                    0

> dotprod(a,b,´orthogonal´);              ⟶                   2 I
```

SEE ALSO: evalc, linalg[crossprod],linalg[innerprod]

3.2.24   linalg[eigenvals] – compute the eigenvalues of a matrix

Calling Sequence:

    eigenvals(A)
    eigenvals(A, C)
    eigenvals(A, ´implicit´)

Parameters:
    A –   a square matrix
    C –   a matrix of the same shape as A

Synopsis:
- The call `eigenvals(A)` returns a sequence of the eigenvalues of A. The eigenvalues are computed by solving the characteristic polynomial det(lambda I - A) = 0 for the scalar variable lambda, where I is the identity matrix. If the characteristic polynomial is of degree greater than four, then it may not be possible to find all the eigenvalues. If a second parameter ´implicit´ is given, the eigenvalues are expressed using Maple´s `RootOf` notation for algebraic extensions.

- The call `eigenvals(`A,C`)` solves the "generalized eigenvalue problem", that is, finds the roots of the polynomial `det(` lambda C `-` A`)`.

- This function is part of the `linalg` package, and so can be used in the form `eigenvals(..)` only after performing the command `with(linalg)` or `with(linalg, eigenvals)`. The function can always be accessed in the long form `linalg[eigenvals](..)`.

Examples:
```
> with (linalg):
  A := matrix(3,3,[1,2,3,1,2,3,2,5,6]);
```
$$\longrightarrow \qquad A := \begin{bmatrix} 1 & 2 & 3 \\ 1 & 2 & 3 \\ 2 & 5 & 6 \end{bmatrix}$$

```
> eigenvals(A);
```
$$\longrightarrow \qquad 0,\ 9/2 + 1/2\ 93^{1/2},\ 9/2 - 1/2\ 93^{1/2}$$

```
> eigenvals(A,´implicit´);
```
$$\longrightarrow \qquad 0,\ RootOf(- 3 - 9\ \_Z + \_Z^2)$$

SEE ALSO: `linalg[eigenvects]`, `Eigenvals`, `linalg[charpoly]`

3.2.25   linalg[eigenvects] – find the eigenvectors of a matrix

Calling Sequence:
    `eigenvects(`A`)`

Parameters:
    A –   a square matrix

Synopsis:
- The procedure `eigenvects` computes exact symbolic eigenvectors of a square matrix A over a field. That is, for each eigenvalue `lambda` of A, it solves the linear system `(I*lambda - ` A`) X = 0` for `X`. For numerical solutions, see `Eigenvects`.

- The result returned is a sequence of lists of the form $[e_i, m_i, \{v[1,i],... v[n_i,i]\}]$, where the e_i are the eigenvalues, m_i their algebraic multiplicities and $\{v[1,i], ..., v[n_i,i]\}$ is the set of basis vectors for the eigenspace corresponding to e_i, and $1 \le n_i \le m_i$ is the dimension of the eigenspace.

- Although eigenvectors can in principle be computed for a matrix over any field F, one must be able to test for zero in F in order to get the correct result. For this reason, this routine is limited at present to the field of algebraic numbers. Therefore, the input A must be a matrix of rationals, or a matrix of algebraic numbers represented using Maple's `RootOf` notation.

- The method used to compute the eigenvectors is to compute and factor the characteristic polynomial of A over the field F, obtaining the factorization `product(f`$_i$`^m`$_i$`, i=1..k)`. For non-linear factors the eigenvalue will be represented in terms of an algebraic extension of F, namely `e`$_i$ = `RootOf(f`$_i$`, lambda)` (meaning the field `F[lambda]/f`$_i$). In other words, `e`$_i$ is a placeholder for one or more eigenvalues. Finally `eigenvects` computes the nullspace of `(I*e`$_i$ `- A) for i =` `1..k`, to obtain the eigenspace.

- This function is part of the `linalg` package, and so can be used in the form `eigenvects(..)` only after performing the command `with(linalg)` or `with(linalg, eigenvects)`. The function can always be accessed in the long form `linalg[eigenvects](..)`.

Examples:
```
> with(linalg):
> A := matrix(3, 3, [1,-3,3,3,-5,3,6,-6,4]);
```
$$A := \begin{bmatrix} 1 & -3 & 3 \\ 3 & -5 & 3 \\ 6 & -6 & 4 \end{bmatrix}$$

```
> eigenvals(A);
```
$$4, -2, -2$$

```
> eigenvects(A);
```
$$[-2, 2, \{[1, 1, 0], [-1, 0, 1]\}], [4, 1, \{[1, 1, 2]\}]$$

```
> nullspace(charmat(A,-2));
```
$$\{[1, 1, 0], [-1, 0, 1]\}$$

```
> B := array([[1,2,3], [2,2,4], [3,3,9]]);
```
$$B := \begin{bmatrix} 1 & 2 & 3 \\ 2 & 2 & 4 \\ 3 & 3 & 9 \end{bmatrix}$$

```
> eigenvects(B);
```
$$[\%1, 1, \{[1, 4/31 + \frac{38}{31}\%1 - 3/31\ \%1^2, 2/31\ \%1^2 - \frac{15}{31}\%1 - \frac{13}{31}]\}]$$

$$\%1 := \qquad RootOf(6 + 4\ \_Z - 12\ \_Z^2 + \_Z^3)$$

```
> alias(sqrt2 = RootOf(x^2-2)):
> C := matrix([[1,-sqrt2], [0,sqrt2]]);
                            [ 1   - sqrt2 ]
                      C := [             ]
                            [ 0    sqrt2  ]
> eigenvects(C);
            [1, 1, {[ 1, 0 ]}], [sqrt2, 1, {[ 1, 1/2 sqrt2 - 1 ]}]
```

SEE ALSO: `linalg[eigenvals]`, `linalg[charmat]`, `linalg[nullspace]`, `Eigenvals`, `Eigenvects`, `RootOf`, `vector`

3.2.26   linalg[equal] – determine whether two matrices are equal

Calling Sequence:
  equal(*A*, *B*)

Parameters:
  A,*B* – matrices

Synopsis:
- The procedure **equal** returns true if matrices *A* and *B* are of the same shape and corresponding entries are equal, false or an error otherwise.

- This function is part of the **linalg** package, and so can be used in the form **equal(..)** only after performing the command **with(linalg)** or **with(linalg, equal)**. The function can always be accessed in the long form **linalg[equal](..)**.

Examples:
```
> with (linalg):
  A := array([[2,1],[1,2]]):
  B := matrix(2,2,[2,1,1,2]):
  equal(A, array([[2,1],[1,2]]));              ⟶                    true
```

SEE ALSO: `linalg[equal]`

3.2.27   linalg[exponential] – matrix exponential

Calling Sequence:
  exponential(*A*)
  exponential(*A*, *t*)

Parameters:
  A – a square matrix
  t – an (optional) scalar parameter of type name

Synopsis:
- The matrix exponential, $\exp(t^*A)$, is a matrix with the same shape as A and is defined as follows: $\exp(A^*t) = I + A^*t + 1/2!^*A\hat{}2^*t\hat{}2 + \ldots$ where I is the identity matrix.

- If the second parameter is not given, then the first indeterminate (if any) in the matrix is removed and used as a parameter.

- The `exponential` function can only return a symbolic answer if the eigenvalues of A can be found. To get a floating-point approximation, use at least one floating-point entry in A.

- This function is part of the `linalg` package, and so can be used in the form `exponential(..)` only after performing the command `with(linalg)` or `with(linalg, exponential)`. The function can always be accessed in the long form `linalg[exponential](..)`.

Examples:
```
> with(linalg,exponential):
> A := array([[t,0,0],[0,t,0],[0,0,t]]);
```

$$A := \begin{bmatrix} t & 0 & 0 \\ 0 & t & 0 \\ 0 & 0 & t \end{bmatrix}$$

```
> exponential(A);
```

$$\begin{bmatrix} \exp(t) & 0 & 0 \\ 0 & \exp(t) & 0 \\ 0 & 0 & \exp(t) \end{bmatrix}$$

```
> B := array([[-13,-10],[21,16]]);
```

$$B := \begin{bmatrix} -13 & -10 \\ 21 & 16 \end{bmatrix}$$

```
> exponential(B,t);
```

$$\begin{bmatrix} 15 \exp(t) - 14 \exp(2\ t) & 10 \exp(t) - 10 \exp(2\ t) \\ -21 \exp(t) + 21 \exp(2\ t) & -14 \exp(t) + 15 \exp(2\ t) \end{bmatrix}$$

3.2.28   linalg[extend] – enlarge a matrix

Calling Sequence:
 extend(A , m, n, x)
 extend(A, m, n)

Parameters:
 A – a matrix

m, n –   non-negative integers

x –   value (optional)

Synopsis:

- The call `extend(`A,m,n,x`)` returns a new matrix which is a copy of the matrix A with m additional rows and n additional columns. The new entries are initialized to x.

- If the parameter x is omitted, then the matrix is extended by m rows and n columns, but the entries in those rows and columns are not assigned.

- This function is part of the `linalg` package, and so can be used in the form `extend(..)` only after performing the command `with(linalg)` or `with(linalg, extend)`. The function can always be accessed in the long form `linalg[extend](..)`.

Examples:

```
> with(linalg):
  A := matrix (2,2,[1,2,3,4]);
```
$\longrightarrow$
```
        [ 1   2 ]
A  :=  [       ]
        [ 3   4 ]
```

```
> extend(A,1,1,0);
```
$\longrightarrow$
```
        [ 1   2   0 ]
        [           ]
        [ 3   4   0 ]
        [           ]
        [ 0   0   0 ]
```

SEE ALSO: `linalg[copyinto]`, `linalg[augment]`, `linalg[stack]`

3.2.29   linalg[ffgausselim] – fraction-free Gaussian elimination on a matrix

Calling Sequence:

    `ffgausselim(`A`)`

    `ffgausselim(`A`, ´`r`´)`

    `ffgausselim(`A`, ´`r`´, ´`d`´)`

    `ffgausselim(`A`, `$rmar$`)`

Parameters:

    A –   a rectangular matrix

    ´r´   –   for returning rank of A (optional)

    ´d´   –   for returning the determinant of A (optional)

    $rmar$ –   non-negative integer

Synopsis:

- Fraction-free Gaussian elimination with row pivoting is performed on A, an n by m matrix of multivariate polynomials over the rationals. The result is an upper triangular matrix of multivariate polynomials.

- If an optional second parameter is specified, and it is a name, it is assigned the rank of A. The rank of A is the number of non-zero rows in the resulting matrix.

- If an optional third parameter is also specified, and the rank of $A = $ n, then it is assigned the determinant of `submatrix(`A`,1..n,1..n)`.

- If an optional second parameter is specified, and it is an integer, the elimination is terminated at this column position.

- This function is part of the `linalg` package, and so can be used in the form `ffgausselim(..)` only after performing the command `with(linalg)` or `with(linalg, ffgausselim)`. The function can always be accessed in the long form `linalg[ffgausselim](..)`.

Examples:

```
> with(linalg):
  A := matrix(3,3,[x,1,0,0,0,1,1,y,1]);
```
$\longrightarrow$
```
                                    [ x   1   0 ]
                                    [           ]
                               A := [ 0   0   1 ]
                                    [           ]
                                    [ 1   y   1 ]
```

```
> ffgausselim(A,'r','d');
```
$\longrightarrow$
```
              [ x      1         0      ]
              [                         ]
              [ 0   y x - 1      x       ]
              [                         ]
              [ 0      0       y x - 1  ]
```

```
> r;
```
$\longrightarrow$ 3

```
> d;
```
$\longrightarrow$ - y x + 1

SEE ALSO: `linalg[gausselim]`, `linalg[gaussjord]`, `linalg[rowspan]`

3.2.30   linalg[fibonacci] – fibonacci matrix

Calling Sequence:

> `fibonacci(`n`)`

Parameters:

> n – a non-negative integer

Synopsis:

- The procedure `fibonacci(`n`)` returns the nth fibonacci matrix, $F(n)$, defined as follows:

```
F(0) = matrix (1,1,[1])
F(1) = matrix (1,1,[1])
F(2) = matrix (2,2,[1,1,1,0])
F(n) = [[F(n-1),F(n-1)],[F(n-1),0]],
          where        dim F(n) = dim F(n-1) + dim F(n-2)
```

- This function is part of the `linalg` package, and so can be used in the form `fibonacci(..)` only after performing the command `with(linalg)` or `with(linalg, fibonacci)`. The function can always be accessed in the long form `linalg[fibonacci](..)`.

Examples:
```
> with (linalg):
  fibonacci(3);
```
$\longrightarrow$

```
[ 1  1  1 ]
[         ]
[ 1  0  1 ]
[         ]
[ 1  1  0 ]
```

3.2.31   linalg[frobenius] – compute the Frobenius form of a matrix

Calling Sequence:

> frobenius(*A*)
>
> frobenius(*A*, *'P'*)

Parameters:

> *A*   –   a square matrix
>
> *'P'*   –   (optional) assigned the transformation matrix

Synopsis:

- The function `frobenius(A)` computes and returns the Frobenius form F of a matrix *A*.

- F has the following structure: F = diag(C1,C2,..Ck) where the C*i*s are companion matrices associated with polynomials p1, p2,.. pk with the property that pi divides p(i-1), for i = 2..k .

- If called in the form `frobenius(A, 'P')`, then *P* will be assigned the transformation matrix corresponding to the Frobenius form, that is, the matrix P such that inverse(P) * *A* * P = F.

- The Frobenius form defined in this way is unique (if we require that pi divides p(i-1) as above).

- If the sizes of the blocks C1,C2,...,Ck are n1,n2,...,nk respectively, then the columns of the matrix *P* are the vectors [f1,*A**f1,..,*A*^(n1-1)*f1,f2,...,fk,*A**fk,...*A*^(nk-1)*fk]

- This function is part of the `linalg` package, and so can be used in the form `frobenius(..)` only after performing the command `with(linalg)` or `with(linalg, frobenius)`. The function can always be accessed in the long form `linalg[frobenius](..)`.

Examples:

```
> with(linalg,frobenius):
  A := array([[-9,21,-15,4,2,0],[-10,21,
  -14,4,2,0],[-8,16,-11,4,2,0],
  [-6,12,-9,3,3,0],[-4,8,-6,0,5,0],[-2,4,
  -3,0,1,3]]);
```
$\longrightarrow$

$$
A := \begin{bmatrix} -9 & 21 & -15 & 4 & 2 & 0 \\ -10 & 21 & -14 & 4 & 2 & 0 \\ -8 & 16 & -11 & 4 & 2 & 0 \\ -6 & 12 & -9 & 3 & 3 & 0 \\ -4 & 8 & -6 & 0 & 5 & 0 \\ -2 & 4 & -3 & 0 & 1 & 3 \end{bmatrix}
$$

```
> frobenius(A, P);
```
$\longrightarrow$

$$
\begin{bmatrix} 0 & 0 & 0 & 0 & 15 & 0 \\ 1 & 0 & 0 & 0 & -47 & 0 \\ 0 & 1 & 0 & 0 & 56 & 0 \\ 0 & 0 & 1 & 0 & -32 & 0 \\ 0 & 0 & 0 & 1 & 9 & 0 \\ 0 & 0 & 0 & 0 & 0 & 3 \end{bmatrix}
$$

```
> print(P);
```
$\longrightarrow$

$$
\begin{bmatrix} 1 & -7 & -25 & -53 & -47 & 0 \\ 0 & -8 & -24 & -44 & -16 & 0 \\ 0 & -6 & -16 & -22 & 32 & 0 \\ 0 & -3 & -6 & 3 & 84 & 0 \\ 1 & 1 & 5 & 29 & 137 & 0 \\ 1 & 2 & 7 & 28 & 109 & 1 \end{bmatrix}
$$

3.2.32   linalg[gausselim] – Gaussian elimination on a matrix

Calling Sequence:

gausselim(A)

gausselim(A, $'r'$)

gausselim(A, $'r'$, $'d'$)

gausselim(A, $rmar$)

Parameters:

A	–	a rectangular matrix
$'r'$	–	for returning rank of A (optional)
$'d'$	–	for returning the determinant of A (optional)
$rmar$	–	non-negative integer

Synopsis:

- Gaussian elimination with row pivoting is performed on A, an n by m matrix over a field. At present, the matrix coefficients must lie in the field of rational numbers Q, or the field of rational functions over Q .

- The result is a an upper triangular matrix B. If A is an n by n matrix then det(B) = +- product(B[i,i], i=1..n) .

- If an optional second parameter is specified, and it is a name, it is assigned the rank of A. The rank of A is the number of non-zero rows in the resulting matrix.

- If an optional third parameter is also specified, and the rank of A = n, then it is assigned the determinant of `submatrix(`A`,1..n,1..n)`.

- If an optional second parameter is specified, and it is an integer, the elimination is terminated at this column position.

- This function is part of the `linalg` package, and so can be used in the form `gausselim(..)` only after performing the command `with(linalg)` or `with(linalg, gausselim)`. The function can always be accessed in the long form `linalg[gausselim](..)`.

Examples:

```
> with(linalg):
  A := matrix(3,3,[x,1,0,0,0,1,1,y,1]);
```
$\longrightarrow$

$$A := \begin{bmatrix} x & 1 & 0 \\ 0 & 0 & 1 \\ 1 & y & 1 \end{bmatrix}$$

```
> gausselim (A,'r','d');
```
$\longrightarrow$

$$\begin{bmatrix} x & 1 & 0 \\ 0 & \dfrac{y\,x - 1}{x} & 1 \\ 0 & 0 & 1 \end{bmatrix}$$

```
> r;
```
$\longrightarrow$ 3

```
> d;
```
$\longrightarrow$ $-\,y\,x + 1$

SEE ALSO: `linalg[ffgausselim]`, `linalg[gaussjord]`, `linalg[rank]`, `linalg[backsub]`

3.2.33   linalg[genmatrix] – generate the coefficient matrix from equations

Calling Sequence:

  genmatrix(*eqns*, *vars*)

  genmatrix(*eqns*, *vars*, *flag*)

Parameters:

  eqns – set or list of equations

  vars – set or list of variables

  flag – (optional) a flag which can be any value

Synopsis:

- The function **genmatrix** generates the coefficient matrix from the linear system of equations *eqns* in the unknowns *vars*.

- If an optional third argument is present, the negative of the "right-hand side" vector will be included as the last column of the matrix.

- This function is part of the `linalg` package, and so can be used in the form **genmatrix(..)** only after performing the command **with(linalg)** or **with(linalg, genmatrix)**. The function can always be accessed in the long form `linalg[genmatrix](..)`.

Examples:
```
> with (linalg):
> genmatrix( [x+2*y-3,3*x-5*y=0], [x,y] );
                        [ 1    2 ]
                        [        ]
                        [ 3   -5 ]
> genmatrix( [x+2*y-3,3*x-5*y=0], [x,y], flag );
                        [ 1    2   -3 ]
                        [            ]
                        [ 3   -5    0 ]
```

SEE ALSO: linalg[linsolve]

3.2.34   linalg[grad] – vector gradient of an expression

Calling Sequence:

  grad(*expr*, *v*)

Parameters:

  expr – scalar expression

  v   – vector or list of variables

Synopsis:

- The function **grad** computes the gradient of *expr* with respect to *v*.

- That is, it computes the following vector of partial derivatives:

vector([diff($expr,v[1]$), diff($expr,v[2]$), ...]) .

- This function is part of the `linalg` package, and so can be used in the form `grad(..)` only after performing the command `with(linalg)` or `with(linalg, grad)`. The function can always be accessed in the long form `linalg[grad](..)`.

Examples:
```
> with (linalg):
  grad( x*y*z, [x,y,z] );              ⟶          [ y z, x z, x y ]
> grad(3*x^2 +2*y*z, vector([x,y,z]) );  ⟶        [ 6 x, 2 z, 2 y ]
```

SEE ALSO: `linalg[diverge]`, `linalg[curl]`, `diff`

3.2.35   linalg[GramSchmidt] – compute orthogonal vectors

Calling Sequence:
> GramSchmidt([v_1, v_2, ... , v_n])
> GramSchmidt({ v_1, v_2, ... , v_n })

Parameters:
> v_1, v_2, ..., v_n –   linearly independent vectors

Synopsis:

- The function `GramSchmidt` computes a list or set of orthogonal vectors from a given list or set of linearly independent vectors, using the Gram-Schmidt orthogonalization process.

- The vectors given must be linearly independent, otherwise the vectors returned will also be dependent.

- The vectors returned are not normalized.

- This function is part of the `linalg` package, and so can be used in the form `GramSchmidt(..)` only after performing the command `with(linalg)` or `with(linalg, GramSchmidt)`. The function can always be accessed in the long form `linalg[GramSchmidt](..)`.

Examples:
```
> with (linalg):
  v1 := vector([1,0,0]);              ⟶        v1 := [ 1, 0, 0 ]
> v2 := vector([1,1,0]);              ⟶        v2 := [ 1, 1, 0 ]
> v3 := vector([1,1,1]);              ⟶        v3 := [ 1, 1, 1 ]
> GramSchmidt({v1,v2,v3});            ⟶  {[ 1, 0, 0 ], [ 0, 1, 0 ], [ 0, 0, 1 ]}
> u1 := vector([2,2,2]);              ⟶        u1 := [ 2, 2, 2 ]
> u2 := vector([0,2,2]);              ⟶        u2 := [ 0, 2, 2 ]
```

```
> u3 := vector([0,0,2]);
```
$\longrightarrow$ u3 := [0, 0, 2]

```
> GramSchmidt([u1,u2,u3]);
```
$\longrightarrow$ [[2, 2, 2], [-4/3, 2/3, 2/3],

[0, -1, 1]]

SEE ALSO: `linalg[basis]`, `linalg[intbasis]`, `linalg[norm]`

3.2.36   linalg[hadamard] – bound on coefficients of the determinant of a matrix

Calling Sequence:
   `hadamard(A)`

Parameters:
   A –   a square matrix

Synopsis:
- The function **hadamard** computes a bound on the **maxnorm** of $\det(A)$ where A is an n by n matrix whose coefficients lie in the domain Z, Q, R, Z[x], Q[x] or R[x].

- For Z, the classical inequality of Hadamard is :

  $|\det(A)| \le$ sqrt(product(sum(A[i,j]^2,j=1..n), i=1..n))

- This function is part of the **linalg** package, and so can be used in the form **hadamard(..)** only after performing the command **with(linalg)** or **with(linalg, hadamard)**. The function can always be accessed in the long form `linalg[hadamard](..)`.

Examples:
```
> with(linalg):
  A := matrix (3,3,[1,-1,0,0,2,-1,-2,0,1])
  ;
```
$\longrightarrow$
```
        [  1  -1   0 ]
        [            ]
   A := [  0   2  -1 ]
        [            ]
        [ -2   0   1 ]
```
```
> hadamard(A);
```
$\longrightarrow$ 1/2
50

SEE ALSO: `linalg[det]`, `linalg[norm]`

3.2.37   linalg[hermite] – Hermite normal form (reduced row echelon form)

Calling Sequence:
   `hermite(A, x)`

Parameters:

    A — a rectangular matrix of polynomials in x

    x — name

Synopsis:

- The function **hermite** computes the Hermite normal form (reduced row echelon form) of an m by n rectangular matrix of univariate polynomials in x over the field of rational numbers Q, or rational expressions over Q.

- In principle this should work for polynomials in x over any field F, i.e. the Euclidean domain F[x], but in practice the code is only as powerful as Maple's **normal** function.

- The Hermite normal form is obtained by doing elementary row operations on A. This includes interchanging rows, multiplying through a row by a unit, and subtracting a multiple of one row from another.

- This function is part of the **linalg** package, and so can be used in the form **hermite(..)** only after performing the command **with(linalg)** or **with(linalg, hermite)**. The function can always be accessed in the long form **linalg[hermite](..)**.

Examples:
```
> with(linalg):
> H := inverse(hilbert(2,x));
```

$$H := \begin{bmatrix} -(-3+x)^2\,(-2+x) & (-3+x)\,(-2+x)\,(-4+x) \\ (-3+x)\,(-2+x)\,(-4+x) & -(-3+x)^2\,(-4+x) \end{bmatrix}$$

```
> hermite(H,x);
```

$$\begin{bmatrix} x^2 - 5\,x + 6 & 0 \\ 0 & x^2 - 7\,x + 12 \end{bmatrix}$$

SEE ALSO: `linalg[smith]`, `linalg[ihermite]`, Hermite

3.2.38   linalg[hessian] – compute the Hessian matrix of an expression

Calling Sequence:

    hessian(*expr*, *vars*)

Parameters:

    expr — a scalar expression

    vars — a vector or list of variables

Synopsis:

- The procedure **hessian**(*expr*, *vars*) computes the Hessian matrix of *expr* with respect to *vars*.

- The matrix result is n x n, where n is the length of *vars*. The (i,j)th entry of the matrix result is **diff**(*expr*, *vars*[i], *vars*[j]).

- This function is part of the **linalg** package, and so can be used in the form **hessian(..)** only after performing the command **with(linalg)** or **with(linalg, hessian)**. The function can always be accessed in the long form **linalg[hessian](..)**.

Examples:
```
> with (linalg):
  hessian( x*y*z, [x,y,z] );
```
$\longrightarrow$
```
                                          [ 0   z   y ]
                                          [           ]
                                          [ z   0   x ]
                                          [           ]
                                          [ y   x   0 ]
```
```
> hessian ( x^2*y + 3*x*y^2, [x,y]);
```
$\longrightarrow$
```
                                    [    2 y        2 x + 6 y ]
                                    [                         ]
                                    [ 2 x + 6 y       6 x     ]
```
```
>
```

SEE ALSO: **diff**, **linalg[grad]**

3.2.39   linalg[hilbert] – create a Hilbert matrix

Calling Sequence:
> hilbert(*n*)
> hilbert(*n*, *x*)

Parameters:
> *n* – a positive integer
> *x* – optional expression

Synopsis:

- The function **hilbert** returns the n x n generalized Hilbert matrix.

- This matrix is symmetric and has $1/(i+j-x)$ as its (i,j)-th entry. If x is not specified, $x = 1$ is used.

- This function is part of the **linalg** package, and so can be used in the form **hilbert(..)** only after performing the command **with(linalg)** or **with(linalg, hilbert)**. The function can always be accessed in the long form **linalg[hilbert](..)**.

Examples:
```
> with (linalg):
  hilbert(3);
```
$\longrightarrow$

```
[  1   1/2  1/3 ]
[               ]
[ 1/2  1/3  1/4 ]
[               ]
[ 1/3  1/4  1/5 ]
```

```
> hilbert(3,x+1);
```
$\longrightarrow$

```
[   1      1      1   ]
[ -----  -----  ----- ]
[ 1 - x  2 - x  3 - x ]
[                     ]
[   1      1      1   ]
[ -----  -----  ----- ]
[ 2 - x  3 - x  4 - x ]
[                     ]
[   1      1      1   ]
[ -----  -----  ----- ]
[ 3 - x  4 - x  5 - x ]
```

3.2.40   linalg[htranspose] – compute the Hermitian transpose of a matrix

Calling Sequence:

    htranspose(A)

Parameters:

   A – an m x n matrix

Synopsis:
- The Hermitian transpose of A is computed. The result is an n x m matrix.

- The [i,j]-th element of the result is equal to the conjugate of the [j,i]-th element of A.

- The result inherits the indexing (for example, diagonal or sparse) function of A, if A has such an indexing.

- If A is a vector, then it is treated as if it were a column (n x 1) matrix.

- This function is part of the `linalg` package, and so can be used in the form `htranspose(..)` only after performing the command `with(linalg)` or `with(linalg, htranspose)`. The function can always be accessed in the long form `linalg[htranspose](..)`.

Examples:
```
> with (linalg):
  A := array([[1,2],[4,I]]);
```
$\longrightarrow$

```
          [ 1  2 ]
    A := [        ]
          [ 4  I ]
```

```
> htranspose(A);                    ⟶                    [ 1    4 ]
                                                          [        ]
                                                          [ 2  - I ]

> B := matrix (3,2,[1+I,-3,-I,2*I,4,5]);   ⟶             [ 1 + I   -3 ]
                                                          [            ]
                                                    B := [  - I    2 I ]
                                                          [            ]
                                                          [  4       5 ]

> htranspose(B);                   ⟶             [ 1 - I   I    4 ]
                                                 [                ]
                                                 [   -3  - 2 I  5 ]
```

SEE ALSO: linalg[transpose]

3.2.41   linalg[ihermite] – integer-only Hermite normal form

Calling Sequence:
>   ihermite(*A*)

Parameters:
>   *A*   –   a rectangular matrix of integers

Synopsis:
- The function ihermite computes the Hermite normal form (reduced row echelon form) of a rectangular matrix of integers.

- The Hermite normal form of *A* is an upper triangular matrix H with rank(*A*) = the number of nonzero rows of H. If *A* is an n by n matrix of full rank then |det(*A*)| = product(H[i,i],i=1..n).

- This is not an efficient method for computing the rank or determinant except that this may yield a partial factorization of det(*A*) without doing any explicit factorizations.

- The Hermite normal form is obtained by doing elementary row operations. This includes interchanging rows, multiplying through a row by -1, and adding an integral multiple of one row to another.

- This function is part of the linalg package, and so can be used in the form ihermite(..) only after performing the command with(linalg) or with(linalg, ihermite). The function can always be accessed in the long form linalg[ihermite](..).

Examples:
```
> with(linalg,ihermite):
  H := array( [[9,-36,30], [-36,192,-180],
  [30,-180,180]] );                        ⟶             [  9     -36    30  ]
                                                         [                   ]
                                                  H := [ -36    192   -180 ]
                                                         [                   ]
                                                         [  30   -180   180  ]
```

```
> ihermite(H);
```
$\longrightarrow$

```
[ 3    0   30 ]
[             ]
[ 0   12    0 ]
[             ]
[ 0    0   60 ]
```

SEE ALSO: linalg[hermite], linalg[ismith]

3.2.42   linalg[indexfunc] – determine the indexing function of an array

Calling Sequence:
indexfunc(A)

Parameters:
A – an array

Synopsis:
- The procedure call indexfunc(A) returns the indexing function for the array A.

- This could be one of the standard indexing functions such as **identity**, **symmetric**, or **diagonal**, or the name of a user-defined indexing function.

- If there is no indexing function specified for the array A, then indexfunc will return NULL.

- This function is part of the linalg package, and so can be used in the form indexfunc(..) only after performing the command with(linalg) or with(linalg, indexfunc). The function can always be accessed in the long form linalg[indexfunc](..).

Examples:
```
> with (linalg):
> indexfunc( array(1..2,1..2,symmetric) );
                            symmetric
> indexfunc( array(1..2,1..2,[[1,0],[0,1]]) );
```

SEE ALSO: indexfcn

3.2.43   linalg[innerprod] – calculate the inner product

Calling Sequence:
innerprod (u, $A_1, A_2, \ldots, A_n$, v)

Parameters:
u, v – vectors

$A_1, \ldots, A_n$ – matrices

Synopsis:

- The function innerprod calculates the inner product of a sequence of vectors and matrices. The dimension of each matrix and vector must be compatible for multiplication in the order given.

- This function is part of the linalg package, and so can be used in the form innerprod(..) only after performing the command with(linalg) or with(linalg, innerprod). The function can always be accessed in the long form linalg[innerprod](..).

Examples:

```
> with (linalg):
  u := vector(2,[1,2]);                   ⟶          u := [ 1, 2 ]

> v := vector(3,[1,2,3]);                 ⟶          v := [ 1, 2, 3 ]

> A := matrix(2,3,[1,1,1,2,2,2]);         ⟶               [ 1  1  1 ]
                                                     A := [         ]
                                                          [ 2  2  2 ]

> innerprod(u,A,v);                       ⟶                30

> w := vector(3,[3,2,1]);                 ⟶          w := [ 3, 2, 1 ]

> innerprod(v,w);                         ⟶                10
```

SEE ALSO: evalm, linalg[crossprod], linalg[dotprod]

3.2.44   linalg[intbasis] – determine a basis for the intersection of spaces

Calling Sequence:

   intbasis(S_1, S_2, ..., S_n)

Parameters:

   S_i –   a vector, set of vectors or list of vectors

Synopsis:

- A basis for the intersection of the vector spaces spanned by the vectors in S_i is returned.

- A basis for the zero-dimensional space is an empty set or list.

- This function is part of the linalg package, and so can be used in the form intbasis(..) only after performing the command with(linalg) or with(linalg, intbasis). The function can always be accessed in the long form linalg[intbasis](..).

Examples:

```
> with(linalg):
  v1 := vector([1,0,1,0]);                ⟶          v1 := [ 1, 0, 1, 0 ]

> v2 := vector([0,1,0,1]);                ⟶          v2 := [ 0, 1, 0, 1 ]

> v3 := vector([1,2,1,1]);                ⟶          v3 := [ 1, 2, 1, 1 ]
```

```
> v4 := vector([-1,-2,1,0]);                    ⟶              v4 := [ -1, -2, 1, 0 ]
> intbasis({v1,v2},{v3,v4});                     ⟶                       {}
> intbasis([v1,v2,v3],[v3,v4],[v2,v3]);          ⟶               {[ 1, 2, 1, 1 ]}
> u := evalm(v1-v2);                             ⟶              u := [ 1, -1, 1, -1 ]
> v := evalm(v1+v2);                             ⟶               v := [ 1, 1, 1, 1 ]
> w := evalm(v1-v3);                             ⟶              w := [ 0, -2, 0, -1 ]
> intbasis({u,v,w},{v1,v2,v3});                  ⟶      {[ 0, -2, 0, -1 ], [ 1, 3, 1, 2 ],
                                                              [ -2, -4, -2, -2 ]}
```

SEE ALSO: `linalg[basis]`, `linalg[sumbasis]`

3.2.45   linalg[inverse] – compute the inverse of a matrix

Calling Sequence:

   inverse(*A*)

Parameters:

   A   –   a square matrix

Synopsis:

- The function **inverse** computes the matrix inverse of A. An error occurs if the matrix is singular.

- This function uses Cramer´s rule for matrices of dimension less than or equal to 4 by 4, and for matrices with the ´sparse´ indexing function.

- For other matrices, the inverse is computed by applying the operations for the Gauss-Jordan reduction of A to an identity matrix of the same shape.

- This function is part of the `linalg` package, and so can be used in the form `inverse(..)` only after performing the command `with(linalg)` or `with(linalg, inverse)`. The function can always be accessed in the long form `linalg[inverse](..)`.

Examples:
```
> with (linalg):
  A := array( [[1,x],[2,3]] );                   ⟶                   [ 1   x ]
                                                             A := [       ]
                                                                   [ 2   3 ]

> inverse( A );                                  ⟶      [     3           x      ]
                                                        [ - --------   -------- ]
                                                        [   - 3 + 2 x   - 3 + 2 x ]
                                                        [                         ]
                                                        [     2           1       ]
                                                        [  --------   - -------- ]
                                                        [  - 3 + 2 x     - 3 + 2 x ]
```

SEE ALSO: `linalg[det]`, `linalg[gaussjord]`

3.2.46   linalg[ismith] – integer-only Smith normal form

Calling Sequence:
    ismith(A)

Parameters:
    A – a rectangular matrix of integers

Synopsis:
- The function ismith computes the Smith normal form S of an n by m rectangular matrix of integers.

- If two n by n matrices have the same Smith normal form, they are equivalent.

- The Smith normal form is a diagonal matrix where

```
rank(A) = number of nonzero rows (columns) of S
sign(S[i,i]) = 1  for 0 < i <= rank(A)
S[i,i] divides S[i+1,i+1] for 0 < i < rank(A)
product(S[i,i],i=1..r) divides det(M) for all minors M of rank 0 < r <= rank(A)
```

  Hence if n = m and rank(A) = n then abs(det(A)) = product(S[i,i],i=1..n).

- The Smith normal form is obtained by doing elementary row and column operations. This includes interchanging rows (columns), multiplying through a row (column) by -1, and adding integral multiples of one row (column) to another.

- Although the rank and determinant can be easily obtained from S this is not an efficient method for computing these quantities except that this may yield a partial factorization of det(A) without doing any explicit factorizations.

- This function is part of the linalg package, and so can be used in the form ismith(..) only after performing the command with(linalg) or with(linalg, ismith). The function can always be accessed in the long form linalg[ismith](..).

Examples:
```
> with(linalg,ismith):
  H := array( [[9,-36,30], [-36,192,-180],
  [30,-180,180]] );
```
$$H := \begin{bmatrix} 9 & -36 & 30 \\ -36 & 192 & -180 \\ 30 & -180 & 180 \end{bmatrix}$$

```
> ismith(H);
```
$$\begin{bmatrix} 3 & 0 & 0 \\ 0 & 12 & 0 \\ 0 & 0 & 60 \end{bmatrix}$$

SEE ALSO: linalg[smith], linalg[ihermite]

3.2.47   linalg[iszero] – determine whether a matrix is zero

Calling Sequence:

    `iszero(A)`

Parameters:

    A –   matrix

Synopsis:

- The call `iszero(A)` returns true if all of the entries of matrix A are zero, false or an error otherwise.

- This function is part of the `linalg` package, and so can be used in the form `iszero(..)` only after performing the command `with(linalg)` or `with(linalg, iszero)`. The function can always be accessed in the long form `linalg[iszero](..)`.

Examples:
```
> with(linalg):
  A := array([[2,1],[1,2]]);                                       [ 2  1 ]
                                            ⟶         A := [        ]
                                                              [ 1  2 ]

> iszero(subs(x=A,charpoly(A,x)));         ⟶               true
> A := matrix(3,3,[1,0,0,0,0,0,0,0,0]);    ⟶            [ 1  0  0 ]
                                                        [          ]
                                                 A := [ 0  0  0 ]
                                                        [          ]
                                                        [ 0  0  0 ]

> iszero (A);                              ⟶              false
```

SEE ALSO: `linalg[equal]`

3.2.48   linalg[jacobian] – compute the Jacobian matrix of a vector function

Calling Sequence:

    `jacobian(f, v)`

Parameters:

    f –   a vector or list of expressions

    v –   a vector or list of variables

Synopsis:

- The procedure `jacobian(f, v)` computes the Jacobian matrix of f with respect to v. The (i,j)-th entry of the matrix result is diff(f[i],v[j]).

- This function is part of the `linalg` package, and so can be used in the form `jacobian(..)` only after performing the command `with(linalg)` or `with(linalg, jacobian)`. The function can always be accessed in the long form `linalg[jacobian](..)`.

Examples:
```
> with (linalg):
  A := vector( [x^2, x*y, x*z] );
```
$\longrightarrow$

$$A := [\ x^2,\ x\,y,\ x\,z\ ]$$

```
> jacobian( A,[x,y,z] );
```
$\longrightarrow$

$$\begin{array}{c}[\ 2\,x\quad 0\quad 0\ ]\\ [\qquad\qquad\quad]\\ [\ y\quad x\quad 0\ ]\\ [\qquad\qquad\quad]\\ [\ z\quad 0\quad x\ ]\end{array}$$

SEE ALSO: linalg[grad], linalg[hessian], diff

3.2.49   linalg[JordanBlock] – return a Jordan block matrix

Calling Sequence:

> JordanBlock(t, n)

Parameters:

> t –   any constant, algebraic number, or rational number
>
> n –   an integer: the size of the Jordan block matrix

Synopsis:

- The function JordanBlock returns a matrix J of a special form: the Jordan block. The matrix J will be such that $J[i,i] = t$, $J[i,i+1] = 1$ for i=1..n-1 and $J[i,j] = 0$ otherwise.

- This function is part of the linalg package, and so can be used in the form JordanBlock(..) only after performing the command with(linalg) or with(linalg, JordanBlock). The function can always be accessed in the long form linalg[JordanBlock](..).

Examples:
```
> with (linalg):
  linalg[JordanBlock](3,5);
```
$\longrightarrow$

$$\begin{array}{c}[\ 3\quad 1\quad 0\quad 0\quad 0\ ]\\ [\qquad\qquad\qquad\qquad\ ]\\ [\ 0\quad 3\quad 1\quad 0\quad 0\ ]\\ [\qquad\qquad\qquad\qquad\ ]\\ [\ 0\quad 0\quad 3\quad 1\quad 0\ ]\\ [\qquad\qquad\qquad\qquad\ ]\\ [\ 0\quad 0\quad 0\quad 3\quad 1\ ]\\ [\qquad\qquad\qquad\qquad\ ]\\ [\ 0\quad 0\quad 0\quad 0\quad 3\ ]\end{array}$$

```
> linalg[JordanBlock](x,7);
```
$$\longrightarrow$$

```
[ x  1  0  0  0  0  0 ]
[                     ]
[ 0  x  1  0  0  0  0 ]
[                     ]
[ 0  0  x  1  0  0  0 ]
[                     ]
[ 0  0  0  x  1  0  0 ]
[                     ]
[ 0  0  0  0  x  1  0 ]
[                     ]
[ 0  0  0  0  0  x  1 ]
[                     ]
[ 0  0  0  0  0  0  x ]
```

SEE ALSO: linalg[diag], linalg[companion]

3.2.50 linalg[jordan] – compute the Jordan form of a matrix

Calling Sequence:
    jordan(A)
    jordan(A, $'P'$)

Parameters:

   A –   a square matrix

   $'P'$ –   (optional) used to return the transition matrix

Synopsis:

- The call jordan(A) computes and returns the Jordan form J of a matrix A.

- J has the following structure: J = diag(j1,j2,..jk) where the ji's are Jordan block matrices. The diagonal entries of these Jordan blocks are the eigenvalues of A (and also of J).

- If the optional second argument is given, then P will be assigned the transformation matrix corresponding to this Jordan form, that is, the matrix P such that inverse(P) * J * P = A.

- The Jordan form is unique up to permutations of the Jordan blocks.

- If the roots of the characteristic polynomial cannot be solved exactly, floating point approximations will be used.

- This function is part of the linalg package, and so can be used in the form jordan(..) only after performing the command with(linalg) or with(linalg, jordan). The function can always be accessed in the long form linalg[jordan](..).

Examples:
```
> with(linalg):
  B := array ([[1,0],[3,2]]);
```
$$\longrightarrow$$

```
        [ 1  0 ]
  B := [      ]
        [ 3  2 ]
```

```
> jordan(B,'P');                      ⟶              [ 1   0 ]
                                                      [       ]
                                                      [ 0   2 ]

> eval(P);                            ⟶              [ 1    0  ]
                                                      [         ]
                                                      [ 1   1/3 ]

> jordan(P);                          ⟶              [ 1    0  ]
                                                      [         ]
                                                      [ 0   1/3 ]
```

SEE ALSO: linalg[eigenvals], Eigenvals, linalg[frobenius]

3.2.51   linalg[kernel] – compute a basis for the null space
linalg[nullspace] – a synonym for kernel

Calling Sequence:
kernel(A)
kernel(A, ´*nulldim*´)
nullspace(A)
nullspace(A, ´*nulldim*´)

Parameters:

A	–	a matrix
´*nulldim*´	–	an optional unevaluated name

Synopsis:

- The procedure **kernel** computes a basis for the null space (kernel) of the linear transformation defined by the matrix A. The result is a set of vectors, possibly empty.

- When given an optional second parameter, **kernel**(A,´*nulldim*´), in addition to the above, assigns the variable *nulldim* the vector space dimension of the kernel (a non-negative integer).

- This function is part of the **linalg** package, and so can be used in the form **kernel**(..) only after performing the command **with(linalg)** or **with(linalg, kernel)**. The function can always be accessed in the long form **linalg[kernel]**(..).

Examples:
```
> with (linalg):
  A := array( [[1,2,3],[1,x,3],[0,0,0]] )
  ;                                   ⟶              [ 1   2   3 ]
                                                      [           ]
                                               A := [ 1   x   3 ]
                                                      [           ]
                                                      [ 0   0   0 ]

> kernel( A );                        ⟶              {[ -3, 0, 1 ]}
```

SEE ALSO: linalg[colspace], linalg[rowspace], Nullspace, linalg[range]

3.2.52 linalg[laplacian] – compute the Laplacian

Calling Sequence:

   laplacian(*expr*, *v*)

Parameters:

   expr – a scalar expression

   v – a vector or list of variables

Synopsis:

- The `laplacian(`*expr*`,`*v*`)` function computes the laplacian of *expr* with respect to *v*.

- The laplacian is defined to be the sum of the second derivatives `diff(`*expr*`, x$2)` for x in *v* .

- This function is part of the `linalg` package, and so can be used in the form `laplacian(..)` only after performing the command `with(linalg)` or `with(linalg, laplacian)`. The function can always be accessed in the long form `linalg[laplacian](..)`.

Examples:
```
> with (linalg):
  laplacian(x^2*y*z,[x,y,z]);                    --->            2 y z
```

SEE ALSO: `linalg[grad]`

3.2.53 linalg[leastsqrs] – least-squares solution of equations

Calling Sequence:

   leastsqrs(A, b)

   leastsqrs(S, v)

Parameters:

   A – a matrix

   b – a vector

   S – a set of equations or expressions

   v – a set of names

Synopsis:

- The call `leastsqrs(`A`,`b`)` returns the vector that best satisfies $A \mathbf{x} = b$ in the least-squares sense. The result returned is the vector x which minimizes `norm(`A `x - ` b`, 2)`.

- The call `leastsqrs(`S`,`v`)` finds the values for the variables in v which minimize the equations or expressions in S in the least-squares sense. The result returned is a set of equations whose left-hand sides are from v .

- This function is part of the `linalg` package, and so can be used in the form `leastsqrs(..)` only after performing the command `with(linalg)` or `with(linalg, leastsqrs)`. The function can always be accessed in the long form `linalg[leastsqrs](..)`.

Examples:
```
> with (linalg):
> A := array( [[1,1,1],[1,2,4],[1,0,0],[1,-1,1]] );
```

$$A := \begin{bmatrix} 1 & 1 & 1 \\ 1 & 2 & 4 \\ 1 & 0 & 0 \\ 1 & -1 & 1 \end{bmatrix}$$

```
> b := array( [3,10,3,9/10] );
```

$$b := [\, 3, \ 10, \ 3, \ 9/10 \,]$$

```
> leastsqrs(A,b);
```

$$[\, \frac{327}{200}, \ \frac{301}{200}, \ \frac{49}{40} \,]$$

```
> S := {c0+c1+c2-3, c0+2*c1+4*c2-10, c0-9/10, c0-c1+c2-3};
  S := {c0 - 9/10, c0 + c1 + c2 - 3, c0 + 2 c1 + 4 c2 - 10, c0 - c1 + c2 - 3}
```

```
> leastsqrs(S,{c0,c1,c2});
```

$$\{c_2 = \frac{91}{40}, \ c_1 = 7/200, \ c_0 = \frac{159}{200}\}$$

SEE ALSO: `linalg[linsolve]`

3.2.54   linalg[linsolve] – solution of linear equations

Calling Sequence:

> linsolve(A, b)
> linsolve(A, B)

Parameters:

> A – a matrix
> b – a vector
> B – a matrix

Synopsis:

- The function `linsolve`(A, b) finds the vector x which satisfies the matrix equation $A\,x = b$. If A has n rows and m columns, then `vectdim`(b) must be n and `vectdim(x)` will be m, if a solution exists.

- If $A\,x = b$ has no solution, then the null sequence **NULL** is returned. If $A\,x = b$ has many solutions, then the result will use the global names t1, t2, t3, ... to describe the family of solutions parametrically.

- The call `linsolve(`A`,`B`)` finds the matrix X which solves the matrix equation A X $= B$ where each column of X satisfies A col(X,i) $=$ col(B,i) . If A X $= B$ has does not have a unique solution, then `NULL` is returned.

- This function is part of the `linalg` package, and so can be used in the form `linsolve(..)` only after performing the command `with(linalg)` or `with(linalg, linsolve)`. The function can always be accessed in the long form `linalg[linsolve](..)`.

Examples:
```
> with (linalg):
  A := array([[1,2],[1,3]]);
```
$\longrightarrow$
```
                              [ 1  2 ]
                       A := [        ]
                              [ 1  3 ]
```
```
> b := array([1,-2]);
```
$\longrightarrow$
```
                       b := [ 1, -2 ]
```
```
> linsolve(A,b);
```
$\longrightarrow$
```
                          [ 7, -3 ]
```
```
> B := array([[1,1],[-2,1]]);
```
$\longrightarrow$
```
                              [  1  1 ]
                       B := [         ]
                              [ -2  1 ]
```
```
> linsolve(A,B);
```
$\longrightarrow$
```
                          [  7  1 ]
                          [       ]
                          [ -3  0 ]
```

SEE ALSO: `linalg[leastsqrs]`, `solve`

3.2.55 linalg[matrix] – create a matrix

Calling Sequence:
>　　`matrix(`L`)`
>　　`matrix(`m`, `n`)`
>　　`matrix(`m`, `n`, `L`)`
>　　`matrix(`m`, `n`, `f`)`
>　　`matrix(`m`, `n`, `lv`)`

Parameters:

L	–	list of lists or vectors of elements
m,n	–	positive integers (row and column dimensions)
f	–	a function used to create the matrix elements
lv	–	a list or vector of elements

Synopsis:
- The `matrix` function is part of the `linalg` package. It provides a simplified syntax for creating matrices. A general description of matrices in Maple is available under the heading `matrix`.

- The call `matrix(m, n, L)` creates an m by n matrix where the first row of the matrix is defined by the list/vector $L[1]$, the second row by $L[2]$, and so forth. The call `matrix(L)` is equivalent to `matrix(m, n, L)` where m = `nops(L)` and n = max(`seq(nops(L[i]),i=1..m)`) .

- The call `matrix(m, n)` creates an m by n matrix with unspecified elements.

- The call `matrix(m, n, f)` creates an m by n matrix whose elements are the result of the function f (possibly a constant) acting on the row and column index of the matrix. Thus, `matrix(m, n, f)` is equivalent to `matrix([[f(1,1), ..., f(1,n)], ..., [f(m,1), ..., f(m,n)]])` .

- The call `matrix(m, n, lv)` creates an m by n matrix whose elements are read off from lv row by row, where lv is a list or vector of elements of type algebraic.

- Since matrices are represented as two-dimensional arrays, see **array** for further details about how to work with matrices.

- This function is part of the `linalg` package, and so can be used in the form `matrix(..)` only after performing the command `with(linalg)` or `with(linalg, matrix)`. The function can always be accessed in the long form `linalg[matrix](..)`.

Examples:
```
> with(linalg):
  matrix(2,2,[5,4,6,3]);
```
$\longrightarrow$
```
[ 5   4 ]
[        ]
[ 6   3 ]
```

```
> matrix([[5,4],[6,3]]);
```
$\longrightarrow$
```
[ 5   4 ]
[        ]
[ 6   3 ]
```

```
> matrix(2,2,0);
```
$\longrightarrow$
```
[ 0   0 ]
[        ]
[ 0   0 ]
```

```
> f := (i,j) -> x^(i+j-1):
  A := matrix(2,2,f);
```
$\longrightarrow$
```
            [       2 ]
            [ x     x ]
      A :=  [         ]
            [  2      3 ]
            [ x      x ]
```

```
> map(diff,A,x);
```
$\longrightarrow$
```
      [  1      2 x ]
      [            ]
      [          2 ]
      [ 2 x    3 x ]
```

```
> A := matrix(2,2,[x,y]);
```
$\longrightarrow$
```
            [    x           y    ]
      A :=  [                     ]
            [ A[2, 1]   A[2, 2] ]
```

```
> matrix([col(A,1..2)]);
```
$\longrightarrow$
```
      [ x   A[2, 1] ]
      [             ]
      [ y   A[2, 2] ]
```

SEE ALSO: `main[matrix]`, `array`, `vector`, `linalg`

3.2.56 linalg[minor] – compute a minor of a matrix

Calling Sequence:

  minor(A, r, c)

Parameters:

  A – a matrix

  r,c – positive integers

Synopsis:

- The function `minor` returns the (r,c)th minor of the matrix A. This is the matrix obtained by removing the rth row and the cth column of A.

- This function is part of the `linalg` package, and so can be used in the form `minor(..)` only after performing the command `with(linalg)` or `with(linalg, minor)`. The function can always be accessed in the long form `linalg[minor](..)`.

Examples:

```
> with (linalg):
  A := matrix(3,3,[1,5,2,6,3,7,4,8,5]);
```
$\longrightarrow$
$$A := \begin{bmatrix} 1 & 5 & 2 \\ 6 & 3 & 7 \\ 4 & 8 & 5 \end{bmatrix}$$

```
> minor (A,2,2);
```
$\longrightarrow$
$$\begin{bmatrix} 1 & 2 \\ 4 & 5 \end{bmatrix}$$

SEE ALSO: `linalg[row]`, `linalg[col]`, `linalg[delrows]`, `linalg[delcols]`, `linalg[submatrix]`

3.2.57 linalg[minpoly] – compute the minimum polynomial of a matrix

Calling Sequence:

  minpoly(A, x)

Parameters:

  A – a square matrix

  x – a name

Synopsis:

- The procedure `minpoly(`A`,`x`)` computes the minimum polynomial of the matrix A in x. The minimum polynomial is the polynomial of lowest degree which annihilates A.

- The minimum polynomial will always divide the characteristic polynomial.

- This function is part of the `linalg` package, and so can be used in the form `minpoly(..)` only after performing the command `with(linalg)` or `with(linalg, minpoly)`. The function can always be accessed in the long form `linalg[minpoly](..)`.

Examples:
```
> with(linalg,minpoly):
  with(linalg,charpoly):
  A := array([[2,1,0,0],[0,2,0,0],[0,0,1,1]
  ,[0,0,-2,4]]):
  m := minpoly(A,x);
```
$$m := -12 + 16\,x - 7\,x^2 + x^3$$

```
> p := expand(charpoly(A,x));
```
$$p := x^4 - 9\,x^3 + 30\,x^2 - 44\,x + 24$$

```
> divide(p,m);
```
$$\text{true}$$

SEE ALSO: `linalg[charpoly]`

3.2.58   linalg[mulcol] – multiply a column of a matrix by an expression
linalg[mulrow] – multiply a row of a matrix by an expression

Calling Sequence:
    mulcol(A, c, $expr$)
    mulrow(A, r, $expr$)

Parameters:

A	–	a matrix
r,c	–	positive integers
$expr$	–	a scalar expression

Synopsis:
- The call `mulcol`$(A,c,expr)$ returns a matrix which has the same entries as A with the cth column multiplied by $expr$. Likewise `mulrow`$(A,r,expr)$ returns a matrix with the rth row multiplied by $expr$.

Examples:
```
> with (linalg):
  A := matrix([[1,2],[3,4]]);
```
$$A := \begin{bmatrix} 1 & 2 \\ 3 & 4 \end{bmatrix}$$

```
> mulrow(A,2,2);
```
$$\begin{bmatrix} 1 & 2 \\ 6 & 8 \end{bmatrix}$$

```
> mulcol(A,2,x);
```
$$\begin{bmatrix} 1 & 2\,x \\ 3 & 4\,x \end{bmatrix}$$

3.2.59   linalg[multiply] – matrix-matrix or matrix-vector multiplication

Calling Sequence:

   multiply(A, B, ...)

Parameters:

   A, B, ... – matrices

Synopsis:

* The function `multiply(`$A,B,...$`)` calculates the matrix product A B ... . The dimensions of each matrix must be consistent with the rules of matrix multiplication.

* The call `multiply(`A`,v)`, for a matrix A and vector v, calculates the matrix-vector product A v. The number of entries in v must be equal to the number of columns of A. Thus if A is an n x m matrix, vectdim(v) must be m. The result is a vector with n entries.

* The `evalm` command provides an easy method of accessing `multiply`.

* This function is part of the `linalg` package, and so can be used in the form `multiply(..)` only after performing the command `with(linalg)` or `with(linalg, multiply)`. The function can always be accessed in the long form `linalg[multiply](..)`.

Examples:
```
> with (linalg):
  A := array( [[1,2],[3,4]] ):
  B := array( [[0,1],[1,0]] ):
  C := array( [[1,2],[4,5]] ):
  multiply(A,B,C);
```
$\longrightarrow$
```
                                                    [  6    9 ]
                                                    [         ]
                                                    [ 16   23 ]
```
```
> v := vector([3,4]):
  multiply(A,v);
```
$\longrightarrow$
```
                                                    [ 11, 25 ]
```
```
> multiply(v,transpose(v));
```
$\longrightarrow$
```
                                                    [  9   12 ]
                                                    [         ]
                                                    [ 12   16 ]
```

SEE ALSO: `evalm`, `linalg[innerprod]`

3.2.60   linalg[norm] – norm of a matrix or vector

Calling Sequence:

   norm(A)

   norm(A, *normname*)

Parameters:

   A – a matrix or vector

   normname – (optional) a matrix/vector norm

Synopsis:

- The function norm(*A*, *normname*) computes the specified matrix or vector norm for the matrix or vector *A*.

- For matrices, *normname* should be one of: 1, 2, ´infinity´, ´frobenius´.

- For vectors, *normname* should be one of: any positive integer, ´infinity´, ´frobenius´

  . The default norm used throughout the linalg package is the infinity norm. Thus norm(A) computes the infinity norm of *A* and is equivalent to norm(*A*,infinity) .

- For vectors, the infinity norm is the maximum magnitude of all elements. The infinity norm of a matrix is the maximum row sum, where the row sum is the sum of the magnitudes of the elements in a given row.

- The frobenius norm of a matrix or vector is defined to be the square root of the sum of the squares of the magnitudes of each element.

- The ´1´-norm of a matrix is the maximum column sum, where the column sum is the sum of the magnitudes of the elements in a given column. The ´2´-norm of a matrix is the square root of the maximum eigenvalue of the matrix A * transpose(A) .

- For a positive integer k, the k-norm of a vector is the *k*th root of the sum of the magnitudes of each element raised to the *k*th power.

- This function is part of the linalg package, and so can be used in the form norm(..) only after performing the command with(linalg) or with(linalg, norm). The function can always be accessed in the long form linalg[norm](..).

Examples:
```
> with (linalg):
  norm( array([[1,-2],[3,-4]]), infinity )
  ;
> norm( array([1,-1,2]), 2 );
```
$\longrightarrow$ 7

$\longrightarrow$ 1/2
                      6

SEE ALSO: linalg[cond], norm

3.2.61   linalg[orthog] – test for orthogonal matrices

Calling Sequence:
    orthog(*A*)

Parameters:
    A – a matrix

Synopsis:

- The function orthog returns true if it can show that the matrix *A* is orthogonal, false if it can show that the matrix is not orthogonal, and FAIL otherwise.

- A matrix is orthogonal if the inner product of any column of A with itself is 1, and the inner product of any column of A with any other column is 0.

- The Maple function **testeq** is used to test whether expressions are equivalent to 0 or 1.

- This function is part of the **linalg** package, and so can be used in the form **orthog(..)** only after performing the command **with(linalg)** or **with(linalg, orthog)**. The function can always be accessed in the long form **linalg[orthog](..)**.

Examples:
```
> with (linalg):
> orthog( array([[-1/2,sqrt(3)/2],[sqrt(3)/2,1/2]]) );
```
$$\text{true}$$

SEE ALSO: **testeq**

3.2.62   linalg[permanent] – compute the permanent of a matrix

Calling Sequence:
>   permanent(A)

Parameters:
> A –   a square matrix

Synopsis:
- The permanent of a matrix is computed in a way similar to the determinant. The permanent of an arbitrary 2x2 matrix, **matrix(2,2,[a,b,c,d])**, is **ad+bc**. To compute the permanent of larger matrices, minor expansion is used in the same manner as when computing determinants, only without alternating sign while expanding along a row or column.

- This function is part of the **linalg** package, and so can be used in the form **permanent(..)** only after performing the command **with(linalg)** or **with(linalg, permanent)**. The function can always be accessed in the long form **linalg[permanent](..)**.

Examples:
```
> with (linalg):
  A := matrix([[1,2,3],[4,5,6],[7,8,9]]);        ⟶
```
$$A := \begin{bmatrix} 1 & 2 & 3 \\ 4 & 5 & 6 \\ 7 & 8 & 9 \end{bmatrix}$$

```
> permanent(A);                                  ⟶
```
$$450$$

```
> B := matrix([[a,b,c],[d,e,f],[g,h,i]]);        ⟶
```
$$B := \begin{bmatrix} a & b & c \\ d & e & f \\ g & h & i \end{bmatrix}$$

>
>   permanent(B); $\longrightarrow$ a e i + a f h + d b i + d c h + g b f
>   + g c e

SEE ALSO: linalg[det]

3.2.63   linalg[pivot] – pivot about a matrix entry

Calling Sequence:

    pivot(A, i, j)

    pivot(A, i, j, $r..s$)

Parameters:

    A –   a matrix

    i,j –   positive integers

    $r..s$ –   range of rows to be pivoted

Synopsis:

* The function **pivot** pivots the matrix A about $A[i,j]$ which must be non-zero.

* The function **pivot**(A,i,j) will add multiples of the ith row to every other row in the matrix, with the result that the (k,j)th entry of the matrix A is set to zero for all k not equal to i. That is, the jth column of the matrix will be all zeros, except for the (i,j)th element.

* The call **pivot**$(A,i,j,r..s)$ acts like **pivot**(A,i,j) except that only rows r through s are set to zero in the jth column. Rows not in the range $r..s$ are not affected.

* This function is part of the **linalg** package, and so can be used in the form **pivot**(..) only after performing the command **with(linalg)** or **with(linalg, pivot)**. The function can always be accessed in the long form **linalg[pivot]**(..).

Examples:

```
> with (linalg):
  A := matrix(4,4,[1,2,3,4,5,6,7,8,9,0,1,2,
  3,4,5,6]);
```
$\longrightarrow$

$$A := \begin{bmatrix} 1 & 2 & 3 & 4 \\ 5 & 6 & 7 & 8 \\ 9 & 0 & 1 & 2 \\ 3 & 4 & 5 & 6 \end{bmatrix}$$

```
> A := pivot(A,2,1);
```
$\longrightarrow$

$$A := \begin{bmatrix} 0 & 4/5 & 8/5 & 12/5 \\ 5 & 6 & 7 & 8 \\ 0 & -54/5 & -58/5 & -62/5 \\ 0 & 2/5 & 4/5 & 6/5 \end{bmatrix}$$

```
> A := pivot(A,3,2);
```
$\longrightarrow$

$$
A := \begin{bmatrix}
0 & 0 & \dfrac{20}{27} & \dfrac{40}{27} \\
5 & 0 & 5/9 & 10/9 \\
0 & -54/5 & -58/5 & -62/5 \\
0 & 0 & \dfrac{10}{27} & \dfrac{20}{27}
\end{bmatrix}
$$

SEE ALSO: `linalg[gausselim]`

3.2.64   linalg[potential] – compute the potential of a vector field

Calling Sequence:

    potential (*f*, *var*, *'V'*)

Parameters:

    f   –   a vector field

    var   –   list of variables

    'V'   –   name in which the potential is returned

Synopsis:

- The function **potential** determines whether a given vector function is derivable from a scalar potential, and determines that potential if it exists.

- The function returns **true** if the function *f* has a scalar potential, and **false** if it does not.

- If a scalar potential for *f* exists, it will be assigned to the name given in the third argument *V*. If potential returns true, then *V* will be assigned a scalar function such that grad $V = f$.

- This function is part of the **linalg** package, and so can be used in the form **potential(..)** only after performing the command **with(linalg)** or **with(linalg, potential)**. The function can always be accessed in the long form **linalg[potential](..)**.

Examples:

```
> with (linalg):
  f := [ 2*x*y + y^3, x^2 + 3*x*y^2];
```
$\longrightarrow$
$$f := [2\,x\,y + y^3,\ x^2 + 3\,x\,y^2]$$

```
> potential (f, [x,y], 'F');
```
$\longrightarrow$ true

```
> F;
```
$\longrightarrow$
$$x\,y^3 + x^2\,y$$

```
> g := [2*x, 2*y, 2*z];
```
$\longrightarrow$
$$g := [2\,x,\ 2\,y,\ 2\,z]$$

```
> potential (g, [x,y,z], 'G');
```
$\longrightarrow$ true

> G; $\longrightarrow$ $$z^2 + y^2 + x^2$$

SEE ALSO: linalg[vecpotent], linalg[grad]

3.2.65   linalg[randmatrix] – random matrix generator

Calling Sequence:
>   randmatrix(*m*, *n*, *options*)

Parameters:
>   *m, n*   –   positive integers
>   *options* –   (optional) names or equations

Synopsis:

- The **randmatrix** function generates a random matrix of dimension *m* by *n*. It is intended to be used for generating examples for debugging, testing and demonstration purposes. Several options are available. The options determine the form of the matrix and the entries in the matrix.

- The first two arguments specify the row dimension and column dimension. The remaining arguments are interpreted as options and are input as equations or names in any order. The names **sparse**, **dense** (default), **symmetric**, **antisymmetric**, and **unimodular** specify the structure of the matrix. The equation **entries = f** specifies that the nullary function **f** is to be used to generate the matrix entries. The default is **rand(-99..99)**. Thus the matrix entries are random two digit integers.

- This function is part of the **linalg** package, and so can be used in the form **randmatrix(..)** only after performing the command **with(linalg)** or **with(linalg, randmatrix)**. The function can always be accessed in the long form **linalg[randmatrix](..)**.

Examples:
```
> with(linalg, randmatrix):
  randmatrix(3, 3);
```
$\longrightarrow$
```
                    [ -85   -55   -37 ]
                    [                 ]
                    [ -35    97    50 ]
                    [                 ]
                    [  79    56    49 ]
```

```
> randmatrix(3, 3, unimodular);
```
$\longrightarrow$
```
                    [ 1    63    57 ]
                    [               ]
                    [ 0     1   -59 ]
                    [               ]
                    [ 0     0     1 ]
```

```
> poly := proc() Randpoly(3, x) mod 3
  end;
```
$\longrightarrow$ poly := proc() Randpoly(3,x) mod 3 end

```
> randmatrix(2, 2, entries = poly);
```
$$
\longrightarrow
\begin{bmatrix}
2x^3 + x^2 & 2x^3 + 1 \\[4pt]
x^3 + 2x + 2 & x^3 + x^2
\end{bmatrix}
$$

SEE ALSO: `linalg[matrix]`

3.2.66   linalg[rank] – rank of a matrix

Calling Sequence:

   rank(A)

Parameters:

   A – a matrix

Synopsis:

- The function **rank** computes the rank of a matrix, by performing Gaussian elimination on the rows of the given matrix. The rank of the matrix A is the number of non-zero rows in the resulting matrix.

- This function is part of the `linalg` package, and so can be used in the form `rank(..)` only after performing the command `with(linalg)` or `with(linalg, rank)`. The function can always be accessed in the long form `linalg[rank](..)`.

Examples:

```
> with (linalg):
  A := matrix(3,3,[x,1,0,0,0,1,x*y,y,1]);
```
$$
\longrightarrow
A := \begin{bmatrix}
x & 1 & 0 \\
0 & 0 & 1 \\
xy & y & 1
\end{bmatrix}
$$

```
> rank(A);
```
$$
\longrightarrow \qquad 2
$$

3.2.67   linalg[row] – extract row(s) of a matrix as vector(s)
linalg[col] – extract column(s) of a matrix as vectors(s)

Calling Sequence:

   row(A, i)
   row(A, $i..k$)
   col(A, i)
   col(A, $i..k$)

Parameters:

   A – matrix

i, k – positive integers

Synopsis:
- The function `col(A,i)` (`row(A,i)`) extracts the ith column (row) of the matrix A. The result is returned as a vector.

- More generally, `col(A,i..k)` (`row(A,i..k)`) extracts a sequence of columns (rows) i..k of the matrix A. The result is returned as a sequence of vectors.

Examples:
```
> with(linalg):
  A := matrix(3,3,[1,2,3,4,5,6,7,8,9]);
```
$$\longrightarrow \qquad A := \begin{bmatrix} 1 & 2 & 3 \\ 4 & 5 & 6 \\ 7 & 8 & 9 \end{bmatrix}$$

```
> row(A,2);
```
$$\longrightarrow \qquad [\ 4,\ 5,\ 6\ ]$$

```
> col(A,2..3);
```
$$\longrightarrow \qquad [\ 2,\ 5,\ 8\ ],\ [\ 3,\ 6,\ 9\ ]$$

SEE ALSO: `linalg[rowdim]`, `linalg[swaprow]`, `linalg[delrows]`, `linalg[stack]`, `linalg[concat]`, `linalg[addrow]`, `linalg[mulrow]`

3.2.68   **linalg[rowdim] – determine the row dimension of a matrix**
linalg[coldim] – determine the column dimension of a matrix

Calling Sequence:
`rowdim(A)`
`coldim(A)`

Parameters:
A – a matrix

Synopsis:
- The functions `rowdim(A)` and `coldim(A)` return an integer, the number of rows and columns respectively, in the matrix A.

Examples:
```
> with (linalg):
  A := matrix(3,4);
```
$$\longrightarrow \qquad A := array(1\ ..\ 3,\ 1\ ..\ 4,\ [])$$

```
> rowdim(A);
```
$$\longrightarrow \qquad 3$$

```
> coldim(A);
```
$$\longrightarrow \qquad 4$$

3.2.69   linalg[rowspace] – compute a basis for the row space
            linalg[colspace] – compute a basis for the column space
            linalg[range] – a synonym for colspace

Calling Sequence:
    rowspace(*A*)
    colspace(*A*)
    range(*A*)
    rowspace(*A*,´*dim*´)
    colspace(*A*,´*dim*´)
    range(*A*,´*dim*´)

Parameters:
    A   –   a matrix
    ´*dim*´   –   an optional unevaluated name

Synopsis:
- The functions rowspace(*A*) and colspace(*A*) return a set of vectors that form a basis for the vector space spanned by the rows and columns of the matrix *A*, respectively. The vectors are returned in a canonical form with leading entries 1. The function **range** is a synonym for colspace.

- The optional second parameter ´*dim*´ is assigned the rank of *A*, which is the dimension of the row and column space.

Examples:
```
> with (linalg):
  A := matrix(3,2,[2,0,3,4,0,5]);
```
$\longrightarrow$
$$A := \begin{bmatrix} 2 & 0 \\ 3 & 4 \\ 0 & 5 \end{bmatrix}$$

```
> rowspace(A);
```
$\longrightarrow$ {[1, 0], [0, 1]}

```
> colspace(A);
```
$\longrightarrow$ {[0, 1, 5/4], [1, 0, -15/8]}

```
> B :=array([[0,0,0,1,0],[0,x,0,0,0],[0,0,
  y,y,0],[0,x,0,1,0]]);
```
$\longrightarrow$
$$B := \begin{bmatrix} 0 & 0 & 0 & 1 & 0 \\ 0 & x & 0 & 0 & 0 \\ 0 & 0 & y & y & 0 \\ 0 & x & 0 & 1 & 0 \end{bmatrix}$$

```
> rowspace(B,´d´);
```
$\longrightarrow$ {[0, 0, 0, 1, 0], [0, 0, 1, 0, 0],
            [0, 1, 0, 0, 0]}

```
> d;
```
$\longrightarrow$ 3

SEE ALSO: `linalg[rowspan]`, `linalg[colspan]`, `linalg[nullspace]`, `linalg[gaussjord]`,
        `linalg[hermite]`, `type[range]`, `plot[range]`, `main[range]`

3.2.70   linalg[rowspan] – compute spanning vectors for the row space
linalg[colspan] – compute spanning vectors for the column space

Calling Sequence:
>   rowspan(A)
>   rowspan(A, $'dim'$)
>   colspan(A)
>   colspan(A, $'dim'$)

Parameters:
> A –   a matrix of multivariate polynomials over the rationals
> $'dim'$ –   assigned the dimension of the row space of A

Synopsis:
- The functions `rowspan` and `colspan` compute a spanning set for the row space and column space respectively of the matrix A. The matrix A must be a matrix of multivariate polynomials over the rationals. The spanning set returned is a set of vectors (one-dimensional arrays) of polynomials.

- If the optional second argument is given, it will be assigned the rank of A, which is the dimension of the row space and column space.

- These functions use so-called "fraction-free" Gaussian elimination to triangularize the matrix. Hence, unlike `linalg[rowspace]` and `linalg[colspace]`, no rational expressions are introduced during the elimination; as a result, if variables are later assigned particular values, division by zero cannot occur.

- These functions are part of the `linalg package`, and so can be used in the form `rowspan(..)` only after performing the command `with(linalg)` or `with(linalg,rowspan)`. These functions can always be accessed in the long form `linalg[rowspan](..)`.

Examples:
```
> with(linalg):
  A := matrix (2,2,[a,b,c,d]);
```
$\longrightarrow$
$$A := \begin{bmatrix} a & b \\ c & d \end{bmatrix}$$

```
> rowspan(A);
```
$\longrightarrow$ $\{[\ a,\ b\ ],\ [\ 0,\ d\ a - b\ c\ ]\}$

```
> colspan(A);
```
$\longrightarrow$ $\{[\ 0,\ d\ a - b\ c\ ],\ [\ a,\ c\ ]\}$

```
> B := matrix (3,2,[x+y,y+2*z,x,y,y,2*z])
  ;
```
                                              $\longrightarrow$
$$B := \begin{bmatrix} x+y & y+2z \\ x & y \\ y & 2z \end{bmatrix}$$

```
> rowspan(B,'d');
```
                                              $\longrightarrow$
$$\{[\ x,\ y\ ],\ [\ 0,\ 2\,x\,z-y^2\ ]\}$$

```
> d;
```
                                              $\longrightarrow$ 2

SEE ALSO: linalg[rowspace], linalg[colspace], linalg[ffgausselim]

3.2.71   linalg[rref] – reduced row echelon form
linalg[gaussjord] – a synonym for rref (Gauss-Jordan elimination)

Calling Sequence:
   gaussjord(A)
   gaussjord(A, $'r'$)
   gaussjord(A, $'r'$, $'d'$)
   gaussjord(A, $rmar$)

Parameters:

A	–	a rectangular matrix
$'r'$	–	for returning the rank of A (optional)
$'d'$	–	for returning the determinant of A (optional)
$rmar$	–	non-negative integer

Synopsis:
- Elementary row operations are performed on A, an n by m matrix over a field, to reduce it to row Echelon (Gauss-Jordan) form. At present, the matrix coefficients must lie in the field of rational numbers Q, or the field of rational functions over Q.

- The resulting matrix is upper triangular with leading nonzero entries 1 . If the matrix contains integers only, then in general rational numbers will appear in the result. Likewise if the matrix contains polynomials, in general rational functions will appear in the result.

- If an optional second parameter is specified, and it is a name, it is assigned the rank of A. The rank of A is the number of nonzero rows in the resulting matrix.

- If an optional third parameter is also specified, and the rank of $A = n$, then it is assigned the determinant of submatrix(A,1..n,1..n).

- If an optional second parameter is specified, and it is an integer, the elimination is terminated at this column position.

- This function is part of the `linalg` package, and so can be used in the form `gaussjord(..)` only after performing the command `with(linalg)` or `with(linalg, gaussjord)`. The function can always be accessed in the long form `linalg[gaussjord](..)`.

Examples:
```
> with(linalg):
  A := array([[4,-6,1,0],[-6,12,0,1],[-2,6,
  1,1]]);
```
$\longrightarrow$

$$A := \begin{bmatrix} 4 & -6 & 1 & 0 \\ -6 & 12 & 0 & 1 \\ -2 & 6 & 1 & 1 \end{bmatrix}$$

```
> gaussjord(A,'r');
```
$\longrightarrow$

$$\begin{bmatrix} 1 & 0 & 1 & 1/2 \\ 0 & 1 & 1/2 & 1/3 \\ 0 & 0 & 0 & 0 \end{bmatrix}$$

```
> r;
```
$\longrightarrow$ 2

SEE ALSO: `linalg[backsub]`, `linalg[gausselim]`, `linalg[hermite]`, `linalg[rowspace]`, `linalg[backsub]`, `linalg[pivot]`

3.2.72   linalg[scalarmul] – multiply a matrix or vector by an expression

Calling Sequence:
    scalarmul(A, *expr*)

Parameters:
    A –   a matrix or vector
    expr –   a scalar expression

Synopsis:
- The call `scalarmul`(A, *expr*) returns the matrix or vector which is the result of multiplying every entry in A by the scalar value *expr*.

- This function is part of the `linalg` package, and so can be used in the form `scalarmul(..)` only after performing the command `with(linalg)` or `with(linalg, scalarmul)`. The function can always be accessed in the long form `linalg[scalarmul](..)`.

Examples:
```
> with (linalg):
  scalarmul(array([[1,2],[3,4]]),2);
```
$\longrightarrow$

$$\begin{bmatrix} 2 & 4 \\ 6 & 8 \end{bmatrix}$$

```
> scalarmul(array([1,2,3,4]),3);
```
$\longrightarrow$ $[\ 3,\ 6,\ 9,\ 12\ ]$

```
> v:=vector(4,[2,x,6,-3]);          ⟶              v := [ 2, x, 6, -3 ]
> scalarmul(v,x);                   ⟶                        2
                                                   [ 2 x, x , 6 x, - 3 x ]
> scalarmul(transpose(v),x);        ⟶                        2
                                               transpose([ 2 x, x , 6 x, - 3 x ])
```

SEE ALSO: `linalg[multiply]`, `linalg[mulrow]`, `linalg[mulcol]`, `evalm`

3.2.73   linalg[singularvals] – compute the singular values of a matrix

Calling Sequence:
singularvals(A)

Parameters:
A –   a square matrix

Synopsis:
- The function `singularvals` returns a list of the singular values of the given matrix A.
- The singular values of a matrix are the square roots of the eigenvalues of `transpose`(A) * A.
- This function is part of the `linalg` package, and so can be used in the form `singularvals(..)` only after performing the command `with(linalg)` or `with(linalg, singularvals)`. The function can always be accessed in the long form `linalg[singularvals](..)`.

Examples:
```
> with (linalg):
  A := array( [[1,0,1],[1,0,1],[0,1,0]] );   ⟶              [ 1  0  1 ]
                                                            [         ]
                                                    A :=    [ 1  0  1 ]
                                                            [         ]
                                                            [ 0  1  0 ]
>
  singularvals(A);                           ⟶              [0, 2, 1]
```

SEE ALSO: `linalg[eigenvals]`, `Svd`

3.2.74   linalg[smith] – compute the Smith normal form of a matrix

Calling Sequence:
smith(A, x)

Parameters:
A –   square matrix of univariate polynomials in x
x –   the variable name

Synopsis:
- The Smith normal form of a matrix with univariate polynomial entries in x over a field F is computed. Thus the polynomials are then regarded as elements of the Euclidean domain $F[x]$.

- This routine is only as powerful as Maple´s normal function, since at present it only understands the field Q of rational numbers and rational functions over Q.

- The Smith normal form of a matrix is the diagonal matrix S obtained by doing elementary row and column operations where S[i,i] is the greatest common divisor of all i by i minors of A and `product(S[i,i],i=1..rank(A)) = det(A)/lcoeff(det(A),x)`.

- This function is part of the `linalg` package, and so can be used in the form `smith(..)` only after performing the command `with(linalg)` or `with(linalg, smith)`. The function can always be accessed in the long form `linalg[smith](..)`.

Examples:
```
> with(linalg):
> A := matrix([[1-x,y-x*y],[0,1-x^2]]);
                          [ 1 - x   y - x y ]
                          [                 ]
                  A := [                 2 ]
                          [   0       1 - x ]
> smith(A,x);
                      [ - 1 + x       0     ]
                      [                     ]
                      [                   2 ]
                      [    0      - 1 + x   ]
> H := inverse(hilbert(2,x));
          [             2                                      ]
          [  - (- 3 + x)  (- 2 + x)     (- 3 + x) (- 2 + x) (- 4 + x) ]
      H := [                                                    ]
          [                                          2         ]
          [ (- 3 + x) (- 2 + x) (- 4 + x)   - (- 3 + x)  (- 4 + x)   ]
> smith(H,x);
            [ - 3 + x                0              ]
            [                                       ]
            [                        2              ]
            [    0      (- 2 + x) (x  - 7 x + 12) ]
```

SEE ALSO: `linalg[hermite]`, `linalg[ismith]`, Smith

3.2.75   linalg[stack] – join two or more matrices together vertically

Calling Sequence:

   stack(A, B,...)

Parameters:

$A, B, \ldots$ – matrices or vectors

Synopsis:
- The function `stack` joins two or more matrices or vectors together vertically, where a vector is interpreted as a row vector. The matrices and vectors must have the same number of columns.

- This function is part of the `linalg` package, and so can be used in the form `stack(..)` only after performing the command `with(linalg)` or `with(linalg, stack)`. The function can always be accessed in the long form `linalg[stack](..)`.

Examples:
```
> with(linalg):
  a := matrix(2,2,[1,2,3,4]);
```
$$\longrightarrow \qquad a := \begin{bmatrix} 1 & 2 \\ 3 & 4 \end{bmatrix}$$

```
> b := matrix(2,2,[5,6,7,8]);
```
$$\longrightarrow \qquad b := \begin{bmatrix} 5 & 6 \\ 7 & 8 \end{bmatrix}$$

```
> stack(a,b);
```
$$\longrightarrow \qquad \begin{bmatrix} 1 & 2 \\ 3 & 4 \\ 5 & 6 \\ 7 & 8 \end{bmatrix}$$

```
> v := vector(2,[1,2]);
```
$$\longrightarrow \qquad v := [\, 1, \ 2 \,]$$

```
> stack(v,b,v);
```
$$\longrightarrow \qquad \begin{bmatrix} 1 & 2 \\ 5 & 6 \\ 7 & 8 \\ 1 & 2 \end{bmatrix}$$

SEE ALSO: `linalg[augment]`, `linalg[extend]`

3.2.76 linalg[submatrix] – extract a specified submatrix from a matrix

Calling Sequence:

submatrix(A, *Rrange*, *Crange*)

submatrix(A, *Rlist*, *Clist*)

Parameters:

A — a matrix

Rrange, Crange — integer ranges of rows/columns of A

Rlist, Clist   —   lists of integer row/column indices

Synopsis:

- The call **submatrix**(*A*, *Rrange*, *Crange*) returns the submatrix of *A* selected by the row range *Rrange* and the column range *Crange*.

- The call **submatrix**(*A*, *Rlist*, *Clist*) returns the matrix whose (i,j)th element is *A*[*Rlist*[i], *Clist*[j]].

- This function is part of the **linalg** package, and so can be used in the form **submatrix(..)** only after performing the command **with(linalg)** or **with(linalg, submatrix)**. The function can always be accessed in the long form **linalg[submatrix](..)**.

Examples:
```
> with (linalg):
  A := array( [[1,2,3],[4,x,6]] );                                [ 1  2  3 ]
                                                       A := [          ]
                                                                [ 4  x  6 ]

> submatrix(A,1..2,2..3);                      ⟶                 [ 2  3 ]
                                                                 [      ]
                                                                 [ x  6 ]

> submatrix(A,[2,1],[2,1]);                    ⟶                 [ x  4 ]
                                                                 [      ]
                                                                 [ 2  1 ]
```

SEE ALSO: linalg[subvector], linalg[row], linalg[col]

3.2.77   linalg[subvector] – extract a specified vector from a matrix

Calling Sequence:

      subvector(*A*, *r*, *c*)

Parameters: ◁

      A   – a matrix

      r,c – a list, range, or an integer

Synopsis:

- The function **subvector**(*A*, *r*, *c*) returns the vector of *A* selected by the single index in either the row or column entry and the range or list in the other entry.

- Note that either the row or column entry must be a single index specifying the desired row or column.

- This function is part of the **linalg** package, and so can be used in the form **subvector(..)** only after performing the command **with(linalg)** or **with(linalg, subvector)**. The function can always be accessed in the long form **linalg[subvector](..)**.

Examples:
```
> with (linalg):
  A := array( [[1,2,3],[4,x,6]] );
```
$$A := \begin{bmatrix} 1 & 2 & 3 \\ 4 & x & 6 \end{bmatrix}$$

```
> subvector(A,1..2,2);                  ⟶              [ 2, x ]
> subvector(A,2,[2,1]);                 ⟶              [ x, 4 ]
```

SEE ALSO: linalg[submatrix]

3.2.78   linalg[sumbasis] – determine a basis for the sum of vector spaces

Calling Sequence:
> sumbasis(S_1,S_2, ...)

Parameters:
> S_i –   a vector, set of vectors or list of vectors

Synopsis:
- A basis for the sum of the vector spaces spanned by the vectors in S_i is returned.

- A basis for the zero-dimensional space is an empty set or list.

- This function is part of the **linalg** package, and so can be used in the form **sumbasis(..)** only after performing the command **with(linalg)** or **with(linalg, sumbasis)**. The function can always be accessed in the long form **linalg[sumbasis](..)**.

Examples:
```
> with(linalg):
  v1 := vector([1,0,1,0]);              ⟶         v1 := [ 1, 0, 1, 0 ]
> v2 := vector([0,1,0,1]);              ⟶         v2 := [ 0, 1, 0, 1 ]
> v3 := vector([1,2,1,1]);              ⟶         v3 := [ 1, 2, 1, 1 ]
> v4 := vector([-1,-2,1,0]);            ⟶         v4 := [ -1, -2, 1, 0 ]
> sumbasis({v1,v2},{v3,v4});            ⟶            {v1, v2, v3, v4}
>
  sumbasis({v1,v2,v3},{v3,v4},{v2,v3}); ⟶            {v1, v2, v3, v4}
>
```

SEE ALSO: linalg[intbasis], linalg[basis]

3.2.79   linalg[swaprow] – swap two rows in a matrix
linalg[swapcol] – swap two columns in a matrix

Calling Sequence:

    swaprow(A, $r1$, $r2$)
    swapcol(A, $c1$, $c2$)

Parameters:

    A –   a matrix

    $r1,r2,c1,c2$ –   positive integers

Synopsis:

- The swaprow(A,$r1$,$r2$) function creates a new matrix in which row $r1$ has been swapped with row $r2$. Similarly, swapcol(A,$c1$,$c2$) creates a new matrix in which column $c1$ has been swapped with column $c2$.

Examples:
```
> with (linalg):
  A := array( [[1,2,x],[3,4,y]] );
```
$$A := \begin{bmatrix} 1 & 2 & x \\ 3 & 4 & y \end{bmatrix}$$

```
> swaprow(A,1,2);
```
$$\begin{bmatrix} 3 & 4 & y \\ 1 & 2 & x \end{bmatrix}$$

```
> swapcol(A,2,3);
```
$$\begin{bmatrix} 1 & x & 2 \\ 3 & y & 4 \end{bmatrix}$$

3.2.80   linalg[sylvester] – create Sylvester matrix from two polynomials

Calling Sequence:

    sylvester(p, q, x)

Parameters:

    p, q –   expanded polynomials in the variable x

    x –   a variable (name)

Synopsis:

- The call sylvester(p,q,x) returns the Sylvester matrix of the polynomials p and q with respect to x. Note that the determinant of this matrix is equal to resultant(p,q,x) .

- If p is of degree m in x and q is of degree n in x then the output is a square matrix of dimension $m+n$.

- The polynomials p and q must be expanded in x, because the coeff function is used on p and q to determine the entries of the matrix.

- This function is part of the `linalg` package, and so can be used in the form `sylvester(..)` only after performing the command `with(linalg)` or `with(linalg, sylvester)`. The function can always be accessed in the long form `linalg[sylvester](..)`.

Examples:
```
> with(linalg):
  p := a+b*x:
  q := c+d*x+e*x^2:
  sylvester(p,q,x);
```
$\longrightarrow$
```
[ b   a   0 ]
[           ]
[ 0   b   a ]
[           ]
[ e   d   c ]
```

SEE ALSO: `linalg[bezout]`, `linalg[det]`, `collect`, `resultant`

3.2.81   linalg[toeplitz] – create a Toeplitz matrix

Calling Sequence:
>   toeplitz(*L*)

Parameters:
>   *L*   –   a list of expressions

Synopsis:
- The procedure `toeplitz(`*L*`)` returns the symmetric toeplitz matrix corresponding to the list *L*.

- The matrix result has as many rows (and columns) as there are elements in the list *L*. The first element of *L* is placed all along the diagonal, and the i-th element of *L* is placed all along the (i-1)th sub and super diagonals of the result.

- This function is part of the `linalg` package, and so can be used in the form `toeplitz(..)` only after performing the command `with(linalg)` or `with(linalg, toeplitz)`. The function can always be accessed in the long form `linalg[toeplitz](..)`.

Examples:
```
> with (linalg):
  toeplitz( [a,b,c] );
```
$\longrightarrow$
```
[ a   b   c ]
[           ]
[ b   a   b ]
[           ]
[ c   b   a ]
```

3.2.82   linalg[trace] – the trace of a matrix

Calling Sequence:
>   trace(*A*)

Parameters:

    A – a square matrix

Synopsis:

- The function `trace(`A`)` computes the trace of the matrix A. The trace of a matrix A is defined to be the sum of the diagonal elements of A.

- This function is part of the `linalg` package, and so can be used in the form `trace(..)` only after performing the command `with(linalg)` or `with(linalg, trace)`. The function can always be accessed in the long form `linalg[trace](..)`.

Examples:
```
> with(linalg):
  trace(array([[1,2],[1,4]]));
```
$\longrightarrow$
$$5$$

```
> A := matrix(3,3,[a,b,c,d,e,f,g,h,i]);
```
$\longrightarrow$
$$A := \begin{bmatrix} a & b & c \\ d & e & f \\ g & h & i \end{bmatrix}$$

```
> trace(A);
```
$\longrightarrow$
$$a + e + i$$

3.2.83   linalg[transpose] – compute the transpose of a matrix

Calling Sequence:

    `transpose(`A`)`

Parameters:

    A – an m x n matrix or a vector

Synopsis:

- The matrix transpose of A is computed. The result is an n x m matrix. The [i,j]-th element of the result is equal to the [j,i]-th element of A. The result inherits the indexing function (for example, diagonal or sparse) of A, if it has one.

- If A is a vector, then it is treated as if it were a column vector; `transpose(`A`)` would therefore be a row vector.

- This function is part of the `linalg` package, and so can be used in the form `transpose(..)` only after performing the command `with(linalg)` or `with(linalg, transpose)`. The function can always be accessed in the long form `linalg[transpose](..)`.

Examples:
```
> with (linalg):
  A := array([[1,2,3],[4,5]]);
```
$\longrightarrow$
$$A := \begin{bmatrix} 1 & 2 & 3 \\ 4 & 5 & A[2, 3] \end{bmatrix}$$

```
> transpose(A);                    ⟶          [ 1      4    ]
                                              [             ]
                                              [ 2      5    ]
                                              [             ]
                                              [ 3   A[2, 3] ]

> B := array(diagonal,1..2,1..2,[(1,1)=5])
  ;                                ⟶               [ 5      0    ]
                                           B :=  [             ]
                                                 [ 0   B[2, 2] ]

> transpose(B);                    ⟶          [ 5      0    ]
                                              [             ]
                                              [ 0   B[2, 2] ]

> C := array([1,2,3]);             ⟶      C := [ 1, 2, 3 ]

> transpose(C);                    ⟶         transpose(C)
```

3.2.84   linalg[vandermonde] – create a Vandermonde matrix

Calling Sequence:

> vandermonde(L)

Parameters:

> L –   a list of expressions

Synopsis:

- The function vandermonde(L) returns the Vandermonde matrix formed from the elements of the list. This square matrix has as its (i,j)-th entry $L[i]^{(j-1)}$.

- This function is part of the linalg package, and so can be used in the form vandermonde(..) only after performing the command with(linalg) or with(linalg, vandermonde). The function can always be accessed in the long form linalg[vandermonde](..).

Examples:

```
> with (linalg):
  vandermonde([1,2,3,4]);          ⟶         [ 1  1   1   1 ]
                                             [              ]
                                             [ 1  2   4   8 ]
                                             [              ]
                                             [ 1  3   9  27 ]
                                             [              ]
                                             [ 1  4  16  64 ]
```

```
> vandermonde([x,y,z]);
```

$$\longrightarrow \quad \begin{bmatrix} & & 2 \\ 1 & x & x \\ & & \\ & & 2 \\ 1 & y & y \\ & & \\ & & 2 \\ 1 & z & z \end{bmatrix}$$

3.2.85   linalg[vecpotent] – compute the vector potential

Calling Sequence:

    vecpotent(f, var, $'V'$)

Parameters:

    f –   a vector function of length three

    var –   list of three variables

    V –   name in which the vector potential is returned

Synopsis:

- The function **vecpotent** determines whether a given vector function has a vector potential, and determines that vector potential if it exists.

- The function returns true if the function f has a vector potential, and false if it does not. The vector potential exists if and only if the divergence of f is zero.

- If a vector potential for f exists, it will be assigned to the name given in the third argument V. If **vecpotent** returns true, then V will be assigned a vector function such that curl $V = f$.

- This function is part of the **linalg** package, and so can be used in the form **vecpotent(..)** only after performing the command **with(linalg)** or **with(linalg, vecpotent)**. The function can always be accessed in the long form **linalg[vecpotent](..)**.

Examples:

```
> with (linalg):
  f := [x^2*y, -1/2*x*y^2, -x*y*z];
```

$$\longrightarrow \quad f := [x^2 y, -1/2\, x\, y^2, -x\, y\, z]$$

```
> vecpotent (f, [x,y,z], 'V');
```

$$\longrightarrow \quad \text{true}$$

```
> print(V);
```

$$\longrightarrow \quad [-1/2\, x\, y^2 z, -x^2 y\, z, 0]$$

```
> g := [x^2, y^2, z^2];
```

$$\longrightarrow \quad g := [x^2, y^2, z^2]$$

```
> vecpotent  (g, [x,y,z], 'G');
```

$$\longrightarrow \quad \text{false}$$

SEE ALSO: `linalg[potential]`

3.2.86　linalg[vectdim] − determine the dimension of a vector

Calling Sequence:
> vectdim(*v*)
> vectdim(*L*)

Parameters:

> *v* − a vector
> *L* − a list

Synopsis:

- The function `vectdim` determines the dimension of the given vector.

- If `vectdim` is given a vector *v*, then the dimension of (number of elements in) *v* is returned; if a list *L* is given to `vectdim`, then the number of elements in the list is returned.

- This function is part of the `linalg` package, and so can be used in the form `vectdim(..)` only after performing the command `with(linalg)` or `with(linalg, vectdim)`. The function can always be accessed in the long form `linalg[vectdim](..)`.

Examples:
```
> with (linalg):
  v := vector([x,y,z]);                    ⟶            v := [ x, y, z ]

> vectdim(v);                              ⟶                   3

> vectdim([1,2,3,4]);                      ⟶                   4

> u := vector(2);                          ⟶           u := array(1 .. 2, [])

> vectdim(u);                              ⟶                   2
```

SEE ALSO: `linalg[rowdim]`, `linalg[coldim]`

3.2.87　linalg[vector] − create a vector

Calling Sequence:

> vector($[x_1, \ldots, x_n]$)
> vector(n, $[x_1, \ldots, x_n]$)
> vector(n)
> vector(n, *f*)

Parameters:

> $x_1, \ldots, x_n$ − vector elements of type algebraic
> n　　　　　　 − length of the vector
> *f*　　　　　　 − a function used to create the vector elements

Synopsis:
- The vector function is part of the **linalg** package. It provides a simplified syntax for creating vectors. For a general description of vectors in Maple, see the information under **vector**.

- The calls `vector([`$x_1, \ldots, x_n$`])` and `vector(`$n,$ `[`$x_1, \ldots, x_n$`])` will produce a vector (a one-dimensional array) of length n containing the given elements.

- The call `vector(`n`)` will produce a vector of length n with unspecified elements.

- The call `vector(`n, f`)` will produce a vector of length n whose elements are the result of the function f acting on the index of the vector. Thus `vector(`n, f`)` is equivalent to `vector(1..`n`,` `[`$f(1)$`,` $f(2)$`,` $\ldots$`,` $f(n)$`])`.

- Since vectors are represented as one-dimensional arrays, see information under **array** for further details about how to work with vectors.

- This function is part of the **linalg** package, and so can be used in the form `vector(..)` only after performing the command **with(linalg)** or **with(linalg, vector)**. The function can always be accessed in the long form `linalg[vector](..)`.

Examples:

```
> with(linalg):
  vector([5,4,6,3]);              ⟶              [ 5, 4, 6, 3 ]

> vector(4);                      ⟶          [ ?[1], ?[2], ?[3], ?[4] ]

> vector(4,0);                    ⟶              [ 0, 0, 0, 0 ]

> f := x -> x^2:
  v := vector(4,f);               ⟶          v := [ 1, 4, 9, 16 ]

> v[2];                           ⟶                    4
```

SEE ALSO: `array`, `matrix`, `vector`, `linalg`

3.3   The Plots Package

3.3.1   Introduction to the `plots` Package

Calling Sequence:

> `plots[`*function*`]` `(args)`
> *function* `(args)`

Synopsis:
- To use a **plots** function, either define that function alone by typing **with(plots,** *function***)**, or define all plots functions by typing **with(plots)**.

- The functions available are:

conformal	cylinderplot	display	display3d
matrixplot	pointplot	polarplot	replot
spacecurve	sparsematrixplot	sphereplot	tubeplot

- For help with a particular function use `help(plots, `*function*`)`

- The package functions are always available without applying `with`, using the long-form notation: `plots[`*function*`] (`*args*`) .`

SEE ALSO: `with`, `plot`, `plot3d`

3.3.2   plots[conformal] – conformal plot of a complex function

Calling Sequence:

    `conformal(F,r1,r2,`*options*`);`

Parameters:

    F   –   a complex procedure or expression

    r1, r2   –   ranges of the form a..b, or name=a..b

Synopsis:

- A conformal plot of a complex function `F(z)` from `a+bi` to `c+di` maps a two-dimensional grid $a \leq x \leq c$, $b \leq y \leq d$ from the plane into a second (curved) grid determined by the images of the original grid lines under `F`. The size of the original grid (the number of lines in either the `x` or `y` direction) is controlled by the user (see below). The result is a set of curves in the plane, which has the property that they also intersect at right angles at the points where `F` is analytic.

- The `conformal` function defines a conformal plot of a complex function `F`. `F` can be an expression or a procedure. The first range, `r1`, defines the grid lines in the plane that are to be conformally mapped via the complex function `F`. If not specified, this first range defaults to `0..1+(-1)^(1/2)`. The second range, `r2`, describes the dimensions of the window in which the conformal plot is to lie. If not present, the defaults are the maximum and minimum of the resulting conformal lines in the range.

- Remaining arguments are interpreted as options which are specified as equations of the form `option = value`.

- An option of the form `grid=[m,n]` with `m` and `n` integers, specifies the number of grid lines in both `x` and `y` directions that are to be mapped conformally. The default is 11 lines in either direction, making an 11 by 11 grid.

- An option of the form `numxy=[m,n]`, with `m` and `n` integers, specifies the number of points that are to be plotted in each gridline, with `m` points in the `x` direction and `n` points in the `y` direction. The default is 15 points in each direction.

- There are also a number of standard two-dimensional plot options that are applicable with conformal plot. These include specifications for style, and the number of horizontal and vertical ticks marks. See also `help(plot,`options`)` .

- The result of a call to `conformal` is a `PLOT` structure which can be rendered by the plotting device. The user may assign a `PLOT` value to a variable, save it in a file, then read it back in for redisplay. See `help(plot,structure)` .

- `conformal` may be defined by `with(plots)` or `with(plots,conformal)`. It can also be used by the name `plots[conformal]` .

Examples:
```
with(plots):
conformal(z^3);
conformal(z^2,z=0..2+2*I);
conformal(1/z,-1-I..1+I,-6-6*I..6+6*I);
conformal(cos(z),z=0..2*Pi+ Pi*I,grid=[8,8],numxy=[11,11]);
conformal(z^3,xtickmarks=3,ytickmarks=6,style=POINT);
```

SEE ALSO: `plot`, `plot[`options`]`, `plot[structure]`, `plots[display]`

3.3.3   plots[cylinderplot] – plot a 3D surface in cylindrical coordinates

Calling Sequence:

   `cylinderplot(L,r1,r2,`*options*`);`

Parameters:

   L   –   a procedure or expression having two variables, or a list of three
               such procedures or expressions

   r1,r2 –   ranges of the form `var=a..b`

Synopsis:

- The `cylinderplot` function gives a three-dimensional plot of a surface or parametric surface in cylindrical coordinates. The individual functions can be in the form of either expressions or procedures.

- If L is not a list, then L represents radius given in terms of the coordinates `theta` and `z`. In this case, `r1` and `r2` are the ranges for `theta` and `z`.

- If L is a list, then the three components of L are parametric representations of the coordinates `radius`, `theta`, and `z`, respectively. In this case, `r1` and `r2` are ranges for the two parameters of the surface.

- Remaining arguments are interpreted as options which are specified as equations of the form `option = value`. These options are the same as those found in `plot3d`. For example, the option `grid = [m,n]` specifies that the set of points representing the three-dimensional plot are evaluated over an `m` by `n` grid of points. See also `help(plot,`options`)` .

- The result of a call to cylinderplot is a PLOT3D structure which can be rendered by the plotting device. The user may assign a PLOT3D value to a variable, save it in a file, then read it back in for redisplay. See help(plot3d,structure) .

- cylinderplot may be defined by with(plots) or with(plots,cylinderplot). It can also be used by the name plots[cylinderplot] .

Examples:
```
with(plots):
cylinderplot(1,theta=0..2*Pi,z=-1..1);
cylinderplot(z+ 3*cos(2*theta),theta=0..Pi,z=0..3);
cylinderplot((5*cos(y)^2 -1)/3,x=0..2*Pi,y=-Pi..Pi,style=PATCH);
cylinderplot([z*theta,theta,cos(z^2)],theta=0..Pi,z=-2..2);
```

SEE ALSO: read, save, plot3d, plots[sphereplot], plot3d[`options`],
plot3d[structure]

3.3.4 plots[display] – display a set of 2D plot structures

Calling Sequence:

display(L)

display(L, *options*)

Parameters:

L – a set or list of PLOT structures to be displayed

Synopsis:

- A typical call to the display function would be display({F,G}), where F and G are previously defined PLOT structures. To see how to generate a single PLOT structure see help(plot). The result of the call to display is a single PLOT structure, that, when rendered, gives a picture of the two plots. Algebraically, it does a union of the PLOT structures.

- Display is not smart about things like collisions with label names. If two plot structures have different labeling, display chooses the first set of labels. Similarly, in the case of multiple titles, the first title found is chosen as the display title.

- The display function does not handle mixed infinity and non-infinity plots. To see what an infinity plot is see plot[infinity].

- Remaining arguments are interpreted as options which are specified as equations of the form option = value. These are interpreted as global specifications. For example, options of the form r=a..b are taken as dimensions for the plotting window in which the plots are to be displayed. The first range specification represents the horizontal range of the viewing window, while the second range specification gives the vertical bounds of the viewing window. These horizontal and vertical bounds can also be specified as view = [a..b,c..d] or view = [x=a..b,y=c..d].

Other options include `title=t` for a string `t`. The global specifications override the ones defined locally in the individual PLOT structures.

Examples:
```
with(plots);
F:=plot(cos(x),x=-Pi..Pi):
G:=plot(tan(x),x=-Pi..Pi,y=-Pi..Pi):
H:=plot([sin(t),cos(t),t=-Pi..Pi]):
display({F,G,H});
display({F,H},view=[x=-10..10,y=-8..3],title=`Two plots`);
display({G,H},x=-1..1,height=-1..1);
```

SEE ALSO: `plot`, *spec* where *spec* is one of `infinity`, `polar`, `parametric`, `spline`, `point`, `line`, `multiple`, `ranges`, `functions`, `` `options` ``, `structure`, `setup`, `device`

3.3.5   plots[display3d] – display a set of 3D plot structures

Calling Sequence:
> `display3d(L)`
> `display3d(L, ` *options*`)`

Parameters:
> *L*   –   a set or list of PLOT3D structures to be displayed

Synopsis:
- A typical call to the display3d function would be `display3d({F,G})`, where F and G are previously defined PLOT3D structures. To see how to generate a single PLOT3D structure see `help(plot3d)`. The functions that generate a PLOT3D structure are `cylinderplot`, `display3d`, `matrixplot`, `plot3d`, `pointplot`, `spacecurve`, `sphereplot`, and `tubeplot`. The result of the call to `display3d` is a single PLOT3D structure, that, when rendered, gives a picture of the set of plots. Algebraically, it does a union of the PLOT3D structures.

- `Display3d` is not smart about things like collisions. If two `plot3d` structures have different titles, for example, then `display3d` choses the first title it comes across. This is also true for other options including axes labelling and style specification.

- Remaining arguments are interpreted as options which are specified as equations of the form `option=value`. These are interpreted as global specifications. These are the same as those options found in `plot3d` (with the exception of `coords`). See `help(plot3d,`´options´`)`. The global specifications override the ones defined locally in the individual PLOT3D structures.

- `display3d` may be defined by `with(plots)` or `with(plots,display3d)`. It can also be used by the name `plots[display3d]` .

Examples:
```
with(plots);
```

```
F:=plot3d(sin(x*y),x=-Pi..Pi,y=-Pi..Pi):
G:=plot3d(x + y,x=-Pi..Pi,y=-Pi..Pi):
H:=plot3d([2*sin(t)*cos(s),2*cos(t)*cos(s),2*sin(s)],s=0..Pi,t=-Pi..Pi):
display3d({F,G,H});
display3d({F,H},view=[-1..1,-1..1,-1..1],title=`Two plots`);
```

SEE ALSO: `plot3d`, `plot3d[structure]`, `plot3d[`options`]`

3.3.6   plots[matrixplot] – 3D plot with z values determined by a matrix

Calling Sequence:

    `matrixplot(A, `*options*`);`

Parameters:

    A – a matrix

Synopsis:

- The `matrixplot` function defines a three-dimensional graph with the x and y coordinates representing row and column indices, respectively. The z values are then given by the input matrix.

- Remaining arguments are interpreted as options which are specified as equations of the form `option = value`. The option `heights=HISTOGRAM` draws the plot in the form of a three-dimensional histogram. Other options include most of those found with `plot3d`. These include `axes`, `view`, `title`, `labels`, `style`, `projection`, and `orientation`.

- The result of a call to `matrixplot` is a PLOT3D structure which can be rendered by the plotting device. The user may assign a PLOT3D value to a variable, save it in a file, then read it back in for redisplay. See `help(plot3d,structure)` .

- `matrixplot` may be defined by `with(plots)` or `with(plots,matrixplot)`. It can also be used by the name `plots[matrixplot]` .

Examples:
```
with(plots):
with(linalg):
A:= hilbert(8): B:= toeplitz([1,2,3,4,-4,-3,-2,-1]):
matrixplot(A);
matrixplot(A &* B);
matrixplot(A+B,heights=HISTOGRAM, axes =BOXED);
```

SEE ALSO: `read`, `save`, `plot3d`, `plot3d[`options`]`, `plot3d[structure]`

3.3.7   plots[pointplot] – create a 3D point plot

Calling Sequence:

    `pointplot(L,`*options*`);`

Parameters:
> L – a set or list of three-dimensional points

Synopsis:
- The pointplot function is used to create a three-dimensional plot of points. The points that are to be plotted come from the set or list L.

- Remaining arguments are interpreted as options which are specified as equations of the form option = value. These options are the same as those found in plot3d. These include the options axes = f, where f is one of BOXED or AXES, title = t, where t is a line of text, labels = [x,y,z] where x, y, and z are each of type string. Additional plot3d options include orientation, projection and view. See also help(plot3d,`options`) .

- The result of a call to pointplot is a PLOT3D structure which can be rendered by the plotting device. The user may assign a PLOT3D value to a variable, save it in a file, then read it back in for redisplay. See help(plot3d,structure) .

- pointplot may be defined by with(plots) or with(plots,pointplot). It can also be used by the name plots[pointplot] .

Examples:
```
with(plots):
pointplot({[0,1,1],[1,-1,2],[3,0,5] },axes=BOXED);
pointplot({[cos(Pi*t/40),sin(Pi*t/40),t/40]$t=0..40});
```

SEE ALSO: plot3d[`options`], plot3d[structure]

3.3.8   plots[polarplot] – plot a 2D curve in polar coordinates

Calling Sequence:
>   polarplot(L, *options*);

Parameters:
> L – a set of two-dimensional curves

Synopsis:
- The polarplot function defines one or more curves in 2-space given in polar coordinates. The first argument can be a procedure, an expression, a list, or a set containing any combination of these three. In the case of a set, the result is a multiple plot of all the curves in the set.

- A single procedure or expression represents a plot of radius given in terms of angle. A list represents a parametric polar curve, with the first component representing radius, the second component angle, and any additional arguments in the list representing local arguments for the given parametric curve (for example, numpoints = n or t=a..b).

- If no range is specified, a default angle range $-\mathrm{Pi} \leq \mathrm{theta} \leq \mathrm{Pi}$ is used.

- Remaining arguments are interpreted as options which are specified as equations of the form `option = value`. These can be used to specify global default values such as `t=a..b` or `numpoints=n`.

- Remaining options are the same as those found in plot. For example, the option `numpoints = 40` specifies that the curves be drawn with a starting value of 40 points. See also `help(plot,`options`)`

- The result of a call to polarplot is a `PLOT` structure which can be rendered by the plotting device. The user may assign a `PLOT` value to a variable, save it in a file, then read it back in for redisplay. See `help(plot,structure)` .

- polarplot may be defined by `with(plots)` or `with(plots,polarplot)`. It can also be used by the name `plots[polarplot]` .

Examples:
```
with(plots):
polarplot(1);
polarplot([cos(t),sin(t),t=0..4*Pi]);
polarplot({1,t/10});
polarplot({t,[2*cos(t),sin(t),t=-Pi..Pi]},t=-Pi..Pi,numpoints=50);
```

SEE ALSO: `read`, `save`, `plot`, `plot[`options`]`, `plot[structure]`

3.3.9   plots[replot] – redo a plot

Calling Sequence:

  `replot( p, h, v, `*options*`);`

Parameters:

  p – a plot data structure

  h – horizontal range (optional)

  v – vertical range (optional)

  ... – optional arguments (see below)

Synopsis:

- This function allows one to change the parameters of a plot without the plot being recalculated. This allows one to zoom in on a plot or change the title, labels, or style. The options to `replot` are the same as the options to `plot`. See `help(plot,`options`)` .

- Note that since `replot` does not recompute the function, it will not provide more data in a new plot than was in the original plot. Therefore changing the scale will not show more of the function or increase any detail. This requires computing the plot again using `plot`. In addition, `replot` cannot switch between infinity and bounded plots.

- Note: If a vertical range is specified, the horizontal range must be specified too, since the first range is assumed to be the horizontal range.

- replot may be defined by `with(plots)` or `with(plots,replot)`. It can also be used by the name `plots[replot]` .

Examples:
```
with(plots):
plot(tan(x),0..Pi);
replot(",x=0..Pi,y=-5..5);
replot(",x=0..Pi/2,title=tan);
p := plot(sin(x),0..2*Pi):
q := replot(p,x=0..Pi,y=-1..1):
q;
```

SEE ALSO: plot

3.3.10   plots[sparsematrixplot] – 2D plot of nonzero values of a matrix

Calling Sequence:
> sparsematrixplot(A, *options*);

Parameters:
> A –   a matrix

Synopsis:

- The sparsematrixplot function defines a two-dimensional graph with the x and y coordinates representing row and column indices, respectively. It plots a point for every nonzero entry of the matrix. It is useful for displaying sparsity patterns of matrices.

- One can also set the number of tickmarks for both x and y coordinates via xticksmarks = n or ytickmarks = m in the second and third arguments.

- The result of a call to sparsematrixplot is a PLOT structure which can be rendered by the plotting device. The user may assign a PLOT value to a variable, save it in a file, then read it back in for redisplay. See plot[structure].

- sparsematrixplot may be defined by `with(plots)` or `with(plots,sparsematrixplot)`. It can also be used by the name `plots[sparsematrixplot]` .

Examples:
```
with(plots):
with(linalg):
A := randmatrix(15,15,sparse):
B := gausselim(A):
sparsematrixplot(A);
sparsematrixplot(B);
C := randmatrix(18,15,sparse);
```

```
sparsematrixplot(C,xtickmarks=8,ytickmarks=5);
```

SEE ALSO: `read`, `save`, `plot`, `plot[structure]`

3.3.11   plots[spacecurve] – plotting of 3D space curves

Calling Sequence:
> `spacecurve(`*L*`, `*options*`) ;`

Parameters:
> *L*   –   a set of space curves

Synopsis:

- The `spacecurve` function defines one or more curves in 3 space. The first argument can be a list of size three or more. The first three components are considered to be the parametric representations of the `x`, `y`, and `z` coordinates. Additional arguments include specifying a range for the particular curve via `t=a..b`, or the number of points that are to be used in drawing the curve via `numpoints=n` (the default is 50). The first argument can also be a set of such lists, in which case several three-dimensional curves are drawn.

- Remaining arguments are interpreted as options which are specified as equations of the form `option = value`. These can be used to specify global default values such as `t=a..b` or `numpoints=n`.

- Remaining options are the same as those found in `plot3d` with the exception of specifying a grid size. For example, the option `axes = BOXED` specifies that the space curves include a boxed axis bounding the plot. See also `help(plot3d,`options`)` .

- The result of a call to spacecurve is a `PLOT3D` structure which can be rendered by the plotting device. The user may assign a `PLOT3D` value to a variable, save it in a file, then read it back in for redisplay. See `help(plot3d,structure)` .

- `spacecurve` may be defined by `with(plots)` or `with(plots,spacecurve)`. It can also be used by the name `plots[spacecurve]` .

Examples:
```
with(plots):
spacecurve([cos(t),sin(t),t],t=0..4*Pi);
spacecurve({[sin(t),0,cos(t),t=0..2*Pi],[cos(t)+1,sin(t),0,numpoints=100]},
 t=-Pi..Pi);
spacecurve({[t*sin(t),t,t*cos(t)],[cos(t),sin(t),0]},t=-Pi..2*Pi,axes=BOXED);
spacecurve( [ -10*cos(t) - 2*cos(5*t) + 15*sin(2*t),
              -15*cos(2*t) + 10*sin(t) - 2*sin(5*t),
               10*cos(3*t) ], t= 0..2*Pi);
```

SEE ALSO: `read`, `save`, `plots[tubeplot]`, `plot3d[`options`]`, `plot3d[structure]`

3.3.12   plots[sphereplot] – plot a 3D surface in spherical coordinates

Calling Sequence:

> sphereplot(L,r1,r2, *options*);

Parameters:

> L –   a procedure or expression having two variables, or a list of three such procedures or expressions
>
> $r1,r2$ –   ranges of the form var=a..b

Synopsis:

- The `sphereplot` function gives a three-dimensional plot of a surface or parametric surface in spherical coordinates. The individual functions can be in the form of either expressions or procedures.

- If L is not a list, then L represents radius given in terms of **theta** and **phi**. Otherwise, if L is a list, then the three components represent **radius**, **theta** and **phi**, respectively.

- Remaining arguments are interpreted as options which are specified as equations of the form `option = value`. These options are the same as those found in `plot3d`. For example, the option `grid = [m,n]` specifies that the set of points representing the three-dimensional plot is evaluated over an **m** by **n** grid of points.curves See also `plot['options']`.

- The result of a call to `sphereplot` is a `PLOT3D` structure which can be rendered by the plotting device. The user may assign a `PLOT3D` value to a variable, save it in a file, then read it back in for redisplay. See `plot3d[structure]`.

- `sphereplot` may be defined by `with(plots)` or `with(plots,sphereplot)`. It can also be used by the name `plots[sphereplot]`:

Examples:
```
with(plots):
sphereplot(1,theta=0..2*Pi,phi=0..Pi);
sphereplot((1.3)^z * sin(theta),z=-1..2*Pi,theta=0..Pi);
sphereplot([z*theta,exp(theta/10),z^2],theta=0..Pi,z=-2..2);
sphereplot((5*cos(y)^2 -1)/2,x=0..2*Pi,y=-Pi..Pi,style=PATCH);
sphereplot((3*sin(x)^2 -1)/2,x=-Pi..Pi,y=0..Pi,orientation=[30,49]);
```

SEE ALSO: `read`, `save`, `plot3d`, `plots[cylinderplot]`, `plot3d['options']`,
        `plot3d[structure]`

3.3.13   plots[tubeplot] – three-dimensional tube plotting

Calling Sequence:

> tubeplot(C, *options*);

Parameters:
  C — a set of spacecurves

Synopsis:
- The `tubeplot` function defines a tube about one or more three-dimensional space curves. A given space curve is a list of three or more components. The initial three components define parametrically the **x**, **y**, and **z** components. Additional components of a given space curve specify various local attributes of the curve.

- Remaining components of an individual space curve are interpreted as local options which are specified as equations of the form option = value. These include equations of the form `numpoints` = n or `tubepoints` = m with n and m integers. These allow the user to designate the number of points evaluated on the space curve and the number of points on the tube, respectively. The default values used by Maple are `numpoints=50` and `tubepoints=10`. An equation of the form `radius` = f, where f is some expression, defines the radius of the tube about the given space curve. If no radius is specified, then the default used is `radius=1`. An equation of the form t=a..b, where a and b evaluate to constants, specifies the range of the parameter of the curve.

- Remaining arguments to tubeplot include such specifications as `numpoints` = n, `tubepoints` = m, t= a..b, and `radius` = f. These are to be used in the case where an individual space curve does not have the option specified.

- Additional options are the same as those found in **spacecurve** (and similar to options for **plot3d**). For example, the option **axes= BOXED** specifies that the **tubeplot** is to include a boxed axis bounding the plot. See also `help(plot3d,`options`)` .

- The result of a call to **tubeplot** is a PLOT3D structure which can be rendered by the plotting device. The user may assign a PLOT3D value to a variable, save it in a file, then read it back in for redisplay. See `help(plot3d,structure)` .

- **tubeplot** may be defined by `with(plots)` or `with(plots,tubeplot)`. It can also be used by the name `plots[tubeplot]` .

Examples:
```
with(plots):
tubeplot([cos(t),sin(t),0],t=0..2*Pi,radius=0.5);
tubeplot([cos(t),sin(t),0,t=Pi..2*Pi,radius=0.25*(t-Pi)]);
tubeplot([sin(t),t,cos(t)],t=0..2*Pi,radius=1.2+sin(t),numpoints=40);
tubeplot([sin(t),t,exp(t)],t=-1..1,radius=cos(t),tubepoints=20);
tubeplot( [ -10*cos(t) - 2*cos(5*t) + 15*sin(2*t),
            -15*cos(2*t) + 10*sin(t) - 2*sin(5*t),
            10*cos(3*t) ], t= 0..2*Pi,radius=3*cos(t*Pi/3));
# Multiple tubeplots are also allowed
tubeplot({[cos(t),sin(t),0],[0,sin(t)-1,cos(t)]},t=0..2*Pi,radius=1/4);
tubeplot({[cos(t),sin(t),0],[0,sin(t)-1,cos(t)]},t=0..2*Pi,radius=1/10*t);
tubeplot({[cos(t),sin(t),0,t=Pi..2*Pi,numpoints=15,radius=0.25*(t-Pi)],
          [0,cos(t)-1,sin(t),t=0..2*Pi,numpoints=45,radius=0.25]});
```

SEE ALSO: plot3d[`options`], plot3d[structure], plots[spacecurve]

3.4 The Statistics Package

3.4.1 Introduction to the stats Package

Calling Sequence:

stats[*function*] (*args*)

function(*args*)

Synopsis:

- To use a **stats** function, either define that function alone by typing the command **with(stats,** *function*), or define all statistics functions by typing the command **with(stats)**.

- The **stats** package allows simple statistical manipulation of data. This package includes operations such as averaging, standard deviation, correlation coefficients, variance, and regression analysis.

- The following functions are available in the statistics package:

addrecord	average	ChiSquare	correlation
covariance	evalstat	Exponential	Fdist
Ftest	getkey	linregress	median
mode	multregress	N	projection
putkey	Q	RandBeta	RandChiSquare
RandExponential	RandFdist	RandGamma	RandNormal
RandPoisson	RandStudentsT	RandUniform	regression
removekey	Rsquared	sdev	serr
statplot	StudentsT	Uniform	variance

- Additional information is available on the following topics which are concerned with the statistical matrix and user-defined distributions.

addrecord	data	distribution	evalstat
getkey	putkey	removekey	

- The package functions are also available using the long-form notation: **stats**[*function*] (*args*). This notation is necessary whenever there is a conflict between a package function name and another function used in the same session.

SEE ALSO: **with, stats**[*function*] (where *function* is from the above list)

3.4.2   stats[addrecord] – add records to a statistical matrix

Calling Sequence:
    addrecord(*dat*, *new*)

Parameters:
    dat   –   a statistical matrix
    new   –   a matrix or list of values

Synopsis:
• Addrecord appends new observations to a statistical matrix. The new statistical matrix is returned.

Examples:
```
> with(stats):
  dat:=array([[x,y],[1,2],[2,3]]):
  addrecord(dat,[3,4]);
```

$\longrightarrow$

```
[ x   y ]
[       ]
[ 1   2 ]
[       ]
[ 2   3 ]
[       ]
[ 3   4 ]
```

SEE ALSO: stats[getkey], stats[removekey], stats[putkey]

3.4.3   stats[average] – calculate the average of a list, matrix, or array of numbers

Calling Sequence:
    average(*x*)
    average(*x1*, *x2*, ...)
    average(*dat*, *k1*, *k2*, ...)

Parameters:
    x   –   a list, list of lists, matrix, or array
    dat   –   statistical matrix
    x1, x2, ...   –   sequence to be averaged
    k1, k2, ...   –   keys of statistical matrix

Synopsis:
• average takes the average of some numbers.

• If average receives a matrix, list of lists, or a statistical matrix with more than one key specified,
average will return a list of values corresponding to the averages of the columns.

Examples:
```
> with(stats):
  k:=array([[1,2,2],[5,5,5],[11,7,8]]):
  average(k);                              ⟶          [17/3, 14/3, 5]
> average(");                              ⟶               46/9
```

SEE ALSO: stats[variance], stats[mode], stats[median]

3.4.4   stats[ChiSquare] – evaluate the Chi squared function

Calling Sequence:
> ChiSquare(F, v)

Parameters:
> F –   ChiSquare value
>
> v –   degrees of freedom

Synopsis:
- ChiSquare returns the value of x such that:

  ```
  int(u^(v/2 - 1)*exp( - u/2),u = 0..x)/(GAMMA(v/2)*(2^(v/2))) = F .
  ```

Examples:
```
> with(stats):
  ChiSquare(.10,5);                        ⟶           1.610307987
```

SEE ALSO: stats[distribution]

3.4.5   stats[correlation] – evaluate the correlation coefficient

Calling Sequence:
> correlation(dat, $key1$, $key2$)
> correlation($list1$, $list2$)

Parameters:
> dat – standard statistical matrix
>
> $key1$, $key2$ – two keys from dat
>
> $list1$, $list2$ – two lists of values

Synopsis:
- The correlation coefficient shows the strength of the relationship between two continuous variables.

Examples:
```
> with(stats):
> correlation([1,2,3,4,5,6,7],[1,2,5,8,10,13,16]):
> evalf(");
```
$$.9954022751$$

SEE ALSO: `stats[Rsquared]`, `stats[covariance]`

3.4.6   stats[covariance] – calculate the covariance

Calling Sequence:
> covariance(*dat, key1, key2*)
> covariance(*list1, list2*)

Parameters:

dat	–	statistical matrix
key1, key2	–	two keys from dat
list1, list2	–	two lists of values

Synopsis:
- This function calculates the covariance between two lists of data.

- A single number representing the covariance will be returned.

Examples:
```
> with(stats):
  covariance([1,2,3],[1.2,2.3,3.7]);            ⟶            1.250000000

> dat:=array([[x,y,z],[1,7,1.2],[2,10,2.3],
  [3,20,3.7]]):
  covariance(dat,x,z);                          ⟶            1.250000000
```

SEE ALSO: `stats[variance]`

3.4.7   stats/statistical matrix

Synopsis:
- The main structure in this statistical package is the "statistical matrix". This is simply an array of data with the first row being the names of each column - the "key". A normal data array can be converted into a statistical array using the procedure putkey.

- In these matrices the first row is the "key" and each additional row is a different observation of the "key" qualities.

- To get information into a Maple readable format you must edit the data in the environment outside of Maple. Use either a program or an editor to make the data readable by Maple. One of the simplest structures for data to take is a list of lists such as [[..], [..], ...]. The following is an `awk` program to take data in a "cards" format and change it into a list of lists:

```
BEGIN {print "["; bob=0}
{if (NF!=0) {
  if (bob!=0)print ","; bob=1;
  print "[", $1;for (i=2;i<=NF;++i)print ",",$i; print "]"
  }
}
END {print "]",";"}
```

• There are a few procedures to manipulate the statistical matrix. These are: `putkey`, `getkey`, `removekey`, and `addrecord`.

SEE ALSO: `stats`, any of the above procedure names

3.4.8   stats/statistical distributions

Synopsis:

• There are standard statistical distributions in the stats package: `N` (Normal), `ChiSquare`, `StudentsT`, `Fdist`, and `Ftest`. These return a single value – the same as you would find in a table.

• There is also another type of distribution. These can be created using the procedures `Exponential` and `Uniform`. If given the proper parameters these procedures return another procedure that is the distribution specified by the given parameters.

Examples:
```
> with(stats):
  exp5:=Exponential(5):
  exp5(6);                          ⟶              5 exp(-30)
```

SEE ALSO: `stats`, any of the above procedure names

3.4.9   stats[evalstat] – manipulation of statistical data

Calling Sequence:

    evalstat(*dat*, *eqn1*, *eqn2*,...)

Parameters:

    dat   – basic statistical matrix

    eqn1, eqn2, ... – new equations involving current data

Synopsis:

- The `evalstat` function constructs a new column in the statistical array with values determined by the given equation.

- This procedure returns the new array which can be assigned to another variable name.

Examples:
```
> with(stats):
  dat:=array([[x,y],[1,2],[3,4]]):
  evalstat(dat,k=5*x+y^2);
```
$\longrightarrow$

```
[ x   y    k ]
[            ]
[ 1   2    9 ]
[            ]
[ 3   4   31 ]
```

SEE ALSO: `stats[data]`

3.4.10   stats[Exponential] – the exponential distribution

Calling Sequence:

Exponential(*lambda*, *bound*)

Exponential(*lambda*)

Parameters:

lambda – the Exponential lambda value

bound – the upper bound of the integral

Synopsis:

- If `Exponential` is passed only a lambda value, then the procedure will return an unnamed procedure which is the exponential distribution for that specific lambda value. This procedure can then be evaluated for any given value.

- If the `Exponential` procedure is passed a lambda value and a bound, then the procedure will return the value of the Exponential distribution evaluated to the given upper bound.

Examples:
```
> with(stats):
  temp:=Exponential(5):
  temp(4);
```
$\longrightarrow$ 5 exp(-20)

```
> Exponential(5,4);
```
$\longrightarrow$ 5 exp(-20)

SEE ALSO: `stats[Uniform]`, `stats[distribution]`

3.4.11   stats[Fdist] – variance ratio distribution

Calling Sequence:

Fdist(F, n, m)

Parameters:

F –   percentile

n, m –   numerator and denominator degrees of freedom

Synopsis:

- This procedure finds an x such that:

```
F = GAMMA((n+m)/2) / ( GAMMA(n/2)*GAMMA(m/2) )  *
      int( (n/m*u)^(n/2 - 1) * (1+n/m*u)^( - (n+m)/2), u = 0..x ) .
```

Examples:
```
> with(stats):
  Fdist(.9,2,5);
```
$\longrightarrow$ 3.779716079

SEE ALSO: stats[Ftest]

3.4.12   stats[Ftest] – test prob(F > x)

Calling Sequence:

Ftest(x, n, m)

Parameters:

x –   some value

n, m –   numerator and denominator degrees of freedom

Synopsis:

- This procedure will evaluate the following integral and take its probabilistic complement (subtract it from 1):

```
F = GAMMA((n+m)/2) / ( GAMMA(n/2)*GAMMA(m/2) )  *
      int( (n/m*u)^(n/2 - 1) * (1+n/m*u)^( - (n+m)/2), u = 0..x ) .
```

Examples:
```
> with(stats):
  Ftest(32,3,5):
  evalf(");
```
$\longrightarrow$ .0010910074

SEE ALSO: stats[Fdist]

3.4.13 stats[getkey] – returns the key from a statistical matrix

Calling Sequence:

getkey(*dat*)

Parameters:

dat – statistical matrix

Synopsis:

• The getkey function returns a list which is the key of the given statistical matrix.

Examples:
```
> with(stats):
> dat:=array([[x,y,z],[1,2,3],[4,5,6],[7,8,9]]):
> getkey(dat);
```
$$[x, y, z]$$

SEE ALSO: stats[putkey], stats[removekey]

3.4.14 stats[linregress] – linear regression

Calling Sequence:

linregress(*dat*, *eqn*)

linregress(*yvals*, *xvals*)

Parameters:

dat   – standard statistical matrix

eqn   – relation of variables

yvals, xvals   – lists of values

Synopsis:

• This routine uses the traditional estimate Sxy/Sxx = b for the equation y = a + b*x. The a value is estimated by a = ybar - b*xbar.

• The parameter *eqn* must be of the form key1 = key2 which are two of the keys in the array.

• The procedure returns a list containing the intercept value and the slope value.

Examples:
```
> with(stats):
  linregress([3,4,5,6],[10,15,17,19]);        ⟶        [-.4413407821, .3240223464]
```

SEE ALSO: stats[multregress], stats[regression]

3.4.15   stats[median] – the median of the given data

Calling Sequence:

    median(*data*)

    median(*dat, k1, k2, ...*)

Parameters:

data	–	either a list, matrix, or list of lists
dat	–	statistical matrix
k1, k2, ...	–	keys of the statistical matrix

Synopsis:

- The median of a list of **n** numbers is the round(n/2) number when the list is sorted.

- If a list of list, matrix, or statistical matrix with more than one key is specified then median will return the list of medians of the columns.

Examples:

```
> with(stats):
  median(3,4,5,6,7);
```
                      $\longrightarrow$ 5

```
> dat:=array([[1,2,3,4],[5,6,7,8],[9,2,4,1]
  ]):
  median(dat);
```
                      $\longrightarrow$ [5, 2, 4, 4]

SEE ALSO: stats[average], stats[mode]

3.4.16   stats[mode] – the mode of the given data

Calling Sequence:

    mode(*data, lowbnd, incr*)

Parameters:

data	–	either a list, matrix, or list of lists
lowbnd	–	(optional) the lower bound of the data
incr	–	(optional) the desired range increment

Synopsis:

- The mode of a set of **n** numbers is the number(s) with the highest frequency in the data set.

- If only *data* is specified then each unique number is considered to be an observation.

- If a lower bound and a range increment are specified then the range of numbers with the maximum frequency of observations is considered the mode. The lower bound must be less than the minimum of the data or erroneous results will occur.

- If more than one number or range has the highest frequency then a list of modes will be returned.

• If the *data* specified is a matrix or a list of lists then a list of lists will be returned containing the lists of modes for each column.

Examples:
```
> with(stats):
  dat:=[1,1,2,2,2,2,5,5,9,9,9,9]:
  mode(dat);                         ⟶                    [2, 9]
> mode(dat,1,2);                     ⟶                    [1 .. 3]
> dat:=array([[1,2,3,4],[1,2,3,3],[1,2,3,2]
  ]):
  mode(dat);                         ⟶          [[1], [2], [3], [2, 3, 4]]
```

SEE ALSO: stats[average], stats[median]

3.4.17 stats[multregress] – multiple regression

Calling Sequence:

> multregress(*dat*, *eqn*, *const*)

Parameters:

> *dat*   –   standard statistical array
>
> *eqn*   –   relation of the form y = [x1, x2, ...]
>
> *const* –   (optional) constant term will be added

Synopsis:

• multregress uses the matrix equation

$$
\begin{array}{ccc}
\text{t} & -1 & \text{t} \\
(\text{X} \quad \text{X}) & \text{X} & \text{Y}
\end{array}
$$

• An array containing the evaluated coefficients is returned.

• The form of the equation should be y = [x1, x2, ...] where y, x1, x2, ... are keys of the statistical matrix. A list of the coefficients for each of x1, x2,.... is calculated

• If the option *const* is specified as the third parameter then a constant term is added as the first entry in the return array.

Examples:
```
> with(stats):
> dat := array([[y,x1,x2,x3,x4,x5,x6], [1,4,2,7,1,6,8], [6,9,1,12,5,3,7],
>         [9,3,4,7,1,8, 6], [4,7,1/2,4/3,6/7,5,1], [9,1,4/3,1,7,6,2],
>         [2,6,1,8,9,3,4], [11,14,2,7,1,3,8]]):
> multregress(dat,x5=[x2,x4,x6],const);
            [ 5.059941787, 1.458974589, -.1287837185, -.4300813849 ]
```

SEE ALSO: stats[regression], stats[projection], stats[linregress]

3.4.18   stats[N] – values for a normal distribution

Calling Sequence:

   N(x, m, v)

Parameters:

   x –   upper bound of integration

   m –   (optional) mean of the distribution

   v –   (optional) variance of the distribution

Synopsis:

- The N function integrates the following function:

$$N = int(\ exp(-1/2*(u-m)^2/v) \ / \ sqrt(2*Pi*v), \ u = - \ infinity..x \ )$$

- The default values are m = 0 and v = 1 (the N(0, 1) distribution). If m and v are specified then the distribution is N(m, v).

Examples:
```
> with(stats):
  N(2.44);                                    ⟶            .9926563690
> N(2.44,2,5);                                ⟶            .5779977938
```

SEE ALSO: stats[distribution]

3.4.19   stats[projection] – create a projection matrix

Calling Sequence:

   projection(dat, eqn, $const$)

Parameters:

   dat –   statistical matrix

   eqn –   relation of the form y = [x1, x2, ...]

   $const$ –   (optional) constant term will be added

Synopsis:

- This procedure constructs a projection matrix according to

$$X \ (X^t \ X)^{-1} \ X^t$$

- The equation must be of the form y = [x1, x2, ...] .

- The projection matrix projects the actual Y values onto the predicted Y values.

- If the option *const* is added as a third parameter then the projection matrix will be constructed as if a constant existed in the equation.

Examples:
```
> with(stats):
> dat := array([[x1,x4,x5], [3/2,9.2,1/7], [4/9,1,2.3], [1,1/6,8]]):
> pro := projection(dat, y = [x1,x4,x5], const);
                    [ .7177328572   .9325857143   3.963428571 ]
                    [                                          ]
            pro := [ 2.270080000   .1358555555   .8428888888 ]
                    [                                          ]
                    [ 4.711783333   .7753000000   1.693333333 ]
```

SEE ALSO: stats[regression]

3.4.20   stats[putkey] – add a key to a matrix

Calling Sequence:
   putkey(*dat*, *key*)

Parameters:
   dat – matrix full of data

   key – list of names making up the key for the data

Synopsis:
- This procedure adds a row to the given data matrix, making it a statistical matrix.

- The argument *key* must be a list of syntactically correct names.

- This procedure returns a new unnamed array with the key added.

Examples:
```
> with(stats):
  dat:=array([[2,3],[5,7]]):
  putkey(dat,[y,x]);
```
$\longrightarrow$
```
                                                    [ y   x ]
                                                    [       ]
                                                    [ 2   3 ]
                                                    [       ]
                                                    [ 5   7 ]
```

SEE ALSO: stats[data], stats[removekey]

3.4.21   stats[Q] – Tail of a N(0,1) distribution

Calling Sequence:
   Q(*x*)

Parameters:

x –   any real number

Synopsis:

- Q evaluates the following expression: `evalf(1/2*(1-erf(x/sqrt(2))))`

Examples:
```
> with(stats):
  Q(0);                              ⟶              .5000000000
> Q(1);                              ⟶              .1586552539
```

SEE ALSO: stats[N]

3.4.22   stats[RandBeta] – random number generator for the Beta distribution

Calling Sequence:

RandBeta(a, b)
RandBeta(a, b, d)

Parameters:

a, b –   positive real numbers

d –   integer specifying number of digits of precision

Synopsis:

- The function RandBeta returns a random number generator for the Beta distribution of order a and b .

Examples:
```
> with(stats):
> f := RandBeta(1,2):
> seq(f(),i=1..5);
        .08506363199, .6040811399, .4344403978, .1062882242, .5627954882
```

SEE ALSO: stats[RandGamma], stats[RandFdist]

3.4.23   stats[RandExponential] – random number generator for the exponential distribution

Calling Sequence:

RandExponential(u)
RandExponential(u, d)

Parameters:

u –   positive real number

    d – (optional) integer specifying number of digits

Synopsis:

- The `RandExponential` function returns a real random number generator for the exponential distribution with mean time between arrivals *u*.

Examples:
```
> with(stats):
> f := RandExponential(3):
> seq(f(),i=1..5);
        .8953519047, 6.592802112, 3.037518162, 2.562664828, .5021620560
```

3.4.24 stats[RandFdist] – random number generator for the F-distribution

Calling Sequence:
    RandFdist(*v1*, *v2*)
    RandFdist(*v1*, *v2*, *d*)

Parameters:

    v1, v2 – positive real numbers

    d – (optional) integer specifying number of digits

Synopsis:

- The function `RandFdist` returns a random number generator for the F-distribution with *v1* and *v2* degrees of freedom.

Examples:
```
> with(stats):
> F := RandFdist(1,3):
> seq(F(),i=1..5);
        .02670979112, 3.064663161, .6850516830, 7.020169617, 2.085687473
```

SEE ALSO: `stats[RandGamma]`, `stats[RandBeta]`

3.4.25 stats[RandGamma] – random number generator for the Gamma distribution

Calling Sequence:
    RandGamma(*a*)
    RandGamma(*a*, *d*)

Parameters:

    a – positive real number

    d – integer specifying number of digits

Synopsis:
- The `RandGamma` function returns a uniform random number generator for the Gamma distribution of order a.

Examples:
```
> with(stats):
> f := RandGamma(1):
> seq(f(),i=1..5);
        .2984506349, 2.197600704, 1.012506054, .8542216094, .1673873520
```

SEE ALSO: stats[RandBeta], stats[RandFdist]

3.4.26   stats[RandNormal] – random number generator for the normal distribution

Calling Sequence:
>   RandNormal(u, s)
>   RandNormal(u, s, d)

Parameters:
> u – real number, the mean
> s – positive real number, the standard deviation
> d – (optional) integer specifying number of digits

Synopsis:
- The function `RandNormal` returns a pseudo-random number generator for the normal distribution $N(u, s)$.

Examples:
```
> with(stats):
> N := RandNormal(0,1):
> seq(N(),i=1..5);
        -.1549980895, -.5339724225, -.8587859939, .5740786490, .5865929574
```

SEE ALSO: stats[RandUniform]

3.4.27   stats[RandPoisson] – random number generator for the Poisson distribution

Calling Sequence:
>   RandPoisson($lambda$)
>   RandPoisson($lambda$, d)

Parameters:

lambda   –   positive real number

d   –   (optional) integer specifying number of digits

Synopsis:

- The `RandPoisson` function returns a pseudo-random number generator for the Poisson distribution where *lambda* is the mean number of occurrences of an event per unit time.

- Note that the method being used is only efficient for small *lambda* .

Examples:
```
> with(stats):
  f := RandPoisson(.9):
  seq(f(),i=1..5);                    ⟶              1, 0, 1, 1, 0
```

SEE ALSO: stats[RandExponential]

3.4.28   stats[RandStudentsT] – random number generator for the Students T distribution

Calling Sequence:
```
RandStudentsT(v)
RandStudentsT(v, d)
```

Parameters:

v   –   positive real number

d   –   (optional) integer specifying number of digits

Synopsis:

- The function `RandStudentsT` returns a random number generator for the Student T distribution with *v* degrees of freedom.

Examples:
```
> with(stats):
> T := RandStudentsT(2):
> seq(T(),i=1..5);
      -.07701893963, -.2888707411, .2334227339, -.1022134775, -3.280223936
```

SEE ALSO: stats[RandChiSquare]

3.4.29   stats[RandUniform] – uniform random number generator

Calling Sequence:
```
RandUniform(a..b)
RandUniform(a..b, d)
```

Parameters:

     $a..b$ –   real numbers $a < b$

     d –   (optional) integer specifying number of digits

Synopsis:

- RandUniform returns a uniform random number generator for the range for $[a, \ b)$.

Examples:
```
> with(stats):
> U := RandUniform(1..2):
> seq(U(),i=1..5);
        1.741966908, 1.111069327, 1.363307370, 1.425614356, 1.845871898
```

SEE ALSO: stats[RandNormal]

3.4.30   stats[RandChiSquare] – random number generator for the Chi-squared distribution

Calling Sequence:

     RandChiSquare(v)

     RandChiSquare(v, d)

Parameters:

     v –   positive real number

     d –   (optional) integer specifying number of digits

Synopsis:

- RandChiSquare returns a random number generator for the X-square distribution with v degrees of freedom.

Examples:
```
> with(stats):
> X := RandChiSquare(3):
> seq(X(),i=1..5);
        2.771207464, .3629432068, 2.107282672, 2.314454502, 1.229701805
```

3.4.31   stats[regression] – generalized regression routine

Calling Sequence:

     regression(dat, eqn)

Parameters:

     dat –   statistical array

     eqn –   equation for regression fit

Synopsis:

- This procedure accepts data and an equation linear in the unknown coefficients and determines values for the coefficients.

- This routine solves the system of normal equations to determine the values of the coefficients.

Examples:
```
> with(stats):
> dat := array([[y,x],[1/2,1/5],[4,1],[6,1],[1,7]]):
> regression(dat,y=a+b*x+c*x^2+d*x^3);
    {b = 14.93827160, c = -8.746688749, a = -2.145417448, d = .9538345915}
```

SEE ALSO: stats[multregress], stats[linregress]

3.4.32   stats[removekey] – take the key away from a statistical matrix

Calling Sequence:
>   removekey(*dat*)

Parameters:
>   *dat*   –   statistical matrix

Synopsis:

- Removekey removes the first row of a statistical matrix and returns a normal matrix full of the data.

Examples:
```
> with(stats):
  dat:=array([[y,x],[3,4],[6,7],[7,8]]):
  removekey(dat);
```
$\longrightarrow$
```
[ 3  4 ]
[      ]
[ 6  7 ]
[      ]
[ 7  8 ]
```

SEE ALSO: stats[putkey], stats[getkey], stats[data]

3.4.33   stats[Rsquared] – square of the correlation coefficient

Calling Sequence:
>   Rsquared(*list1*, *list2*)
>   Rsquared(*dat*, *key1*, *key2*)

Parameters:
>   *dat*   –   a statistical array

> *key1, key2* – two keys defined in the data
>
> *list1, list2* – two lists of values of the same size

Synopsis:

- The `Rsquared` function determines the square of the correlation coefficient.

- The `Rsquared` statistic is the proportion of variance in one of the variables that can be explained by variation in the other variables.

Examples:
```
> with(stats):
> Rsquared([1,2,3,4,5],[3.5,9.7,2.3,10.4,11.66]);
                          .3936148819
> dat:=array([[y,x1,x2],[2,5,7],[3,6,8],[4,7,9]]):
> Rsquared(dat,x1,x2);
                              1
```

SEE ALSO: `stats[correlation]`, `stats[variance]`, `stats[data]`

3.4.34   stats[sdev] – standard deviation

Calling Sequence:

> sdev(x)
>
> sdev(*dat*, *k1*, *k2*, ...)

Parameters:

> x – either a list, list of lists, matrix, or vector
>
> *dat*   – statistical matrix
>
> *k1, k2, ...* – keys of the statistical matrix

Synopsis:

- This procedure calculates the standard deviation of a list of numbers.

- Standard deviation is a measure of how widely the numbers are spread.

- If a list of lists, matrix, or a statistical matrix with more than one key is specified, then a list of values is returned.

Examples:
```
> with(stats):
  sdev(3,4,5,6,7,8,100);                    ⟶            1/2  1/2
                                                     2/3 959    3

> dat:=array([[y,x],[3,4],[9.5,6.7],[0.001,
  0.005]]):
  sdev(dat,y,x);                            ⟶    [4.855838445, 3.368309417]
```

SEE ALSO: `stats[variance]`, `stats[serr]`

3.4.35   stats[serr] – standard error

Calling Sequence:

> serr(x)
> serr(dat, $k1$, $k2$, ...)

Parameters:

x	–	can be list, list of lists, matrix, or vector
dat	–	statistical matrix
$k1$, $k2$, ...	–	keys of the statistical matrix

Synopsis:

- The standard error of a list of numbers tells us how far off the mean estimate might be.

- If a list of lists, matrix, or a statistical matrix with more than one key is specified, then a list of values is returned.

Examples:
```
> with(stats):
  serr([3,4,0.001,9.9]);                    ⟶                    2.073625062
> dat:=array([[x,y],[2,3],[6.7,31],[0.001,
  0.005]]):
  serr(dat,x,y);                            ⟶                         1/2                      1/2
                                            [1.146351717 3    , 5.698700386 3    ]
```

SEE ALSO: stats[sdev], stats[variance]

3.4.36   stats[statplot] – statistical plotting

Calling Sequence:

> statplot(dat, eqn)
> statplot(dat, $prod$)

Parameters:

dat	–	statistical array
eqn, $prod$	–	an equation or a product of the keys

Synopsis:

- A specially designed plotting routine to help with statistical plotting.

- If an equation is specified as the second parameter then the line is plotted on top of the x and y points. This is very handy in plotting regression results.

- If a product is specified then this is interpreted as "x vs y".

- The plotdevice must be changed outside of this routine.

Examples:
```
# see if linear regression is appropriate results in a plot of y vs x
```

```
statplot(dat,y*x);
regression(dat,y=A+B*x); # where y,x are keys to dat
statplot(dat,y=A+B*x);   # regression assigns the calculated
                         # values to A,B
#   results in a plot of y vs x with the regression line
#   overlaid.
```

SEE ALSO: stats[data] stats[regression]

3.4.37   stats[StudentsT] – the Student's T distribution

Calling Sequence:

   StudentsT(F, v, $area$)

Parameters:

   F –   cumulative probability

   v –   degrees of freedom

   $area$ –   optional parameter

Synopsis:

• StudentsT returns the x value to satisfy the following integral:

```
F = GAMMA((v+1)/2) / ( sqrt(Pi*v)*GAMMA(v/2) )  *
    int( (1+u^2/v)^(-(v+1)/2), u = -infinity..x )
```

• If a third argument $area$ is specified then StudentT will compute the area from -infinity to F under a Student's central t-distribution with v degrees of freedom.

Examples:
```
> with(stats):
  StudentsT(.75,4);                  ⟶              .7406970841
> StudentsT(.75,4,'area');           ⟶              .7406970841
```

SEE ALSO: stats[distribution]

3.4.38   stats[Uniform] – the Uniform distribution

Calling Sequence:

   Uniform(a, b)

Parameters:

   a, b –   the range of the Uniform distribution

Synopsis:

• A Uniform distribution has value 0 everywhere except a..b where its value is $1/(b-a)$.

- The procedure Uniform returns an unnamed procedure which is the Uniform distribution of the given interval.

Examples:
```
> with(stats):
  k:=Uniform(a,b):
  k();
```
$\longrightarrow$

$$\{[\frac{1}{b-a}, a .. b]\}$$

```
> j:=Uniform(5,7):
  j(6);
```
$\longrightarrow$ $\qquad$ 1/2

```
> j(10);
```
$\longrightarrow$ $\qquad$ 0

SEE ALSO: stats[Exponential], stats[distribution]

3.4.39 stats[variance] – compute the variance of a set of data

Calling Sequence:
> variance(x)
> variance(dat, $k1$, $k2$, ...)

Parameters:

x	–	can be a list, matrix, or vector
dat	–	a statistical matrix
$k1$, $k2$, ...	–	keys of the statistical matrix

Synopsis:

- The variance is a measure of spread in the data.

- If a list of lists, matrix, or a statistical matrix with more than one key is specified a list of values is returned.

Examples:
```
> with(stats):
  variance([3,4,5,6]);
```
$\longrightarrow$ $\qquad$ 5/3

```
> dat:=array([[x,y],[3,5],[6.7,8.9],[0.001,
  0.005]]):
  variance(dat,y,x);
```
$\longrightarrow$ $\qquad$ [19.88017500, 11.26010033]

SEE ALSO: stats[covariance], stats[sdev]

3.5   The Simplex Linear Optimization Package

3.5.1   Introduction to the simplex package

Calling Sequence:

>   simplex[*function*](args)
>   *function*(args)

Synopsis:

- The **simplex** package is a collection of routines for linear optimization using the simplex algorithm as a whole, and using only certain parts of the simplex algorithm.

- The particular implementation of the simplex algorithm used here is based on the initial chapters of *Linear Programming* by Chvatal, 1983, W.H. Freeman and Company, New York.

- In addition to the routines **feasible**, **maximize** and **minimize**, the simplex package provides routines to assist the user in carrying out the steps of the algorithm one at a time: setting up problems, finding a pivot element, and executing a single pivot operation.

- The functions available are:

basis	convexhull	cterm	dual
feasible	maximize	minimize	pivot
pivoteqn	pivotvar	ratio	setup
standardize			

- See **simplex[*function*]** for more information on any of the functions listed above.

- To maximize a linear function **f** subject to the set of linear constraints **c**, use **with(simplex)** followed by **maximize(f,c)**.

- To use a simplex function, either define that function alone using the command **with(simplex, *function*)**, or define all simplex functions using the command **with(simplex)**. Alternatively, invoke the function using the long form **simplex[*function*]**. This long form notation is necessary whenever there is a conflict between a package function name and another function used in the same session.

Examples:
```
> with(simplex):
> cnsts := {3*x+4*y-3*z <= 23, 5*x-4*y-3*z <= 10,
>   7*x+4*y+11*z <= 30}:
> obj := -x + y + 2*z:
> maximize(obj,cnsts union {x>=0,y>=0,z>=0});
                        {x = 0, y = 49/8, z = 1/2}
```

```
> maximize(obj,cnsts,NONNEGATIVE);
                        {x = 0, y = 49/8, z = 1/2}

> maximize(obj,cnsts);
```

SEE ALSO: `with`, `simplex[`*function*`]` (where *function* is from the above list)

3.5.2   simplex[basis] – computes a list of variables, corresponds to the basis

Calling Sequence:

> basis(*C*)

Parameters:

> *C* – a set of linear equations

Synopsis:

- The function `basis(`*C*`)` computes a list of variables, one per equation, which corresponds to the basis used by the simplex algorithm.

- The set of linear equations *C* passed to `basis` should be in the special form produced by `simplex[setup]`. The function `simplex[setup]` ensures that each of these variables occurs in exactly one equation, and only on the left-hand side.

- This function is part of the `simplex` package, and so can be used in the form `basis(..)` only after performing the command `with(simplex)` or `with(simplex, basis)`. The function can always be accessed in the long form `simplex[basis](..)`.

Examples:
```
> with(simplex):
> basis( [ x = 3*y + z , w = 2*y - z ] );
                        [x, w]
```

SEE ALSO: `simplex[setup]`

3.5.3   simplex[convexhull] – finds convex hull enclosing the given points

Calling Sequence:

> convexhull(*ps*)

Parameters:

> *ps* – a list or set of points

Synopsis:

- The result returned by `convexhull` is a list of points in counter-clockwise order.

- An `n*log(n)` algorithm computing tangents of pairs of points is used.

- Every point in *ps* is a list of an x and y coordinate, both of which must be numeric.

- This function is part of the `simplex` package, and so can be used in the form `convexhull(..)` only after performing the command `with(simplex)` or `with(simplex, convexhull)`. The function can always be accessed in the long form `simplex[convexhull](..)`.

Examples:
```
> with(simplex):
> convexhull( { [0,0], [1,1], [2,0], [1,0], [1,1/2] } );
                    [[0, 0], [2, 0], [1, 1]]
```

3.5.4   simplex[cterm] – computes the list of constants from the system

Calling Sequence:
>   cterm(*C*)

Parameters:
>   *C*   –   a set or list of linear equations or inequalities

Synopsis:
- The function cterm(*C*) computes a list of constants, one per equation or inequality. These are computed as if the constants were collected on the right hand side of the equation or inequality.

- This function is part of the `simplex` package, and so can be used in the form `cterm(..)` only after performing the command `with(simplex)` or `with(simplex, cterm)`. The function can always be accessed in the long form `simplex[cterm](..)`.

Examples:
```
> with(simplex):
> cterm( [ 3*x + y <= 5 , 4*y - z - 3 = 3 ] );
                         [5, 6]
```

3.5.5   simplex[dual] – computes the dual of a linear program

Calling Sequence:
>   dual(*f*, *C*, *y*)

Parameters:
>   *f*   –   a linear expression
>   *C*   –   a set of linear inequalities
>   *y*   –   a name

Synopsis:
- The procedure dual(*f, C, y*) computes the dual of a linear program which is in standard inequality form.

- The expression *f* is the linear objective function to be maximized, subject to the linear inequalities *C*. These inequalities are in the special form produced by `simplex[convert/stdle]`. The name `y` is used to construct the names `y1`, `y2`, ... for the dual variables.

- The resulting dual is returned as an expression sequence: objective, constraints.

- This function is part of the `simplex` package, and so can be used in the form `dual(..)` only after performing the command `with(simplex)` or `with(simplex, dual)`. The function can always be accessed in the long form `simplex[dual](..)`.

Examples:
```
> with(simplex):
> dual( x+y, {3*x+4*y <= 4, 4*x+3*y <= 5}, z );
            4 z1 + 5 z2, {1 <= 3 z1 + 4 z2, 1 <= 4 z1 + 3 z2}
```

SEE ALSO: `simplex[stdle]`

3.5.6   simplex[feasible] – determine if system is feasible or not

Calling Sequence:
>   feasible(*C*)
>   feasible(*C*, *vartype*)
>   feasible(*C*, *vartype*, ′*NewC*′, ′*Transform*′)

Parameters:

C	–	a set of linear constraints
vartype	–	(optional) NONNEGATIVE or UNRESTRICTED
NewC	–	(optional) a name
Transform	–	(optional) a name

Synopsis:
- The function `feasible` returns `true` if a feasible solution to the linear system *C* exists, and `false` otherwise.

- Non-negativity constraints on all the variables can be indicated by use of a second argument, NONNEGATIVE, or by explicitly listing the constraints. No restriction on the signs of the variable may be indicated by using UNRESTRICTED as the second argument to `feasible`.

- The final two arguments are used to return, as sets, the final system found by `feasible`, and any variable transformations which occurred. The new system may have global `artificial` and `slack` variables present (such as `_AR` or `_SL1`).

- This function is part of the `simplex` package, and so can be used in the form `feasible(..)` only after performing the command `with(simplex)` or `with(simplex, feasible)`. The function can always be accessed in the long form `simplex[feasible](..)`.

Examples:
```
> with(simplex):
> feasible({4*x+3*y <= 5, 3*x+4*y = 4}, NONNEGATIVE);
                              true

> feasible({4*x-3*y <= 5, 3*x-4*y = 4, x>=0, y>=0 });
                              false
```

3.5.7   simplex[maximize] – maximize a linear program

Calling Sequence:
>   maximize(*f*, *C*)
>   maximize(*f* , *C*, *vartype*)
>   maximize(*f* , *C*, *vartype*, ´*NewC*´, ´*transform*´)

Parameters:

f	–	a linear expression
C	–	a set or list of linear constraints
vartype	–	(optional) NONNEGATIVE or UNRESTRICTED
NewC	–	(optional) a name
transform	–	(optional) a name

Synopsis:

- The expression *f* is the linear objective function to be maximized subject to the linear constraints *C*. The function `maximize` returns either a set of equations describing the optimal solution to the specified linear program, or the empty set in the case where no feasible solution to C exists, or NULL in the case where the solution is unbounded.

- The equations returned by `maximize` can be substituted back into the objective function *f* to obtain the value of the objective function at the optimal solution.

- A third parameter may be used to specify that all variables are constrained to be NONNEGATIVE; such constraints may also be listed explicitly. Similarly, UNRESTRICTED indicates that no sign constraint is to be placed on the variables.

- A fourth and a fifth parameter may be included to specify names for returning the optimal description, and any variable transformations used to set up the problem.

- This function is part of the `simplex` package, and so can be used in the form `maximize(..)` only after performing the command `with(simplex)` or `with(simplex, maximize)`. The function can always be accessed in the long form `simplex[maximize](..)`.

Examples:
```
> with(simplex):
> maximize( x+y, {4*x+3*y <= 5, 3*x+4*y <= 4 } );
                       {x = 8/7, y = 1/7}
```

```
> maximize( x-y, {3*x+4*y <= 4, 4*x+3*y <= -3} );
> maximize( x-y, {3*x+4*y <= 4, 4*x+3*y <= -3}, NONNEGATIVE );
                            {}
```

SEE ALSO: `simplex[minimize]`

3.5.8   simplex[minimize] – minimize a linear program

Calling Sequence:

> minimize(f, C)
> minimize(f , C, *vartype*)
> minimize(f , C, *vartype*, ´*NewC*´, ´*transform*´)

Parameters:

f	–	a linear expression
C	–	a set or list of linear constraints
vartype	–	(optional) NONNEGATIVE or UNRESTRICTED
NewC	–	(optional) a name
transform	–	(optional) a name

Synopsis:

- The expression f is the linear objective function to be minimized subject to the linear constraints C. The function `minimize` returns either a set of equations describing the optimal solution to the specified linear program, or the empty set in the case where no feasible solution to C exists, or NULL in the case where the solution is unbounded.

- The equations returned by `minimize` can be substituted back into the objective function f to obtain the value of the objective function at the optimal solution.

- A third parameter may be used to specify that all variables are constrained to be NONNEGATIVE; such constraints may also be listed explicitly. Similarly, UNRESTRICTED indicates that no sign constraint is to be placed on the variables.

- A fourth and a fifth parameter may be included to specify names for returning the optimal description, and any variable transformations used to set up the problem.

- This function is part of the `simplex` package, and so can be used in the form `minimize(..)` only after performing the command `with(simplex)` or `with(simplex, minimize)`. The function can always be accessed in the long form `simplex[minimize](..)`.

Examples:

```
> with(simplex):
> minimize( x+y, {4*x+3*y <= 5, 3*x+4*y <= 4}, NONNEGATIVE );
                        {x = 0, y = 0}
```

SEE ALSO: `simplex[maximize]`

3.5.9   simplex[pivot] – construct a new set of equations given a pivot

Calling Sequence:
   pivot(C, x, eqn)

Parameters:

C	–	a set or list of equations
x	–	a variable name
eqn	–	an equation or list of equations

Synopsis:
- The call `pivot`(C, x, eqn) constructs a new set of equations in a form compatible with the forms used in `simplex[setup]`, by solving the specified equation eqn for x, then substituting the result into C. This is equivalent to the standard simplex pivot on an array of coefficients.

- To re-express the objective function `f` in terms of this new description, one can substitute the result of `pivot` into `f`.

- If eqn is a list of equations, then $eqn[1]$ is used. This is to facilitate easy user interaction with the function `simplex[pivoteqn]` used to choose an equation to use for the next pivot.

- This function is part of the `simplex` package, and so can be used in the form `pivot(..)` only after performing the command `with(simplex)` or `with(simplex, pivot)`. The function can always be accessed in the long form `simplex[pivot](..)`.

Examples:
```
> with(simplex):
> pivot( {_SL1 = 5-4*x-3*y, _SL2 = 4-3*x-4*y}, x, [_SL1 = 5-4*x-3*y] );
              {_SL2 = 4 - 3 x - 4 y, x = - 1/4 _SL1 + 5/4 - 3/4 y}
```

SEE ALSO: `simplex[pivoteqn]`

3.5.10   simplex[pivoteqn] – returns a sublist of equations given a pivot

Calling Sequence:
   pivoteqn(C, var)

Parameters:

C	–	a set of linear equations
var	–	the variable chosen to carry out a pivot operation

Synopsis:
- The function `pivoteqn`(C, var) returns a sublist of equations from C. Each equation eq returned achieves the minimum non-negative ratio of

$$(- \text{cterm}(\{\backslash\text{var eq}\})) / \text{coeff}(\{\backslash\text{var eq}\}, \{\backslash\text{var var}\}, 1)$$

- In the special case where all ratios are negative, the function returns **FAIL**.

- The linear equations C are in the special form produced by `simplex[setup]`.

- This function is part of the `simplex` package, and so can be used in the form `pivoteqn(..)` only after performing the command `with(simplex)` or `with(simplex, pivoteqn)`. The function can always be accessed in the long form `simplex[pivoteqn](..)`.

Examples:
```
> with(simplex):
> pivoteqn( {_SL1 = 5-4*x-3*y, _SL2 = 4-3*x-4*y}, x );
                    [_SL1 = 5 - 4 x - 3 y]
```

3.5.11   simplex[pivotvar] – returns a variable with positive coefficient

Calling Sequence:
>   pivotvar(f, *List*)
>   pivotvar(f)

Parameters:

> f –   an objective function expressed in terms of non-basic variables relative to some description
>   *List*   –   a list of the variables which occur in the problem

Synopsis:

- The function pivotvar(f, *List*) returns either a variable which has a positive coefficient or **FAIL**. The optional *List* (of variable names) is used to determine the order in which the variables should be examined. The first success is used. If no *List* is provided, then `pivotvar` chooses its own order.

- This function is part of the `simplex` package, and so can be used in the form `pivotvar(..)` only after performing the command `with(simplex)` or `with(simplex, pivotvar)`. The function can always be accessed in the long form `simplex[pivotvar](..)`.

Examples:
```
> with(simplex):
> pivotvar( x1 + 3*x3 - x4 );
                              x1
> pivotvar( x1 + 3*x3 - x4 , [x4,x3,x1] );
                              x3
```

3.5.12   simplex[ratio] – returns a list of ratios

Calling Sequence:
>   ratio(C, x)

Parameters:

    C – a set of linear equations

    x – a variable to be used in calculating ratios

Synopsis:

- The call `ratio(`C, x`)` returns a list of ratios which can be used to determine which equation in C is most restrictive with respect to the next simplex pivot. For each equation, the ratio computed, if possible, is

$$- \text{cterm(eq) / coeff(eq, x, 1)}$$

- All negative ratios are reported as `infinity`, as are those for which the denominator is 0.

- The equations C are in the special form produced by `simplex[setup]`.

- This function is part of the `simplex` package, and so can be used in the form `ratio(..)` only after performing the command `with(simplex)` or `with(simplex, ratio)`. The function can always be accessed in the long form `simplex[ratio](..)`.

Examples:
```
> with(simplex):
> ratio( [_SL1 = 5-4*x-3*y, _SL2 = 4-3*x-4*y], x );
                        [5/4, 4/3]
```

3.5.13   simplex[setup] – constructs a set of equations with variables on the lhs

Calling Sequence:

    `setup(`C`)`

    `setup(`C`, NONNEGATIVE)`

    `setup(`C`, NONNEGATIVE, `$'t'$`)`

Parameters:

    C – a set of linear equations

    $'t'$ – a name

Synopsis:

- The function `setup(`C`)` constructs a set of equations in a form with isolated variables on the left-hand side. Those variables form a basis for the corresponding linear system, and do not occur on the right hand side of any equation. Slack variables of the form `_SL.i` are introduced to deal with inequalities. Unrestricted variables are transformed to be the difference of two variables.

- The resulting system is equivalent to the original system C, in that solutions to the new system can be transformed into solutions of the original system. The resulting system need not correspond to a feasible basic solution.

- If the optional second parameter ´NONNEGATIVE´ is present, all variables are assumed to be non-negative. If a third parameter is present, then it is assigned the transformations used for any variable deemed to be unrestricted.

- This function is part of the **simplex** package, and so can be used in the form **setup(..)** only after performing the command **with(simplex)** or **with(simplex, setup)**. The function can always be accessed in the long form **simplex[setup](..)**.

Examples:
```
> with(simplex):
> setup( {3*x+4*y <= 4, 4*x+3*y = 5} );
                {y = - 4/3 x + 5/3, _SL1 = - 8/3 + 7/3 x}
```

3.5.14   simplex[standardize] – converts a set of equations to type ≤

Calling Sequence:
standardize(*C*)

Parameters:
C   –   a set or list of linear constraints

Synopsis:
- The function **standardize**(*C*) returns a set or list of inequalities of the form ≤, equivalent to the original set of constraints, with all constants on the right-hand side.

- This function differs from **convert(*C*,std)** in that equations are transformed into pairs of inequalities.

- This function is equivalent to **convert(*C*,stdle)**.

- This function is part of the **simplex** package, and so can be used in the form **standardize(..)** only after performing the command **with(simplex)** or **with(simplex, standardize)**. The function can always be accessed in the long form **simplex[standardize](..)**.

Examples:
```
> with(simplex):
> standardize( {3*x+4*y <= 4, 4*x+3*y = 5} );
           {3 x + 4 y <= 4, 4 x + 3 y <= 5, - 4 x - 3 y <= -5}
```

SEE ALSO: simplex[convert[std]], simplex[convert[stdle]]

3.5.15   convert/equality – convert inequalities to equations

Calling Sequence:
convert(*s*, equality)

Parameters:
> s – a list or set of equalities (and/or) inequalities

Synopsis:

- All equations and inequalities are converted to type `` `=` ``. Note that the feasible region described by the original set of relations is usually not preserved by this operation.

- This function is part of the simplex package and can be used only after performing the command `with(simplex)`.

Examples:
```
> with(simplex):
> convert( {3*x+4*y <= 4, 4*x+3*y <= 5}, equality );
                    {3 x + 4 y = 4, 4 x + 3 y = 5}
```

SEE ALSO: `convert[equality]`

3.5.16   convert/std – convert to standard form for simplex manipulation

Calling Sequence:
>   convert(C, std)

Parameters:
> C – a set or list of linear equations and inequalities

Synopsis:

- The `convert(`C`,std)` function returns a set (or list) of constraints obtained by moving all constants to the right-hand side of the equation (inequality) for each constraint in C.

- No attempt is made to represent equations by inequalities.

- Note that inequalities of the form $\geq$ are automatically transformed by Maple into the form $\leq$.

- This function is part of the simplex package and can only be used after performing the command `with(simplex)`.

Examples:
```
> with(simplex):
> convert( {4 <= 3*x+4*y, 5 <= 4*x+3*y}, std );
                    {- 3 x - 4 y <= -4, - 4 x - 3 y <= -5}
```


3.5.17   convert/stdle – convert inequalities to type $\leq$

Calling Sequence:
>   convert(C, stdle)

Parameters:

   C –   a set or list of linear constraints

Synopsis:

- the function `convert(`C`,stdle)` returns a set or list of inequalities of the form $\leq$, equivalent to the original set of constraints, with all constants on the right hand side.

- This function is part of the simplex package and can only be used after performing the command `with(simplex)`.

Examples:
```
> with(simplex):
> convert( {3*x+4*y <= 4, 4*x+3*y = 5} , stdle );
            {3 x + 4 y <= 4, 4 x + 3 y <= 5, - 4 x - 3 y <= -5}
```

SEE ALSO: `simplex[standardize]`

3.5.18   type/nonneg – test if constraint is of the form a $\geq$ 0

Calling Sequence:

   `type(`*expr*`, nonneg)`

Parameters:

   expr   –   any expression

Synopsis:

- This is a Boolean function to test if the expression is a constraint of the form a $\geq$ 0 .

- This function is part of the simplex package and can only be used after performing the command `with(simplex)`.

Examples:
```
> with(simplex):
  type( -a <= 0 , nonneg );          ⟶          true

> type(  0 <= a , nonneg );          ⟶          true

> type(  a = 0 , nonneg );           ⟶          false

> type(  0 <= a+b , nonneg );        ⟶          false
```

3.6 The Grobner Basis Package

3.6.1 Introduction to the grobner package

Calling Sequences:

   function`(args)`
   `grobner[`*function*`](args)`

Synopsis:

- The **grobner** package is a collection of routines for doing Gröbner basis calculations.

- To use a **grobner** function, either define that function alone by the command **with(grobner,** *function*), or define all grobner functions by the command **with(grobner)**. To access a **grobner** function without using the **with** command, use the function in the form **grobner**[*function*].

- The functions available are the following:

finduni	finite	gbasis	gsolve
leadmon	normalf	solvable	spoly

- The available term orderings are as follows:

plex	pure lexicographic, where X = [x1, x2, ..., xn] induces x1 > x2 > ... > xn
tdeg	total degree (graduated), with ties broken by inverse lexicographical order.

- For help with a particular function see the information under **grobner**[*function*], where *function* is one from the above list.

- Here is an example of using a **grobner** function. To find the Gröbner basis of a set of polynomials F over the variables X = [x_1, x_2, ..., x_n], with respect to total degree ordering, one would first define F and X, and then execute the sequence of commands that appears below.

Examples:
```
with(grobner, gbasis);
gbasis(F, X, tdeg);
```

SEE ALSO: **with**, **grobner**[*function*]

3.6.2   grobner[finduni] – attempt to find the smallest univariate polynomial in the ideal generated by the given polynomials

Calling Sequence:

    finduni(x, F)

    finduni(x, F, X)

Parameters:

- x –   an indeterminate appearing in F
- F –   set or list of polynomials
- X –   set or list of indeterminates, which must include x (not including parameters; default is indets(F))

Synopsis:

- The command `finduni`(x, F, X) will attempt to construct the univariate polynomial (in x) of least degree in `Ideal`(F); this gives all possible roots (of F) in x. The construction makes use of the total degree Gröbner basis.

- The existence of such a polynomial is only guaranteed if the system F has finitely many solutions; therefore, `grobner[finite]` should be called first to avoid infinite loops!

- When called, `grobner[finduni]` may attempt the construction even if `grobner[finite]` yields `false`.

- If X is a list, the given permutation of variables is used in all associated Gröbner basis computations.

- If X is omitted, the set `indets`(F) is used as a default.

- This function is part of the **grobner** package, and so can be used in the form `finduni(..)` only after performing the command `with(grobner)` or `with(grobner, finduni)`. The function can always be accessed in the long form `grobner[finduni](..)`.

Examples:

```
> with(grobner):
  F := [x^2 - 2*x*z + 5, x*y^2 + y*z^3, 3*
  y^2 - 8*z^3]:
  finite(F);                        ⟶                true
> finduni(z,F);                     ⟶
```
$$1600\, z^3 + 240\, z^6 - 96\, z^8 + 9\, z^9$$

SEE ALSO: `grobner[solvable]`, `grobner[finite]`, `grobner[gsolve]`

3.6.3   grobner[finite] – decide if a given algebraic system has (at most) finitely many solutions

Calling Sequence:

   finite(F)
   finite(F, X)

Parameters:

- F –   set or list of polynomials
- X –   set or list of indeterminates (not including parameters; default is indets(F))

Synopsis:

- The command `finite(F, X)` decides, by using the total degree Gröbner basis and a criterion of Buchberger, if a set/list of polynomials F with respect to the indeterminates X has (at most) finitely many solutions.

- If X is a list, the given permutation of variables is used in all associated Gröbner basis computations.

- If X is omitted, the set `indets(F)` is used as a default.

- This function is part of the **grobner** package, and so can be used in the form `finite(..)` only after performing the command `with(grobner)` or `with(grobner, finite)`. The function can always be accessed in the long form `grobner[finite](..)`.

Examples:
```
> with(grobner):
  F := [x^2 - 2*x*z + 5, x*y^2 + y*z^3, 3*
  y^2 - 8*z^3]:
  finite(F);                    ⟶            true
> finite([F[1],F[2]]);          ⟶            false
```

SEE ALSO: `grobner[solvable]`, `grobner[finduni]`, `grobner[gsolve]`

3.6.4   grobner[gbasis] – compute reduced, minimal Gröbner basis

Calling Sequence:

> gbasis(F, X)
> gbasis(F, X, *termorder*)
> gbasis(F, X, *termorder*, $'Y'$)

Parameters:

F	–	set or list of polynomials
X	–	set or list of indeterminates (not including parameters)
Y	–	(a name) list of reordered indeterminates (output only required if X is a set)
termorder	–	term ordering: $'plex'$ (pure lexicographic) or $'tdeg'$ (total degree is default)

Synopsis:

- The command `gbasis`$(F, X, termorder)$ computes the reduced, minimal Gröbner basis of the polynomials F with respect to the indeterminates X and the given term ordering. If X is a set, reordering will occur and a fourth argument MUST appear.

- A list of indeterminates $X := [x_1, x_2, \ldots, x_n]$ induces the ordering $x_1 > x_2 > \ldots > x_n$.

- If the indeterminates are given as a set $X:=\{x_1, x_2, \ldots, x_n\}$, they are automatically reordered to be "heuristically optimal" (in the sense of Boge, Gebauer, Kredel), and the new ordering is assigned to the fourth parameter; a list of indeterminates will not be reordered.

- This function is part of the `grobner` package, and so can be used in the form `gbasis(..)` only after performing the command `with(grobner)` or `with(grobner, gbasis)`. The function can always be accessed in the long form `grobner[gbasis](..)`.

Examples:

```
> with(grobner):
> F := [x^2 - 2*x*z + 5, x*y^2 + y*z^3, 3*y^2 - 8*z^3]:
> gbasis(F,[y,x,z],plex);
         2     3        3      8       7      5    2
     [3 y  - 8 z , 80 y z  - 3 z  + 32 z  - 40 z , x  - 2 x z + 5,

          7      8       5        3        6        3      8      9
      - 96 z  + 9 z  + 120 z  + 640 z  x, 240 z  + 1600 z  - 96 z  + 9 z ]

> gbasis(F,{x,y,z},plex,'Y');
      2                      7      8       5        3      2      3
     [x  - 2 x z + 5, - 96 z  + 9 z  + 120 z  + 640 z  x, 3 y  - 8 z ,

           3      8       7      5        6        3      8      9
      80 y z  - 3 z  + 32 z  - 40 z , 240 z  + 1600 z  - 96 z  + 9 z ]

> Y;
                          [x, y, z]

> gbasis(F,[x,y,z],tdeg);
      2                    2      3      4        3      2      2      3
     [x  - 2 x z + 5, - 3 y  + 8 z , 9 y  + 48 z y  + 320 y , 8 x y  + 3 y ]
```

SEE ALSO: `grobner[normalf]`

3.6.5   grobner[gsolve] – prepare the given algebraic system for solving

Calling Sequence:

    `gsolve(F)`

    `gsolve(F, nonzero)`

    `gsolve(F, nonzero, X)`

Parameters:

F	–	set or list of polynomials in indeterminates X
X	–	set or list of indeterminates (not including parameters; default is $\mathrm{indets}(F)$)
nonzero	–	set of polynomials in indeterminates X

Synopsis:

- The command `gsolve(F, nonzero, X)` computes a collection of reduced (lexicographic) Gröbner bases corresponding to F.

- First, the system corresponding to F is subdivided by factorization.

- Then, each subsystem is passed to a variant of Buchberger's algorithm which factors all intermediate results.

- The result is a list of reduced subsystems whose roots are those of the original system, but whose variables have been successively eliminated and separated as far as possible.

- If desired, the **solve()** function may then be applied to each sub-list.

- The set *nonzero* may be used to prevent certain quantities from being considered in roots; however, this may not stop all such solutions.

- If X is omitted, the set **indets(F)** is used as a default.

- If X is a list, the given permutation of variables is used in all associated Gröbner basis computations; otherwise, each subsystem may use a different permutation.

- This function is part of the **grobner** package, and so can be used in the form **gsolve(..)** only after performing the command **with(grobner)** or **with(grobner, gsolve)**. The function can always be accessed in the long form **grobner[gsolve](..)**.

Examples:
```
> with(grobner):
> F := [x^2 - 2*x*z + 5, x*y^2 + y*z^3, 3*y^2 - 8*z^3]:
> gsolve(F);
              2
    [[y, x  + 5, z],

                    5      4      2      4            5       2
        [80 y - 3 z  + 32 z  - 40 z , - 96 z  + 640 x + 9 z  + 120 z ,

                 3          5      6
           240 z  + 1600 - 96 z  + 9 z ],

          2
    [z, y, x  + 5]]
```

SEE ALSO: grobner[solvable], grobner[finite], grobner[finduni]

3.6.6   grobner[leadmon] – compute the leading monomial of a polynomial

Calling Sequence:
    leadmon(*poly*, X, *termorder*)
    leadmon(*poly*, X)

Parameters:
    X – a list of indeterminates (variable names)
    poly   – a polynomial over X

termorder – (optional) term ordering: either ´plex´ (pure lexicographic), or
´tdeg´ (total degree — default)

Synopsis:

- The command `leadmon(`*poly*, *X*, *termorder*`)` computes the leading monomial of *poly* with respect to the chosen term ordering and given indeterminates. It returns a list of two elements; the first is the leading coefficient, and the second is the leading term of *poly*.

- This function is part of the `grobner` package, and so can be used in the form `leadmon(..)` only after performing the command `with(grobner)` or `with(grobner, leadmon)`. The function can always be accessed in the long form `grobner[leadmon](..)`.

Examples:
```
> with(grobner):
  p := 320*x*y^2 + 9*x*y^4 - 96*z^2*y^4*x +
    1600*y^3 - 18*y^5*x*z
           - 592*x*z*y^3 + 45*y^5 + 240*z*
  y^4:
  h := leadmon(p,[x,y],plex);
```
$\longrightarrow$
$$h := [-18\,z,\ x\,y^5]$$

```
> h := leadmon(p,[x,y,z],plex);
```
$\longrightarrow$
$$h := [-18,\ y^5\,x\,z]$$

```
> h := leadmon(p,[z,y,x]);
```
$\longrightarrow$
$$h := [-96,\ z^2\,y^4\,x]$$

SEE ALSO: `grobner`

3.6.7 grobner[normalf] – reduced form of a polynomial modulo an ideal

Calling Sequence:
```
normalf(poly, F, X)
normalf(poly, F, X, termorder)
```

Parameters:

F	– list of polynomials (normally, a Gröbner basis)
X	– list of indeterminates (not including free parameters)
poly	– the polynomial to be reduced
termorder	– (optional) term ordering: either ´plex´ (pure lexicographic) or ´tdeg´ (total degree — default)

Synopsis:

- The command `normalf(`*poly*, *F*, *X*, *termorder*`)` computes the fully reduced form of *poly* with respect to the ideal basis *F*, indeterminates *X*, and term ordering *termorder* (either ´plex´ for pure lexicographic ordering, or ´tdeg´ for total degree ordering).

- Usually, one first computes a Gröbner basis for a set of polynomials (algebraic equations, or side relations) via grobner[gbasis], for use as the reducing basis F. Note: F must be a Gröbner basis with respect to X, *termorder* (and not with respect to some permutation of X, for example).

- The function normalf yields a canonical form for polynomials modulo the ideal generated by F if F is a Gröbner basis.

- This function is part of the grobner package, and so can be used in the form normalf(..) only after performing the command with(grobner) or with(grobner, normalf). The function can always be accessed in the long form grobner[normalf](..).

Examples:
```
> with(grobner):
> F := [x^2 - 2*x*z + 5, x*y^2 + y*z^3, 3*y^2 - 8*z^3]:    X := [x,y,z]:
> G := gbasis(F,X);
            2                    2     3     2     3     4      3        2
  G := [x  - 2 x z + 5, 8 x y  + 3 y , - 3 y  + 8 z , 9 y  + 48 z y  + 320 y ]
> p := 320*x*y^2 + 9*x*y^4 - 96*z^2*y^4*x + 1600*y^3 - 18*y^5*x*z
>            - 592*x*z*y^3 + 45*y^5 + 240*z*y^4:
> normalf(p,G,X);
                                    0

> q := 3*x^3*y*z^2 - x*z^2 + y^3 + y*z:
> normalf(q,G,X);
              2 3        2          3       2     2
             9 z  y  - 15 y z  x - 41/4 y  + 60 y  z - z  x + z y
```

SEE ALSO: grobner[gbasis]

3.6.8   grobner[solvable] – decide if given algebraic system is solvable

Calling Sequence:
    solvable(F)
    solvable(F, X)
    solvable(F, X, *termorder*)

Parameters:

F	–	set or list of polynomials
X	–	set or list of indeterminates (not including parameters; default is indets(F))
termorder	–	term ordering: ´plex´ (pure lexicographic) or ´tdeg´ (total degree — default)

Synopsis:
- solvable(F, X, *termorder*) decides, using Gröbner basis methods, if a set/list of polynomials F with respect to the indeterminates X is algebraically consistent.

- If X is a list, the given permutation of variables is used in all associated Gröbner basis computations.

- If X is omitted, the set `indets`(F) is used as a default.

- If the term ordering is not specified, the total degree Gröbner basis is used for the test.

- This function is part of the **grobner** package, and so can be used in the form `solvable(..)` only after performing the command `with(grobner)` or `with(grobner, solvable)`. The function can always be accessed in the long form `grobner[solvable](..)`.

Examples:
```
> with(grobner):
  F := [x^2 - 2*x*z + 5, x*y^2 + y*z^3, 3*
  y^2 - 8*z^3]:
  solvable(F);                    ⟶              true
> solvable([op(F),x]);            ⟶              false
```

SEE ALSO: `grobner[finite]`, `grobner[finduni]`, `grobner[gsolve]`

3.6.9 grobner[spoly] – compute the S polynomial of two polynomials

Calling Sequence:

    spoly($p1$, $p2$, X, *termorder*)

    spoly($p1$, $p2$,X)

Parameters:

X	–	a list of indeterminates (variable names)
$p1$, $p2$	–	polynomials over X
termorder	–	(optional) term ordering: either ´plex´ (pure lexicographic), or ´tdeg´ (total degree — default)

Synopsis:

- The command `spoly`($p1, p2, X, termorder$) computes the S-polynomial of $p1$ and $p2$ with respect to the chosen term ordering and given indeterminates.

- This function is part of the **grobner** package, and so can be used in the form `spoly(..)` only after performing the command `with(grobner)` or `with(grobner, spoly)`. The function can always be accessed in the long form `grobner[spoly](..)`.

Examples:
```
> with(grobner):
  p := x - 13*y^2 - 12*z^3:
  q := x^2 - x*y + 92*z:
  spoly(p,q,[x,y,z],plex);
```
$$\longrightarrow \quad -13\, x\, y^2 - 12\, x\, z^3 + x\, y - 92\, z^3$$

> spoly(p,q,[x,y,z]); $\longrightarrow$ $x^3 - 13\,x^2\,y^2 - 12\,z^3\,x\,y + 1104\,z^4$

SEE ALSO: **grobner**, **grobner[leadmon]**

4

Packages for Discrete Mathematics

4.1 The Combinatorial Functions Package

4.1.1 Introduction to the combinat package

Calling Sequence:

    function(**args**)

    combinat[*function*](**args**)

Synopsis:

- The combinat package contains combinatorial functions such as permutations, combinations, and partitions.

- The functions available are:

bell	binomial	cartprod	character
Chi	combine	composition	decodepart
encodepart	fibonacci	firstpart	inttovec
lastpart	multinomial	nextpart	numbcomb
numbcomp	numbpart	numbperm	partition
permute	powerset	prevpart	randcomb
randpart	randperm	stirling1	stirling2
subsets	vectoint		

- For help with a particular function see combinat[*function*], where *function* is taken from the above list.

- An example of using a function from the combinat package is the following. To compute the nth Fibonacci number, one would use the command with(combinat, fibonacci) followed by fibonacci(n).

- To use a combinatorial function, either define that specific function by using the command `with(combinat, `*function*`)`, or define all combinatorial functions by the command `with(combinat)`. The package functions are always available without applying `with`, by using the long form of the function name: `combinat[`*function*`]`(*args*). This latter notation is necessary whenever there is a conflict between a package function name and another function used in the same session.

SEE ALSO: `with`

4.1.2   combinat[bell] – compute Bell numbers

Calling Sequence:
>   `bell(`*n*`)`

Parameters:
>   *n*   –   an expression

Synopsis:
- The procedure `bell` computes the nth Bell number if the argument **n** is an integer; otherwise, it returns the unevaluated function call.

- The Bell numbers are defined by the exponential generating function:

$$\texttt{exp(exp(x)-1)} = \texttt{sum(bell(n)/n!*x\^{}n, n=0..infinity)}$$

- The Bell numbers are computed using the umbral definition :

$$\texttt{bell(n+1)} = \texttt{(bell()+1)\^{}n}$$

where `bell()^n` represents `bell(n)`.

- For example:

```
bell(3) = bell(2) + 2*bell(1) + 1
bell(4) = bell(3) + 3*bell(2) + 3*bell(1) + 1
bell(n+1) = sum(binomial(n, i)*bell(i), i=0..n)   if n>0
```

- The **n**th Bell number has several interesting interpretations, including

```
the number of rhyming schemes in a stanza of n lines
the number of ways n unlike objects can be placed in n like boxes
the number of ways a product of n distinct primes may be factored
```

- This function is part of the `combinat` package, and so can be used in the form `bell(..)` only after performing the command `with(combinat)` or `with(combinat, bell)`. The function can always be accessed in the long form `combinat[bell](..)`.

Examples:
```
> with(combinat, bell);
```
$\longrightarrow$ `[bell]`

> bell(1);	$\longrightarrow$	1
> bell(4);	$\longrightarrow$	15
> bell(-1);	$\longrightarrow$	1
> bell(n);	$\longrightarrow$	bell(n)

SEE ALSO: combinat, binomial

4.1.3 combinat[cartprod] – iterate over a list of lists

Calling Sequence:

cartprod(LL)

Parameters:

LL – a list of lists of anything

Synopsis:

- The command cartprod is a special iterating function. It returns a table, the entries of which are: finished, delete, nextvalue, listoflists, and counter.

- The nextvalue entry, a function, returns a list. The first element in this list is an element from the first list in LL, the second element in this list is an element from the second list in LL, and so forth. Calling the function nextvalue also updates the counter and sets the finished entry.

- The finished entry is either true or false. So long as the finished flag is false, one may call the nextvalue function and the delete function. The finished entry is set by nextvalue or delete.

- The counter entry is an array of length nops(LL). It keeps track of the values that have or have not been returned. One can think of counter as an odometer which rotates through all possible values. It is updated by nextvalue and delete.

- The delete function is a function that allows you to remove any one element from each list in LL, thus it is called with a list of length nops(LL).

- This function is part of the combinat package, and so can be used in the form cartprod(..) only after performing the command with(combinat) or with(combinat, cartprod). The function can always be accessed in the long form combinat[cartprod](..).

Examples:

> with(combinat, cartprod);	$\longrightarrow$	[cartprod]
> T:=cartprod([[1,2,3], [a,b]]): T[nextvalue]();	$\longrightarrow$	[1, a]
> T[delete]([1, a]);	$\longrightarrow$	T[delete]([1, a])
> T[nextvalue]();	$\longrightarrow$	[1, b]

> T[finished];	$\longrightarrow$	false
> T[delete]([2, b]);	$\longrightarrow$	T[delete]([2, b])
> T[finished];	$\longrightarrow$	false

SEE ALSO: combinat

4.1.4   combinat[character] – compute character table for a symmetric group
combinat[Chi] – compute Chi function for partitions of symmetric group

Calling Sequence:

    character(n)

    Chi(*lambda*, *rho*)

Parameters:

    n –   a positive integer

    lambda –   a list representing a partition of n

    rho   –   another list representing a partition of n

Synopsis:

- Given a group (G,*), a group of matrices (H,&*) homomorphic to G is termed a **representation** of G. A representation is said to be **reducible** if there exists a similarity transformation

$$S : H \ -> \ X \ H \ X\_inverse$$

that maps all elements of H to the same non-trivial block diagonal structure. If a representation is not reducible, it is termed an **irreducible** representation.

- Given two elements of the same conjugacy class in G, the traces of their corresponding matrices in any representation are equal. The character function Chi is defined such that Chi of a conjugacy class of an irreducible representation of a group is the trace of any matrix corresponding to a member of that conjugacy class.

- Taking G to be the symmetric group on n elements, S_n, there is a one-to-one correspondence between the partitions of n and the non-equivalent irreducible representations of G. There is also a one-to-one correspondence between the partitions of n and the conjugacy classes of G.

- The Maple function Chi works on symmetric groups. Chi(*lambda*, *rho*) will compute and return the trace of the matrices in the conjugacy class corresponding to the partition *rho* in the irreducible representation corresponding to the partition *lambda*. Clearly, both *rho* and *lambda* must be partitions of the same number.

- The function character(n) computes Chi(*lambda*, *rho*) for all partitions *lambda* and *rho* of n. Thus, it computes the character of all conjugacy classes for all irreducible representations of the symmetric group on n elements.

- This function is part of the combinat package, and so can be used in the form character(..) only after performing the command with(combinat) or with(combinat, character). The function can always be accessed in the long form combinat[character](..).

Examples:

```
> with(combinat):
  Chi([1,2], [1,1,1]);                    ⟶                          2
> character(3);                            ⟶            [ 1    1    1 ]
                                                        [            ]
                                                        [ 2    0   -1 ]
                                                        [            ]
                                                        [ 1   -1    1 ]
```

SEE ALSO: combinat, combinat[partition]

4.1.5 combinat[combine] – construct the combinations of a list

Calling Sequence:

    combine(n)

    combine(n, m)

Parameters:

    n – a list of objects or an integer

    m – (optional) integer

Synopsis:

- If n is a list, then combine returns a list of the combinations of the list elements. If n is a non-negative integer, it is interpreted in the same way as a list of the first n integers.

- If m is given, then only combinations of size m are generated; otherwise, all combinations are generated, including the empty combination, that is, the power set is generated. Note that duplicates in the list n are taken into account.

- This function is part of the combinat package, and so can be used in the form combine(..) only after performing the command with(combinat) or with(combinat, combine). The function can always be accessed in the long form combinat[combine](..).

Examples:

```
> with(combinat, combine);                ⟶              [combine]
> combine(3, 2);                           ⟶          [[1, 2], [1, 3], [2, 3]]
> combine([a, a, b]);                      ⟶    [[], [a], [b], [a, b], [a, a], [a, a, b]
                                                 ]
> combine([a, b, c]);                      ⟶      [[], [a], [b], [a, b], [c], [a, c],
                                                     [b, c], [a, b, c]]
```

> combine([a, b, b, c], 2); $\longrightarrow$ [[a, b], [a, c], [b, b], [b, c]]

SEE ALSO: combinat[numbcomb], combinat[permute], combine

4.1.6   combinat[composition] – k-compositions of an integer

Calling Sequence:
   composition(n, k)

Parameters:
   n, k –   non-negative integers

Synopsis:
- The procedure composition computes and returns a list containing all distinct ordered k-tuples of positive integers whose elements sum to n. These are known as the compositions of n.

- In the list of k-tuples returned, each k-tuple is a list representing a particular composition of n.

- See also the function numbcomp, which computes the number of different compositions of an integer.

- This function is part of the combinat package, and so can be used in the form composition(..) only after performing the command with(combinat) or with(combinat, composition). The function can always be accessed in the long form combinat[composition](..).

Examples:
> with(combinat, composition); $\longrightarrow$ [composition]

> composition(5, 2); $\longrightarrow$ {[4, 1], [3, 2], [2, 3], [1, 4]}

> composition(3, 3); $\longrightarrow$ {[1, 1, 1]}

> composition(4, 3); $\longrightarrow$ {[2, 1, 1], [1, 2, 1], [1, 1, 2]}

SEE ALSO: combinat[numbcomp], @, @@

4.1.7   combinat[encodepart] – compute canonical integer representing partition
combinat[decodepart] – compute canonical partition represented by integer

Calling Sequence:
   encodepart(l)
   decodepart(n, m)

Parameters:
   l –   a list of positive integers
   n –   a positive integer

m – a positive integer

Synopsis:

- The functions `encodepart` and `decodepart` provide a canonical labeling of all the partitions of n.

- Given a partition l of n, `encodepart(`l`)` computes and returns the integer m between 1 and `numbpart(n)` uniquely representing this partition.

- Given a positive integer n and a positive integer m between 1 and `numbpart(`n`)` the function `decodepart(`m`, `n`)` computes and returns the unique partition l represented by this positive integer.

- This function is part of the `combinat` package, and so can be used in the form `encodepart(..)` only after performing the command `with(combinat)` or `with(combinat, encodepart)`. The function can always be accessed in the long form `combinat[encodepart](..)`.

Examples:
```
> with(combinat):
  partition(3);                    ⟶         [[1, 1, 1], [1, 2], [3]]
> decodepart(3, 1);               ⟶              [1, 1, 1]
> decodepart(3, 2);               ⟶                [1, 2]
> decodepart(3, 3);               ⟶                 [3]
> decodepart(3, 4);               ⟶         Error, (in decodepart)
                                            2nd argument is out of range
> encodepart([1, 2]);             ⟶                  2
```

SEE ALSO: `combinat[vectoint]`, `combinat[inttovec]`, `combinat[partition]`, `combinat[numbpart]`, `combinat[nextpart]`, `combinat[prevpart]`

4.1.8   combinat[fibonacci] – compute Fibonacci numbers or polynomials

Calling Sequence:

> `fibonacci(`n`)`
> `fibonacci(`n`, `x`)`

Parameters:

n, x – algebraic expressions

Synopsis:

- The call `fibonacci(`n`)` computes the nth Fibonacci number $F(n)$, if n is an integer; otherwise it returns unevaluated.

- The call `fibonacci(`n`, `x`)` computes the nth Fibonacci polynomial in x if n is an integer; otherwise it returns unevaluated.

- The Fibonacci numbers are defined by the linear recurrence

$$F(n) = F(n-1) + F(n-2) \text{ where } F(0) = 0 \text{ and } F(1) = 1$$

- The Fibonacci polynomials are defined similarly by

$$F(n, x) = x F(n-1, x) + F(n-2, x) \text{ where } F(0, x) = 0 \text{ and } F(1, x) = 1$$

Note that `F(n) = F(n, 1)`.

- The method used to compute `F(n)` is, however, based on the following identity: Let `A` be the two by two matrix `[[1, 1], [1, 0]]`. Observe that `[ F(n+1), F(n) ] = A [ F(n), F(n-1) ]` Thus `F(n)` can be computed quickly (in time `O(log(n)^3)` instead of `O(n^2)`) by computing `A^n` using binary powering.

- The generating function for `F(n, x)` is

$$t/(1-x*t-t^2) = sum(F(n, x)*t^n, n=0..infinity)$$

- This function is part of the `combinat` package, and so can be used in the form `fibonacci(..)` only after performing the command `with(combinat)` or `with(combinat, fibonacci)`. The function can always be accessed in the long form `combinat[fibonacci](..)`.

Examples:
```
> with(combinat, fibonacci):
  fibonacci(5);                        ⟶                    5
> seq(fibonacci(i), i=0..10);         ⟶      0, 1, 1, 2, 3, 5, 8, 13, 21, 34, 55
> seq(fibonacci(i), i=-10..0);        ⟶     -55, 34, -21, 13, -8, 5, -3, 2, -1, 1, 0
> seq(fibonacci(i, x), i=1..5);       ⟶            2        3        4     2
                                             1, x, x + 1, x + 2 x, x + 3 x + 1
> fibonacci(n);                       ⟶                fibonacci(n)
```

SEE ALSO: `combinat`, `linalg[fibonacci]`

4.1.9   combinat[firstpart] – first partition in canonical partition sequence
combinat[nextpart] – next partition in canonical partition sequence
combinat[lastpart] – last partition in canonical partition sequence
combinat[prevpart] – previous partition in canonical partition sequence

Calling Sequence:
```
firstpart(n)
nextpart(l)
lastpart(n)
prevpart(l)
```

Parameters:

> l – a list of positive integers
>
> n – a positive integer

Synopsis:

- All four of the functions use the canonical partition sequence defined by combinat[encodepart].

- Given a positive integer n, firstpart(n) computes and returns the first partition of n in the canonical partition sequence.

- Given a partition l of n, nextpart(l) computes and returns the next partition of n in the canonical partition sequence.

- Given a positive integer n, lastpart(n) computes and returns the last partition of n in the canonical partition sequence.

- Given a partition l of n, prevpart(l) computes and returns the previous partition of n in the canonical partition sequence.

- This function is part of the combinat package, and so can be used in the form firstpart(..) only after performing the command with(combinat) or with(combinat, firstpart). The function can always be accessed in the long form combinat[firstpart](..).

Examples:

```
> with(combinat):
  partition(3);                    ⟶        [[1, 1, 1], [1, 2], [3]]
> firstpart(3);                    ⟶               [1, 1, 1]
> nextpart(");                     ⟶                [1, 2]
> nextpart(");                     ⟶                 [3]
> prevpart(");                     ⟶                [1, 2]
> lastpart(3);                     ⟶                 [3]
```

SEE ALSO: combinat[encodepart], combinat[partition], combinat[numbpart], combinat[randpart], combinat[inttovec]

4.1.10   combinat[multinomial] – compute the multinomial coefficients

Calling Sequence:

> multinomial(n, k_1, k_2, ..., k_m)

Parameters:

> n – a positive integer
>
> k_i – positive integers whose sum is n

Synopsis:

- The procedure `multinomial`(n, k_1, k_2, ..., k_m) computes the multinomial coefficient denoted (n | k_1, k_2, ..., k_m) equal to $n!/(k_1!\ k_2!\ ...\ k_m!)$ where it is assumed that $n = k_1 + k_2 + ... + k_m$ and that `m` > 0; that is, there are at least 2 arguments.

- This function is part of the `combinat` package, and so can be used in the form `multinomial(..)` only after performing the command `with(combinat)` or `with(combinat, multinomial)`. The function can always be accessed in the long form `combinat[multinomial](..)`.

Examples:

`> with(combinat, multinomial);`	$\longrightarrow$	[multinomial]
`> multinomial(8, 2, 6);`	$\longrightarrow$	28
`> binomial(8, 2);`	$\longrightarrow$	28
`> multinomial(8, 2, 3, 3);`	$\longrightarrow$	560

SEE ALSO: `combinat`, `binomial`

4.1.11   combinat[numbcomb] – count the number of combinations

Calling Sequence:

> `numbcomb`(n)
>
> `numbcomb`(n, m)

Parameters:

> n –   a list of objects or an integer
>
> m –   (optional) integer

Synopsis:

- If n is a list, then `numbcomb` counts the combinations of the elements of n taken m at a time. If m is not given, then all combinations are considered. If n is a non-negative integer, it is interpreted in the same way as a list of the first n integers.

- Note that the result of `numbcomb`(n, m) is equivalent to `nops(combine(`n, m`))`.

- The count of combinations takes into account duplicates in n. In the case where there are no duplicates, the count is given by the formula `2^`n if m is not specified, or by the formula `binomial(`n, m`)` if m is specified. If there are duplicates in the list, then the generating function is used.

- This function is part of the `combinat` package, and so can be used in the form `numbcomb(..)` only after performing the command `with(combinat)` or `with(combinat, numbcomb)`. The function can always be accessed in the long form `combinat[numbcomb](..)`.

Examples:

`> with(combinat, numbcomb):` `  numbcomb(3, 2);`	$\longrightarrow$	3

> numbcomb([a, a, b]);	$\longrightarrow$	6
> numbcomb([a, b, c]);	$\longrightarrow$	8
> numbcomb([a, b, b, c], 2);	$\longrightarrow$	4

SEE ALSO: combinat[combine]

4.1.12   combinat[numbcomp] – the number of k-compositions of an integer

Calling Sequence:
  numbcomp(n, k)

Parameters:
  n, k –   any integers

Synopsis:
- The procedure numbcomp counts the number of distinct ordered k-tuples of positive integers whose elements sum to n.

- See the composition function which constructs the k-compositions. Note that numbcomp(n, k) = nops(composition(n, k)) for n, k > 0.

- This function is part of the combinat package, and so can be used in the form numbcomp(..) only after performing the command with(combinat) or with(combinat, numbcomp). The function can always be accessed in the long form combinat[numbcomp](..).

Examples:
> with(combinat, numbcomp);	$\longrightarrow$	[numbcomp]
> numbcomp(7, 3);	$\longrightarrow$	15
> numbcomp(5, 4);	$\longrightarrow$	4

SEE ALSO: composition

4.1.13   combinat[numbpart] – the number of partitions of an integer

Calling Sequence:
  numbpart(n)

Parameters:
  n –   a non-negative integer

Synopsis:
- This procedure counts the number of partitions of an integer n, that is, the number of ways n can be split into sums without regard to order.

- See the `partition` function that constructs the partitions. Note that `numbpart`(n) = `nops(partition`(n)`)` for $n \geq 0$.

- This function is part of the `combinat` package, and so can be used in the form `numbpart(..)` only after performing the command `with(combinat)` or `with(combinat, numbpart)`. The function can always be accessed in the long form `combinat[numbpart](..)`.

Examples:
```
> with(combinat):
> numbpart(5);
```
7

```
> partition(5);
  [[1, 1, 1, 1, 1], [1, 1, 1, 2], [1, 2, 2], [1, 1, 3], [2, 3], [1, 4], [5]]
```

SEE ALSO: `combinat[partition]`

4.1.14   combinat[numbperm] – count the number of permutations

Calling Sequence:

> numbperm(n)

> numbperm(n, r)

Parameters:

> n – a list of objects or an integer

> r – (optional) integer

Synopsis:

- If n is a list, then `numbperm` counts the permutations of the elements of n taken r at a time. If n is a non-negative integer, it is interpreted in the same way as a list of the first n integers. If r is not specified, it is taken to be r = `nops`(n).

- The count of permutations takes into account duplicates in n. In the case where there are no duplicates, the count is given by the formula $n! \; / \; (n - r)!$. Otherwise the generating function is used.

- The function `permute` will compute the number of permutations. Thus `numbperm`$(n, \; r)$ = `nops(permute`$(n, \; r)$`)`.

- This function is part of the `combinat` package, and so can be used in the form `numbperm(..)` only after performing the command `with(combinat)` or `with(combinat, numbperm)`. The function can always be accessed in the long form `combinat[numbperm](..)`.

Examples:
```
> with(combinat, numbperm);                    ⟶              [numbperm]

> numbperm(3);                                  ⟶                  6
```

> numbperm(3, 2);	$\longrightarrow$	6
> numbperm([a, b]);	$\longrightarrow$	2
> numbperm([a, a, b], 2);	$\longrightarrow$	3

SEE ALSO: combinat[permute]

4.1.15   combinat[partition] – partition an integer

Calling Sequence:

partition(n)

Parameters:

n –   a non-negative integer

Synopsis:

- The procedure **partition** partitions an integer n into all possible sums without regard to order. The output is a list of lists of integers where the sum of the elements in each list is n.

- See the function **numbpart**, which computes the number of partitions.

- This function is part of the **combinat** package, and so can be used in the form **partition(..)** only after performing the command **with(combinat)** or **with(combinat, partition)**. The function can always be accessed in the long form **combinat[partition](..)**.

Examples:

> with(combinat, partition);	$\longrightarrow$	[partition]
> partition(0);	$\longrightarrow$	[[]]
> partition(3);	$\longrightarrow$	[[1, 1, 1], [1, 2], [3]]

SEE ALSO: combinat[numbpart]

4.1.16   combinat[permute] – construct the permutations of a list

Calling Sequence:

permute(n)

permute(n, r)

Parameters:

n –   a list of objects or an integer

r –   (optional) integer

Synopsis:

- If n is a list, then **permute** returns a list of all the permutations of the elements taken r at a time. If n is a non-negative integer, it is interpreted in the same way as a list of the first n integers. If r is not specified, then it is taken to be equal to the number of elements in n.

- The permutations are generated in order. Duplicates in n are respected.

- The function **numbperm** will compute the number of possible permutations: $\text{numbperm}(n, \ r) = \text{nops}(\text{permute}(n, \ r))$.

- This function is part of the **combinat** package, and so can be used in the form **permute(..)** only after performing the command **with(combinat)** or **with(combinat, permute)**. The function can always be accessed in the long form **combinat[permute](..)**.

Examples:

`> with(combinat, permute);`	$\longrightarrow$	[permute]
`> permute(3);`	$\longrightarrow$	[[1, 2, 3], [1, 3, 2], [2, 1, 3], [2, 3, 1], [3, 1, 2], [3, 2, 1]]
`> permute(3, 2);`	$\longrightarrow$	[[1, 2], [1, 3], [2, 1], [2, 3], [3, 1], [3, 2]]
`> permute([a, b]);`	$\longrightarrow$	[[a, b], [b, a]]
`> permute([a, a, b], 2);`	$\longrightarrow$	[[a, a], [a, b], [b, a]]

SEE ALSO: combinat[numbperm]

4.1.17   combinat[powerset] – construct the power set of a set

Calling Sequence:

    powerset(s)

Parameters:

   s – a set or list, or a non-negative integer

Synopsis:

- If s is a set, then **powerset** returns a set of all the subsets of s. The power set has 2^n entries, where n is the number of elements in s. If s is a non-negative integer, it is interpreted in the same way as a set of the first s integers.

- If s is a list, a list of all the sublists of s is returned. The number of sublists depends on the multiplicities of s.

- This function is part of the **combinat** package, and so can be used in the form **powerset(..)** only after performing the command **with(combinat)** or **with(combinat, powerset)**. The function can always be accessed in the long form **combinat[powerset](..)**.

Examples:
```
> with(combinat, powerset);
                        [powerset]

> powerset(3);
          {{}, {1, 2, 3}, {2, 3}, {3}, {1, 3}, {1}, {2}, {1, 2}}
```

```
> powerset({a, b});
```
$$\{\{\}, \ \{a, \ b\}, \ \{b\}, \ \{a\}\}$$
```
> powerset([a, a, b]);
```
$$[[], \ [a], \ [b], \ [a, \ b], \ [a, \ a], \ [a, \ a, \ b]]$$

SEE ALSO: `combinat[combine]`

4.1.18   combinat[randcomb] – construct a random combination

Calling Sequence:

   randcomb(n, m)

Parameters:

   n – a list or set of objects, or a positive integer

   m – a non-negative integer

Synopsis:

- If n is a list or set, then `randcomb` returns a random combination of m elements of n. If n is a positive integer, a random combination of m of the first n positive integers is returned.

- This function is part of the `combinat` package, and so can be used in the form `randcomb(..)` only after performing the command `with(combinat)` or `with(combinat, randcomb)`. The function can always be accessed in the long form `combinat[randcomb](..)`.

Examples:

```
> with(combinat, randcomb);
```
                                  $\longrightarrow$ [randcomb]
```
> randcomb(5, 3);
```
                                  $\longrightarrow$ {3, 4, 5}
```
> randcomb([a, b, c, d], 2);
```
                                  $\longrightarrow$ [b, d]
```
> randcomb({W, X, Y, Z}, 1);
```
                                  $\longrightarrow$ {Z}

SEE ALSO: `combinat[combine]`

4.1.19   combinat[randpart] – construct a random partition

Calling Sequence:

   randpart(n)

Parameters:

   n – a positive integer

Synopsis:

- The function `randpart` computes and returns a random partition of the positive integer n. This consists of a random list of positive integers whose sum is n.

- This function is part of the combinat package, and so can be used in the form randpart(..) only after performing the command with(combinat) or with(combinat, randpart). The function can always be accessed in the long form combinat[randpart](..).

Examples:

> with(combinat, randpart);	$\longrightarrow$	[randpart]
> randpart(10);	$\longrightarrow$	[2, 3, 5]
> randpart(10);	$\longrightarrow$	[1, 1, 8]
> randpart(10);	$\longrightarrow$	[2, 8]

SEE ALSO: combinat[partition]

4.1.20   combinat[randperm] – construct a random permutation

Calling Sequence:

    randperm(n)

Parameters:

    n – a list or set of objects, or a positive integer

Synopsis:

- If n is a list or set, then randperm returns a random permutation of the elements of n. If n is a positive integer, then a random permutation of the first n positive integers is returned.

- This function is part of the combinat package, and so can be used in the form randperm(..) only after performing the command with(combinat) or with(combinat, randperm). The function can always be accessed in the long form combinat[randperm](..).

Examples:

> with(combinat, randperm);	$\longrightarrow$	[randperm]
> randperm(3);	$\longrightarrow$	[1, 2, 3]
> randperm([a, b, c, d]);	$\longrightarrow$	[d, c, b, a]
> randperm({W, X, Y, Z});	$\longrightarrow$	[X, Y, Z, W]

SEE ALSO: combinat[permute], combinat[numbperm]

4.1.21   combinat[stirling1] – Stirling numbers of the first kind

Calling Sequence:

    stirling1(n, m)

Parameters:

   n, m –   any integers

Synopsis:

- The function **stirling1** computes the Stirling numbers of the first kind, denoted S(n, m), via the generating function for S:

```
  n
 -----                                                    (n + 1)
  \              m                               (-1)            GAMMA(n - x)
   )   S(n, m) x  = x (x - 1) .. (x - n + 1) = - ------------------------
  /                                                          GAMMA(- x)
 -----
 m = 0
```

- Note that $(-1)\char`^(n - m)$ S(n, m) is the number of permutations of n symbols which have exactly m cycles.

- This function is part of the **combinat** package, and so can be used in the form **stirling1(..)** only after performing the command **with(combinat)** or **with(combinat, stirling1)**. The function can always be accessed in the long form **combinat[stirling1](..)**.

Examples:

> with(combinat, stirling1);	$\longrightarrow$	[stirling1]
> stirling1(10, 5);	$\longrightarrow$	-269325
> stirling1(5, k);	$\longrightarrow$	stirling1(5, k)

SEE ALSO: combinat[stirling2]

4.1.22   combinat[stirling2] – Stirling numbers of the second kind

Calling Sequence:

   stirling2(n, m)

Parameters:

   n, m –   any integers

Synopsis:

- The function **stirling2** computes the Stirling numbers of the second kind S(n,m) from the well-known formula in terms of the binomial coefficients:

$$S(n,m) = \frac{1}{m!} \sum_{k=0}^{m} (-1)^{(m-k)} \binom{m}{k} k^{n}$$

- Note that $S(n, m)$ is the number of ways of partitioning a set of n elements into m non-empty subsets.

- This function is part of the **combinat** package, and so can be used in the form **stirling2(..)** only after performing the command **with(combinat)** or **with(combinat, stirling2)**. The function can always be accessed in the long form **combinat[stirling2](..)**.

Examples:

`> with(combinat, stirling2);`	$\longrightarrow$	`[stirling2]`
`> stirling2(10, 5);`	$\longrightarrow$	`42525`
`> stirling2(5, k);`	$\longrightarrow$	`stirling2(5, k)`

SEE ALSO: combinat[stirling1]

4.1.23   combinat[subsets] – iterate over the power set of a set or list

Calling Sequence:
    subsets(L)

Parameters:
    L – a set or list of elements

Synopsis:

- The function **subsets** is an iterator for generating the power set of a set one set at a time. It returns a table with two entries: **finished** and **nextvalue**.

- The **finished** entry will be either **true** or **false**, depending on whether the end of the set has been reached or not. It is initially set to **false**.

- The **nextvalue** entry is a procedure that traverses the power set. The result of a call to **nextvalue** will be the next set in the power set. When the entire set has been traversed, **nextvalue** will set the value of **finished** to **true**.

- If the input L is a set, the subsets are returned as sets. If the input is a list, the subsets are returned as lists. Otherwise the only difference is in the order in which the subsets are generated.

- This function is part of the **combinat** package, and so can be used in the form **subsets(..)** only after performing the command **with(combinat)** or **with(combinat, subsets)**. The function can always be accessed in the long form **combinat[subsets](..)**.

Examples:
```
> with(combinat):
> S := subsets({1, 2}):
> while not S[finished] do S[nextvalue]() od;
                            {}

                           {1}

                           {2}

                          {1, 2}
```

SEE ALSO: `combinat`, `combinat[powerset]`, `combinat[combine]`

4.1.24   combinat[vectoint] – index of vector in canonical ordering
combinat[inttovec] – vector referenced by integer in canonical ordering

Calling Sequence:
> vectoint(l)
> inttovec(m, n)

Parameters:

l –   a list of non-negative integers

m –   a non-negative integer

n –   a non-negative integer

Synopsis:

- These two functions provide a one-to-one correspondence between the non-negative integers and all vectors composed of **n** non-negative integers.

- The one-to-one correspondence is defined as follows. View all vectors of **n** non-negative integers as exponent vectors on **n** variables. Therefore, for each vector, there is a corresponding monomial. Collect all such monomials and order them by increasing total degree. Resolve ties by ordering monomials of the same degree in lexicographic order. This gives a canonical ordering.

- Given a vector l of **n** non-negative integers, the corresponding integer **m** is its index in this canonical ordering. The function `vectoint($l$)` computes and returns this integer **m**.

- Given a non-negative integer m, the corresponding vector l is the mth vector in this canonical ordering of vectors of length n. The function `inttovec($m$, $n$)` computes and returns this vector l.

- Here is a sample canonical ordering where **n** is 3:

Vector	Number	Monomial
[0,0,0]	0	1
[1,0,0]	1	x
[0,1,0]	2	y
[0,0,1]	3	z
[2,0,0]	4	x^2
[1,1,0]	5	x*y
[1,0,1]	6	x*z
[0,2,0]	7	y^2
...	...	...

- This function is part of the `combinat` package, and so can be used in the form `vectoint(..)` only after performing the command `with(combinat)` or `with(combinat, vectoint)`. The function can always be accessed in the long form `combinat[vectoint](..)`.

Examples:
```
> with(combinat):
  vectoint([1,0,1]);                    ⟶                        6
> inttovec(6,3);                        ⟶                   [1, 0, 1]
```

SEE ALSO: `combinat[encodepart]`, `combinat[decodepart]`

4.2 The Permutation Group and Finitely-Presented Group Package

4.2.1 Introduction to the group package

Calling Sequence:

    function(**args**)

    group[*function*] (**args**)

Synopsis:

- To use a **group** function, either define that specific function alone by the command `with(group,function)` or define all **group** functions by the command `with(group)`.

- The functions in this package are the following:

centralizer	cosets	cosrep	groupmember
grouporder	inter	invperm	isnormal
mulperms	normalizer	permrep	pres

- In addition, when this package has been loaded, the **type** function will accept **disjcyc**, and the **convert** function will accept **disjcyc** and **permlist** as second arguments. These provide type-checking and conversion routines for the data types used within the **group** package. See the information on the **type** and **convert** functions for details.

- For help with a particular function see the information under **group**[*function*], where *function* is one from the above list.

- For help with how to represent groups, subgroups, and group elements for this package, see the information under **group[grelgroup]**, **group[permgroup]**, and **group[subgrel]**.

- An example of using the **group** package to find the order of a permutation group is the following:

Examples:
```
> with(group):
> grouporder(permgroup(8, {a=[[1,2]], b=[[1,2,3,4,5,6,7,8]]}));
                              40320
```

SEE ALSO: **with**, **group[convert]**, **group[type]**, **define**, **combinat[permute]**

4.2.2   group[centralizer] – find the centralizer of a set of permutations

Calling Sequence:
> centralizer(*pg*, *s*)

Parameters:
> *pg* – group in which the centralizer is to be found
> *s* – set of permutations or a single permutation

Synopsis:
- This function finds the largest subgroup of the permutation group *pg* in which every element commutes with every element of *s*. *s* need not be contained in *pg*. The result is returned as an unevaluated **permgroup** call.

- This function is part of the **group** package, and so can be used in the form **centralizer(..)** only after performing the command **with(group)** or **with(group, centralizer)**.

Examples:
```
> with(group):
> centralizer(permgroup(7, {[[1,2]], [[1,2,3,4,5,6,7]]}), {[[3,6]]});
        permgroup(7, {[[1, 2]], [[1, 2, 5], [3, 6], [4, 7]], [[2, 4]]})

> centralizer(permgroup(5, {[[1,2,3]], [[3,4,5]]}), [[1,2]]);
                permgroup(5, {[[3, 4, 5]], [[1, 2], [4, 5]]})
```

SEE ALSO: **group[permgroup]**

4.2.3   group[cosets] – find a complete list of right coset representatives for a subgroup of a permutation group or a group given by generators and relations

Calling Sequence:

cosets(*sbgrl*)

cosets(*pg*, *sbpg*)

Parameters:

sbgrl   –   a subgroup of a group given by generators and relations

pg, *sbpg* –   permutation groups of same degree

Synopsis:

- For groups given by generators and relations, the argument *sbgrl* should be a `subgrel`. A set of words in the generators of the group is returned.

- For permutation groups, both arguments should be `permgroups` and *sg* should be a subgroup of *pg*. A set of permutations in disjoint cycle notation is returned.

- This function is part of the `group` package, and so can be used in the form cosets(..) only after performing the command with(group) or with(group, cosets).

Examples:
```
> with(group):
> g := grelgroup({a,b,c}, {[a,b,c,a,1/b],[b,c,a,b,1/c],[c,a,b,c,1/a]}):
> cosets(subgrel({y=[a,b,c]}, g));
                          {[], [a], [a, b]}
> pg1 := permgroup(7, {[[1,2]], [[1,2,3,4,5,6,7]]}):
> pg2 := permgroup(7, {[[1,2,3]], [[3,4,5,6,7]]}):
> cosets(pg1,pg2);
                          {[], [[6, 7]]}
```

SEE ALSO: group[permgroup], group[grelgroup], group[subgrel]

4.2.4   group[cosrep] – express group element as product of an element in a subgroup multiplied by a right coset representative for that subgroup

Calling Sequence:

cosrep(*elem*, *sub*)

Parameters:

elem –   a permutation or a word in the group generators

sub   –   a permutation group or a `subgrel`

Synopsis:
- If *sub* is a `subgrel`, then *elem* should be a word in the group generators. A two-element list will be returned. The first element is the subgroup element expressed as a word in the subgroup generators, the second is the right coset representative. The coset representative will be an element of the set returned by `cosets(sub)`.

- If *sub* is a `permgroup`, then *elem* should be a permutation in disjoint cycle notation. A two-element list is returned. The first element is a permutation contained in *sub*, the second is a right coset representative permutation for *sub* in the symmetric group of the same degree. The coset representative will be an element of the set returned by `cosets(Sn, sub)`, where `Sn` is the symmetric group of the same degree as *sub*.

- This function is part of the `group` package, and so can be used in the form `cosrep(..)` only after performing the command `with(group)` or `with(group, cosrep)`.

Examples:
```
> with(group):
> g := grelgroup({a,b,c}, {[a,b,c,a,1/b],[b,c,a,b,1/c],[c,a,b,c,1/a]}):
> cosrep([c], subgrel({y=[a,b,c]}, g));
                        [[y, y, y, y, y], [a]]
> pg := permgroup(7, {[[1,2,3]], [[3,4,5,6,7]]}):
> cosrep([[3,4,5,6]],pg);
                    [[[3, 4, 5, 7, 6]], [[6, 7]]]
```

SEE ALSO: `group[permgroup]`, `group[subgrel]`, `group[cosets]`

4.2.5   group[grelgroup] – represent a group by generators and relations

Calling Sequence:
>   grelgroup(*gens*, *rels*)

Parameters:
>   *gens*   –   a set of names taken to be the generators of the group
>   *rels*   –   a set of relations among the generators which define the group

Synopsis:
- The function `grelgroup` is used as a procedure and an unevaluated procedure call. As a procedure, `grelgroup` checks its arguments and then either exits with an error or returns the unevaluated `grelgroup` call.

- The first argument is a set of Maple names which stand for the generators of the group. The second argument is a set of "words" in the generators. A "word" is a list of generators and/or inverses of generators representing a product. The inverse of a generator `g` is represented by `1/g`. An empty list represents the identity element. The words `w1`, `w2`, ... `wn` in *rels* are such that the relations `w1 = w2 = ... = wn = 1` define the group.

- This function is part of the **group** package, and so can be used in the form **grelgoup(..)** only after performing the command **with(group)** or **with(group, grelgoup)**.

Examples:
```
> with(group):
> grelgroup({a,b}, {[a,a,a], [b,b], [a,b,1/a,1/b]});
            grelgroup({a, b}, {[a, a, a], [b, b], [a, b, 1/a, 1/b]})

> grelgroup({a,b}, {[a,1/c,a], [b,a]});
Error, (in grelgroup) invalid arguments
```

SEE ALSO: group[subgrel]

4.2.6   group[groupmember] – test if a permutation is in a given group

Calling Sequence:
>   groupmember(*perm*, *pg*)

Parameters:
>   *perm*   –   the permutation to test
>   *pg*   –   the permutation group

Synopsis:
- The function returns **true** or **false**, indicating whether *perm* is in *pg*.

- This function is part of the **group** package, and so can be used in the form **groupmember(..)** only after performing the command **with(group)** or **with(group, groupmember)**.

Examples:
```
> with(group):
  pg := permgroup(7, {[[1,2,3]], [[3,4,5,6,
  7]]}):
  groupmember([[1,5],[3,6]], pg);            ⟶            true

> groupmember([[3,2,4,7]], pg);              ⟶            false
```

SEE ALSO: group[permgroup]

4.2.7   group[grouporder] – compute the order of a group

Calling Sequence:
>   grouporder(*g*)

Parameters:
>   *g*   –   a **permgroup** or **grelgroup**

Synopsis:

- The number of elements in the given group is computed and returned.

- This function is part of the **group** package, and so can be used in the form `grouporder(..)` only after performing the command `with(group)` or `with(group, grouporder)`.

Examples:
```
> with(group):
> grouporder(permgroup(7, {[[1,2,3]], [[3,4,5,6,7]]}));
                            2520
> grouporder(grelgroup({x,y}, {[x,x,y,x,y,y,y],[y,y,x,y,x,x,x]}));
                            56
```

SEE ALSO: `group[permgroup]`, `group[grelgroup]`

4.2.8   group[inter] – find the intersection of two permutation groups

Calling Sequence:

    inter(*pg1*, *pg2*)

Parameters:

    pg1, pg2   –   two **permgroups** of the same degree

Synopsis:

- The intersection of two permutation groups is computed, and returned in the form of a **permgroup**.

- This function is part of the **group** package, and so can be used in the form `inter(..)` only after performing the command `with(group)` or `with(group, inter)`.

Examples:
```
> with(group):
> pg1 := permgroup(7, {[[2,3,4]], [[3,4,5,6,7]]}):
> pg2 := permgroup(7, {[[1,2]], [[1,2,3,4,5,6]]}):
> inter(pg1,pg2);
            permgroup(7, {[[2, 3, 4]], [[4, 5, 6]], [[3, 5, 4]]})
```

SEE ALSO: `group[permgroup]`

4.2.9   group[invperm] – find inverse of a permutation

Calling Sequence:

    invperm(*perm*)

Parameters:

    –   a permutation in disjoint cycle notation

Synopsis:
- The inverse is computed, and the result is output in disjoint cycle notation.

- This function is part of the **group** package, and so can be used in the form `invperm(..)` only after performing the command `with(group)` or `with(group, invperm)`.

Examples:
```
> with(group):
  invperm([[4,6,5], [2,11]]);                    ⟶              [[2, 11], [4, 5, 6]]
```

SEE ALSO: group[permgroup]

4.2.10   group[isnormal] – determine whether a subgroup is normal

Calling Sequence:
> isnormal(*subgrl*)
> isnormal(*pg*, *sg*)

Parameters:
> *subgrl*   –   a **subgrel**
> *pg, sg*   –   two **permgroups** of the same degree

Synopsis:
- With two arguments, this function determines whether *sg* is a normal subgroup of the permutation group generated by the union of *sg* and *pg*. If it is, then **true** is returned; otherwise **false** is returned.

- With one argument, this function tests whether *subgrl* is a normal subgroup. The function returns **true** or **false**.

- This function is part of the **group** package, and so can be used in the form `isnormal(..)` only after performing the command `with(group)` or `with(group, isnormal)`.

Examples:
```
> with(group):
> pg := permgroup(8, {[[1,2]], [[1,2,3,4,5,6,7,8]]}):
> sg1 := permgroup(8, {[[1,2,3,4]], [[1,2]], [[5,6,7,8]], [[5,6]]}):
> isnormal(pg,sg1);
                                false
> sg2 := permgroup(8, {[[1,2,3]], [[2,3,4,5,6,7,8]]}):
> isnormal(pg,sg2);
                                true
> g := grelgroup({a,b}, {[a,a,a,a,a], [b,b,b], [a,b,1/a,1/b]}):
> isnormal(subgrel({x=[a]}, g));
                                true
```

SEE ALSO: group[permgroup], group[subgrel], group[grelgroup]

4.2.11   group[mulperms] – multiply two permutations in disjoint cycle notation

Calling Sequence:

    mulperms(*perm1*, *perm2*)

Parameters:

    perm1, perm2   –   the permutations in disjoint cycle notation

Synopsis:

- The product is expressed in disjoint cycle notation.

- This function is part of the **group** package, and so can be used in the form `mulperms(..)` only after performing the command `with(group)` or `with(group, mulperms)`.

Examples:
```
> with(group):
  mulperms([[2,3,4], [1,6]], [[4,6]]);        ⟶        [[1, 4, 2, 3, 6]]
```

SEE ALSO: group[permgroup]

4.2.12   group[normalizer] – find the normalizer of a subgroup

Calling Sequence:

    normalizer(*pg*, *sg*)

Parameters:

    pg, sg   –   two **permgroups** of the same degree

Synopsis:

- This function finds the largest subgroup of *pg* in which *sg* is a normal subgroup. The group *sg* should be a subgroup of *pg*. The result is returned as an unevaluated **permgroup** call.

- This function is part of the **group** package, and so can be used in the form `normalizer(..)` only after performing the command `with(group)` or `with(group, normalizer)`.

Examples:
```
> with(group):
> pg := permgroup(7, {[[1,2,3]], [[3,4,5,6,7]]}):
> sg := permgroup(7, {[[1,2,3]],[[3,4,5]]}):
> normalizer(pg,sg);
          permgroup(7, {[[1, 2, 3]], [[3, 4, 5]], [[4, 5], [6, 7]]})
```

SEE ALSO: group[permgroup]

4.2.13   group[permgroup] – represent a permutation group

Calling Sequence:

permgroup(*deg, gens*)

Parameters:

deg – degree of the permutation group

gens – set of generators for the permutation group

Synopsis:

- The function **permgroup** is used as a procedure and an unevaluated procedure call. As a procedure, **permgroup** checks its arguments and then either exits with an error or returns the unevaluated **permgroup** call.

- The first argument is the degree of the group, and should be an integer. The second argument is a set of group generators. Each generator is represented in disjoint cycle notation. The generators may be named or unnamed. A named generator is an equation; the left operand is the generator´s name, the right operand is the permutation in disjoint cycle notation.

- A permutation in disjoint cycle notation is a list of lists. Each sub-list represents a cycle; the permutation is the product of these cycles. The cycle [a[1], a[2], ..., a[n]] represents the permutation which maps a[1] to a[2], a[2] to a[3], ..., a[n-1] to a[n], and a[n] to a[1]. The identity element is represented by the empty list [].

- This package follows the convention that "permutations act on the right". In other words, if p1 and p2 are permutations, then the product of p1 and p2 (p1 &* p2) is defined such that (p1 &* p2)(i) = p2(p1(i)) for i=1..*deg*.

- This function is part of the **group** package, and so can be used in the form **permgroup(..)** only after performing the command **with(group)** or **with(group, permgroup)**.

Examples:

```
> with(group):
> permgroup(5, {a=[[1,2], [4,5]], b=[[5,4,3,2,1]]});
        permgroup(5, {a = [[1, 2], [4, 5]], b = [[5, 4, 3, 2, 1]]})
> permgroup(6, {[[1,2]], [[1,2,3,4,5,6]]});
            permgroup(6, {[[1, 2]], [[1, 2, 3, 4, 5, 6]]})
> permgroup(5, {x=[[3,4], y=[7,2]]});
Error, (in permgroup) invalid parameters
```

4.2.14   group[permrep] – find a permutation representation of a group

Calling Sequence:

permrep(*sbgrl*)

Parameters:

    sbgrl – subgroup of a group given by generators and relations (i.e. a
         `subgrel`)

Synopsis:

- This function finds all the right cosets of the given subgroup in the given group, assigns integers consecutively to these cosets, constructs a permutation on these coset numbers for each group generator, and returns the permutation group generated by these permutations. Thus the permutation group will be a homomorphic image of (but not necessarily isomorphic to) the original group. A `permgroup` is returned whose generators are named the same as the original group generators.

- This function is part of the **group** package, and so can be used in the form `permrep(..)` only after performing the command `with(group)` or `with(group, permrep)`.

Examples:
```
> with(group):
> g := grelgroup({x,y}, {[x,x,y,x,y,y,y],[y,y,x,y,x,x,x]}):
> sg := subgrel({y=[y]},g):
> permrep(sg);
    permgroup(8, {y = [[2, 5, 8, 6, 7, 3, 4]], x = [[1, 2, 3, 7, 5, 6, 4]]})
```

SEE ALSO: `group[permgroup]`, `group[grelgroup]`, `group[subgrel]`

4.2.15   group[pres] – find a presentation for a subgroup of a group

Calling Sequence:

    `pres`(*sbgrl*)

Parameters:

    sbgr – subgroup of a group given by generators and relations (i.e. a
         `subgrel`)

Synopsis:

- This function finds a set of relations among the given subgroup's generators sufficient to define the subgroup. The result is returned as a `grelgroup`.

Examples:
```
> with(group):
> g := grelgroup({a,b,c,d}, {[a,b,c,1/d],[b,c,d,1/a],[c,d,a,1/b],[d,a,b,1/c]}):
> sg := subgrel({x=[a,b],y=[a,c]},g):
> pres(sg);
    grelgroup({x, y},

        {[1/x, y, y, 1/x, 1/y, x, x, 1/y], [1/x, y, 1/x, 1/y, x, y, x, 1/y],

          [y, 1/x, 1/x, 1/x, y, x, 1/y, x, 1/y, x]}                        )
```

SEE ALSO: `group[grelgroup]`, `group[subgrel]`

4.3   The Boolean Logic Package

4.3.1 Introduction to the `logic` package

Calling Sequence:

    `logic[` *function* `] (` args `)]`
    function `(` args `)`

Synopsis:

- To use a function in the logic package, either define that function alone by typing `with( logic,` *function* `)`, or define all logic functions by typing `with( logic )`.

- Once the function has been defined, you may use the function directly by invoking `logic[` *function* `] (` args `)` or just *function*.

- The functions available are:

bequal	bsimp	canon	convert/frominert
convert/MOD2	convert/toinert	distrib	dual
environ	randbool	satisfy	tautology

- The following Boolean operators are used in this package: `&and`, `&or`, `&not`, `&iff`, `&nor`, `&nand`, `&xor`, and `&implies`.

- For help with a particular function try: `?logic,` *function*

- Example: to simplify the Boolean expression b, one would do the following: `with(logic);` `bsimp( b );`

SEE ALSO: `with, boolean`

4.3.2   logic[bequal] – logical equivalence of two expressions

Calling Sequence:

    `bequal(`*b1,b2*`)`
    `bequal(`*b1,b2,′p′*`)`

Parameters:

    b1,b2 – Boolean expressions
    p   – (optional) unevaluated name

Synopsis:

- This function must be defined by `with(logic);` before it can be used.

- `bequal` tests whether or not the two given Boolean expressions are logically equivalent. The result returned is `true` in the case that the two expressions are logically equivalent, `false` if they are not.

- If the third parameter is present and the two expressions are not logically equivalent, a valuation is assigned to p in which the two expressions have different results. If the two expressions are equivalent, p is assigned `NULL`. Note that the equivalence test may be significantly faster if the third argument is not present.

Examples:

```
> with(logic);                     ⟶   [bequal, bsimp, canon, convert/MOD2,
                                         convert/frominert, convert/toinert,
                                         distrib, dual, environ, randbool,
                                         satisfy, tautology]
> bequal(a &and (a &or b),a);      ⟶             true
> bequal(a &iff (a &or b),b &implies a);  ⟶       true
> bequal(a &implies b,b &implies a,'p');  ⟶       false
> p;                               ⟶        {a = true, b = false}
> bequal(&not(a &and b),(&not a) &or (&not
  b),'p');                         ⟶             true
> p;
```

SEE ALSO: `logic`

4.3.3   logic[bsimp] – Boolean expression simplification

Calling Sequence:

    bsimp(b)

Parameters:

    b –   a Boolean expression

Synopsis:

- This function must be defined by `with(logic)` before it can be used.

- `bsimp` returns a minimal sum of products expansion of the given Boolean expression.

- The expression returned is a `minimal` (irreducible) sum of prime implicants: note that `bsimp` does not guarantee that the returned expansion is a `minimum` sum of prime implicants.

Examples:

```
> with(logic);
  [bequal, bsimp, canon, convert/MOD2, convert/frominert, convert/toinert,
```

          `distrib, dual, environ, randbool, satisfy, tautology]`

```
> bsimp(a &or (a &and b));
                                    a
> bsimp((a &and b) &or (a &and (&not b)));
                                    a
> bsimp(a &iff (a &or b));
                            a &or (&not b)
> bsimp(&or(a &and b,(&not a) &and c,b &and c));
                    (a &and b) &or ((&not a) &and c)
```

SEE ALSO: `logic` , `logic[distrib]` , `simplify`

4.3.4 logic[canon] – canonical representation of expression

Calling Sequence:
>   canon(*b,alpha*)
>   canon(*b,alpha,form*)

Parameters:

b	–	a Boolean expression
alpha	–	set or list of names
form	–	(optional) name of canonical form

Synopsis:

- This function must be defined by `with(logic);` before it can be used.

- `canon` converts the given boolean expression to a canonical form. The canonical form is given with respect to the symbol names given in *alpha*. By default, if no third parameter is given, the input expression is converted to its disjunctive normal form.

- There are three canonical expansions available :

MOD2	The Boolean expression is converted to its equivalent modulo 2 canonical form. The second argument *alpha* has no affect but still must be supplied.
CNF	The conjunctive normal form is returned.
DNF	The disjunctive normal form is returned.

Examples:

```
> with(logic);              ⟶  [bequal, bsimp, canon, convert/MOD2,
                                    convert/frominert, convert/toinert,
                                    distrib, dual, environ, randbool,
```

satisfy, tautology]

> canon(a &and b,{a,b,c});	$\longrightarrow$	&and(a, b, c) &or &and(a, b, &not c)
> canon(a &xor b,{a,b},MOD2);	$\longrightarrow$	a + b
> canon(a &xor b,{a,b},CNF);	$\longrightarrow$	((&not a) &or (&not b)) &and (b &or a)
> canon(a &xor b,{a,b},DNF);	$\longrightarrow$	(b &and (&not a)) &or (a &and (&not b))

SEE ALSO: logic , logic[convert]

4.3.5 logic[convert] – convert an expression to a different form

Calling Sequence:

   convert(b, *form*, *args*);

Parameters:

   b –   a Boolean expression

   form   –   a name

   args   –   optional arguments

Synopsis:

* The options for **convert** described here must be defined by **with(logic);** before they can be used.

* **convert** allows one to change the format of a Boolean expression.

* There are three conversions available (*form* must be one of these):

* MOD2: The Boolean expression is converted to its equivalent modulo 2 format. The optional parameter **expanded** indicates that the modulo 2 representation should be returned fully expanded.

* **frominert**: The inert operators **&and**, **&or**, and **&not** are replaced by the system operators **and**, **or**, and **not** respectively.

* **toinert**: The conversion from system-defined operators to inert ones is performed.

Examples:

> with(logic): convert(a &or b,MOD2);	$\longrightarrow$	1 + (1 + a) (1 + b)
> convert(a &or b,MOD2,´expanded´);	$\longrightarrow$	b + a + a b
> convert(&and(a,b,c) &or b,frominert);	$\longrightarrow$	a and b and c or b
> convert(&or(a,b,c) &and b,frominert);	$\longrightarrow$	(a or b or c) and b
> convert(a and b or c,toinert);	$\longrightarrow$	c &or (a &and b)

SEE ALSO: logic, boolean, convert, convert[mod2]

4.3.6   logic[distrib] – expand a Boolean expression

Calling Sequence:
  distrib(b)
  distrib(b,alpha)

Parameters:

  b –   a Boolean expression

  $alpha$ –   (optional) set of names

Synopsis:

- This function must be defined by `with(logic);` before it can be used.

- `distrib` takes the given Boolean expression and expands it into a sum of products form (a disjunction of conjunctions). The expansion is performed by simply applying the distributive law and DeMorgan´s law to the given expression; therefore, the expanded expression is not guaranteed to be in either minimized form nor in canonical form.

- The presence of the optional second parameter indicates that the expansion should be the canonical disjunctive normal form of the expression with respect to the symbols in $alpha$.

Examples:
```
> with(logic);
   [bequal, bsimp, canon, convert/MOD2, convert/frominert, convert/toinert,

       distrib, dual, environ, randbool, satisfy, tautology]
> distrib(&not(a &and b));
                              (&not a) &or (&not b)
> distrib(&and(a,b &or c));
                           (a &and b) &or (a &and c)
> distrib(&not(a &or b),{a,b,c});
            &and(&not a, &not b, c) &or &and(&not a, &not b, &not c)
```

SEE ALSO: `logic` , `logic[canon]` , `logic[bsimp]` , `expand`

4.3.7   logic[dual] – construct the dual of a Boolean expression

Calling Sequence:
  dual(b)

Parameters:
  b – a Boolean expression

Synopsis:

- This function must be defined by `with(logic);` before it can be used.

- `dual` returns the dual of the given Boolean expression. All occurrences of **true** and **false** are replaced by **false** and **true**, respectively. In addition, **&and**'s are replaced by **&or**'s, and **&or**'s are replaced by **&and**'s.

- The other Boolean operators are left unchanged; however, if automatic simplifications are being made via `logic[environ]`, then the output expression will be the dual of the automatically simplified input expression.

Examples:
```
> with(logic);
    [bequal, bsimp, canon, convert/MOD2, convert/frominert, convert/toinert,

        distrib, dual, environ, randbool, satisfy, tautology]
> dual(a &and (&not a)  = false);
                      a &or (&not a) = true
> dual(a &implies b);
                         a &implies b
> environ(2):
> dual(a &implies b);
                       b &and (&not a)
```

SEE ALSO: `logic` , `logic[environ]`

4.3.8 logic[environ] – set level of automatic logical simplification

Calling Sequence:
> environ(n)

Parameters:
> n – an integer between 0 and 3 inclusive

Synopsis:
- This function must be defined by `with(logic);` before it can be used.

- `environ` sets an environment that causes automatic simplification of Boolean expressions. In other words, the user can specify to what degree laws of associativity and the like be applied automatically.

- There are three levels to choose from:

- n=0 — No simplifications.

- n=1 — The associative property is applied to remove redundant brackets. The expression is expressed in terms of the operators: **&and**, **&or**, and **&not**.

- n=2 — In addition to the simplifications of level 2, additional properties are understood. They include: **a &and a --> a**, **a &or a --> a**, and a knowledge of **true** and **false**.

- n=3 — The expression is converted to its unexpanded modulo 2 form.

Examples:

`> with(logic);`	$\longrightarrow$	`[bequal, bsimp, canon, convert/MOD2,`
		`convert/frominert, convert/toinert,`
		`distrib, dual, environ, randbool,`
		`satisfy, tautology]`

```
> environ(0);
  a &and b &and a;
```
$\longrightarrow$ (a &and b) &and a

```
> a &iff a;
```
$\longrightarrow$ a &iff a

```
> environ(1);
  a &and b &and a;
```
$\longrightarrow$ &and(a, a, b)

```
> a &iff a;
```
$\longrightarrow$ (a &and a) &or ((&not a) &and (&not a))

```
> environ(2);
  a &and b &and a;
```
$\longrightarrow$ a &and b

```
> a &iff a;
```
$\longrightarrow$ a &or (&not a)

```
> environ(3);
  a &and b &and a;
```
$\longrightarrow$ $\begin{array}{cc} & 2 \\ a & b \end{array}$

```
> a &iff a;
```
$\longrightarrow$ 2 a + 1

SEE ALSO: `logic` , `operators`

4.3.9   logic[randbool] – construct a random Boolean function

Calling Sequence:

   `randbool(`*alpha* `,` *c*`);`

Parameters:

   alpha   –   a list or set of symbols representing the alphabet

   c   –   (optional) name of canonical form

Synopsis:

- This function must be defined by `with(logic);` before it can be used.

- `randbool` returns a random Boolean expression in the canonical form `ivar CNF` (conjunctive normal form) , or `MOD2` (modulo 2 normal form) . By default, a random Boolean expression in disjunctive normal form is returned. The Boolean expression is in normal form with respect to the symbols in *alpha*.

Examples:
```
> with(logic);
   [bequal, bsimp, canon, convert/MOD2, convert/frominert, convert/toinert,
```

```
             distrib, dual, environ, randbool, satisfy, tautology]
> randbool({a,b});
             &or(a &and b, (&not a) &and (&not b), b &and (&not a))
> randbool([a,b,c],DNF);
    &or(&and(a, b, c), &and(b, &not a, &not c), &and(c, &not a, &not b),
        &and(a, c, &not b), &and(b, c, &not a))
> randbool([a,b],CNF);
                   (a &or b) &and ((&not a) &or (&not b))
> randbool({a,b,c},MOD2);
                      a b c + b + a + a b
```

SEE ALSO: `logic`

4.3.10   logic[satisfy] – return a valuation satisfying an expression

Calling Sequence:
> satisfy(b)
> satisfy($b, alpha$)

Parameters:

b –   a Boolean expression

$alpha$ –   alphabet set (optional second parameter)

Synopsis:

- This function must be defined by **with(logic);** before it can be used.

- **satisfy** returns an assignment to the variables of b such that b is **true**. If the given Boolean expression b is not satisfiable, NULL is returned.

- If the optional second parameter is present, the valuation includes all variable names in the given expression and in *alpha*.

Examples:
```
> with(logic);
    [bequal, bsimp, canon, convert/MOD2, convert/frominert, convert/toinert,

        distrib, dual, environ, randbool, satisfy, tautology]
> satisfy(a &or b);
                         {b = true, a = false}
> satisfy(a &or b,{a,b,c});
                    {b = true, a = false, c = false}
> satisfy(a &and (&not a));
```

SEE ALSO: `logic`

4.3.11   logic[tautology] – test for tautology

Calling Sequence:

    tautology(b)

    tautology($b, \acute{p}\acute{}$)

Parameters:

    b – a Boolean expression

    p – (optional) unevaluated name

Synopsis:

- This function must be defined by with(logic); before it can be used.

- tautology tests whether or not the given Boolean expression b is a tautology. **true** is returned if b is a tautology, otherwise **false** is returned.

- If the optional second parameter is present, a valuation is assigned to the name specified by the second parameter proving a negative result. If the second parameter $\acute{p}\acute{}$ is present and b is a tautology, then NULL is assigned to p. Note that the tautology test may be significantly faster if the second parameter is not present.

Examples:
```
> with(logic);
  [bequal, bsimp, canon, convert/MOD2, convert/frominert, convert/toinert,

      distrib, dual, environ, randbool, satisfy, tautology]
> tautology(&and(a,b) &or (&not a) &or (&not b));
                               true
> tautology((a &iff b) &or b, ´p´);
                               false

> p;
                        {a = true, b = false}
> tautology(a &or (&not a), ´p´);
                               true

> p;
```

SEE ALSO: logic

4.4   The Number Theory Package

4.4.1 Introduction to the numtheory package

Calling Sequence:

    numtheory[*function*] (*args*)

function (*args*)

Synopsis:

- To use a numtheory function, either define that function alone by typing with(numtheory, *function*), or define all numtheory functions by typing with(numtheory).

- The functions available are:

B	bernoulli	cfrac	cyclotomic
divisors	E	euler	F
factorset	fermat	GIgcd	ifactor
imagunit	isolve	isprime	issqrfree
ithprime	J	jacobi	L
lambda	legendre	M	mcombine
mersenne	mipolys	mlog	mobius
mroot	msqrt	nextprime	nthpow
order	phi	pprimroot	prevprime
primroot	rootsunity	safeprime	sigma
tau			

- For more information on a particular function see numtheory[*function*].

- The package functions are always available without applying with, using the long-form notation numtheory[*function*](*args*).

- This notation is necessary whenever there is a conflict between a package function name and another function used in the same session.

SEE ALSO: with

4.4.2   numtheory[cfrac] – continued fraction convergents

Calling Sequence:
    cfrac(r)
    cfrac(r, *maxit*)

Parameters:
    r –   a rational or float number
    maxit   –   (optional) integer (the number of convergents)

Synopsis:

- The cfrac function prints the sequence of (maximum *maxit*) quotients and convergents of the number r.

Examples:
```
> with(numtheory):
> cfrac(evalf(Pi),5);
the    1   th quotient is:    3        the convergent is:    3
the    2   th quotient is:    7        the convergent is:    22/7
the    3   th quotient is:    15        the convergent is:    333/106
the    4   th quotient is:    1        the convergent is:    355/113
the    5   th quotient is:    293        the convergent is:    104348/33215
```

SEE ALSO: `convert[confrac]`

4.4.3   numtheory[cyclotomic] – calculate cyclotomic polynomial

Calling Sequence:

cyclotomic(n, t)

Parameters:

n – an integer

t – variable

Synopsis:

- The function `cyclotomic`(n, t) will return the nth cyclotomic polynomial in t.

- The nth cyclotomic polynomial is the polynomial, over the rationals, whose roots are the primitive nth roots of unity.

Examples:
```
> with(numtheory):
  cyclotomic(1,x);                              ⟶              x - 1
> cyclotomic(20,z);                             ⟶       z^8 - z^6 + z^4 - z^2 + 1
```

4.4.4   numtheory[divisors] – positive divisors of an integer

Calling Sequence:

divisors(n)

Parameters:

n – an integer

Synopsis:

- The function `divisors` will compute the set of positive divisors of its integer argument n.

Examples:
```
> with(numtheory):
  divisors(-9);                                 ⟶              {1, 3, 9}
```

> `divisors(0);`	$\longrightarrow$	`{}`
> `divisors(1);`	$\longrightarrow$	`{1}`
> `divisors(10);`	$\longrightarrow$	`{1, 2, 5, 10}`

4.4.5   numtheory[factorset] – prime factors of an integer

Calling Sequence:
  `factorset(n);`

Parameters:
  n – an integer

Synopsis:
• The function `factorset` will compute the set of prime factors of its integer argument n.

Examples:
> `with(numtheory):` `factorset(10);`	$\longrightarrow$	`{2, 5}`
> `factorset(96);`	$\longrightarrow$	`{2, 3}`

SEE ALSO: `ifactor`

4.4.6   numtheory[fermat] – nth Fermat number

Calling Sequence:
  `fermat(n)`

Parameters:
  n – a non-negative integer

Synopsis:
• The function `fermat` will compute the nth Fermat number, which is `2^(2^n)+1`.

Examples:
> `with(numtheory):` `fermat(2);`	$\longrightarrow$	17
> `fermat(3);`	$\longrightarrow$	257

4.4.7   numtheory[GIgcd] – gcd of Gaussian integers

Calling Sequence:
  `GIgcd(x1, x2, ... xn)`

Parameters:

   $x1$, $x2$, ..., xn –   Gaussian integers

Synopsis:

* This function computes and returns the first quadrant associate of a gcd of the Gaussian integers $x1$ = `a1` + `b1*I`, ... ,xn = `an` + `bn*I`, where the `ai`s and the `bi`s are integers. The first quadrant associate of a Gaussian integer is defined as `i^j*x` where `x` is a Gaussian integer and `j(1..4)` is the quadrant containing `x`.

Examples:
```
> with(numtheory):
  GIgcd(-345+515*I,1574+368*I);          ⟶          41 + 117 I

> GIgcd(24,12);                          ⟶               12
```

SEE ALSO: `gcd`, `igcd`, `gcdex`.

4.4.8   numtheory[imagunit] – square root of -1 mod n

Calling Sequence:

   `imagunit(`n`)`

Parameters:

   n –   an integer

Synopsis:

* The function `imagunit` will find a square root of -1 (mod n) , if possible, otherwise it returns FAIL.

Examples:
```
> with(numtheory):
  imagunit(5);                           ⟶                2

> imagunit(7);                           ⟶              FAIL
```

4.4.9   numtheory[issqrfree] – test if integer is square free

Calling Sequence:

   `issqrfree(`n`)`

Parameters:

   n –   integer

Synopsis:

* The function `issqrfree(`n`)` returns **true** if n is square free, and **false** otherwise. The integer n is square free if it is not divisible by a perfect square.

Examples:

```
> with(numtheory):
  issqrfree(20);                    ⟶               false
> issqrfree(21);                    ⟶               true
```

4.4.10   numtheory[jacobi] – Jacobi symbol

Calling Sequence:

jacobi(a, b)

Parameters:

a – an integer relatively prime to b

b – a positive odd integer

Synopsis:

- The function `jacobi` will compute the Jacobi symbol J(a/b) of a and b. If the factorization of b is $b[1]*b[2]* \ldots *b[n]$, then J(a, b) = `product( L(`a`, b[i]), i = 1..n)`, where L(a,b) is the Legendre symbol of a and b.

Examples:

```
> with(numtheory):
  jacobi(6,11);                     ⟶                -1
> jacobi(7,9);                      ⟶                 1
```

SEE ALSO: `numtheory[legendre]`

4.4.11   numtheory[lambda] – Carmichael's lambda function

Calling Sequence:

lambda(n)

Parameters:

n – a positive integer

Synopsis:

- The size of the largest cyclic group generated by `g^i` (mod n) is given by `lambda(`n`)`.

- Carmichael's theorem states that `a^lambda(`n`) = 1` (mod n) if `gcd(a, `n`) = 1`.

Examples:

```
> with(numtheory):
  lambda(13);                       ⟶                12
> lambda(200);                      ⟶                20
```

4.4.12   numtheory[legendre] – Legendre symbol

Calling Sequence:

   legendre(a, b)

Parameters:

   a, b – integers

Synopsis:

• The function legendre will compute the Legendre symbol L(a/b) of a and b, which is defined to be 1 if a is a quadratic residue (mod b) and -1 if a is a quadratic non-residue (mod b) . The number a is a quadratic residue of b if it has a square root (mod b); i.e., an integer c exists such that c^2 is congruent to a (mod b).

Examples:
```
> with(numtheory):
  legendre(7,9);                    ⟶                    1
> legendre(3,5);                    ⟶                   -1
```

4.4.13   numtheory[mcombine] – Chinese remaindering

Calling Sequence:

   mcombine(a, ra, b, rb)

Parameters:

   a, ra, b, rb – integers

Synopsis:

• The function mcombine will compute an integer x such that x = ra (mod a) and x = rb (mod b), for a and b relatively prime, otherwise it returns **FAIL**.

Examples:
```
> with(numtheory):
  mcombine(6,5,9,7);                ⟶                  FAIL
> mcombine(7,4,11,5);               ⟶                   60
```

4.4.14   numtheory[mersenne] – nth Mersenne prime

Calling Sequence:

   mersenne(n)
   mersenne([i])

Parameters:

   n, i –   positive integers

Synopsis:

- The function **mersenne** will compute the nth Mersenne prime. If the argument n is an integer then **mersenne**(n) will return 2^n-1 if 2^n-1 is prime; otherwise it will return **FAIL**. If the argument is a list with one integer element [i], then **mersenne**([i]) will return the ith Mersenne prime.

Examples:

```
> with(numtheory):
  mersenne(3);
```
                                $\longrightarrow$ 7

```
> mersenne([3]);
```
                                $\longrightarrow$ 31

```
> mersenne(4);
```
                                $\longrightarrow$ FAIL

4.4.15   numtheory[mipolys] – number of monic irreducible univariate polynomials

Calling Sequence:

   mipolys(n, p)

   mipolys(n, p, m)

Parameters:

   n –   non-negative integer

   p –   prime integer (characteristic of a finite field)

   m –   (optional) positive integer

Synopsis:

- In the first form, **mipolys**(n, p), the number of monic irreducible univariate polynomials of degree n over the finite field **Z mod** p is computed.

- In the second form, **mipolys**(n, p, m), the number of monic irreducible univariate polynomials of degree n over the Galois field **GF**(p^m) is computed.

- The first form is a special case of the second, where m defaults to 1. In this context, the general mathematical definition of **mipolys** is

$$1/n * sum(mobius(n/d)*(p^m)^d, \text{ for } d \text{ in } divisors(n))$$

Examples:

```
> with(numtheory):
  mipolys(3,5);
```
                                $\longrightarrow$ 40

```
> mipolys(1,2,4);
```
                                $\longrightarrow$ 16

SEE ALSO: mobius, divisors

4.4.16   numtheory[mlog] – discrete logarithm

Calling Sequence:
   mlog(x, a, n)

Parameters:
   x, a, n – integers

Synopsis:
- The function mlog will compute the discrete logarithm of x to the base a (mod n). It finds an integer y such that $a \char94 y = x$ (mod n) if possible, otherwise it returns FAIL.

Examples:
```
> with(numtheory):
  mlog(9,4,11);
```
$\longrightarrow$ 3
```
> mlog(5,2,7);
```
$\longrightarrow$ FAIL

4.4.17   numtheory[mobius] – Mobius function

Calling Sequence:
   mobius(n)

Parameters:
   n – a positive integer

Synopsis:
- The function mobius(n) gives the Mobius function of n (lattice of divisors).

Examples:
```
> with(numtheory):
  mobius(20);
```
$\longrightarrow$ 0
```
> mobius(21);
```
$\longrightarrow$ 1

4.4.18   numtheory[mroot] – modular root

Calling Sequence:
   mroot(x, r, p)

Parameters:
   x, r, p – integers

Synopsis:
- The function mroot will compute the rth root of x (mod p). It finds an integer y such that $y \char94 r$ = x (mod p) if possible, otherwise it returns FAIL.

• The order, r, of the root must be prime.

Examples:
```
> with(numtheory):
  mroot(5,11,13);                    ⟶              8
> mroot(2,3,6);                      ⟶              2
```

4.4.19 numtheory[msqrt] – modular square root

Calling Sequence:
>   msqrt(x, n)

Parameters:
> x, n – integers

Synopsis:
• The function `msqrt` will compute the square root of x (mod n). It finds an integer y such that
  y^2 = x (mod n) if possible, otherwise it returns **FAIL**.

Examples:
```
> with(numtheory):
  msqrt(3,11);                       ⟶              5
> msqrt(3,7);                        ⟶            FAIL
```

4.4.20 numtheory[nthpow] – find largest nth power in a number

Calling Sequence:
>   nthpow(m, n)

Parameters:
> m – non-zero integer
> n – natural number

Synopsis:
• The function `nthpow` will return (b)^n where b is the greatest natural number such that b^n
  divides m.

Examples:
```
> with(numtheory):
  nthpow(4,2);                       ⟶              2
                                                   (2)
> nthpow(250,3);                     ⟶              3
                                                   (5)
```

4.4.21   numtheory[order] – order of a number

Calling Sequence:

    order(n, m)

Parameters:

n – an integer

m – a positive integer

Synopsis:

- The smallest integer i such that n^i = 1 (mod m) is returned.

- More formally, the order of n in the multiplicative group **mod** m is returned. If n and m are not coprime, then **NULL** is returned.

Examples:
```
> with(numtheory):
  order(13,100);                          ⟶              20

> order(5,8);                             ⟶               2

> order(8,12);
```

4.4.22   numtheory[phi] – totient function

Calling Sequence:

    phi(n)

Parameters:

n – an integer

Synopsis:

- The function **phi** will compute the totient function of n, which is the number of positive integers not exceeding n and relatively prime to n.

Examples:
```
> with(numtheory):
  phi(6);                                 ⟶               2

> phi(15);                                ⟶               8
```

4.4.23   numtheory[pprimroot] – compute a pseudo primitive root

Calling Sequence:

    pprimroot(g, n)

Parameters:
> g – an integer
> n – an integer greater than 2

Synopsis:
- The function pprimroot(g, n) computes the next primitive root larger than g or, if n does not have primitive roots, computes a number which is not a root of order of any of the factors of phi(n).

- Thus (in all cases), find an integer y, such that there is no x for which x^r = y (mod n) when r is a divisor of phi(n) greater than 1 and igcd(y, n) = 1.

Examples:
```
> with(numtheory):
  pprimroot(1,41);                          ⟶          6
> pprimroot(2,8);                           ⟶          3
```

SEE ALSO: numtheory[primroot]

4.4.24   numtheory[primroot] – compute a primitive root

Calling Sequence:
> primroot(g, n)

Parameters:
> g – an integer
> n – an integer greater than 2

Synopsis:
- The function primroot will compute the first primitive root of n that is greater than g, if possible, otherwise it returns FAIL. The integers that are relatively prime to n form a group of order phi(p) under multiplication (mod n). If this group is cyclic then a generator of the group is called a primitive root of n.

Examples:
```
> with(numtheory):
  primroot(1,41);                           ⟶          6
> primroot(2,8);                            ⟶          FAIL
```

SEE ALSO: numtheory[pprimroot]

4.4.25   numtheory[rootsunity] – roots of unity

Calling Sequence:

    `rootsunity(p, r)`

Parameters:

    p – prime
    r – integer

Synopsis:

- This function will calculate all the pth roots of unity **mod** r and return the result as an expression sequence.

- The order, p, of the root must be prime.

- Note that there will always be at least the root 1.

Examples:
```
> with(numtheory):
  rootsunity(5,11);                    ⟶              1, 3, 4, 5, 9
> rootsunity(3,11);                    ⟶                   1
```

4.4.26   numtheory[safeprime] – compute a safe prime

Calling Sequence:

    `safeprime(n)`

Parameters:

    n – an integer

Synopsis:

- The function `safeprime` will compute the smallest safe prime that is greater than n. A safe prime is a number p such that p is prime and `(p-1)/2` is prime.

Examples:
```
> with(numtheory):
  safeprime(8);                        ⟶                  11
> safeprime(12);                       ⟶                  23
```

4.4.27   numtheory[sigma] – sum of divisors

Calling Sequence:

    `sigma(n)`

Parameters:

n – an integer

Synopsis:

- The sum of the positive divisors of n is given by **sigma**(n).

Examples:
```
> with(numtheory):
  sigma(2);                    ⟶              3
> sigma(9);                    ⟶              13
> sigma(100);                  ⟶              217
```

4.4.28   numtheory[tau] – number of divisors

Calling Sequence:

tau(n)

Parameters:

n – a positive integer

Synopsis:

- The number of positive divisors of n is given by **tau**(n).

Examples:
```
> with(numtheory):
  tau(100);                    ⟶              9
> tau(101);                    ⟶              2
> tau(6);                      ⟶              4
```

5
Packages for Applied Mathematics

5.1 The Differential Forms Package

5.1.1 Introduction to the `difforms` package

Calling Sequence:

function(`args`)

`difforms`[*function*]`(args)`

Synopsis:

- To use a `difforms` function, either define that function alone by typing `with(difforms,` *function*`)`, or define all `difforms` functions by typing `with(difforms)`. Alternatively, invoke the function using the long form `difforms`[*function*]`.

- The functions available are:

`&^`	`d`	`defform`	`formpart`
`mixpar`	`parity`	`scalarpart`	`simpform`
`wdegree`			

- For example, to take the exterior derivative of an expression:

```
with(difforms):
defform(w1=1,w2=1,w3=1,f=scalar,g=scalar,C=const);
d( f*w1^2+g*&^(w2,w1)+f*&^(w2,w3) );
      &^(d(f), w1^2) + &^(d(g), w2, w1) + g &^(d(w2), w1)
          - g &^( w2, d(w1)) + &^(d(f), w2, w3)
          + f &^(d(w2), w3) - f &^(w2, d(w3))
```

- For help with a particular function see `difforms`[*function*]`.

- For help with the basic types used by the package (`const`, `scalar`, `form`), see `difforms[const]`, `difforms[scalar]` or `difforms[form]`.

SEE ALSO: `with`

5.1.2   difforms[d] – exterior differentiation

Calling Sequence:

> d(*expr*)
> d(*expr*, *forms*)

Parameters:

expr – expression or list of expressions

forms – (optional) list of 1-forms, i.e. wdegree=one

Synopsis:

- The function **d** computes the exterior derivative of an expression. If the expression is a list, then d is applied to each element in the list.

- When **d** is called with an expression and a list of 1-forms, any name of type **scalar** in the expression will be expanded in these 1-forms. It is assumed that the 1-forms are independent. For each 1-form, a new scalar is created to be the component for that 1-form.

- This function is part of the `difforms` package, and so can be used in the form `d(..)` only after performing the command `with(difforms)` or `with(difforms, d)`. The function can always be accessed in the long form `difforms[d](..)`.

Examples:

```
> with(difforms):
> defform(f=0,w1=1,w2=1,w3=1,v=1,x=0,y=0,z=0);
> d(x^2*y);
```

$$2 x y \, d(x) + x^2 \, d(y)$$

```
> d(f(x,y,z));
    D[1](f)(x, y, z) d(x) + D[2](f)(x, y, z) d(y) + D[3](f)(x, y, z) d(z)
> d(f, [w1,w2,w3]);
```

$$fw1 \; w1 + fw2 \; w2 + fw3 \; w3$$

```
> d(f, [d(x),d(y[1]),v[1]]);
```

$$fx \; d(x) + fy[1] \; d(y[1]) + fv[1] \; v[1]$$

```
> d(f*&^(w1,w2));
```

$$\&^{\wedge}(d(f), w1, w2) + f \; (d(w1) \; \&^{\wedge} \; w2) - f \; (w1 \; \&^{\wedge} \; d(w2))$$

```
> d([x*y,y^2,w2]);
```

$$[y \; d(x) + x \; d(y), \; 2 y \; d(y), \; d(w2)]$$

SEE ALSO: `D`, `diff`, `difforms`, `difforms[mixpar]`

5.1.3   difforms[defform] – define a constant, scalar, or form

Calling Sequence:
   defform(*n1* = *e1*, *n2* = *e2*, . . .)

Parameters:
   n1, n2, ... – names or calls to difforms[d]
   e1, e2, ... – Maple expressions

Synopsis:
- The function defform is used to define the basic variables used in a computation, or to define the exterior derivative of an expression.

- The function defform clears the remember tables of all functions in the forms package, as changing the definition of a form can make the remembered results invalid. However, definitions made through defform are not cleared; they are remembered permanently.

- The function defform takes an arbitrary number of equations, where each equation is *name* = *expr*. There are certain expressions - const, scalar, form, odd, even,-1, and 0 that have special meanings. Except for these, *name* = *expr* means *name* is a form and wdegree(*name*) = *expr*.

- The expression const or -1 means type(*name*,const) is true. The name even or odd means type(*name*,const) is true, but also that parity(*name*) is 0 or 1, as appropriate. The names even and odd are useful for specifying a form with even or odd, but otherwise unknown wdegree.

- The expression scalar or 0 means type(*name*,scalar) is true. The name form means that type(*name*,form) is true, but does not give a value to wdegree(*name*).

- The function defform can be used to define the exterior derivative of an expression, and these derivatives are remembered permanently.

- This function is part of the difforms package, and so can be used in the form defform(..) only after performing the command with(difforms) or with(difforms, defform). The function can always be accessed in the long form difforms[defform](..).

Examples:
```
> with(difforms):
  defform(a=const,b=scalar,e=nonhmg,j=3,
  f=odd,l=f^3,d(l)=e);
  type(a,const);                    ⟶              true

> type(b,scalar);                   ⟶              true

> type(e,form),wdegree(e);          ⟶         true, nonhmg

> type(j,form),wdegree(j);          ⟶           true, 3

> type(f,const),parity(f);          ⟶           true, 1

> d(l);                             ⟶               e
```

5.1.4   difforms[formpart] – find part of an expression which is a form

Calling Sequence:

    formpart(*expr*)

Parameters:

    expr –   any Maple expression, but usually a product

Synopsis:

- The function **formpart** is most effective on products. It will try to simplify other expressions to products.

- This function removes scalars and constants from a product, leaving only that part which is a form, or is undeclared.

- This function and **scalarpart** can be used to separate a product into form and scalar parts.

- This function is part of the **difforms** package, and so can be used in the form formpart(..) only after performing the command with(difforms) or with(difforms, formpart). The function can always be accessed in the long form difforms[formpart](..).

Examples:
```
> with(difforms): defform(f=scalar);
  formpart(2*&^(f*v,u));                    ⟶                    v &^ u
```

5.1.5   difforms[mixpar] – ensure equality of mixed partial derivatives

Calling Sequence:

    mixpar(*expr*)

Parameters:

    expr –   a Maple expression

Synopsis:

- The function **mixpar** will take nested calls to **diff**, and sort the sequence of differentiations so that they are sorted by lexicographical ordering on the variables of differentiation.

- The purpose of this function is to ensure that equal mixed partials are recognized as equal.

- This function is part of the **difforms** package, and so can be used in the form mixpar(..) only after performing the command with(difforms) or with(difforms, mixpar). The function can always be accessed in the long form difforms[mixpar](..).

Examples:
```
> with(difforms): defform(f=0,x=0,y=0,z=0);
> d(f(x,y,z));
    D[1](f)(x, y, z) d(x) + D[3](f)(x, y, z) d(z) + D[2](f)(x, y, z) d(y)
```

```
> d( " );
 d(D[1](f)(x, y, z) d(x)) + d(D[3](f)(x, y, z) d(z)) + d(D[2](f)(x, y, z) d(y))

> mixpar( " );
 d(D[1](f)(x, y, z) d(x)) + d(D[3](f)(x, y, z) d(z)) + d(D[2](f)(x, y, z) d(y))
```

SEE ALSO: `diff`, `sort`

5.1.6   difforms[parity] – extension of mod 2

Calling Sequence:

   parity(*expr*)

Parameters:

   expr   –   a Maple expression

Synopsis:

- The function `parity` computes the parity of an expression, by assuming that unspecified exponents are integers. Given `n` is an integer, `p^n mod 2 = p`. For unassigned names, the function `parity` returns the name.

- It is with the `parity` function that `defform` recognizes certain names as even or odd.

- This function is part of the `difforms` package, and so can be used in the form `parity(..)` only after performing the command `with(difforms)` or `with(difforms, parity)`. The function can always be accessed in the long form `difforms[parity](..)`.

Examples:
```
> with(difforms): defform(w=p,p=even);
  parity(3*p);                              ⟶              0

> parity(p^m+k);                            ⟶              k
```

SEE ALSO: `` `mod` ``

5.1.7   difforms[scalarpart] – find part of an expression which is a scalar

Calling Sequence:

   scalarpart(*expr*)

Parameters:

   expr   –   a Maple expression, but usually a product

Synopsis:

- The function `scalarpart` is most effective on products. It will try to simplify other expressions to products.

- This function collects scalars and constants from a product, removing those parts which are forms or are undeclared.

- This function and `formpart` may be used together to separate a product into scalar and form parts.

- This function is part of the `difforms` package, and so can be used in the form `scalarpart(..)` only after performing the command `with(difforms)` or `with(difforms, scalarpart)`. The function can always be accessed in the long form `difforms[scalarpart](..)`.

Examples:
```
> with(difforms): deform(f=scalar,c=const);
> scalarpart(2*&^(v*f,c*u));
                              2 f c
```

5.1.8   difforms[simpform] – simplify an expression involving forms

Calling Sequence:

    simpform(*expr*)

Parameters:

    expr – a Maple expression

Synopsis:

- The function `simpform` will simplify an expression involving forms. Its operations include collecting like terms, simplifying wedge products, and pulling out scalar factors.

- This function is part of the `difforms` package, and so can be used in the form `simpform(..)` only after performing the command `with(difforms)` or `with(difforms, simpform)`. The function can always be accessed in the long form `difforms[simpform](..)`.

Examples:
```
> with(difforms): deform(f=scalar, g=scalar);
> &^(f*v,u)+&^(u,g*v);
                         f (v &^ u) + g (u &^ v)
> simpform(");
                              (wdegree(u) wdegree(v))
                 (f + (-1)                            g) (v &^ u)
> f*(&^(u,v)+&^(u,w))+g*&^(u,v):
> simpform(");
                      (f + g) (u &^ v) + f (u &^ w)
```

5.1.9   difforms[wdegree] – degree of a form

Calling Sequence:

    wdegree(*expr*)

Parameters:

    expr – a Maple expression

Synopsis:
- The function **wdegree** computes the degree of an expression, considered as a form.
- The function **wdegree** returns **nonhmg** when the expression is a sum of forms which have different degrees.
- Expressions of type **scalar** or **const** have **wdegree** = 0.
- This function is part of the **difforms** package, and so can be used in the form **wdegree(..)** only after performing the command **with(difforms)** or **with(difforms, wdegree)**. The function can always be accessed in the long form **difforms[wdegree](..)**.

Examples:
```
> with(difforms): defform(x=p, c=const);
  wdegree(c);
```
$\longrightarrow$ 0

```
> wdegree(x);
```
$\longrightarrow$ p

```
> d(x^2);
```
$\longrightarrow$ $(1 + (-1)^p) \ (d(x) \ \&^\wedge \ x)$

```
> wdegree(");
```
$\longrightarrow$ $2\ p + 1$

```
> wdegree(x + d(x));
```
$\longrightarrow$ nonhmg

5.1.10 difforms[&ˆ] – wedge product

Calling Sequence:

    $\&^\wedge(expr_1, \ expr_2, \ \dots)$

    $expr_1 \ \&^\wedge \ expr_2 \ \&^\wedge \ \dots$

Parameters:

    $expr_1, expr_2, \dots$ – Maple expressions

Synopsis:
- The operator **&ˆ** represents the wedge product of differential forms.
- Elementary simplifications are done on wedge products. For example, if **a** is a form of odd degree, then **&ˆ(a,a)** is simplified to 0.
- The operator **&ˆ** will distribute over + whenever possible. The preferred representation of **&ˆ** is a sum of wedge products. Otherwise, it may be necessary to apply **expand**, then **simpform** to an expression to reduce it to simplest form.
- This function is part of the **difforms** package, and so can be used in the form **&ˆ(..)** only after performing the command **with(difforms)** or **with(difforms, &ˆ)**. The function can always be accessed in the long form **difforms[&ˆ](..)**.

Examples:
```
> with(difforms):
  defform(a=1,b=1,c=1,d=2,e=2);
  &^(a,b,c+&^(d,e));                          ⟶              &^(a, b, c) + &^(a, b, d, e)

> &^(a,b,c+&^(d,e,&^(a,d)));                  ⟶                      &^(a, b, c)
```

5.1.11   type/const – extension of type/constant

Calling Sequence:

>   type(*expr*, const)

Parameters:

>   *expr*   –   a Maple expression

Synopsis:

- The type const includes any *expr* which is of type constant, plus those names which are defined to be const through difforms[defform].

- All sums, products, or powers of objects of type const are considered to be const.

- The types const, form and scalar are mutually exclusive. Any expression should belong to at most one of these types.

- This function is part of the difforms package and can only be used after performing the command with(difforms).

Examples:
```
> with(difforms): defform(a=const,b=const,
  c=const,d=const);
  type(a,const);                              ⟶                  true

> type(a+b*c^d,const);                        ⟶                  true

> type(e+a,const);                            ⟶                  false
```

SEE ALSO: type, type/form, type/scalar

5.1.12   type/form – check for forms

Calling Sequence:

>   type(*expr*, form)
>   type(*expr*, form, *int*)

Parameters:

>   *expr*   –   a Maple expression

int   –   (optional) positive integer

Synopsis:
- The command `type(`*expr*`,form,`*int*`)` returns `true` if `type(`*expr*`,form)` is `true` and `wdegree(`*expr*`)` = *int*; in other words, if *expr* is an *int*-form

- The type `form` includes all expressions declared to be forms through `defform`, as well as any product or sum that contains at least one term of type `form`. A power is of type `form` if the base is of type `form`.

- The types `const`, `scalar`, and `form` are mutually exclusive. Any expression can belong to at most one of these types.

- This function is part of the difforms package and can only be used after performing `with(difforms)`.

Examples:
```
> with(difforms): defform(a=const,b=scalar,
  f=form,g=a);
  type(a+b+f, form);                    ⟶            true
> type(a+b, form);                      ⟶            false
> type(f^a, form);                      ⟶            true
> type(a*g, form);                      ⟶            true
```

SEE ALSO: `type`, `type/const`, `type/scalar`

5.1.13   type/scalar – check for scalars

Calling Sequence:
    `type(`*expr*`, scalar)`

Parameters:
    expr   –   a Maple expression

Synopsis:
- The type `scalar` includes all names defined to be `scalar` through `defform`.

- Any sum or product which contains one term which is of type `scalar`, and remaining terms that are all either of type `scalar` or of type `const`, is considered to be `scalar`.

- Any power, where the base type is `scalar`, is considered to be `scalar`.

- The types `const`, `scalar`, and `form` are mutually exclusive. Any expression can belong to at most one of these types.

- This function is part of the difforms package and can only be used after performing `with(difforms)`.

Examples:
```
> with(difforms): defform(a=const,b=const,
  c=scalar,d=scalar);
  type(c+d^3,scalar);                   ⟶            true
```

```
> type(a+d,scalar);                    ⟶           true
> type(a+d+f,scalar);                  ⟶           false
> type(a+b,scalar);                    ⟶           false
```

5.2   The Lie Symmetries Package

5.2.1   Introduction to the `liesymm` package

Synopsis:

- To load the `liesymm` package, use the command `with(liesymm)`.

- This is an implementation of the Harrison-Estabrook procedure (as outlined in the *Journal of Mathematical Physics*, vol 12, American Institute of Physics, New York, 1971, pp. 653-665). It obtains the determining equations leading to the similarity solutions of a system of partial differential equations using a number of important refinements and extensions as developed by J. Carminati.

- To construct the determining equations for the isovector using Cartan´s geometric formulation of partial differential equations in terms of differential ideals use **determine()**. Other routines help to convert the set of equations to an equivalent set of differential forms or vice versa.

- You can compute or check for closure of a given set of forms and annul to a specified sublist of independent coordinates. Modding lists are used to eliminate those parts of a differential form belonging to the ideal.

- The implementation makes use of the exterior derivative (d) and wedge product (&^) but is completely independent of the Maple **difforms** package. It requires a specific coordinate system as defined by setup(). Unknowns default to constants, and automatic simplifications take into account a consistent ordering of the 1-forms and the extraction of coefficients.

- The functions available are:

&mod	&^	annul	close
d	determine	getcoeff	getform
hasclosure	hook	Lie	Lrank
makeforms	mixpar	setup	value
wcollect	wdegree	wedgeset	wsubs

- A brief description of the functionality available follows.

`setup`	to define (or redefine) a list of coordinate variables (0-forms).
`d`	to compute the exterior derivative with respect to the specified coordinates.
`&^`	to compute the wedge product. It automatically simplifies relative to an "address" ordering of the basis variables to sums of expressions of the form `c*(d(x)&^d(y)&^d(z))`.
`Lie`	to compute the Lie derivative of an expression involving forms, relative to a specified vector.
`wcollect`	to express a form as a sum of forms each multiplied by a coefficient of wedge degree 0.
`wsubs`	to substitute an expression for a k-form that is part of an n-form.

- Various other routines such as `choose()`, `getcoeff()`, `mixpar()`, `wdegree()`, `wedgeset()`, and `value()` are used in manipulating the forms and results.

- Let `eqn` be a set or list of partial differential equations involving functions,

    `flist = [f1,f2,...fn]  (x1,..,xj)`

`convert(eqlist, forms, eqlist, w)`
or
`makeforms(eqns, flist, w)`
	Generates a set of forms that when closed characterize the equations in `eqlist` in the sense of Cartan.

`convert(forms, system, vlist)`
or
`annul(forms, vlist)`
	Generates a set of partial differential equations represented by the given forms.

`close(forms)`	Extends the given list of forms to achieve closure under application of d().
`hasclosure(forms)`	Checks if the forms list is closed under applications of `d()`
`&mod`	Reduces a form modulo an exterior ideal (specified by a closed list of forms).
`determine(forms, V)`	Given a list of forms describing a particular set of partial differential equations with coordinates (`x1,...xn`), the routine will produce a set of first order equations for the isovector vector (`V1, ...Vn`). The resulting equations are expressed using **alias** and an inert **Diff** rather than **diff** but evaluation can be forced by using `value()`.
`determine(f,V,h(t,x),w)`	As above, but with f as an equation and with the extra arguments used by makeforms() to construct the initial forms list.

- You need not work with the differential forms directly. When given a list of partial differential equations instead of a forms list, the routine `determine()` sets up the coordinates and differential forms as required.

- Partial derivatives should be expressed in terms of `Diff()` rather than `diff()` or `D()`. The routine `mixpar()` may be used to force mixed partials to a consistent ordering.

- Use `value()` to convert `Diff()` to `diff()` when interpreting or using the result of determine.

Examples:
```
# Non-linear Boltzman's equation.
> with(liesymm):
> setup();
```
$$[]$$
```
> eqn := Diff(u(x,t),x,t) + Diff(u(x,t),x) +u(x,t)^2=0;
```
$$eqn := Diff(u(x, t), x, t) + \left(\frac{d}{dx} u(x, t)\right) + u(x, t)^2 = 0$$
```
> forms := makeforms(eqn,u(x,t),w);
  forms :=
```
$$[d(u) - w1\ d(x) - w2\ d(t), - (d(t)\ \&^\wedge\ d(w2)) + (w1 + u^2)\ (d(x)\ \&^\wedge\ d(t))]$$
```
> eqn := mixpar(eqn);
```
$$eqn := \left(\frac{d^2}{dx\ dt} u(x, t)\right) + \left(\frac{d}{dx} u(x, t)\right) + u(x, t)^2 = 0$$
```
> determine( eqn, V, u(x,t), w ):
> value("):
> wedgeset(0);
```
$$x,\ t,\ u,\ w1,\ w2$$
```
#
> close(forms);
```
$$[d(u) - w1\ d(x) - w2\ d(t), - (d(t)\ \&^\wedge\ d(w2)) + (w1 + u^2)\ (d(x)\ \&^\wedge\ d(t)),$$
$$(d(x)\ \&^\wedge\ d(w1)) + (d(t)\ \&^\wedge\ d(w2))]$$
```
> annul(",[x,t]);
```
$$[\left(\frac{d}{dx} u(x, t)\right) - w1(x, t) = 0,\ \left(\frac{d}{dt} u(x, t)\right) - w2(x, t) = 0,$$
$$\left(\frac{d}{dx} w2(x, t)\right) + w1(x, t) + u(x, t)^2 = 0,$$
$$\left(\frac{d}{dt} w1(x, t)\right) - \left(\frac{d}{dx} w2(x, t)\right) = 0]$$

SEE ALSO: `dsolve`, `liesymm`[*function*] where *function* is any of the functions in the `liesymm` package

5.2.2   liesymm[annul] – annul a set of differential forms

Calling Sequence:

   annul(*forms*, *vlist*)

Parameters:

> *forms* – A list or set of differential forms
>
> *vlist* – A list of those coordinates that are to be treated as independent
>   on the solution manifold

Synopsis:

- This routine is part of the `liesymm` package and is loaded via `with(liesymm)` .

- Given a set of differential forms and a list of coordinates, this command sections the forms and sets them equal to 0. The result is a set of partial differential equations corresponding to the differential forms.

- If the set of forms is closed then the resulting equations include the integrability conditions.

Examples:
```
> with(liesymm):
> setup(t,x,u,w1,w2);
                          [t, x, u, w1, w2]

> a1 := d(u) - w1*d(t) - w2*d(x);
                   a1 := d(u) - w1 d(t) - w2 d(x)

> a2 := (w2+u^2) * (d(x) &^ d(t)) - d(w2) &^ d(x);
               2
      a2 := - (w2 + u ) (d(t) &^ d(x)) + (d(x) &^ d(w2))

> annul([a1,a2],[t,x]);
          / d        \                    / d        \
       [|---- u(t, x)| - w1(t, x) = 0,  |---- u(t, x)| - w2(t, x) = 0,
          \ dt       /                    \ dx       /

                           2  / d         \
            - w2(t, x) - u(t, x)  - |---- w2(t, x)| = 0]
                               \ dt        /

> close([a1,a2]);
                                        2
   [d(u) - w1 d(t) - w2 d(x), - (w2 + u ) (d(t) &^ d(x)) + (d(x) &^ d(w2)),

       (d(t) &^ d(w1)) + (d(x) &^ d(w2))]

> annul(",[t,x]);
          / d        \                    / d        \
       [|---- u(t, x)| - w1(t, x) = 0,  |---- u(t, x)| - w2(t, x) = 0,
          \ dt       /                    \ dx       /

                           2  / d         \
            - w2(t, x) - u(t, x)  - |---- w2(t, x)| = 0,
                               \ dt        /

          / d         \   / d         \
         |---- w1(t, x)| - |---- w2(t, x)| = 0]
          \ dx        /   \ dt        /
```

SEE ALSO: `with, liesymm, setup, makeforms, hasclosure, close, determine`

5.2.3   liesymm[close] – compute the closure of a set of differential forms

Calling Sequence:

> close(*forms*)

Parameters:

> *forms*   –   A list or set of differential forms

Synopsis:

- This routine is part of the `liesymm` package and is loaded via `with(liesymm)` .

- A set of differential forms is closed with respect to the exterior derivative `d()` by adding additional forms to the original set.

Examples:
```
> with(liesymm):
> setup(t,x,u,w1,w2);
                            [t, x, u, w1, w2]
> a1 := d(u) - w1*d(t) - w2*d(x);
                    a1 := d(u) - w1 d(t) - w2 d(x)
> a2 := (w2+u^2) * (d(x) &^ d(t)) - d(w2) &^ d(x);
                    2
        a2 := - (w2 + u ) (d(t) &^ d(x)) + (d(x) &^ d(w2))
> hasclosure([a1,a2]);
                            false
> close([a1,a2]);
                                2
    [d(u) - w1 d(t) - w2 d(x), - (w2 + u ) (d(t) &^ d(x)) + (d(x) &^ d(w2)),
        (d(t) &^ d(w1)) + (d(x) &^ d(w2))]
> hasclosure(");
                            true
```

SEE ALSO: `with, liesymm, setup, makeforms, hasclosure, &mod`

5.2.4   liesymm[d] – the exterior derivative

Calling Sequence:

> d(*form*)

Parameters:

> *form*   –   An expression involving differential forms relative to specific
>   coordinates

Synopsis:

- This routine is part of the `liesymm` package and is loaded via `with(liesymm)` .

- It computes the exterior derivative of the differential form `form` with respect to the coordinates defined by `setup()` .

- For coordinate `x`, `d(x)` is a 1-form, and `d(d(x))=0`.

- Expressions not involving the coordinates are treated as constants. The coordinate list is given by `wedgeset(0)`.

Examples:

`> with(liesymm):`		
`  setup(x,y,z);`	$\longrightarrow$	`[x, y, z]`
`> map(d,");`	$\longrightarrow$	`[d(x), d(y), d(z)]`
`> d(x^2*d(x));`	$\longrightarrow$	`0`
`> d(x + y);`	$\longrightarrow$	`d(x) + d(y)`
`> d( z* (d(x) &^ d(y)));`	$\longrightarrow$	`&^(d(x), d(y), d(z))`
`> d( 1/c* (d(x) &^ d(y)));`	$\longrightarrow$	`0`

SEE ALSO: `with`, `liesymm`, `setup`, `wedgeset`, `` `&^` ``, `Lie` , `hook`

5.2.5   liesymm[determine] – find the determining equations for the isovectors of a pde

Calling Sequence:

 `determine(`*forms,* *Vname*`)`

 `determine(`*eqns,* *Vname,* *fcns,* *Extd*`)`

Parameters:

 forms   –   A list or set of differential forms

 eqns   –   A list or set of partial differential equations

 Vname   –   A name for constructing the names of the components of the isovector

 fcns   –   A list of functions and the dependent variables. For example, u(t,x).

 Extd   –   A name or list of names for constructing the extended variable names.

Synopsis:

- This routine is part of the `liesymm` package and is loaded via `with(liesymm)` .

- Given a set of differential forms and a name `V` this routine constructs the determining equations for the isovectors which are generators of the invariance group (isogroup) of the differential equations. These form a coupled set of linear first order differential equations for the components [V1,...Vn].

- If differential equations are given directly to `determine()` then the required differential forms are constructed automatically using `makeforms()`. Additional arguments are used to identify the dependent and independent variables (e.g. `u(t,x)`), and to describe the extended variables.

- If a set of forms is provided it must be closed. A method of automatically closing is provided through the routine `close()`.

- The components of the isovector `V` correspond to `[wedgeset(0)]`. Thus if `[wedgeset(0)]` = `[x,y,z,w,p]` then `V1` corresponds to `x`, `V2` to `y`, and so forth. This order can be established any time prior to using `determine()` by specifying the coordinates in the desired order as the arguments to `setup()`.

- The determining equations are constructed using an unevaluated `Diff()` and aliases for each of the components (`V1,...Vn`). This is to suppress the functional arguments and to compress output. To force the unevaluated `Diff()` to evaluate, use `value()`.

Examples:

```
> with(liesymm):
# The Heat equation
> eqn := Diff(h(t,x),x,x) = Diff(h(t,x),t);
```

$$eqn := Diff(h(t, x), x, x) = \frac{d}{dt} h(t, x)$$

```
> determine(eqn,V,h(t,x),[p,q]): # a set of eight equations.
## or directly from the forms.
> setup(t,x,h,p,q);
```

$$[t, x, h, p, q]$$

```
> f1 := d(u) - p*d(t) - q*d(x);
```

$$f1 := - p \ d(t) - q \ d(x)$$

```
> f2 := p*(d(x) &^ d(t)) - d(q)&^ d(t);
```

$$f2 := (d(t) \ \&\hat{}\ d(q)) - p \ (d(t) \ \&\hat{}\ d(x))$$

```
> close([f1,f2]);
```

$$[- p \ d(t) - q \ d(x), \ (d(t) \ \&\hat{}\ d(q)) - p \ (d(t) \ \&\hat{}\ d(x)),$$
$$(d(t) \ \&\hat{}\ d(p)) + (d(x) \ \&\hat{}\ d(q))]$$

```
> determine(",V):
```

SEE ALSO: `with`, `liesymm`, `setup`, `makeforms`, `close`, `hasclosure` `,&mod`, `wsubs`

5.2.6   liesymm[getcoeff] – extract the coefficient part of a basis wedge product

Calling Sequence:

   `getcoeff(`*expr*`)`

Parameters:

   expr   –   A term involving one wedge product

Synopsis:
- This routine is part of the `liesymm` package and is loaded via `with(liesymm)` .

- Given an expression `expr` involving a single wedge product `w` find c such that `expr = c*w`.

Examples:
```
> with(liesymm): setup(x,y,z):
  getcoeff( a*d(x));                    ⟶                        a

> getcoeff( a*d(x) + b*d(x));          ⟶                      a + b

> getcoeff( a*d(x) + b*d(y));          ⟶     Error, (in getcoeff)
                                              unable to handle this type
> getcoeff( 1/c* (d(x) &^ d(y)) );     ⟶                       1/c

> getcoeff( c );                       ⟶                        c
```

SEE ALSO: `with`, `liesymm`, `setup`, `` `&^` `` , `d` , `getcoeff`

5.2.7   liesymm[getform] – extract the basis element of a single wedge product

Calling Sequence:
   `getform(`*expr*`)`

Parameters:
   expr   –   A term involving a wedge product

Synopsis:
- This routine is part of the `liesymm` package and is loaded via `with(liesymm)` .

- Given an expression `expr` involving a single wedge product factor out the coefficient.

Examples:
```
> with(liesymm): setup(x,y,z):
  getform( a*d(x));                     ⟶                      d(x)

> getform( a*d(x) + b*d(x));           ⟶                      d(x)

> getform( a*d(x) + b*d(y));           ⟶     Error, (in liesymm[getcoeff])
                                              unable to handle this type
> getform( 1/c* (d(x) &^ d(y)) );      ⟶                 d(x) &^ d(y)

> getform( c );                        ⟶                        1
```

SEE ALSO: `with`, `liesymm`, `setup`, `` `&^` `` , `d` , `getcoeff`

5.2.8   liesymm[hasclosure] – verify closure with respect to d()

Calling Sequence:
>   hasclosure(*forms*)

Parameters:
>   *forms*   –   A list or set of differential forms

Synopsis:
- This routine is part of the `liesymm` package and is loaded via `with(liesymm)` .

- A set of differential forms is tested for closure with respect to the exterior derivative `d()`.

Examples:
```
> with(liesymm):
> setup(t,x,u,w1,w2);
```
$$[t,\ x,\ u,\ w1,\ w2]$$
```
> a1 := d(u) - w1*d(t) - w2*d(x);
```
$$a1 := d(u)\ -\ w1\ d(t)\ -\ w2\ d(x)$$
```
> a2 := (w2+u^2) * (d(x) &^ d(t)) - d(w2) &^ d(x);
```
$$a2 := -\ (w2 + u^2)\ (d(t)\ \&\hat{}\ d(x))\ +\ (d(x)\ \&\hat{}\ d(w2))$$
```
> hasclosure([a1,a2]);
```
$$false$$
```
> close([a1,a2]);
```
$$[d(u)\ -\ w1\ d(t)\ -\ w2\ d(x),\ -\ (w2 + u^2)\ (d(t)\ \&\hat{}\ d(x))\ +\ (d(x)\ \&\hat{}\ d(w2)),$$
$$(d(t)\ \&\hat{}\ d(w1))\ +\ (d(x)\ \&\hat{}\ d(w2))]$$
```
> hasclosure(");
```
$$true$$

SEE ALSO: `with`, `liesymm`, `setup`, `makeforms`, `close` , `&mod`

5.2.9   liesymm[hook] – inner product (hook)

Calling Sequence:
>   hook(*f*, *V*)

Parameters:
>   *f*   –   an expression involving differential forms relative to specific
>   coordinates
>   *V*   –   a vector (or list)

Synopsis:
- This routine is part of the `liesymm` package and is loaded via `with(liesymm)` .

- Compute the inner product of **f** with respect to a vector **V**.

- Use `setup()` to change the underlying coordinate system.

Examples:
```
> with(liesymm):
> setup(x,y,z);
```
$$[x, \ y, \ z]$$

```
> hook(f(x,y,z),V);
```
$$0$$

```
> hook(d(f(x,y,z)),V);
```
$$\left(\frac{d}{dy} \ f(x, \ y, \ z)\right) V[2](x, \ y, \ z) \ + \ \left(\frac{d}{dx} \ f(x, \ y, \ z)\right) V[1](x, \ y, \ z)$$

$$+ \ \left(\frac{d}{dz} \ f(x, \ y, \ z)\right) V[3](x, \ y, \ z)$$

```
> hook(d(f(x,y,z) &^ d(z)),V);
on line 6, syntax error:
hook(d(f(x,y,z) &^ d(z),V);
                      ^
```

SEE ALSO: with, liesymm, setup, wedgeset, `&^`, Lie, d

5.2.10   liesymm[Lie] – the Lie derivative

Calling Sequence:
    Lie(*form*, *V*)

Parameters:
    form –   An expression involving differential forms relative to specific
                coordinates
    V   –   A name or an explicit isovector [V1,V2,...Vn])

Synopsis:
- This routine is part of the `liesymm` package and is loaded via `with(liesymm)`.

- The Lie derivative of the differential form **form** is constructed with respect to **V1**, ...**Vn** where n is the number of coordinates.

Examples:
```
> with(liesymm):
> setup(x,y,z);
```
$$[x, \ y, \ z]$$

```
> Lie(d(x),V);
```
$$\left(\frac{d}{dy} \ V[1](x, \ y, \ z)\right) d(y) \ + \ \left(\frac{d}{dx} \ V[1](x, \ y, \ z)\right) d(x)$$

```
                   /  d          \
              +   |---- V[1](x, y, z)|  d(z)
                   \  dz         /
> Lie(f( x,y,z)*d(x),V);
                   /  d         \               /  d         \
      (V[2](x, y, z) |---- f(x, y, z)|  + V[1](x, y, z) |---- f(x, y, z)|
                   \  dy        /               \  dx        /

              /  d         \
       + V[3](x, y, z) |---- f(x, y, z)|)  d(x) + f(x, y, z) (
              \  dz        /

      /  d          \         /  d          \
     |---- V[1](x, y, z)|  d(y) + |---- V[1](x, y, z)|  d(x)
      \  dy         /         \  dx         /

          /  d          \
      +  |---- V[1](x, y, z)|  d(z))
          \  dz         /
> Lie(f(x,y,z)*(d(x) &^ d(y)),V);
                   /  d         \               /  d         \
      (V[2](x, y, z) |---- f(x, y, z)|  + V[1](x, y, z) |---- f(x, y, z)|
                   \  dy        /               \  dx        /

              /  d         \
       + V[3](x, y, z) |---- f(x, y, z)|)  (d(x) &^ d(y)) + f(x, y, z) (
              \  dz        /

      /  d          \                  /  d          \
     |---- V[1](x, y, z)|  (d(x) &^ d(y)) - |---- V[1](x, y, z)|  (d(y) &^ d(z))
      \  dx         /                  \  dz         /

          /  d          \
      +  |---- V[2](x, y, z)|  (d(x) &^ d(y))
          \  dy         /

          /  d          \
      +  |---- V[2](x, y, z)|  (d(x) &^ d(z)))
          \  dz         /
```

SEE ALSO: `with, liesymm, setup, wedgeset, `&^`, d , hook`

5.2.11   liesymm[Lrank] – the Lie Rank of a set of forms

Calling Sequence:

    Lrank(*forms*)

Parameters:

    form   –   A list or set of differential forms

Synopsis:

- This routine is part of the `liesymm` package and is loaded via `with(liesymm)` .

- It removes forms which are redundant with respect to the generation of the determining equations.

Examples:
```
> with(liesymm):
> setup();
                                    []
> eqn := Diff(u(x,t),x,t) + Diff(u(x,t),x) +u(x,t)^2=0;
                          / d       \        2
        eqn := Diff(u(x, t), x, t) + |---- u(x, t)| + u(x, t)  = 0
                          \ dx      /
> forms := makeforms(eqn,u(x,t),w);
  forms :=

                                                              2
        [d(u) - w1 d(x) - w2 d(t), - (d(t) &^ d(w2)) + (w1 + u ) (d(x) &^ d(t))]
> forms := close(forms);
              forms := [d(u) - w1 d(x) - w2 d(t),

                                        2
                - (d(t) &^ d(w2)) + (w1 + u ) (d(x) &^ d(t)),

                (d(x) &^ d(w1)) + (d(t) &^ d(w2))]
> Lrank(forms);
                                                          2
        [d(u) - w1 d(x) - w2 d(t), - (d(t) &^ d(w2)) + (w1 + u ) (d(x) &^ d(t))]
```

SEE ALSO: `with`, `liesymm`, `setup`, `` `&^` ``, `Lie` , `d` , `makeforms` , `close`

5.2.12   liesymm[makeforms] – construct a set of differential forms from a pde

Calling Sequence:
>   `makeforms(`*eqns, fncs, rootname*`)`
>   `makeforms(`*eqns, fncs,* `[`*Extd*`])`

Parameters:

eqns	–	A pde, or a list or set of pde´s.
fcns	–	A list of functions. For example, u(t,x).
rootname	–	A name used to construct the names for the extended coordinates.
Extd	–	A name or list of names for constructing the extended variable names.

Synopsis:
- This routine is part of the `liesymm` package and is loaded via `with(liesymm)` .

- Given one or more partial differential equations this routine constructs a set of differential forms which after closure is equivalent to the original system of equations in the sense of Cartan. The forms are obtained by first reducing the system of pde´s to a system of first order equations. It will handle systems which are quasi-linear.

- If more than one equation is given, they must be given as a list or a set.

- The second argument specifies the dependent and independent variables as in `h(t,x)`. If there is more than one dependent variable, these must be specified as a set or list. The Maple expression `[h,u](t,x,y)` can be used to construct the required list when more than one dependent variable is involved.

- The third argument is used to construct the names of the extended variables. If it is the name `k` then the names `k1,k2,k3,...kn` are constructed as needed. If a list of names is provided then these names are used in place of `k1, k2,...kn`.

- The dependent variables (as defined by `[h,u](t,x)`) are processed in the order given. For each dependent variable we introduce extended variables corresponding to the partial derivatives with respect to `t` and `x` in that order. These extended variables are in turn treated as dependent variables until the reduction to first order is achieved. The resulting coordinate list is `[t,x,h,k1,k2,u,k3,w4]` with k1 and k2 being the partials of h with respect to t and x and w3 and w4 being the partials of u with respect to t and x.

Examples:
```
> with(liesymm):
> eq1 := Diff(h(t,x),x,x) = Diff(h(t,x),t);
```
$$eq1 := Diff(h(t, x), x, x) = \frac{d}{dt} h(t, x)$$

```
> makeforms(eq1,h(t,x),w);
        [d(h) - w1 d(t) - w2 d(x), (d(t) &^ d(w2)) - w1 (d(t) &^ d(x))]
> makeforms(eq1,h(t,x),[p,q]);
        [d(h) - p d(t) - q d(x), (d(t) &^ d(q)) - p (d(t) &^ d(x))]
> annul(",[t,x]);
```
$$\left[\left(\frac{d}{dt} h(t, x) \right) - p(t, x) = 0, \left(\frac{d}{dx} h(t, x) \right) - q(t, x) = 0, \right.$$
$$\left. \left(\frac{d}{dx} q(t, x) \right) - p(t, x) = 0 \right]$$

```
> eq3 := Diff(U(t,r),r,r) + 1/r* Diff(U(t,r),r) - Diff(U(t,r),t,t)
> = exp(-2*U(t,r))* (Diff(C(t,r),t)^2 - Diff(C(t,r),r)^2);
```
$$eq3 := Diff(U(t, r), r, r) + \frac{\frac{d}{dr} U(t, r)}{r} - Diff(U(t, r), t, t) =$$
$$exp(-2 U(t, r)) \left(\left(\frac{d}{dt} C(t, r) \right)^2 - \left(\frac{d}{dr} C(t, r) \right)^2 \right)$$

```
> eq4 := Diff(C(t,r),r,r) + 1/r* Diff(C(t,r),r) - Diff(C(t,r),t,t)
> = 2*(Diff(C(t,r),r)*Diff(U(t,r),r) - Diff(C(t,r),t)*Diff(U(t,r),t));
```

$$
eq4 := Diff(C(t,\ r),\ r,\ r) + \frac{\frac{d}{dr} C(t,\ r)}{r} - Diff(C(t,\ r),\ t,\ t) =
$$

$$
2 \left(\frac{d}{dr} C(t,\ r)\right)\left(\frac{d}{dr} U(t,\ r)\right) - 2 \left(\frac{d}{dt} C(t,\ r)\right)\left(\frac{d}{dt} U(t,\ r)\right)
$$

```
> makeforms([eq3,eq4],[U(t,r),C(t,r)],[A,B,F,G]);
[d(U) - A d(t) - B d(r), d(C) - F d(t) - G d(r),
```

$$
(d(t)\ \&^\wedge\ d(B)) + (d(r)\ \&^\wedge\ d(A)) + \left(B/r - \frac{F^2}{\exp(U)^2} + \frac{G^2}{\exp(U)^2}\right)(d(t)\ \&^\wedge\ d(r)),
$$

$$
(d(t)\ \&^\wedge\ d(G)) + (d(r)\ \&^\wedge\ d(F)) + (G/r - 2\ G\ B + 2\ F\ A)\ (d(t)\ \&^\wedge\ d(r))]
$$

```
> annul(",[t,r]);
```

$$
[\left(\frac{d}{dt} U(t,\ r)\right) - A(t,\ r) = 0,\quad \left(\frac{d}{dr} U(t,\ r)\right) - B(t,\ r) = 0,
$$

$$
\left(\frac{d}{dt} C(t,\ r)\right) - F(t,\ r) = 0,\quad \left(\frac{d}{dr} C(t,\ r)\right) - G(t,\ r) = 0,
$$

$$
\left(\frac{d}{dr} B(t,\ r)\right) - \left(\frac{d}{dt} A(t,\ r)\right) + \frac{B(t,\ r)}{r} - \frac{F(t,\ r)^2}{\exp(U(t,\ r))^2} + \frac{G(t,\ r)^2}{\exp(U(t,\ r))^2} = 0,
$$

$$
\left(\frac{d}{dr} G(t,\ r)\right) - \left(\frac{d}{dt} F(t,\ r)\right) + \frac{G(t,\ r)}{r} - 2\ G(t,\ r)\ B(t,\ r)
$$

$$
+ 2\ F(t,\ r)\ A(t,\ r) = 0\qquad\qquad]
$$

SEE ALSO: `with`, `liesymm`, `setup`, `makeforms`, `close`, `determine`, `annul`

5.2.13   liesymm[mixpar] – order the mixed partials

Calling Sequence:

   `mixpar(expr)`

Parameters:

   expr   –   A term involving `diff` or `Diff`

Synopsis:
- This routine is part of the `liesymm` package and is loaded via `with(liesymm)` .

- Given an expression involving `Diff` or `diff`, `expr` reorder the partials to an "address" ordering so that mixed partials can be recognized as equivalent.

Examples:
```
> with(liesymm): setup(x,y,z):
  mixpar( Diff(f(x,y),x,y) );
```
$$\longrightarrow \qquad \frac{d^2}{dx\,dy}\,f(x,\ y)$$

```
> mixpar( Diff(Diff(f(x,y),x),y) );
```
$$\longrightarrow \qquad \frac{d^2}{dx\,dy}\,f(x,\ y)$$

```
> mixpar( Diff(Diff(f(x,y),y),x) );
```
$$\longrightarrow \qquad \frac{d^2}{dx\,dy}\,f(x,\ y)$$

SEE ALSO: `with`, `liesymm`, `setup`, `` `&^` `` , `d` , `Lie`

5.2.14   liesymm[setup] – define the coordinates

Calling Sequence:

> setup(*coords*)

Parameters:

> *coords* – An expression sequence, list or set of strings to be used as names for coordinates

Synopsis:
- This routine is part of the `liesymm` package and is loaded via `with(liesymm)` .

- It returns a list of the coordinates to be used in constructing wedge products and defining the actions of functions such as `d()`.

- If no argument is given, then all existing coordinates are removed.

- As part of its action, the remember tables of functions such as `d()`, `` `&^` ``, `Lie()`, `hook()`, `makeforms()`, and `wcollect()` are cleared in preparation for any new definitions. It is advisable that setup be used between problems so as to keep the underlying coordinate list to a minimum size.

Examples:
```
> with(liesymm):
  setup();
```
$$\longrightarrow \qquad []$$

> d(x+y+t);	$\longrightarrow$	0
> setup(x,y,z);	$\longrightarrow$	[x, y, z]
> a := d(x+y+t);	$\longrightarrow$	a := d(x) + d(y)
> setup();	$\longrightarrow$	[]
> a;	$\longrightarrow$	0

SEE ALSO: with, liesymm, `&^`, Lie , d , makeforms , close

5.2.15   liesymm[value] – force evaluation of derivatives

Calling Sequence:

value(f)

Parameters:

f – Any expression involving Diff

Synopsis:

- This routine is part of the liesymm package and is loaded via with(liesymm) .

- For display purposes, routines in this package producing results involving the derivatives of unevaluated function calls are returned using an unevaluated Diff(). The value() command replaces calls to Diff() by calls to diff(), thereby causing normal evaluation to take place.

- Note that under some circumstances diff() and Diff() will display the same way. To determine which of these two is actually present, use lprint().

Examples:

```
> with(liesymm):
  Diff(f(x),x);
```
$\longrightarrow$
$$\frac{d}{dx} f(x)$$

```
> value(");
```
$\longrightarrow$
$$\frac{d}{dx} f(x)$$

```
> Diff(f(x,y),x);
```
$\longrightarrow$
$$\frac{d}{dx} f(x, y)$$

```
> value(");
```
$\longrightarrow$
$$D[1](f)(x, y)$$

SEE ALSO: with, liesymm, dsolve, diff

5.2.16   liesymm[wcollect] – regroup the terms as a sum of products

Calling Sequence:
    wcollect(*expr*)

Parameters:
    expr   –   An expression involving wedge products

Synopsis:
- This routine is part of the `liesymm` package and is loaded via `with(liesymm)`.

- The expression is rewritten in a sum of products form with each distinct wedge product or 1-form occurring exactly once. As the arguments to a wedge product are sorted into "address" order and coefficients are automatically extracted, (d(x) &^ d(t)) is recognized as being the same as - (d(t) &^ d(x)).

Examples:
```
> with(liesymm): setup(x,y,z):
> wcollect( a*(d(x) &^ d(y)) + b*(d(x) &^ d(y)));
                        (a + b) (d(x) &^ d(y))
> wcollect( a*(d(x) &^ d(y)) + b*(d(y) &^ d(x)));
                        (a - b) (d(x) &^ d(y))
```

SEE ALSO: `with`, `liesymm`, `setup`, `` `&^` ``, `d`, `Lie`

5.2.17   liesymm[wdegree] – compute the wedge degree of a form

Calling Sequence:
    wdegree(*expr*)

Parameters:
    expr   –   A differential form

Synopsis:
- This routine is part of the `liesymm` package and is loaded via `with(liesymm)`.

- It returns the wedge degree of a differential form.

- It is defined relative to a given coordinate system, so use of setup() may change its value.

- As the list of coordinates fully defines the one-forms, the value of this routine is just the number of terms in the basic wedge product (that is, `nops( getform( expr ))`).

Examples:
```
> with(liesymm):
  setup(x,y,z);              ⟶              [x, y, z]
> wdegree(0);                ⟶                 -1
```

`> wdegree(x);`	$\longrightarrow$	0
`> wdegree( x^2*d(x));`	$\longrightarrow$	1
`> a := x^2*(d(x) &^ d(y) &^ d(z)) ;`	$\longrightarrow$	2

$$a := x \;\; \&\hat{}(d(x), d(y), d(z))$$

`> wdegree(a);`	$\longrightarrow$	3

SEE ALSO: `with`, `liesymm`, `` `&^` ``, `d`

5.2.18   liesymm[wedgeset] – find the coordinate set

Calling Sequence:
   `wedgeset(n)`

Parameters:
   n – A non-negative integer

Synopsis:
- This routine is part of the `liesymm` package and is loaded via `with(liesymm)` .

- All forms of the indicated degree over the underlying coordinates are returned in an expression sequence.

Examples:

`> with(liesymm): setup(x,y);`	$\longrightarrow$	`[x, y]`
`> wedgeset( 0 );`	$\longrightarrow$	`x, y`
`> wedgeset( 1 );`	$\longrightarrow$	`d(x), d(y)`
`> wedgeset( 2 );`	$\longrightarrow$	`d(x) &^ d(y)`
`> setup();`	$\longrightarrow$	`[]`
`> wedgeset(0);` `  setup(a,b,c);`	$\longrightarrow$	`[a, b, c]`
`> wedgeset(0);`	$\longrightarrow$	`a, b, c`

SEE ALSO: `with`, `liesymm`, `setup`, `` `&^` ``, `d` , `Lie`

5.2.19   liesymm[wsubs] – replace part of a wedge product

Calling Sequence:
   `wsubs(eqn, expr)`
   `wsubs(lst, expr)`

Parameters:

    expr   –   An expression involving wedgeproducts

    eqn   –   An expression of the form wedgeprod = expression

    lst   –   A list or set of expressions like `eqn`

Synopsis:

- This routine is part of the `liesymm` package and is loaded via `with(liesymm)` .

- The routine `wsubs()` is analogous to `powsubs()` but is for wedge products. One or more equations specify replacements that are to be made.

- The replacements occur even if only a subset of the `&^` arguments match.

- If two or more substitutions are specified, they are completed in the order specified, even if they are given as a set. Simultaneous substitution (as in subs()) has not been implemented.

Examples:

```
> with(liesymm): setup(x,y,z,t):
  wsubs( d(x) = d(z), a*(d(x) &^ d(y)) + b*
  (d(x) &^ d(y)));                          ⟶      - a (d(y) &^ d(z)) - b (d(y) &^ d(z))
> getform(a*(d(x) &^ d(y)));                ⟶                    d(x) &^ d(y)
> wsubs( " = d(x) &^ d(t) , d(x) &^ d(y)
  &^ d(z));                                 ⟶              - &^(d(x), d(z), d(t))
```

SEE ALSO: `with`, `liesymm`, `setup`, `` `&^` `` , `d` , `Lie` ,`&mod`

5.2.20 &mod – reduce a form modulo an exterior ideal

Calling Sequence:

    f **&mod** *formlist*

Parameters:

    f   –   A differential form

    formlist   –   A list or set of differential forms

Synopsis:

- This routine is part of the `liesymm` package and is loaded via `with(liesymm)` .

- Reduce the form *f* modulo the exterior ideal generated by `formlist`. Use a closed `formlist` to specify a differential ideal.

Examples:

```
> with(liesymm):
  setup(t,x,u,w1,w2);                       ⟶              [t, x, u, w1, w2]
> a1 := d(u) - w1*d(t) - w2*d(x);           ⟶       a1 := d(u) - w1 d(t) - w2 d(x)
```

```
> a2 := (w2+u^2) * (d(x) &^ d(t)) - d(w2)
  &^ d(x);                                    ⟶                        2
                                                  a2 := - (w2 + u ) (d(t) &^ d(x))

                                                        + (d(x) &^ d(w2))

> (d(u) &^ d(t)) &mod [a1,a2];                ⟶          - w2 (d(t) &^ d(x))

> (d(u) &^ d(t)) &mod close([a1,a2]);         ⟶          - w2 (d(t) &^ d(x))
```

SEE ALSO: with, liesymm, setup, Lie, d, `&^`, close

5.2.21   &^ – the wedge product

Calling Sequence:

  a &^ b

  `&^`(a, b, c)

Parameters:

  a, b, c –   An expression involving differential forms relative to specific
coordinates

Synopsis:

- This routine is part of the liesymm package and is loaded via with(liesymm) .

- It computes the wedge product of differential forms relative to the coordinates defined by setup().

- All 1-forms are generated by applying d() to the coordinates.

- All wedge products are automatically simplified to a wedge product of n 1-forms by extracting coefficients of wedge degree 0.

- All results of a wedge product are reported using an address ordering of the 1-forms to facilitate simplifications. Thus d(y) &^ d(x) may simplify to - (d(x) &^ d(y))   and if so will do so consistently within a given session.

- The ordering used for simplifications of the products of 1-forms is available as { wedgeset(1)}.

Examples:
```
> with(liesymm):
  setup(x,y,z);                              ⟶                [x, y, z]

> d(t) &^ d(x);                              ⟶                    0

> d(x) &^ d(y);                              ⟶               d(x) &^ d(y)

> (5*d(x)) &^ d(y) &^ (3* d(z));             ⟶          15 &^(d(x), d(y), d(z))

> `&^`(a*d(x), b*d(y), c*d(z));              ⟶         a b c &^(d(x), d(y), d(z))
```

SEE ALSO: with, liesymm, setup, wedgeset, `&^`, Lie , hook

5.3 The Newman-Penrose Formalism Package

5.3.1 Introduction to the np package

Calling Sequence:

np[*function*] (args)

function (args)

Synopsis:

- To use an np function, you must first define all the functions in the Newman-Penrose package by typing with(np).

- The functions available are:

conj   D   eqns   suball
V   V_D   X   X_D
X_V   Y   Y_D   Y_V
Y_X

- For more information on the Pfaffian operators see np,Pfaffian.

- For more information on the commutators see np,commutators.

- For help with any function in this package, see np,*function*

- WARNING: This package uses global names for the various NP quantities; it is essential to be wary of unintentional conflicts. For a list of all global names used, after with(np) display the value of `np/globals`.

Examples:
```
> with(np):
> eq := k*r = r^2 + e*p:
> D(eq);
          k D(r) + r D(k) = 2 r D(r) + e D(p) + p D(e)
```

5.3.2 Newman-Penrose commutators – V_D, X_D, Y_D, X_V, Y_V, Y_X

Synopsis:

- The functions V_D, X_D, Y_D, X_V, Y_V, Y_X correspond to the commutation relations associated with the Lie brackets [V,D], [X,D], and so forth. They return an equation whose left-hand side is the Lie bracket, and whose right-hand side is its known linear expansion in terms of the NP derivatives.

Examples:
```
> with(np):
> X_D( p );
 X(D(p)) - D(X(p)) = (ac + b - pc) D(p) + k V(p) - (rc + e - ec) X(p) - s Y(p)
```

5.3.3   Newman-Penrose Pfaffian operators – D, V, X, Y

Synopsis:

- The functions D, V, X, and Y correspond to the Pfaffians associated with the null tetrad vectors
  l, n, m, and $\bar{m}$, respectively. They are simply applied as functions, and behave as differential
  operators on expressions that are commonly encountered in the formalism (such as polynomials
  and rational functions in the NP quantities).

- These functions use a global variable const (initially $\{I\}$), which allows the marking of special
  names as constants.

Examples:
```
> with(np):
> eq1 := D(r) = r^2 + (e + ec)*p:
> V( eq1 );
            V(D(r)) = 2 r V(r) + (e + ec) V(p) + p (V(e) + V(ec))

> const := const union {L}: X( s^2 + R01 + L );
                           2 s X(s) + X(R01)
```

5.3.4   np[conj] – Newman-Penrose complex conjugation operator

Calling Sequence:

> conj(*expr*)

Synopsis:

- The function conj computes the complex conjugate of an object in the NP formalism according
  to the usual rules. The value of conj(*expr*) is output as exprc for readability; therefore one must
  realize that the value will thereafter be treated as an independent quantity (except by the conj
  function!). The (Hermitian) Ricci spinor is treated accordingly.

- In addition, the conj function uses the sets real_fcns and imag_fcns to identify real and imaginary quantities (e.g. the operators D, V, and the constant I).

Examples:
```
> with(np):
  eq1 := D(r) = r^2 + (e + ec)*p:
  conj( eq1 );
```
$$\longrightarrow \qquad D(rc) = rc^2 + (e + ec)\ pc$$

```
> real_fcns := real_fcns union {W2}: conj(                              2
    s^2 + R01 + W2 );                        ⟶           sc  + R10 + W2
```

5.3.5   np[eqns] – initialization and display of NP equations

Synopsis:

- The function eqns is used to initialize and/or display the set of "basic" equations, which are those named eq.k where $1 \leq k \leq$ neqs. An initial call of the form eqns(); causes the equations eq1 through eq29 to be assigned (but not printed), and the variable neqs to be initialized to 29.

- Once the equations have been initialized, a sequential subset of them is printed by typing eqns(a,b); (to display eq.a to eq.b), or eqns(); to display eq1 to eq.neqs (whatever is the current value of neqs). Note that the user must reset neqs when the set of basic equations is extended.

Examples:
```
> with(np): k := 0: kc := 0: l := 0: lc := 0: eqns(): neqs;
                              29

> eqns(2,3);
                (eq2), D(s) = (r + rc) s + (3 e - ec) s + W0

        (eq3), D(t) = (t + pc) r + (tc + p) s + (e - ec) t + W1 + R01
```

5.3.6   np[suball] – substitution utility for Newman-Penrose package

Synopsis:

- The function suball attempts to make substitutions from the set of basic equations and their conjugates. The call suball(expr); will use all equations in the range 1..neqs; a range may also be given as in suball(expr,5,11);. In order to function properly, the user may need to rearrange the equations so that a single, atomic quantity appears on the left-hand side. (See also help(subs).)

- Note: this will be slow if the range of equations is large; it helps to examine the equations first, and use limited ranges for substitution.

Examples:
```
> with(np):
> eqns():  k := 0 : kc := 0 :
> eqns( 1, 1 );
                              2
                (eq1), D(r) = r  + s sc + (e + ec) r + R00

> suball(D(r-rc),1,1): factor( " );
                    - (- r + rc) (rc + e + ec + r)
```

6
Packages for Geometry

6.1 The Euclidean Geometry Package

6.1.1 Introduction to the geometry package

Calling Sequence:

    function(args)

    geometry[*function*](args)

Synopsis:

- The functions in this package deal with two-dimensional Euclidean geometry.

- To use this package, use the command with(geometry) to define all geometry functions.

- The functions available are:

altitude	Appolonius	area	are_collinear
are_concurrent	are_harmonic	are_orthogonal	are_parallel
are_perpendicular	are_similar	are_tangent	bisector
center	centroid	circumcircle	conic
convexhull	coordinates	detailf	diameter
distance	ellipse	Eulercircle	Eulerline
excircle	find_angle	Gergonnepoint	harmonic
incircle	inter	inversion	is_equilateral
is_right	make_square	median	midpoint
Nagelpoint	onsegment	on_circle	on_line
orthocenter	parallel	perpendicular	perpen_bisector
polar_point	pole_line	powerpc	projection
radius	rad_axis	rad_center	randpoint
reflect	rotate	sides	similitude
Simsonline	square	symmetric	tangent
tangentpc			

- For help with a particular function see the information under **geometry**[*function*], where *function* is one from the above list.

- The information available under **geometry[point]**, **geometry[line]**, **geometry[circle]**, and **geometry[triangle]** describes how to define points, lines, circles, and triangles.

- In this package, **x** and **y** are used as global names for the coordinates of points and in the equations of lines and circles.

- For example, to define the point **A** with coordinates 1 and 2, one would execute the following sequence of commands:

Examples:
```
> with(geometry):
  point(A, [1, 2]);
```
                                        $\longrightarrow$ A

SEE ALSO: **projgeom**, **geometry[type]**, **geometry[point]**, **geometry[line]**, **geometry[circle]**, **geometry[triangle]**, **geometry**[*function*]

6.1.2   geometry[altitude] – find the altitude of a given triangle

Calling Sequence:
    altitude(*ABC*, *A*, *hA*)

Parameters:
    ABC – a triangle
    hA   – the *A*-altitude of *ABC*
    A   – a vertex of *ABC*

Synopsis:
- This function is part of the **geometry** package, and so can be used in the form **altitude(..)** only after performing the command **with(geometry)** or **with(geometry, altitude)**.

- The vertices of the triangle must be known.

- Use **op** or **detailf** to have more details on the altitude.

Examples:
```
> with(geometry):
  triangle(ABC, [point(A,0,0), point(B,2,0)
  , point(C,1,3)]):
  altitude(ABC, A, hA);
```
                                        $\longrightarrow$ hA
```
> type(hA, line2d);
```
                                        $\longrightarrow$ true

---

SEE ALSO: **geometry[median]** and **geometry[bisector]**

6.1.3 geometry[Appolonius] – find the Appolonius circles of three given circles

Calling Sequence:
>   Appolonius(*c1*, *c2*, *c3*)

Parameters:
>   *c1, c2, c3* – three circles

Synopsis:
- This function is part of the **geometry** package, and so can be used in the form `Appolonius(..)` only after performing the command `with(geometry)` or `with(geometry, Appolonius)`.

- Return the set of Appolonius circles. In general, there are eight circles.

- The coordinates of the centers and the radii of the circles must be numeric.

6.1.4 geometry[area] – compute the area of a triangle, square, or circle

Calling Sequence:
>   area(*g*)

Parameters:
>   *g* – a triangle, a square, or a circle.

Synopsis:
- This function is part of the **geometry** package, and so can be used in the form `area(..)` only after performing the command `with(geometry)` or `with(geometry, area)`.

- The sign of the area does not depend on the order of the vertices of the triangle.

Examples:
```
> with(geometry):
> triangle(ABC, [point(A,0,0), point(B,2,0), point(C,1,3)]):
> area(ABC);
                            3
```

6.1.5 geometry[are_collinear] – test if three points are collinear

Calling Sequence:
>   are_collinear(*P*, *Q*, *R*)

Parameters:
>   *P, Q, R* – three points

Synopsis:
- This function is part of the **geometry** package, and so can be used in the form `are_collinear(..)` only after performing the command `with(geometry)` or `with(geometry, are_collinear)`.

- The output will be `true`, `false`, or a `condition`.

Examples:
```
> with(geometry):
  point(A, 0, 0), point(B, 2, 0), point(C,
  -3, 0), point(D, 2, 3):
  point(M, mx, my):
  are_collinear(A, B, C);
```
$\longrightarrow$ true

```
> are_collinear(A, C, D);
```
$\longrightarrow$ false

```
> are_collinear(A, D, M);
```
$\longrightarrow$ THE POINTS ARE COLLINEAR IF :

                                         2 my - 3 mx = 0

SEE ALSO: `geometry[are_concurrent]`

6.1.6   geometry[are_concurrent] – test if three lines are concurrent

Calling Sequence:
   `are_concurrent(`*l1*, *l2*, *l3*`)`

Parameters:
   l1, l2, l3   –   three lines

Synopsis:
- This function is part of the `geometry` package, and so can be used in the form `are_concurrent(..)` only after performing the command `with(geometry)` or `with(geometry, are_concurrent)`.

- The output will be `true`, `false`, or a `condition`.

Examples:
```
> with(geometry):
  line(l1,[x = 0]), line(l2,[y=0]),
  line(l3,[x = y]), line(l4,[x+y=1]):
  are_concurrent(l1, l2, l3);
```
$\longrightarrow$ true

```
> are_concurrent(l1, l2, l4);
```
$\longrightarrow$ false

SEE ALSO: `geometry[are_collinear]`

6.1.7   geometry[are_harmonic] – test if a pair of points is harmonic conjugate to another pair of points

Calling Sequence:
   `are_harmonic(`*A*, *B*, *C*, *D*`)`

Parameters:
    A, B, C, D –   four points

Synopsis:
- This function is part of the **geometry** package, and so can be used in the form **are_harmonic(..)** only after performing the command **with(geometry)** or **with(geometry, are_harmonic)**.

- The output will be **true**, **false**, or a **condition**.

- The given points must first be collinear.

Examples:
```
> with(geometry):
  point(A, 0, 0), point(B, 3, 3), point(C,
  7, 7), point(D, 21/11, 21/11):
  are_harmonic(A, B, C, D);              ⟶              true

> are_harmonic(B, A, C, D);             ⟶              true

> are_harmonic(A, C, B, D);             ⟶              false
```

SEE ALSO: geometry[harmonic]

6.1.8   geometry[are_orthogonal] – test if two circles are orthogonal to each other

Calling Sequence:
    are_orthogonal($c1$, $c2$)

Parameters:
    $c1$, $c2$ –   two circles

Synopsis:
- This function is part of the **geometry** package, and so can be used in the form **are_orthogonal(..)** only after performing the command **with(geometry)** or **with(geometry, are_orthogonal)**.

- The output will be **true**, **false**, or a **condition**.

Examples:
```
> with(geometry):
  circle(c1,[x^2 + y^2 =1]), circle(c2,
  [(x-2)^2 + y^2 = 2]):
  circle(c3, [x^2 + y^2 = 2]):
  are_orthogonal(c1,c2);               ⟶              false

> are_orthogonal(c2,c3);              ⟶              true
```

6.1.9   geometry[are_parallel] − test if two lines are parallel to each other

Calling Sequence:

    are_parallel(*l1*, *l2*)

Parameters:

  l1, *l2*   −   two lines

Synopsis:

- This function is part of the **geometry** package, and so can be used in the form **are_parallel(..)** only after performing the command **with(geometry)** or **with(geometry, are_parallel)**.

- The output will be **true**, **false**, or a **condition**.

Examples:
```
> with(geometry):
  line(l1, [x = 0]), line(l2, [y = 0]),
  line(l3, [x = 2]):
  are_parallel(l1, l2);                          ⟶              false
> are_parallel(l1, l3);                          ⟶              true
```

SEE ALSO: geometry[are_perpendicular]

6.1.10   geometry[are_perpendicular] − test if two lines are perpendicular to each other

Calling Sequence:

    are_perpendicular(*l1*, *l2*)

Parameters:

  l1, *l2*   −   two lines

Synopsis:

- This function is part of the **geometry** package, and so can be used in the form **are_perpendicular(..)** only after performing the command **with(geometry)** or **with(geometry, are_perpendicular)**.

- The output will be **true**, **false**, or a **condition**.

Examples:
```
> with(geometry):
  line(l1, [x = 0]), line(l2, [y = 0]),
  line(l3, [x = 2]):
  are_perpendicular(l1, l2);                     ⟶              true
> are_perpendicular(l1, l3);                     ⟶              false
```

SEE ALSO: geometry[are_parallel]

6.1.11   geometry[are_similar] – test if two triangles are similar

Calling Sequence:
   are_similar($T1$, $T2$)

Parameters:
   $T1$, $T2$ –   are given triangles

Synopsis:
- This function is part of the **geometry** package, and so can be used in the form **are_similar(..)** only after performing the command **with(geometry)** or **with(geometry, are_similar)**.

- The output will be **true**, **false**, or a **condition**.

Examples:
```
> with(geometry):
  point(A,0,0),point(B,0,3),point(C,1,0),
  point(D,0,6),point(F,2,0):
  point(G,3,1):
  triangle(T1, [A, B, C]):
  triangle(T2, [A, D, F]):
  triangle(T3, [A, D, G]):
  are_similar(T1, T2);                    ⟶              true
> are_similar(T1, T3);                    ⟶              false
```

6.1.12   geometry[are_tangent] – test if a line and a circle or two circles are tangent to each other

Calling Sequence:
   are_tangent(f, g)

Parameters:
   f, g –   a line and a circle or two circles

Synopsis:
- This function is part of the **geometry** package, and so can be used in the form **are_tangent(..)** only after performing the command **with(geometry)** or **with(geometry, are_tangent)**.

- The output will be **true**, **false**, or a **condition**.

Examples:
```
> with(geometry):
> circle(c1,[x^2 + y^2 =1]), circle(c2,[(x-2)^2 + y^2 =1]):
> circle(c3,[(x-2)^2 + (y-2)^2 = r^2]), line(l, [2*x + 3*y =0]):
> are_tangent(c1, c2);
                              true
```

```
> are_tangent(c1, c3);
      THE CIRCLES ARE TANGENT TO EACH OTHER IF ONE OF THE FOLLOWING IS TRUE
                           2                2
                    [7 - r  - 2 r = 0, 7 - r  + 2 r = 0]
> are_tangent(l, c1);
                                    false
```

6.1.13   geometry[bisector] – find the bisector of a given triangle

Calling Sequence:

    bisector(ABC, A, bA)

Parameters:

ABC	–	a triangle
A	–	a vertex of ABC
bA	–	an A-bisector of ABC

Synopsis:

- This function is part of the **geometry** package, and so can be used in the form `bisector(..)` only after performing the command `with(geometry)` or `with(geometry, bisector)`.

- The vertices of the triangle must be known.

- Use `op` or `detailf` to have more details on the bisector.

Examples:
```
> with(geometry):
> triangle(ABC, [point(A,0,0), point(B,2,0), point(C,1,3)]):
> bisector(ABC, A, bA);
                                    bA
```

SEE ALSO: geometry[altitude] and geometry[median]

6.1.14   geometry[circumcircle] – find the circumcircle of a given triangle

Calling Sequence:

    circumcircle(T, cc)

Parameters:

T	–	a triangle
cc	–	the name of the circumcircle

Synopsis:

- This function is part of the **geometry** package, and so can be used in the form `circumcircle(..)` only after performing the command `with(geometry)` or `with(geometry, circumcircle)`.

- The vertices of the triangle must be known.

- Use **detailf** or **op** to have more details on **circumcircle**.

Examples:
```
> with(geometry):
> triangle(T, [point(A,0,0), point(B,2,0), point(C,1,3)]):
> circumcircle(T, Elc);
```
$$Elc$$

6.1.15   geometry[center] – find the center of a given circle

Calling Sequence:

    center(c)

Parameters:

    c –   a circle.

Synopsis:
- This function is part of the **geometry** package, and so can be used in the form **center(..)** only after performing the command **with(geometry)** or **with(geometry, center)**.

- Use **coordinates** or **detailf** for more details on the center.

Examples:
```
> with(geometry):
  circle(c, [(x-3)^2 + (y-1)^2 =4]);          ⟶                    c

> center(c);                                  ⟶              center_c

> coordinates(");                             ⟶                 [3, 1]

> c[center];                                  ⟶              center_c
```

6.1.16   geometry[centroid] – compute the centroid of a triangle or a set or list of points on a plane

Calling Sequence:

    centroid(g, G)

Parameters:

    g –   a triangle, a set of points or a list of points

    G –   the name of the centroid

Synopsis:
- This function is part of the **geometry** package, and so can be used in the form **centroid(..)** only after performing the command **with(geometry)** or **with(geometry, centroid)**.

- To find the centroid of a triangle, the vertices must be known.

- Use `coordinates` or `detailf` for more details on the centroid.

Examples:
```
> with(geometry):
  ps := [point(A,0,0),point(B,2,0),point(C,
  1,3),point(D,1,6)];
```
$\longrightarrow$ `ps := [A, B, C, D]`
```
> centroid(ps,G);
```
$\longrightarrow$ `G`
```
> coordinates(G);
```
$\longrightarrow$ `[1, 9/4]`

6.1.17   geometry[circle] – define the circles

Calling Sequence:
> circle(*c*, *gv*)

Parameters:

> *c*   –   the name of the circle
>
> *gv* –   a list

Synopsis:

- This function is part of the **geometry** package, and so can be used in the form `circle(..)` only after performing the command `with(geometry)` or `with(geometry, circle)`.

- The given list may consist of three points, the equation of the circle, the center and the radius, or two endpoints of the diameter.

- The variables x and y are used as global for the circle equation, so never use x and y for other purposes.

- For more details on the circles, use `op` or `detailf`.

Examples:
```
> with(geometry):
  circle(c1, [point(A, 0, 0), point(B, 2,
  0), point(C, 1, 2)]);
```
$\longrightarrow$ `c1`
```
> c1[form];
```
$\longrightarrow$ `circle`
```
> c1[given];
```
$\longrightarrow$ `[A, B, C]`
```
> circle(c2, [A, 3]);
```
$\longrightarrow$ `c2`
```
> c2[radius];
```
$\longrightarrow$ `3`
```
> c2[equation];
```
$\longrightarrow$ $x^2 + y^2 - 9 = 0$
```
> circle(c3, [A, B, diameter]);
```
$\longrightarrow$ `c3`
```
> c3[radius];
```
$\longrightarrow$ `1`

SEE ALSO: **geometry[line]**, **geometry[point]**, **geometry[triangle]**

6.1.18   geometry[convexhull] – find convex hull enclosing the given points

Calling Sequence:
    convexhull(*ps*)

Parameters:
    ps – a list or set of points

Synopsis:
- This function is part of the **geometry** package, and so can be used in the form `convexhull(..)` only after performing the command `with(geometry)` or `with(geometry, convexhull)`.

- The result is returned as a list of points in counter-clockwise order.

- Uses an `n*log(n)` algorithm computing tangents of pairs of points.

Examples:
```
> with(geometry):
> point(A,[0,0]),point(B,[1,1]),point(C,[2,0]),
> point(D,[1,0]),point(E1,[1,1/2]):
> convexhull({ A, B, C, D, E1});
```
$$[A, C, B]$$

6.1.19   geometry[conic] – find the conic going through five points

Calling Sequence:
    conic(*e*, *gv*)

Parameters:
    e   – the name of the conic
    gv – a list of five points

Synopsis:
- This function is part of the **geometry** package, and so can be used in the form `conic(..)` only after performing the command `with(geometry)` or `with(geometry, conic)`.

- The given list must consist of five points.

- For details on the conic use **op**.

Examples:
```
> with(geometry):
> ps := point(A,0,0),point(B,2,0),point(C,3,2),point(D,2,4),point(M,1,5):
> conic(e1, [A,B,C,D,M]);
```
$$e1$$
```
> e1[form];
```
$$conic$$

```
> e1[equation];
```
$$160\ x + 592\ y - 88\ x\ y - 80\ x^2 - 104\ y^2 = 0$$

SEE ALSO: geometry[ellipse]

6.1.20   geometry[coordinates] – compute the coordinates of a given point

Calling Sequence:
    coordinates(P)

Parameters:
    P –   a point

Synopsis:
- This function is part of the **geometry** package, and so can be used in the form `coordinates(..)` only after performing the command `with(geometry)` or `with(geometry, coordinates)`.
- The output will be represented as a list.

Examples:
```
> with(geometry):
  point(A, 2, 8), point(M, u, v);                  ⟶            A, M
> A[x];                                            ⟶              2
> A[y];                                            ⟶              8
> coordinates(A);                                  ⟶            [2, 8]
> coordinates(M);                                  ⟶            [u, v]
```

6.1.21   geometry[detailf] – give floating-point information about points, circles, and lines.

Calling Sequence:
    detailf(g)

Parameters:
    g –   a point, a circle or a line.

Synopsis:
- This function is part of the **geometry** package, and so can be used in the form `detailf(..)` only after performing the command `with(geometry)` or `with(geometry, detailf)`.
- Give the coordinates of the center of the circle and its radius.
- Give the equation of the line.

• Give the coordinates of the point.

Examples:
```
> with(geometry):
  point(A,sqrt(5-3^(1/3)), 3/4), circle(c,
  [(x-sqrt(2))^2 + y^2 = 1]):
  detailf(A);                    ⟶        [1.886199997, .7500000000]
> detailf(c);                    ⟶        [center = [1.414213562, 0], radius = 1.]
```

6.1.22   geometry[diameter] – compute the diameter of a given set or list of points on a plane.

Calling Sequence:
    diameter(g)

Parameters:
    g – a list or set of points

Synopsis:
• This function is part of the `geometry` package, and so can be used in the form `diameter(..)` only after performing the command `with(geometry)` or `with(geometry, diameter)`.

• To find the diameter of a list or set of points, the coordinates of the points must be numeric.

• The diameter is found with its two endpoints.

Examples:
```
> with(geometry):
> point(A,0,0),point(B,2,0),point(C,1,3),point(D,1,6):
> point(M,sqrt(2),3):
> ps := [A, B, C, D, M]:
> diameter(ps);
                                1/2
                    [A, D, 37    ]
```

SEE ALSO: geometry[convexhull]

6.1.23   geometry[distance] – find the distance between two points or a point with a line.

Calling Sequence:
    distance(P, l)

Parameters:
    P – a point

l – a point or a line

Synopsis:
- This function is part of the **geometry** package, and so can be used in the form **distance(..)** only after performing the command **with(geometry)** or **with(geometry, distance)**.

Examples:
```
> with(geometry):
  point(A,0,0), point(B,2,0), line(l ,[x +
  y = 3]);                                    ⟶           A, B, l
> distance(A, B);                             ⟶              2
> distance(A, l);                             ⟶             1/2
                                                          3/2 2
```

6.1.24   geometry[ellipse] – define the ellipse

Calling Sequence:
    ellipse(e, gv)

Parameters:
    e – the name of the ellipse

    gv – a list

Synopsis:
- This function is part of the **geometry** package, and so can be used in the form **ellipse(..)** only after performing the command **with(geometry)** or **with(geometry, ellipse)**.

- The given list must contain an equation, five points, or a center and the major and minor axis.

- For details on the ellipse use **op**.

Examples:
```
> with(geometry):
  ps := point(A,0,0),point(B,2,0),point(C,
  3,2),point(D,2,4),point(M,1,5):
  ellipse(e1, [A,B,C,D,M]);                   ⟶             e1
> ellipse(e2, [A, 6, 4, x_axis]);            ⟶             e2
> ellipse(e3, [A, 6, 4, y_axis]);            ⟶             e3
> ellipse(e4, [x^2 + y^2/4 = 1]);            ⟶             e4
```

SEE ALSO: geometry[conic]

6.1.25 geometry[Eulercircle] – find the Euler circle of a given triangle

Calling Sequence:
>   Eulercircle(*T*, *Elc*)

Parameters:
>   *T*   –   a triangle
>   *Elc* –   the name of the Euler circle

Synopsis:

* This function is part of the **geometry** package, and so can be used in the form `Eulercircle(..)` only after performing the command `with(geometry)` or `with(geometry, Eulercircle)`.

* The vertices of the triangle must be known.

* Use `detailf` or `op` to have more details on the Euler circle.

Examples:
```
> with(geometry):
> triangle(T, [point(A,0,0), point(B,2,0), point(C,1,3)]):
> Eulercircle(T, Elc);
```
$$Elc$$

6.1.26 geometry[Eulerline] – find the Euler line of a given triangle

Calling Sequence:
>   Eulerline(*T*, *Ell*)

Parameters:
>   *T*   –   a triangle
>   *Ell* –   the name of the Euler line

Synopsis:

* This function is part of the **geometry** package, and so can be used in the form `Eulerline(..)` only after performing the command `with(geometry)` or `with(geometry, Eulerline)`.

* The vertices of the triangle must be known.

* Use `detailf` or `op` to have more details on the Euler line.

Examples:
```
> with(geometry):
> triangle(T, [point(A,0,0), point(B,2,0), point(C,1,3)]):
> Eulerline(T, Ell);
```
$$Ell$$

6.1.27   geometry[excircle] – find three excircles of a given triangle

Calling Sequence:
    excircle(*T*)

Parameters:
    T –   a triangle

Synopsis:
- This function is part of the **geometry** package, and so can be used in the form excircle(..) only after performing the command with(geometry) or with(geometry, excircle).

- To find the excircle of a triangle, the vertices must be known.

- For more details on the excircle use coordinates or detailf or op.

Examples:
```
> with(geometry):
> ps := point(A,0,0),point(B,2,0),point(C,1,3):
> triangle(T, [ps]);
```
$$T$$
```
> excircle(T);
```
            excircle_of_T_A, excircle_of_T_B, excircle_of_T_C

SEE ALSO: geometry[incircle] and geometry[circumcircle]

6.1.28   geometry[find_angle] – find the angle between two lines or two circles

Calling Sequence:
    find_angle(u, v)

Parameters:
    u, v –   two lines or circles

Synopsis:
- This function is part of the **geometry** package, and so can be used in the form find_angle(..) only after performing the command with(geometry) or with(geometry, find_angle).

- The smallest angle between two lines is reported.

- The angle between two circles takes its value in the range 0..Pi.

Examples:
```
> with(geometry):
  line(l1,[x + y = 1]),
  line(l2,[x - y =1]);                      ⟶              l1, l2
> circle(c1,[x^2 + y^2 = 1]),
  circle(c2,[(x-2)^2+y^2=4]);               ⟶              c1, c2
```

> find_angle(l1, l2);	$\longrightarrow$	1/2 Pi
> find_angle(c1, c2);	$\longrightarrow$	arccos(1/4)

SEE ALSO: geometry[distance]

6.1.29   geometry[Gergonnepoint] – find the Gergonne point of a given triangle

Calling Sequence:
   Gergonnepoint(T, G)

Parameters:
   T –   a triangle
   G –   the name of the Gergonne-point

Synopsis:
- This function is part of the **geometry** package, and so can be used in the form Gergonnepoint(..) only after performing the command with(geometry) or with(geometry, Gergonnepoint).

- The vertices of the triangle must be known.

- Use detailf or coordinates to have more details on the Gergonne-point

Examples:
```
> with(geometry):
> triangle(T, [point(A,0,0), point(B,2,0), point(C,1,3)]):
> Gergonnepoint(T, G);
```
                                   G

6.1.30   geometry[harmonic] – find a point which is harmonic conjugate to another point with respect to two other points

Calling Sequence:
   harmonic(C, A, B, D)

Parameters:
   A, B, C –   three points
   D –   the name of a point harmonic conjugate to C

Synopsis:
- This function is part of the **geometry** package, and so can be used in the form harmonic(..) only after performing the command with(geometry) or with(geometry, harmonic).

- The output will be a point D which is harmonic conjugate to C with respect to A and B.

Examples:

```
> with(geometry):
  point(A, 0, 0), point(B, 3, 3), point(C,
  7, 7);                                        ⟶        A, B, C

> harmonic(C, A, B, D);                          ⟶           D

> D[form];                                       ⟶          point

> coordinates(D);                                ⟶        21     21
                                                         [----, ----]
                                                          11     11
```

SEE ALSO: geometry[tharmonic]

6.1.31 geometry[incircle] – find the incircle of a given triangle

Calling Sequence:
> incircle(*T*, *ic*)

Parameters:
> *T* – a triangle
> *ic* – the name of the incircle

Synopsis:
- This function is part of the **geometry** package, and so can be used in the form incircle(..) only after performing the command with(geometry) or with(geometry, incircle).

- The vertices of the triangle must be known.

- Use detailf or op to have more details on the incircle.

Examples:

```
> with(geometry):
> triangle(T, [point(A,0,0), point(B,2,0), point(C,1,3)]):
> incircle(T, inc);
                              inc
```

6.1.32 geometry[inter] – find the intersections between two lines or a line and a circle or two circles.

Calling Sequence:
> inter(*f*, *g*)

Parameters:
> *f*, *g* – the lines or circles

Synopsis:

- This function is part of the `geometry` package, and so can be used in the form `inter(..)` only after performing the command `with(geometry)` or `with(geometry, inter)`.

- It may return zero, one, or two points.

- For more details on the intersections, use `coordinates` or `op` or `detailf`.

Examples:
```
> with(geometry):
  line(l1, [x = 0]), line(l2, [x + y = 1]),
    circle(c, [x^2 + y^2 = 1]);                    ⟶              l1, l2, c
> G := inter(l1, l2);                              ⟶           G := l1_intersect_l2
> H := inter(l2, c);                               ⟶     H := l2_intersect1_c, l2_intersect2_c
> G[form];                                         ⟶                point
> coordinates(G);                                  ⟶               [0, 1]
> coordinates(H[1]);                               ⟶               [0, 1]
> coordinates(H[2]);                               ⟶               [1, 0]
```

6.1.33   geometry[inversion] – find the inversion of a point, line, or circle with respect to a given circle.

Calling Sequence:

    `inversion(`p`, `c`, `q`)`

Parameters:

    p –   a point, line, or circle

    c –   a circle

    q –   the name of the inversion

Synopsis:

- This function is part of the `geometry` package, and so can be used in the form `inversion(..)` only after performing the command `with(geometry)` or `with(geometry, inversion)`.

- For more details on the inversion, use `op` or `detailf`.

Examples:
```
> with(geometry):
  point(A,2,0), circle(c,[x^2+y^2 = 1]),
  line(l, [x = 2]);                                ⟶              A, c, l
> inversion(A, c, A1);                             ⟶                A1
> coordinates(A1);                                 ⟶              [1/2, 0]
> inversion(l, c, l1);                             ⟶                l1
> l1[form];                                        ⟶               circle
```

6.1.34   geometry[is_equilateral] – test if a given triangle is equilateral

Calling Sequence:

    is_equilateral(*ABC*)

Parameters:

    ABC –   a triangle

Synopsis:

- This function is part of the **geometry** package, and so can be used in the form **is_equilateral(..)** only after performing the command **with(geometry)** or **with(geometry, is_equilateral)**.

- The output will be **true**, **false**, or a **condition**.

Examples:

```
> with(geometry):
  triangle(T, [a, a, a]);                    ⟶              T

> triangle(ABC, [point(A,0,0), point(B,2,0)
  , point(C,1,2)]);                          ⟶             ABC

> is_equilateral(T);                         ⟶             true

> is_equilateral(ABC);                       ⟶             false
```

SEE ALSO: **geometry[is_right]**

6.1.35   geometry[is_right] – test if a given triangle is a right triangle

Calling Sequence:

    is_right(*ABC*)

Parameters:

    ABC –   a triangle

Synopsis:

- This function is part of the **geometry** package, and so can be used in the form **is_right(..)** only after performing the command **with(geometry)** or **with(geometry, is_right)**.

- The output will be **true**, **false**, or a **condition**.

Examples:

```
> with(geometry):
  triangle(T, [2, 2, 3]);                    ⟶              T

> triangle(ABC, [point(A,0,0), point(B,2,0)
  , point(C,0,2)]);                          ⟶             ABC

> is_right(T);                               ⟶             false

> is_right(ABC);                             ⟶             true
```

SEE ALSO: **geometry[is_equilateral]**

6.1.36   geometry[line] – define the lines

Calling Sequence:

   line(l, gv)

Parameters:

   l –   the name of the line

   gv –   a list

Synopsis:

- This function is part of the **geometry** package, and so can be used in the form **line(..)** only after performing the command **with(geometry)** or **with(geometry, line)**.

- The list can be two points or a linear equation (or polynomial) in x and y.

- The names x and y are always used in the line-equation as globals, so never use x or y for other purposes.

- For more details on the line, use **op** or **detailf**.

Examples:

```
> with(geometry):
  point(A, 0, 0), point(B, 2, 0);                    ⟶              A, B

> line(ab,[A,B]), line(l,[x+y = 1]),
  line(t,[x + 2*y -1]);                              ⟶              ab, l, t

> ab[form];                                          ⟶              line

> ab[equation];                                      ⟶              2 y = 0

> l[equation];                                        ⟶              x + y - 1 = 0

> t[equation];                                        ⟶              x + 2 y - 1 = 0
```

SEE ALSO: geometry[point], geometry[circle]

6.1.37   geometry[median] – find the median of a given triangle

Calling Sequence:

   median(ABC, A, mA)

Parameters:

   ABC –   a triangle

   A –   a vertex of ABC

   mA –   an A-median of ABC

Synopsis:

- This function is part of the **geometry** package, and so can be used in the form **median(..)** only after performing the command **with(geometry)** or **with(geometry, median)**.

- The vertices of the triangle must be known.

- Use op or detailf to have more details on the median.

Examples:
```
> with(geometry):
> triangle(ABC, [point(A,0,0), point(B,2,0), point(C,1,3)]):
> median(ABC, A, mA);
```
$$mA$$

SEE ALSO: geometry[altitude] and geometry[bisector]

6.1.38   geometry[midpoint] – find the midpoint of segment joining two points

Calling Sequence:
>   midpoint(A, B, C)

Parameters:

    A, B –   two points

    C –   the name of the midpoint

Synopsis:

- This function is part of the **geometry** package, and so can be used in the form midpoint(..) only after performing the command with(geometry) or with(geometry, midpoint).

- Use coordinates or op or detailf for more details on the midpoint.

Examples:
```
> with(geometry):
  point(A,0,0), point(B,2,0);                ⟶             A, B

> midpoint(A, B, C);                         ⟶                C

> coordinates(C);                            ⟶             [1, 0]
```

6.1.39   geometry[make_square] – construct squares

Calling Sequence:
>   make_square(sqr, $given$)

Parameters:

    sqr –   the name of the square

    $given$ –   a list

Synopsis:

- This function is part of the **geometry** package, and so can be used in the form make_square(..) only after performing the command with(geometry) or with(geometry, make_square).

- The list can consist of two adjacent vertices, two opposite vertices or a vertex and the center.

- For more details on the squares, use op.

Examples:
```
> with(geometry):
  point(A,0,0), point(B,2,0), point(C,2,2),
   point(D,0,2), point(M,1,1):
  make_square(s1, [A, B, adjacent]);          ⟶           s1_1, s1_2

> make_square(s2, [A, C, diagonal]);          ⟶                s2

> make_square(s3, [A, center = M]);           ⟶                s3

> s1_2[form];                                 ⟶              square

> s2[diagonal];                               ⟶               1/2
                                                             2 2

> area(s3);                                   ⟶                4
```

SEE ALSO: geometry[square]

6.1.40 geometry[Nagelpoint] – find the Nagel point of a given triangle

Calling Sequence:
  Nagelpoint(*T*, *N*)

Parameters:
  T – a triangle
  N – the name of the Nagel point

Synopsis:
- This function is part of the **geometry** package, and so can be used in the form Nagelpoint(..) only after performing the command with(geometry) or with(geometry, Nagelpoint).

- The vertices of the triangle must be known.

- Use detailf or coordinates to have more details on the Nagel point.

Examples:
```
> with(geometry):
> triangle(T, [point(A,0,0), point(B,2,0), point(C,1,3)]):
> Nagelpoint(T, N);
                                N
```

6.1.41   geometry[on_circle] – test if a point, a list or set of points are on a given circle.

Calling Sequence:

   on_circle(*f*, *c*)

Parameters:

   c – a circle

   f – a point, a list or a set of points

Synopsis:

- This function is part of the **geometry** package, and so can be used in the form on_circle(..) only after performing the command with(geometry) or with(geometry, on_circle).

- The output will be **true**, **false**, or a **condition**.

Examples:

```
> with(geometry):
  circle(c1,[x^2 + y^2 =1]), circle(c2,
  [(x-2)^2 + y^2 =1]), point(A,-1,0):
  on_circle(A, c1);                          ⟶            true
> on_circle(A, c2);                          ⟶            false
```

SEE ALSO: geometry[on_line]

6.1.42   geometry[on_line] – test if a point, a list or a set of points are on a given line.

Calling Sequence:

   on_line(*f*, *c*)

Parameters:

   c – a line

   f – a point, a list or set of points

Synopsis:

- This function is part of the **geometry** package, and so can be used in the form on_line(..) only after performing the command with(geometry) or with(geometry, on_line).

- The output will be **true**, **false**, or a **condition**.

Examples:

```
> with(geometry):
  line(l1, [y = 0]), line(l2, [x + y = 1]),
   point(A, 1/2, 1/2):
  on_line(A, l1);                            ⟶            false
> on_line(A, l2);                            ⟶            true
```

SEE ALSO: geometry[on_circle]

6.1.43 geometry[onsegment] – find the point which divides the segment joining two given points by some ratio.

Calling Sequence:

    onsegment(A, B, k, C)

Parameters:

   A, B – two points

   k – a ratio

   C – the name of the result

Synopsis:

- This function is part of the **geometry** package, and so can be used in the form **onsegment(..)** only after performing the command **with(geometry)** or **with(geometry, onsegment)**.

- Use **op**, **coordinates**, or **detailf** for more details on the point.

Examples:

```
> with(geometry):
  point(A,0,0), point(B,2,0);            ⟶            A, B
> onsegment(A, B, 1, C1);                ⟶            C1
> coordinates(C1);                       ⟶            [1, 0]
> onsegment(A, B, 1/2, C2);              ⟶            C2
> coordinates(C2);                       ⟶            [2/3, 0]
```

6.1.44 geometry[orthocenter] – compute the orthocenter of a triangle or of a set or list of points in a plane.

Calling Sequence:

    orthocenter(g, H)

Parameters:

   g – a triangle

   H – the name of the orthocenter

Synopsis:

- This function is part of the **geometry** package, and so can be used in the form **orthocenter(..)** only after performing the command **with(geometry)** or **with(geometry, orthocenter)**.

- To find the orthocenter of a triangle, the vertices must be known.

- For more details on the orthocenter use **coordinates** or **detailf**.

Examples:
```
> with(geometry):
  ps := point(A,0,0),point(B,2,0),point(C,
  1,3):
  triangle(ABC, [ps]);                    ⟶              ABC
> orthocenter(ABC, H);                    ⟶               H
> coordinates(H);                         ⟶            [1, 1/3]
```

SEE ALSO: geometry[orthocenter]

6.1.45   geometry[parallel] – the line which goes through a given point and is parallel to a given line.

Calling Sequence:
>   parallel(*P*, *l*, *lp*)

Parameters:

> *P* – a point
>
> *l* – a line
>
> *lp* – the name of the result

Synopsis:

- This function is part of the **geometry** package, and so can be used in the form **parallel(..)** only after performing the command **with(geometry)** or **with(geometry, parallel)**.

- Use **detailf** or **op** to have more details on the line.

Examples:
```
> with(geometry):
  point(P, 2 , 3), line(l,[x + y =1]);   ⟶             P, l
> parallel(P, l, lp);                     ⟶              lp
> lp[form];                               ⟶             line
> lp[equation];                           ⟶         x + y - 5 = 0
```

SEE ALSO: geometry[perpendicular]

6.1.46   geometry[perpen_bisector] – find the line through the midpoint of two given points and perpendicular to the line joining them

Calling Sequence:
>   perpen_bisector(*A*, *B*, *l*)

Parameters:

A, B – two points

l – the name of the result

Synopsis:

- This function is part of the `geometry` package, and so can be used in the form `perpen_bisector(..)` only after performing the command `with(geometry)` or `with(geometry, perpen_bisector)`.

- Use `op` or `detailf` for more details on the `perpen_bisector`.

Examples:

```
> with(geometry):
  point(A,0,0), point(B,2,0);                    ⟶              A, B
> perpen_bisector(A, B, l);                      ⟶                l
> l[form];                                       ⟶               line
> l[equation];                                   ⟶            2 x - 2 = 0
```

6.1.47 **geometry[perpendicular]** – **find a line which goes through a given point and is perpendicular to a given line**

Calling Sequence:

perpendicular(P, l, lp)

Parameters:

P – a point

l – a line

lp – the name of the result

Synopsis:

- This function is part of the `geometry` package, and so can be used in the form `perpendicular(..)` only after performing the command `with(geometry)` or `with(geometry, perpendicular)`.

- Use `detailf` or `op` to have more details on the line.

Examples:

```
> with(geometry):
  point(P, 2 , 3), line(l,[x + y =1]);          ⟶               P, l
> perpendicular(P, l, lp);                       ⟶                lp
> lp[form];                                      ⟶               line
> lp[equation];                                  ⟶            x - y + 1 = 0
```

SEE ALSO: `geometry[parallel]`

6.1.48   geometry[point] – define the points

Calling Sequence:

> point(P, Px, Py)
> point(P, [Px, Py])
> point(P, x=Px, y=Py)
> point(P, [x=Px, y=Py])

Parameters:

P –   the name of the point

Px –   the x-coordinate

Py –   the y-coordinate

Synopsis:

- This function is part of the **geometry** package, and so can be used in the form **point(..)** only after performing the command **with(geometry)** or **with(geometry, point)**.

- Never let the coordinates of the points contain **x** or **y**.

- For more details on the points, use **coordinates**, **op**, or **detailf**.

Examples:

```
> with(geometry):
  point(A,0,0), point(B,2,0);                      ⟶                    A, B
> A[x];                                            ⟶                     0
> A[y];                                            ⟶                     0
> coordinates(B);                                  ⟶                   [2, 0]
> coordinates(A);                                  ⟶                   [0, 0]
```

SEE ALSO: geometry[line], geometry[circle]

6.1.49   geometry[polar_point] – determine the polar line of a given point with respect to a given conic or a given circle

Calling Sequence:

> polar_point(A, e, l)

Parameters:

A –   a point

e –   a conic or circle

l –   the name of the polar line

Synopsis:

- This function is part of the **geometry** package, and so can be used in the form **polar_point(..)** only after performing the command **with(geometry)** or **with(geometry, polar_point)**.

- For more details on the polar-line, use op or detailf.

Examples:
```
> with(geometry):
  circle(c, [x^2 + y^2 =1]), ellipse(e,
  [x^2/4 + y^2 =1]), point(A,3,0):
  polar_point(A, c, l1);                    ⟶                    l1

> l1[equation];                             ⟶                3 x - 1 = 0

> polar_point(A, e, l2);                    ⟶                    l2

> l2[equation];                             ⟶               3/4 x - 1 = 0
```

SEE ALSO: geometry[pole_line]

6.1.50   geometry[pole_line] – determine the pole of a given line with respect to a given conic or a given circle.

Calling Sequence:

   pole_line(*l*, *e*, *P*)

Parameters:

   l   –   a line

   e   –   a conic or a circle

   P   –   the name of the pole of the line

Synopsis:

- This function is part of the **geometry** package, and so can be used in the form pole_line(..) only after performing the command with(geometry) or with(geometry, pole_line).

- For more details on the pole, use coordinates, op, or detailf.

Examples:
```
> with(geometry):
  circle(c, [x^2 + y^2 =1]), ellipse(e,
  [x^2/4 + y^2 =1]);                        ⟶                    c, e

> line(l1, [3*x - 1 = 0]), line(l2, [3/4*x
  - 1 = 0]);                                ⟶                   l1, l2

> pole_line(l1, c, P1);                     ⟶                    P1

> coordinates(P1);                          ⟶                   [3, 0]

> pole_line(l2, e, P2);                     ⟶                    P2

> coordinates(P2);                          ⟶                   [3, 0]
```

SEE ALSO: geometry[polar_point]

6.1.51   geometry[powerpc] – power of a given point with respect to a given circle

Calling Sequence:
  powerpc(P, c)

Parameters:
  P – a point
  c – a circle

Synopsis:
- This function is part of the **geometry** package, and so can be used in the form `powerpc(..)` only after performing the command `with(geometry)` or `with(geometry, powerpc)`.

Examples:
```
> with(geometry):
  circle(c, [x^2 + y^2 =1]), point(A,3,0),
  point(B, 6/7,2);                              ⟶            c, A, B
> powerpc(A, c);                                ⟶                8
> powerpc(B, c);                                ⟶               183
                                                                ---
                                                                 49
```

6.1.52   geometry[projection] – find the projection of a given point on a given line

Calling Sequence:
  projection(P, l, Q)

Parameters:
  P – a point
  l – a line
  Q – the name of the projection

Synopsis:
- This function is part of the **geometry** package, and so can be used in the form `projection(..)` only after performing the command `with(geometry)` or `with(geometry, projection)`.

- Use `coordinates`, `op`, or `detailf` to have more details on the projection.

Examples:
```
> with(geometry):
  point(P, 2 , 3), line(l,[x + y =1]);          ⟶             P, l
> projection(P, l, Q);                          ⟶              Q
```

```
> Q[form];                                  ⟶          point
> coordinates(Q);                           ⟶          [0, 1]
```

SEE ALSO: geometry[perpendicular] and geometry[parallel]

6.1.53 geometry[rad_axis] – find the radical axis of two given circles

Calling Sequence:

    rad_axis(*c1*, *c2*, *l*)

Parameters:

    c1, *c2* – two circles

    l   – the name of the radical axis

Synopsis:

- This function is part of the **geometry** package, and so can be used in the form **rad_axis(..)** only after performing the command **with(geometry)** or **with(geometry, rad_axis)**.

- Use **op** or **detailf** to have more details on the radical axis.

Examples:

```
> with(geometry):
  circle(c1, [x^2 + y^2 =1]), circle(c2,
  [point(A,3,3), 4]):
  rad_axis(c1, c2, l);                      ⟶              l

> l[form];                                  ⟶            line

> l[equation];                              ⟶     - 3 + 6 x + 6 y = 0
```

SEE ALSO: geometry[rad_center]

6.1.54 geometry[rad_center] – find the radical center of three given circles

Calling Sequence:

    rad_center(*c1*, *c2*, *c3*, *O*)

Parameters:

    O   – the name of the radical center

    c1, *c2*, *c3* – three circles

Synopsis:

- This function is part of the **geometry** package, and so can be used in the form **rad_center(..)** only after performing the command **with(geometry)** or **with(geometry, rad_center)**.

- Use **coordinates**, **op**, or **detailf** to have more details on the radical center.

Examples:
```
> with(geometry):
  circle(c1, [x^2 + y^2 =1]), circle(c2,
  [point(A,3,3), 4]):
  circle(c3, [(x-2)^2 + y^2 = 9/4]):
  rad_center(c1, c2, c3, O);                  ⟶                    O

> O[form];                                    ⟶                  point

> coordinates(O);                             ⟶             11
                                                           [----, -3/16]
                                                            16
```

SEE ALSO: geometry[rad_axis]

6.1.55   geometry[radius] – compute the radius of a given circle

Calling Sequence:

    radius(c)

Parameters:

    c –   a circle

Synopsis:

• This function is part of the **geometry** package, and so can be used in the form radius(..) only after performing the command with(geometry) or with(geometry, radius).

• This function is equivalent to c[radius].

Examples:
```
> with(geometry):
  circle(ABC, [point(A,0,0), point(B,2,0),
  point(C,1,3)]):
  circle(c, [x^2 + y^2 = 9]):
  radius(ABC);                           ⟶              1/2   1/2
                                               1/9 25      9

> radius(c);                             ⟶                 3
```

6.1.56   geometry[randpoint] – find a random point on a line or a circle

Calling Sequence:

    randpoint(u, v, w)

Parameters:

    u –   a line, a circle or a range

    v –   a range or the name of a random point

w –   the name of a random point

Synopsis:
- This function is part of the **geometry** package, and so can be used in the form **randpoint(..)** only after performing the command **with(geometry)** or **with(geometry, randpoint)**.
- The output will be a point with coordinates in the given range or lying on the given line or circle.

Examples:
```
> with(geometry):
  line(l,[x - y = 1]), circle(c, [x^2 +
  y^2 = 1]):
  randpoint(-1..2, -1..2, P);                    ⟶              P
> type(P, point);                                ⟶            false
> randpoint(1, -2..3, Q);                        ⟶              Q
> randpoint(c, R);                               ⟶              R
```

SEE ALSO: geometry[point]

6.1.57   geometry[reflect] – find the reflection of a given point with respect to a given line

Calling Sequence:
    reflect(P, l, Q)

Parameters:
P –   a point
l –   a line
Q –   the name of the reflected point

Synopsis:
- This function is part of the **geometry** package, and so can be used in the form **reflect(..)** only after performing the command **with(geometry)** or **with(geometry, reflect)**.
- Use **coordinates**, **op**, or **detailf** to have more details on the reflection.

Examples:
```
> with(geometry):
  point(P, 2 , 3), line(l,[x + y =1]);           ⟶             P, l
> reflect(P, l, Q);                              ⟶              Q
> Q[form];                                       ⟶            point
> coordinates(Q);                                ⟶           [-2, -1]
```

SEE ALSO: geometry[rotate] or geometry[symmetric]

6.1.58 geometry[rotate] – find the point of rotation of a given point with respect to a given point

Calling Sequence:

rotate(P, g, co, Q, R)

Parameters:

P, g	–	a point and the angle of rotation
co	–	the direction of rotation
Q	–	the name of the rotated point
R	–	the center of rotation

Synopsis:

- This function is part of the **geometry** package, and so can be used in the form rotate(..) only after performing the command with(geometry) or with(geometry, rotate).

- If the fifth argument is omitted, then the origin will be the center of rotation.

- The third argument must be either clockwise or counterclockwise.

- Use coordinates, op, or detailf to have more details on the rotated point.

Examples:

```
> with(geometry):
  point(P, 2 , 0), point(Q, 1, 0);        ⟶        P, Q
> rotate(P,Pi,counterclockwise,P1);       ⟶        P1
> coordinates(P1);                        ⟶        [-2, 0]
> rotate(P, Pi/2, clockwise,P2, Q);       ⟶        P2
> coordinates(P2);                        ⟶        [1, -1]
```

SEE ALSO: geometry[reflect] or geometry[symmetric]

6.1.59 geometry[sides] – compute the sides of a given triangle or a given square

Calling Sequence:

sides(g)

Parameters:

g	–	a triangle or a square

Synopsis:

- This function is part of the **geometry** package, and so can be used in the form sides(..) only after performing the command with(geometry) or with(geometry, sides).

- The sides of a triangle are represented by a list.

Examples:
```
> with(geometry):
  triangle(ABC, [point(A,0,0), point(B,3,0)
  , point(C,0,4)]):
  triangle(T, [a, b, c]):
  sides(ABC);                              ⟶        [5, 4, 3]
> sides(T);                                ⟶        [a, b, c]
```

6.1.60   geometry[similitude] – find the insimilitude and outsimilitude of two circles

Calling Sequence:

     similitude($c1$, $c2$)

Parameters:

     $c1$, $c2$ –   two circles

Synopsis:

- This function is part of the **geometry** package, and so can be used in the form `similitude(..)` only after performing the command `with(geometry)` or `with(geometry, similitude)`.

- There might be one or two similitudes.

- Use `coordinates`, `op`, or `detailf` to have more details on the similitudes.

Examples:
```
> with(geometry):
> circle(c1, [x^2 + y^2 =1]), circle(c2, [point(A,3,3), 4]):
> circle(c3, [(x-2)^2 + y^2 = 1]):
> similitude(c1,c2);
                in_similitude_of_c1_c2, ex_similitude_of_c1_c2
> similitude(c1,c3);
                      in_similitude_of_c1_c3
```

6.1.61   geometry[Simsonline] – find the Simson line of a given triangle with respect to a given point on the circumcircle of the triangle

Calling Sequence:

     Simsonl(T, N, sl)

Parameters:

     T –   a triangle

     N –   a point on the circumcircle

sl   –   the name of the Simson line

Synopsis:
- This function is part of the **geometry** package, and so can be used in the form Simsonline(..) only after performing the command with(geometry) or with(geometry, Simsonline).

- The vertices of the triangle must be known.

- Use detailf or coord to have more details on the Simson line.

Examples:
```
> with(geometry):
> triangle(T, [point(A,-1,0), point(B,1,0), point(C,0,1)]):
> point(N,1/sqrt(2),1/sqrt(2)):
> Simsonline(T, N, sl);
```
$$sl$$

6.1.62   geometry[square] – define the squares

Calling Sequence:
    square(*sqr*, [*A*, *B*, *C*, *D*])

Parameters:
    sqr   –   the name of the square
    A, B, C, D   –   four points

Synopsis:
- This function is part of the **geometry** package, and so can be used in the form square(..) only after performing the command with(geometry) or with(geometry, square).

- The vertices must be given in the correct order in the list.

- For more details on a square, use op.

Examples:
```
> with(geometry):
  gv := [point(A,0,0), point(B,1,0),
  point(C,1,1),point(D,0,1)]:
  square(sqr, gv);                        ⟶              sqr

> sqr[form];                              ⟶              square

> sqr[diagonal];                          ⟶              1/2
                                                          2
```

SEE ALSO: **geometry[make_square]**

6.1.63   geometry[symmetric] – find the symmetric point of a point with respect to a given point

Calling Sequence:
> symmetric(*P1*, *Q*, *P2*)

Parameters:
> *P1, Q* – two points
>
> *P2*   – the name of the symmetric point

Synopsis:
- This function is part of the **geometry** package, and so can be used in the form **symmetric(..)** only after performing the command **with(geometry)** or **with(geometry, symmetric)**.

- Use **coordinates**, **op**, or **detailf** to have more details on the symmetric point.

Examples:
```
> with(geometry):
  point(P,2,3), point(Q,0,1);                    ⟶           P, Q
> symmetric(P, Q, R);                            ⟶             R
> R[form];                                       ⟶           point
> coordinates(R);                                ⟶          [-2, -1]
```

SEE ALSO: geometry[rotate] or geometry[reflect]

6.1.64   geometry[tangent] – find the tangents of a point with respect to a circle

Calling Sequence:
> tangent(*P*, *c*, *l1*, *l2*)

Parameters:
> *P*   – a point
>
> *c*   – a circle
>
> *l1, l2* – the names of the two tangents

Synopsis:
- This function is part of the **geometry** package, and so can be used in the form **tangent(..)** only after performing the command **with(geometry)** or **with(geometry, tangent)**.

- For more details on the tangent, use **op** or **detailf**.

Examples:
```
> with(geometry):
  point(A, 1, 1), circle(c, [x^2 + y^2 = 1]
  );                                             ⟶            A, c
```

> tangent(A, c, l1, l2);	$\longrightarrow$	l1, l2
> l1[form];	$\longrightarrow$	line
> l2[form];	$\longrightarrow$	line
> l1[equation];	$\longrightarrow$	x - 1 = 0
> l2[equation];	$\longrightarrow$	- y + 1 = 0

SEE ALSO: geometry[tangentpc]

6.1.65   geometry[tangentpc] – find the tangent of a point on a circle with respect to that circle

Calling Sequence:
tangentpc(P, c, l)

Parameters:

P –   a point

c –   a circle

l –   the name of the tangent line

Synopsis:

- This function is part of the **geometry** package, and so can be used in the form tangentpc(..) only after performing the command with(geometry) or with(geometry, tangentpc).

- For more details on the result, use **op** or **detailf**.

Examples:

```
> with(geometry):
  point(A, 1, 0), circle(c, [x^2 + y^2 = 1]
  );
```
	$\longrightarrow$	A, c
> tangentpc(A, c, l);	$\longrightarrow$	l
> l[form];	$\longrightarrow$	line
> l[equation];	$\longrightarrow$	x - 1 = 0

SEE ALSO: geometry[tangent]

6.1.66   geometry[triangle] – define the triangles

Calling Sequence:
triangle(T, gv)

Parameters:

T – the name of the triangle

gv – a list

Synopsis:

- This function is part of the **geometry** package, and so can be used in the form **triangle(..)** only after performing the command **with(geometry)** or **with(geometry, triangle)**.

- The list can consist of three vertices, three sides, two sides and the angle between them or three lines.

- For more details on a triangle, use **op**.

Examples:

```
> with(geometry):
  triangle(ABC, [point(A,0,0), point(B,2,0)
  , point(C,1,2)]);                          ⟶          ABC
> ABC[form];                                 ⟶          triangle
> triangle(T1, [2, 2, 3]);                   ⟶          T1
> T1[given];                                 ⟶          [2, 2, 3]
> triangle(T2, [3, angle = Pi/3, 3]);        ⟶          T2
> T2[form];                                  ⟶          triangle
> triangle(T3,[line(l1,[x=0]),line(l2,[y=0]
  ),line(l3,[x+y=1])]);                      ⟶          T3
```

SEE ALSO: geometry[line], geometry[point], geometry[circle]

6.1.67   geometry[type] – check for a point or a line or a circle

Calling Sequence:

type(A, *geometrytype*)

Parameters:

A – a point, a line or a circle

geometrytype – the name **point2d**, **line2d** or **circle2d**

Synopsis:
- The function returns **true** or **false**.

- These functions are part of the geometry package and can only be used after performing **with(geometry)**.

Examples:

```
> with(geometry):
  point(A,0,0), line(l, [x-y=1]), circle(c,
  [x^2+y^2=1]):
  type(A, point2d);                          ⟶          true
```

> `type(l, line2d);`	$\longrightarrow$	true
> `type(c, circle2d);`	$\longrightarrow$	true
> `type(A, circle2d);`	$\longrightarrow$	false

SEE ALSO: `geometry[circle]`, `geometry[line]` and `geometry[point]`

6.2   The 3-D Geometry Package

6.2.1   Introduction to the geom3d Package

Calling Sequence:

> *function*(`args`)
>
> `geom3d`[*function*]`(args)`

Synopsis:

- The functions in this package deal with three-dimensional geometry.

- This package may be loaded by the command `with(geom3d)`. The package must be loaded before any functions in this package may be used.

- The functions in this package are the following:

angle	area	are_collinear	are_concurrent
are_parallel	are_perpendicular	are_tangent	center
centroid	coordinates	coplanar	distance
inter	midpoint	onsegment	on_plane
on_sphere	parallel	perpendicular	powerps
projection	radius	rad_plane	reflect
sphere	symmetric	tangent	tetrahedron
triangle3d	volume		

- In addition, the functions `point3d`, `line3d`, `plane`, and `sphere` are used to define points, lines, planes, and spheres, respectively.

- For help with a particular function see `geom3d`[*function*], where *function* is one listed above.

- In this package, `x`, `y`, `z`, and `_t` are used globally as names for the coordinates of points and in the equations of lines, planes, and spheres.

- To define the point A with coordinates (1,2,3), one would use the following sequence of commands:

Examples:
```
> with(geom3d):
  point3d(A, [1, 2, 3]);                  ⟶                        A
```

SEE ALSO: geometry, geom3d[type]

6.2.2   geom3d[angle] − find the angle between two lines or two planes or a line and a plane

Calling Sequence:
    angle(*u*, *v*)

Parameters:
    u, *v* −   lines or planes

Synopsis:
- This function is part of the geom3d package, and so can be used in the form angle(..) only after performing the command with(geom3d) or with(geom3d, angle).

- The smallest angle between two lines (planes) is reported.

Examples:
```
> with(geom3d):
  point3d(A,0,0,0), point3d(B,1,0,0),
  point3d(C,0,1,0), point3d(D,0,0,1):
  plane(p1, [x + y + z = 1]), line3d(A_B,
  [A, B]), line3d(C_D, [C, D]):
  plane(p2, [x = 0]):
  angle(p1, p2);                         ⟶                  1/2
                                                  arccos(1/3 3   )

> angle(A_B, C_D);                        ⟶                1/2 Pi

> angle(A_B, p1);                         ⟶                  1/2
                                                  arcsin(1/3 3   )
```

SEE ALSO: geom3d[distance]

6.2.3   geom3d[area] − compute the area of a triangle or the surface of a sphere

Calling Sequence:
    area(*g*)

Parameters:
    g −   a triangle or a sphere.

Synopsis:

- This function is part of the **geom3d** package, and so can be used in the form **area(..)** only after performing the command **with(geom3d)** or **with(geom3d, area)**.

- If the coordinates of the vertices of the triangle are numeric, then the area is always a positive number.

Examples:

```
> with(geom3d):
  triangle3d(ABC,[point3d(A,0,0,0),
  point3d(B,0,0,1),point3d(C,1,0,0)]):
  sphere(s, [A, 1]):
  area(ABC);                              ⟶                        1/2

> area(s);                                ⟶                        4 Pi
```

SEE ALSO: **geom3d[volume]**

6.2.4   geom3d[are_collinear] − test if three points are collinear

Calling Sequence:

    are_collinear(P, Q, R)

Parameters:

    P, Q, R −   three points

Synopsis:

- This function is part of the **geom3d** package, and so can be used in the form **are_collinear(..)** only after performing the command **with(geom3d)** or **with(geom3d, are_collinear)**.

- The output will be **true**, **false**, or a condition.

Examples:

```
> with(geom3d):
> point3d(A,0,0,0),point3d(B,1,0,0),point3d(C,-2,0,0),point3d(D,dx,dy,dz):
> are_collinear(A, B, C);
                                      true

> are_collinear(A, B, D);
              THE POINTS ARE COLLINEAR IF THE FOLLOWING IS TRUE:
                              {- dz = 0, dy = 0}
```

SEE ALSO: **geom3d[are_concurrent]**

6.2.5   geom3d[are_concurrent] – test if three lines are concurrent

Calling Sequence:

are_concurrent(*l1*, *l2*, *l3*)

Parameters:

l1, l2, l3   –   three lines

Synopsis:

- This function is part of the **geom3d** package, and so can be used in the form **are_concurrent(..)** only after performing the command **with(geom3d)** or **with(geom3d, are_concurrent)**.

- The output will be **true**, **false**, or a condition.

Examples:

```
> with(geom3d):
  point3d(A,[0,0,0]),point3d(B,[0,0,1]),
  point3d(C,[0,1,0]),point3d(D,[1,0,0]);          ⟶          A, B, C, D
> line3d(AB,[A,B]): line3d(AC,[A,C]):
  line3d(AD,[A,D]): line3d(BC,[B,C]):
  are_concurrent(AB, AC, AD);                     ⟶               true
> are_concurrent(AB, AC, BC);                     ⟶               false
```

SEE ALSO: geom3d[are_collinear]

6.2.6   geom3d[are_parallel] – test if two lines or two planes or a line and a plane are parallel to each other

Calling Sequence:

are_parallel(*l1*, *l2*)

Parameters:

l1, l2   –   lines or planes

Synopsis:

- This function is part of the **geom3d** package, and so can be used in the form **are_parallel(..)** only after performing the command **with(geom3d)** or **with(geom3d, are_parallel)**.

- The output will be **true**, **false**, or a condition.

Examples:

```
> with(geom3d):
  plane(p, [x + y + z = 1]):
  line3d(l1, [point3d(A, [0, 0, 0]), [1,0,
  2]]):
  line3d(l2, [A, [1, 0, -1]]):
  are_parallel(l1, p);                            ⟶               false
```

```
> are_parallel(l2, p);                      ⟶                    true
> are_parallel(l1, l2);                      ⟶                    false
```

SEE ALSO: geom3d[are_perpendicular]

6.2.7   geom3d[are_perpendicular] – test if two lines or two planes or a line and a plane are perpendicular to each other

Calling Sequence:
    are_perpendicular(*l1*, *l2*)

Parameters:
    l1, l2 –   are lines or planes

Synopsis:
- This function is part of the **geom3d** package, and so can be used in the form are_perpendicular(..) only after performing the command with(geom3d) or with(geom3d, are_perpendicular).

- The output will be **true**, **false**, or a condition.

Examples:
```
> with(geom3d):
  plane(p, [x + y + z = 1]):
  line3d(l1, [point3d(A, [0, 0, 0]), [1,1,
  1]]):
  line3d(l2, [A, [1, 0, -1]]):
  are_perpendicular(l1, p);                   ⟶                    true
> are_perpendicular(l2, p);                   ⟶                    false
> are_perpendicular(l1, l2);                  ⟶                    true
```

SEE ALSO: geom3d[are_parallel]

6.2.8   geom3d[are_tangent] – test if a plane (line) and a sphere or two spheres are tangent to each other

Calling Sequence:
    are_tangent(*f*, *g*)

Parameters:
    f, g –   is a plane (line) and a sphere or two spheres

Synopsis:
- This function is part of the **geom3d** package, and so can be used in the form are_tangent(..) only after performing the command with(geom3d) or with(geom3d, are_tangent).

- The output will be `true`, `false`, or a condition.

Examples:
```
> with(geom3d):
  sphere(s1, [point3d(A, [0, 0, 0]), 1]):
  sphere(s2, [point3d(B, [2, 0, 0]), r]):
  plane(p, [x + y + z = 3^(1/2)]):
  are_tangent(s1, s2);
```
$\longrightarrow$ THE SPHERES ARE TANGENT TO EACH OTHER I\
F ONE OF THE FOLLOWING IS TRUE:

$$[3 - r^2 - 2\,r = 0, \ 3 - r^2 + 2\,r = 0]$$

```
> are_tangent(p, s1);
```
$\longrightarrow$ true

SEE ALSO: geom3d[tangent]

6.2.9   geom3d[center] – find the center of a given sphere

Calling Sequence:

   center(s)

Parameters:

   s – a sphere.

Synopsis:

- This function is part of the `geom3d` package, and so can be used in the form `center(..)` only after performing the command `with(geom3d)` or `with(geom3d, center)`.

- Find the center of a sphere.

- Use `coordinates` or `op` for more detailed information about the center.

Examples:
```
> with(geom3d):
  sphere(s,[x^2+y^2+z^2=1]);
```
$\longrightarrow$ s
```
> center(s);
```
$\longrightarrow$ center_of_s
```
> coordinates(");
```
$\longrightarrow$ [0, 0, 0]
```
> s[center];
```
$\longrightarrow$ center_of_s

SEE ALSO: geom3d[sphere] and geom3d[point3d]

6.2.10   geom3d[centroid] – compute the centroid of a given tetrahedron or a given triangle or a list of points in the space

Calling Sequence:

   centroid(g, G)

Parameters:

 g – a tetrahedron or a triangle or a list of points

 G – the name of the centroid

Synopsis:

- This function is part of the **geom3d** package, and so can be used in the form **centroid(..)** only after performing the command **with(geom3d)** or **with(geom3d, centroid)**.

- For more details on the centroid *G* use **coordinates** or **op**.

Examples:
```
> with(geom3d):
  ps := [point3d(A,0,0,0),point3d(B,1,0,0),
  point3d(C,0,1,0),point3d(D,0,0,1)]:
  tetrahedron(ABCD, ps), triangle3d(BCD,
  [B, C, D]):
  centroid(ps, G1):
  coordinates(G1);                    ⟶         [1/4, 1/4, 1/4]

> centroid(ABCD, G2):
  coordinates(G1);                    ⟶         [1/4, 1/4, 1/4]

> centroid(BCD, G3):
  coordinates(G3);                    ⟶         [1/3, 1/3, 1/3]
```

SEE ALSO: geom3d[coordinates] and geom3d[tetrahedron]

6.2.11 geom3d[coordinates] – return the coordinates of a given point

Calling Sequence:

 coordinates(*P*)

Parameters:

 P – a point

Synopsis:

- This function is part of the **geom3d** package, and so can be used in the form **coordinates(..)** only after performing the command **with(geom3d)** or **with(geom3d, coordinates)**.

- The output will be represented as a list.

Examples:
```
> with(geom3d):
  point3d(A, 2, 8, 6): point3d(M, u, v, w)
  :
  A[x];                               ⟶              2

> A[y];                               ⟶              8

> A[z];                               ⟶              6
```

```
> coordinates(A);                                    [2, 8, 6]
> coordinates(M);                                    [u, v, w]
```

SEE ALSO: geom3d[point3d]

6.2.12   geom3d[coplanar] – test if four points or two lines are coplanar

Calling Sequence:

    coplanar(u, v, w, s)

Parameters:

    u, v, w, s – four points
    u, v   – two lines

Synopsis:

- This function is part of the **geom3d** package, and so can be used in the form coplanar(..) only after performing the command with(geom3d) or with(geom3d, coplanar).

- The output will be **true**, **false**, or a **condition**.

Examples:
```
> with(geom3d):
  point3d(A,0,0,0),point3d(B,1,0,0),
  point3d(C,0,1,0),point3d(D,0,0,1):
  coplanar(A, B, C, D);                                 false
> line3d(ab, [A, B]), line3d(ac, [A, C]):
  coplanar(ab, ac);                                     true
```

SEE ALSO: geom3d[are_collinear]

6.2.13   geom3d[distance] – find the distance between two points or two lines or between a point and a line (plane)

Calling Sequence:

    distance(u, v)

Parameters:

    u – a point or line
    v – a point, a line or a plane

Synopsis:

- This function is part of the **geom3d** package, and so can be used in the form distance(..) only after performing the command with(geom3d) or with(geom3d, distance).

- The points, lines, or plane can be given in any order.

Examples:
```
> with(geom3d):
  point3d(A,0,0,0), point3d(B,1,0,0):
  point3d(C,0,1,0), point3d(D,0,0,1):
  plane(p, [x + y + z = 1]), line3d(A_B,
  [A, B]), line3d(C_D, [C, D]):
  distance(A,B);                          ⟶                        1

> distance(A_B, C_D);                      ⟶                            1/2
                                                              1/2 2

> distance(A, p);                          ⟶                            1/2
                                                              1/3 3
```

SEE ALSO: geom3d[angle]

6.2.14 geom3d[inter] – find the intersections between two lines or planes

Calling Sequence:
>   inter(u, v, w, s)

Parameters:

u, v	–	two lines or two planes
w	–	the name of the intersection or a plane
s	–	the name of the intersection (in the case of three planes)

Synopsis:

- This function is part of the **geom3d** package, and so can be used in the form **inter(..)** only after performing the command **with(geom3d)** or **with(geom3d, inter)**.

- For more details on the intersections, use **coordinates** or **op**.

Examples:
```
> with(geom3d):
  point3d(A,0,0,0), point3d(B,1,0,0):
  plane(p1, [x + y + z = 1]), plane(p2, [x
  - y + z = 0]):
  line3d(A_B,[A, B]):
  inter(p1, p2, l);                        ⟶                        l

> type(l ,line3d);                         ⟶                      true

> inter(A_B, p1, P);                        ⟶                       P

> type(P, point3d);                        ⟶                      true

> coordinates(P);                          ⟶                   [1, 0, 0]
```

6.2.15 geom3d[line3d] – define the lines in three-dimensional space

Calling Sequence:

> line3d(*l*, *gv*)

Parameters:

> *l*   – the name of the line
>
> *gv* – a list

Synopsis:

- This function is part of the **geom3d** package, and so can be used in the form `line(..)` only after performing the command `with(geom3d)` or `with(geom3d, line)`.

- The list can be two points or a point and a directional vector.

- The names x, y, z, _t are always used as globals, so never use x, y, z, _t for other purposes.

- For more details on the line, use **op**.

Examples:
```
> with(geom3d):
  point3d(A,1,1,1), point3d(B,1,2,3):
  line3d(A_B,[A, B]), line3d(l, [A, [0, 1,
  2]]):
  type(l, line3d);                    ⟶                    true
> l[equation];                        ⟶            [1, 1 + _t, 1 + 2 _t]
> A_B[equation];                      ⟶            [1, 1 + _t, 1 + 2 _t]
> A_B[form];                          ⟶                  line3d
> A_B[direction_vector];              ⟶                 [0, 1, 2]
```

SEE ALSO: geom3d[point3d], geom3d[plane]

6.2.16 geom3d[midpoint] – find the midpoint of the segment joining two points

Calling Sequence:

> midpoint(*A*, *B*, *C*)

Parameters:

> *A, B* – two points
>
> *C*   – the name of the midpoint

Synopsis:

- This function is part of the **geom3d** package, and so can be used in the form `midpoint(..)` only after performing the command `with(geom3d)` or `with(geom3d, midpoint)`.

- Use **coordinates** or **op** for more details on the midpoint.

Examples:
```
> with(geom3d):
  point3d(A,0,0,0):
  point3d(B,2,2,2):
  midpoint(A, B, C);                          ⟶                        C

> type(C, point3d);                           ⟶                       true

> coordinates(C);                             ⟶                     [1, 1, 1]
```

SEE ALSO: geom3d[onsegment] and geom3d[symmetric]

6.2.17   geom3d[on_plane] – test if a point, list, or set of points are on a given plane

Calling Sequence:

on_plane(f, c)

Parameters:

c – a plane

f – a point, a list or set of points

Synopsis:

- This function is part of the **geom3d** package, and so can be used in the form on_plane(..) only after performing the command with(geom3d) or with(geom3d, on_plane).

- The output will be **true**, **false**, or a condition.

Examples:
```
> with(geom3d):
  plane(p, [x + y + z = 1]), point3d(A, 1,
  2,1), point3d(B,1/3,1/3,1/3):
  on_plane(A, p);                             ⟶                      false

> on_plane(B, p);                             ⟶                      true
```

SEE ALSO: geom3d[on_sphere]

6.2.18   geom3d[onsegment] – find the point which divides the segment joining two given points by some ratio

Calling Sequence:

onsegment(A, B, k, C)

Parameters:

A, B – two points

k	–	the ratio
C	–	the name of the result

Synopsis:

- This function is part of the **geom3d** package, and so can be used in the form **onsegment(..)** only after performing the command **with(geom3d)** or **with(geom3d, onsegment)**.

- The ratio must not be -1.

- Use **op** or **coordinates** for more details on the point.

Examples:
```
> with(geom3d):
  point3d(A,0,0,0),
  point3d(B,2,0,1);
```
$\longrightarrow$ A, B
```
> onsegment(A, B, 1, C1);
```
$\longrightarrow$ C1
```
> coordinates(C1);
```
$\longrightarrow$ [1, 0, 1/2]
```
> onsegment(A, B, 1/2, C2);
```
$\longrightarrow$ C2
```
> coordinates(C2);
```
$\longrightarrow$ [2/3, 0, 1/3]

6.2.19   geom3d[on_sphere] – test if a point or a list or set of points are on a given sphere

Calling Sequence:

on_sphere(f, c)

Parameters:

c	–	a sphere
f	–	a point, a list or set of points

Synopsis:

- This function is part of the **geom3d** package, and so can be used in the form **on_sphere(..)** only after performing the command **with(geom3d)** or **with(geom3d, on_sphere)**.

- The output will be **true**, **false**, or a condition.

Examples:
```
> with(geom3d):
  sphere(s,[x^2 + y^2 + z^2 = 1]),
  point3d(A,-1,0,0), point3d(B,1,2,1):
  on_sphere(A, s);
```
$\longrightarrow$ true
```
> on_sphere(A, s);
```
$\longrightarrow$ true

SEE ALSO: geom3d[on_plane] and geom3d[sphere]

6.2.20   geom3d[parallel] – find a plane or line going through a given point or line, and parallel to a given line or plane

Calling Sequence:

   parallel(u, v, w)

Parameters:

   u –   a point

   v –   a line or plane

   w –   the name of the result

Synopsis:

- This function is part of the **geom3d** package, and so can be used in the form `parallel(..)` only after performing the command `with(geom3d)` or `with(geom3d, parallel)`.

- Use **op** to have more details on the line or the plane.

Examples:
```
> with(geom3d):
> point3d(A,0,0,0),point3d(B,1,0,0),point3d(C,0,1,0),point3d(D,0,0,1):
> plane(p, [x + y + z = 1]), line3d(A_B,[A, B]), line3d(C_D, [C, D]):
> parallel(A, p, p1):
> p1[equation];
```
$$x + y + z = 0$$
```
> parallel(A_B, C_D, p2):
> p2[equation];
```
$$- y - z = 0$$

SEE ALSO: geom3d[perpendicular]

6.2.21   geom3d[perpendicular] – find a plane or line going through a given point or line, and perpendicular to a given line or plane

Calling Sequence:

   perpendicular(u, v, w)

Parameters:

   u –   a point or a line

   v –   a line or a plane

   w –   the name of the result

Synopsis:

- This function is part of the **geom3d** package, and so can be used in the form `perpendicular(..)` only after performing the command `with(geom3d)` or `with(geom3d, perpendicular)`.

- Use op to have more details on the line or the plane.

Examples:
```
> with(geom3d):
  point3d(A,0,0,0), point3d(B,1,0,0);                    ⟶        A, B
> plane(p, [x + y + z = 1]), line3d(A_B,[A,
  B]):
  perpendicular(A, p, l):
  type(l, line3d);                                       ⟶        true
> perpendicular(A, A_B, p1):
  type(p1, plane);                                       ⟶        true
> p1[equation];                                          ⟶        x = 0
```

SEE ALSO: geom3d[perpendicular]

6.2.22   geom3d[plane] – define the planes

Calling Sequence:
    plane(*l*, *gv*)

Parameters:
    l   –   the name of the plane
    gv –   a list

Synopsis:

- This function is part of the **geom3d** package, and so can be used in the form **plane(..)** only after performing the command **with(geom3d)** or **with(geom3d, plane)**.

- The given list can consist of three points, a linear equation (or polynomial) in x, y, and z, a point and the normal vector or a point and two vectors.

- Vectors in this geometry package are represented by a list of three elements.

- The names x, y, z, and _t are always used in this three-dimensional geometry as globals so never use x, y, z, or _t for other purposes.

- For more details on the plane, use op.

Examples:
```
> with(geom3d):
  plane(p1,[x+y+z=1]),plane(p2,[point3d(A,
  0,0,0),[1,1,1]]):
  plane(p3,[point3d(B,1,0,0),point3d(C,0,1,
  0),point3d(D,0,0,1)]):
  p2[equation];                              ⟶     x + y + z = 0
> p3[equation];                              ⟶     x + y + z - 1 = 0
```

```
> p1[normal_vector];                    ⟶            [1, 1, 1]
```

SEE ALSO: `geom3d[point3d]`, `geom3d[line3d]`

6.2.23 geom3d[point3d] – define the points in three-dimensional space

Calling Sequence:

point3d(*P*, [*Px*, *Py*, *Pz*])
point3d(*P*, *Px*, *Py*, *Pz*)

Parameters:

P – the name of the point

Px – the x-coordinate

Py – the y-coordinate

Pz – the z-coordinate

Synopsis:

- This function is part of the `geom3d` package, and so can be used in the form `point(..)` only after performing the command `with(geom3d)` or `with(geom3d, point)`.
- Never let the coordinates of the points contain `x`, `y`, or `z` or `_t`.
- For more details on the points, use `coordinates` or `op`.

Examples:

```
> with(geom3d):
  point3d(A, u, v, w);                  ⟶            A
> A[x];                                 ⟶            u
> A[y];                                 ⟶            v
> A[z];                                 ⟶            w
> coordinates(A);                       ⟶            [u, v, w]
```

SEE ALSO: `geom3d[line3d]`, `geom3d[plane]`

6.2.24 geom3d[powerps] – power of a given point with respect to a given sphere

Calling Sequence:

powerps(*P*, *s*)

Parameters:

P – a point

s – a sphere

Synopsis:
- This function is part of the **geom3d** package, and so can be used in the form **powerps(..)** only after performing the command **with(geom3d)** or **with(geom3d, powerps)**.

- The inputs must be in the above order.

Examples:
```
> with(geom3d):
  sphere(s,[x^2+y^2+z^2=1]),point3d(A,1,1,
  1),point3d(B,1/2,1/2,1):
  powerps(A, s);                          ⟶                    2
> powerps(B, s);                          ⟶                   1/2
```

6.2.25   geom3d[projection] – find the projection of a given point/line on a given line/plane

Calling Sequence:
    projection(u, v, w)

Parameters:
    u –   a point or a line
    l –   a line or a plane
    Q –   the name of the projection

Synopsis:
- This function is part of the **geom3d** package, and so can be used in the form **projection(..)** only after performing the command **with(geom3d)** or **with(geom3d, projection)**.

Examples:
```
> with(geom3d):
  point3d(A,0,0,0), point3d(B,1,0,0),
  point3d(C,0,0,1):
  plane(p, [x + y + z = 1]), line3d(A_B, [A,
  B]):
  projection(A, p, A1);                   ⟶                    A1
> coordinates(A1);                        ⟶              [1/3, 1/3, 1/3]
> projection(A_B, p, l);                  ⟶                    l
> type(l, line3d);                        ⟶                   true
> l[direction_vector];                    ⟶            [2/3, -1/3, -1/3]
> projection(C, A_B, C1);                 ⟶                    C1
> coordinates(C1);                        ⟶                [0, 0, 0]
```

SEE ALSO: **geom3d[perpendicular]**

6.2.26   geom3d[radius] – compute the radius of a given sphere

Calling Sequence:
> radius(*c*)

Parameters:
> *c* – a sphere

Synopsis:
- This function is part of the **geom3d** package, and so can be used in the form `radius(..)` only after performing the command `with(geom3d)` or `with(geom3d, radius)`.
- This function is equivalent to the command `c[radius]`.

Examples:
```
> with(geom3d):
  sphere(s, [x^2 + y^2 + z^2 - 2*x + 4*y
  -4*z = 7]):
  radius(s);
> s[radius];
```
$$\longrightarrow \qquad 4$$
$$\longrightarrow \qquad \frac{1/2}{16}$$

6.2.27   geom3d[rad_plane] – find the radical plane of two given spheres

Calling Sequence:
> rad_plane(*s1*, *s2*, *p*)

Parameters:
> *p*   – the name of the radical plane
> *s1*, *s2* – two spheres

Synopsis:
- This function is part of the **geom3d** package, and so can be used in the form `rad_plane(..)` only after performing the command `with(geom3d)` or `with(geom3d, rad_plane)`.
- The parameter *p* is updated to be the radical plane of the two spheres *s1* and *s2*.
- Use subscripts or **op** for details about the radical plane.

Examples:
```
> with(geom3d):
  sphere(s1, [x^2 + y^2 + z^2 = 1]):
  sphere(s2, [point3d(B, [5, 5, 5]), 2]):
  rad_plane(s1 ,s2, p);
> type(p, plane);
> p[equation];
> p[normal_vector];
```
$$\longrightarrow \qquad p$$
$$\longrightarrow \qquad true$$
$$\longrightarrow \qquad -72 + 10\ x + 10\ y + 10\ z = 0$$
$$\longrightarrow \qquad [10, 10, 10]$$

6.2.28 geom3d[reflect] – find the reflection of a given point/line with respect to a line/plane

Calling Sequence:

  reflect(u, v, w)

Parameters:

  u – a point or a line

  v – a line or a plane

  w – the name of the reflected point or line

Synopsis:

- This function is part of the **geom3d** package, and so can be used in the form reflect(..) only after performing the command with(geom3d) or with(geom3d, reflect).

- Use op to have more details on the reflection.

Examples:
```
> with(geom3d):
> point3d(A,0,0,0), point3d(B,1,0,0):
> plane(p, [x + y + z = 1]), line3d(A_B,[A, B]):
> reflect(A, p, A1);
```
$$A1$$
```
> coordinates(A1);
```
$$[2/3, 2/3, 2/3]$$
```
> reflect(A_B, p, l);
```
$$l$$
```
> type(l, line3d);
```
$$true$$
```
> l[equation];
```
$$[2/3 + 1/3 \_t, 2/3 - 2/3 \_t, 2/3 - 2/3 \_t]$$

SEE ALSO: geom3d[symmetric]

6.2.29 geom3d[sphere] – define a sphere

Calling Sequence:

  sphere(s, gv)

Parameters:

  s – the name of the sphere

  gv – a list of parameters to specify the sphere

Synopsis:

- This function is part of the **geom3d** package, and so can be used in the form sphere(..) only after performing the command with(geom3d) or with(geom3d, sphere).

- The list can consist of any of the following pieces of information to specify the sphere:

  - an equation in x, y, z

  - coordinates of the center and the length of the radius

  - four points in space

  - two points and the name **diameter**.

- For details about the sphere, use subscripts or **op**.

Examples:
```
> with(geom3d):
  point3d(A,0,0,0),point3d(B,1,0,0),
  point3d(C,0,1,0),point3d(D,0,0,1):
  sphere(s1, [A, 1]):
  volume(s1);                        ⟶            4/3 Pi

> sphere(s2, [A, B, C, D]):
  type(s2, sphere);                  ⟶             true

> sphere(s3, [B, C, diameter]):
  area(s3);                          ⟶             2 Pi

> sphere(s4, [x^2 + y^2 + z^2 = 1]):
  coordinates(center(s4));           ⟶          [0, 0, 0]
```

SEE ALSO: geom3d[line3d], geom3d[point3d] and geom3d[plane]

6.2.30 geom3d[symmetric] – find the symmetric point of a point with respect to a given point

Calling Sequence:
>   symmetric(*P1*, *Q*, *P2*)

Parameters:
>   *P1, Q* – two points
>   *P2*   – the name of the symmetric point of *P1*

Synopsis:
- This function is part of the **geom3d** package, and so can be used in the form **symmetric(..)** only after performing the command **with(geom3d)** or **with(geom3d, symmetric)**.

- Use **coordinates** or **op** for more details on the symmetric point.

Examples:
```
> with(geom3d):
  point3d(P1,1,2,3), point3d(Q,4,5,6);   ⟶            P1, Q

> symmetric(P1, Q, P2);                   ⟶              P2
```

```
> type(P2, point3d);                          ⟶            true
> coordinates(P2);                            ⟶          [7, 8, 9]
```

SEE ALSO: geom3d[reflect] or geom3d[midpoint]

6.2.31   geom3d[tangent] – find the tangent plane of a point on a given sphere with respect to the sphere

Calling Sequence:
    tangent(P, s, p)

Parameters:

   P – a point

   s – a sphere

   p – the name of the tangent plane

Synopsis:

- This function is part of the **geom3d** package, and so can be used in the form tangent(..) only after performing the command with(geom3d) or with(geom3d, tangent).

- The point must be on the sphere.

- For more detail of the tangent, use op.

Examples:
```
> with(geom3d):
  sphere(s, [x^2 + y^2 + z^2 =1]):
  point3d(A, 1, 0, 0):
  tangent(A, s, p);                           ⟶              p
> type(p, plane);                             ⟶            true
> p[equation];                                ⟶         - 1 + x = 0
```

SEE ALSO: geom3d[are_tangent]

6.2.32   geom3d[tetrahedron] – define a tetrahedron

Calling Sequence:
    tetrahedron(T, gv)

Parameters:

   T – the name of the tetrahedron

   gv – a list

Synopsis:
- This function is part of the **geom3d** package, and so can be used in the form **tetrahedron(..)** only after performing the command **with(geom3d)** or **with(geom3d, tetrahedron)**.

- The given list may consist of four vertices or four planes.

- For more details on the tetrahedron, use **op**.

Examples:
```
> with(geom3d):
  point3d(A,0,0,0),point3d(B,1,0,0),
  point3d(C,0,0,1),point3d(D,0,1,0):
  tetrahedron(T1, [A, B, C, D]);
```
$\longrightarrow$ T1

```
> type(T1, tetrahedron);
```
$\longrightarrow$ true

```
> volume(T1);
```
$\longrightarrow$ 1/6

```
> plane(p1,[x+y+z=1]),plane(p2,[x=0]),
  plane(p3,[y=0]),plane(p4,[z=0]):
  tetrahedron(T2, [p1, p2, p3, p4]);
```
$\longrightarrow$ T2

```
> type(T2, tetrahedron);
```
$\longrightarrow$ true

```
> volume(T2);
```
$\longrightarrow$ 1/6

SEE ALSO: geom3d[line3d], geom3d[point3d], geom3d[plane]

6.2.33 geom3d[triangle3d] – define the triangle in three-dimensional space

Calling Sequence:
triangle3d(*T*, *gv*)

Parameters:

T – the name of the triangle

gv – a list

Synopsis:
- This function is part of the **geometry** package, and so can be used in the form **triangle3d(..)** only after performing the command **with(geometry)** or **with(geometry, triangle3d)**.

- The list consists of three vertices.

- For more details on the triangles, use **op**.

Examples:
```
> with(geom3d):
  triangle3d(ABC, [point3d(A,0,0,0),
  point3d(B,2,0,0),point3d(C,0,2,0)]):
  ABC[form];
```
$\longrightarrow$ triangle3d

```
> type(ABC, triangle3d);                    ⟶            true
> area(ABC);                                 ⟶             1/2
                                                      1/2 16
> centroid(ABC, G);                          ⟶             G
> coordinates(G);                            ⟶        [2/3, 2/3, 0]
```

SEE ALSO: geom3d[line3d], geom3d[point3d], geom3d[sphere]

6.2.34   geom3d[type] – check for a point, line, plane, sphere, triangle, or a tetrahedron in three-dimensional space

Calling Sequence:

     type(A, *geom3dtype*)

Parameters:

     A –   A three-dimensional geometry object

     geom3dtype   –   type of the object

Synopsis:

- The function returns `true` or `false`.

- The second argument can be any of the names `point3d`, `line3d`, `plane`, `sphere`, `triangle3d` or `tetrahedron`.

- These functions are part of the geom3d package and can only be used after performing `with(geom3d)`.

Examples:

```
> with(geom3d):
  point(A,0,0,0), plane(p, [x + y - z =1])
  :
  type(A, point3d);                          ⟶           false
> type(p, plane);                            ⟶           true
> type(p, sphere);                           ⟶           false
```

SEE ALSO: geom3d[point3d], geom3d[line3d] and geom3d[plane]

6.2.35   geom3d[volume] – compute the volume of a tetrahedron or a sphere

Calling Sequence:

     volume(g)

Parameters:
> g – a tetrahedron or a sphere

Synopsis:
- This function is part of the geom3d package, and so can be used in the form volume(..) only after performing the command with(geom3d) or with(geom3d, volume).

- If the coordinates of the vertices of the triangle are numeric, then the volume is always a positive number.

Examples:

```
> with(geom3d):
  point3d(A,0,0,0),point3d(B,0,0,1),
  point3d(C,1,0,0),point3d(D,0,1,0):
  tetrahedron(ABCD, [A, B, C, D]);                 ⟶          ABCD

> volume(ABCD);                                     ⟶          1/6

> sphere(s, [A, 1]);                                ⟶          s

> volume(s);                                        ⟶          4/3 Pi
```

SEE ALSO: geom3d[area]

6.3   The Projective Geometry Package

6.3.1   Introduction to the projgeom package

Calling Sequence:
> *function* (args)
> projgeom[*function*] (args)

Synopsis:
- To use a projective geometry function, either define that specific function alone by typing with(projgeom, *function*), or define all projective geometry functions by typing with(projgeom).

- The functions available are:

collinear	concur	conjugate	ctangent
fpconic	harmonic	inter	join
lccutc	lccutr	lccutr2p	linemeet
midpoint	onsegment	polarp	poleline
ptangent	rtangent	tangentte	tharmonic

- For help with a particular function use help(projgeom, *function*) .

- Use `help(projgeom, point)`, `help(projgeom, line)`, and `help(projgeom, conic)` for the definitions of points, lines, and conics.

- As an example, to find the line joining points `A(1,1,1)` and `B(0,1,2)`, type:

    `with(projgeom,join);  join(point(A,[1,1,1]), point(B,[0,1,2]));`

SEE ALSO: geometry

6.3.2   projgeom[collinear] – test whether three given points are collinear

Calling Sequence:
    `collinear(A, B, C)`

Parameters:
    A, B, C – three given points

Synopsis:
- The output will be **true**, **false**, or a **condition**.

- Use `help(projgeom, point)` for the definition of points.

Examples:
```
> with (projgeom):
> collinear(point(A,[0,1,1]), point(B,[1,1,1]), point(C,[x,y,1]));
            THE POINTS ARE COLLINEAR IF :
```
$$- 1 + y = 0$$

SEE ALSO: `projgeom[concur]`

6.3.3   projgeom[concur] – test whether three given lines are concurrent

Calling Sequence:
    `concur(l1, l2, l3)`

Parameters:
    l1, l2, l3 – three given lines

Synopsis:
- The output will be **true**, **false**, or a **condition**.

- Use `help(projgeom, line)` for the definition of lines.

Examples:
```
> with (projgeom):
> concur(line(l1, [0,1,1]), line(l2, [1,1,1]), line(l3, [x,y,1]));
                THE LINES ARE CONCURRENT IF :
```

$$- 1 + y = 0$$

SEE ALSO: projgeom[collinear]

6.3.4   projgeom[conic] – define the conics

Calling Sequence:
    conic(cc, [a, b, c, d, e, f]);

Parameters:

    cc   –   name of the conic

    a, b, c, d, e, f   –   coefficients of the equation of the conic

Synopsis:
- The coordinates must always be given in a list.

- For more details of the conic cc, use cc[coords].

- The equation of the conic cc is:

    a x^2 + b y^2 + c z^2 + d y z + e z x + f x y .

Examples:
```
> with (projgeom):
  conic(cc, [0,0,1,23,1,6]);              ⟶                    cc
> cc[form];                               ⟶                  conic
> cc[coords];                             ⟶          [0, 0, 1, 23, 1, 6]
```

SEE ALSO: projgeom[line], projgeom[point]

6.3.5   projgeom[conjugate] – check whether two given points are conjugate with respect to a given conic

Calling Sequence:
    conjugate(P, Q, c)

Parameters:

    c   –   a conic

    P, Q   –   two given points

Synopsis:
- The output will be true, false, or a condition.

- Use help(projgeom,point) and help(projgeom,conic) for the definitions of points and conics.

Examples:
```
> with (projgeom):
  point(A,[x,y,z]),point(B,[1,-2,5]);          ⟶                    A, B
> conic(c, [-516, 0, 0, 350, 518, -402]);      ⟶                     c
> conjugate(A,B,c);                            ⟶   THE POINTS ARE CONJUGATE WITH RESPECT T\
                                                   O THE CONIC IF :
```
$$2878\ x\ +\ 1348\ y\ -\ 182\ z\ =\ 0$$

6.3.6   projgeom[ctangent] – find the tangent lines to the conic from any point in the complex plane

Calling Sequence:
> ctangent(P, c)

Parameters:

P	–	a point
c	–	a conic
$l1, l2$	–	name of the tangents

Synopsis:
- There are one or two tangents.

- There must be four arguments in the above order.

- Use help(projgeom, line) and help(projgeom, conic) for the definitions of lines and conics.

Examples:
```
> with (projgeom):
  point(P, [1, 1, 2]), conic(c, [0, 0, 0,
  2, 0, -4]);                                  ⟶                    P, c
> ctangent(P, c, l1, l2);                      ⟶                     l1
> l1[form];                                    ⟶                    line
> l1[coords];                                  ⟶                 [-4, 0, 2]
```

SEE ALSO: projgeom[ptangent] and projgeom[rtangent]

6.3.7   projgeom[fpconic] – find the coefficients of the equation of a conic which goes through five given points

Calling Sequence:
> fpconic(c, p1, p2, p3, p4, p5)

Parameters:

c	–	name of the conic
p1,p2,p3,p4,p5	–	the five given points

Synopsis:
- For more detail of the conic c, use `op` or `c[coords]`.
- Use `help(projgeom,conic)` and `help(projgeom,point)` for the definitions of the conics and points.

Examples:
```
> with (projgeom):
  point(A, [0,0,1]), point(B, [1,0,0]),
  point(C,[0,1,0]),
  point(D, [1,1,2]), point(F, [2,1,4]);        ⟶          A, B, C, D, F
> fpconic(c, A, B, C, D, F);                    ⟶              c
> c[form];                                      ⟶            conic
> c[coords];                                    ⟶      [0, 0, 0, 2, 0, -4]
```

6.3.8   projgeom[harmonic] – find a point which is harmonic conjugate to another point with respect to two other points

Calling Sequence:
```
    harmonic(A, B, C, D)
```

Parameters:

D	–	name of the required point
A, B, C	–	three given points

Synopsis:
- The output will be point D which is harmonic conjugate to A with respect to B, C. It will return NULL if such a point does not exist.
- There must be four arguments in the above order.
- Use `help(projgeom, point)` for the definition of points.

Examples:
```
> with (projgeom):
  point(A, [1,-2,5]), point(B, [1,-2,3]),
  point(C, [0,0,1]);                            ⟶           A, B, C
> harmonic(A, B, C, D);                         ⟶              D
> D[form];                                      ⟶            point
> D[coords];                                    ⟶         [1, -2, 1]
```

SEE ALSO: `projgeom[tharmonic]`

6.3.9   projgeom[inter] – find the intersection of two given lines

Calling Sequence:

    inter(l1, l2, P);

Parameters:

   l1,l2 –   any two lines

   P   –   name of the intersection

Synopsis:

- There must be three arguments in that order.

- Use `op(P)` or `P[coords]` for detail of the intersection P.

Examples:
```
> with (projgeom):
  line(l1, [1,2,0]), line(l2, [0,1,7]);        ⟶           l1, l2
> line(l3, [x,y,0]), line(l4, [0,m,n]);        ⟶           l3, l4
> inter(l1, l2, P);                            ⟶             P
> P[form];                                     ⟶           point
> P[coords];                                   ⟶        [14, -7, 1]
> inter(l3, l4, Q);                            ⟶             Q
> Q[coords];                                   ⟶     [y n, - x n, x m]
```

SEE ALSO: `projgeom[join]`

6.3.10   projgeom[join] – find the line joining two given points

Calling Sequence:

    join(P, Q, l);

Parameters:

   P,Q –   any two points

   l   –   name of the line

Synopsis:

- For detail of the line l, use `op(l)` or `l[coords]`.

- There must be three arguments in the above order.

Examples:
```
> with (projgeom):
  point(A, [1,2,0]), point(B, [0,1,7]);        ⟶           A, B
> point(C, [x,y,0]), point(D, [0,m,n]);        ⟶           C, D
```

> join(A, B, l);	$\longrightarrow$	l
> l[form];	$\longrightarrow$	line
> l[coords];	$\longrightarrow$	[14, -7, 1]
> join(C, D, cd);	$\longrightarrow$	cd
> cd[coords];	$\longrightarrow$	[y n, - x n, x m]

SEE ALSO: projgeom[cut]

6.3.11   projgeom[lccutc] – find the intersection of a conic and a line in a complex plane

Calling Sequence:
    lccutc(c, l, P, Q)

Parameters:

    c – a conic

    l – a line

    P, Q – name of the two intersections

Synopsis:
- There must be four arguments.

- Use help(projgeom, point) for the definition of points.

Examples:
```
> with (projgeom):
  conic(c, [0, 0, 0, -7, -11, 75/2]),
  line(l, [7, -11, 5]):
  lccutc(c, l, P, Q);                    ⟶           P, Q
> P[coords];                             ⟶      [220/7, 440/7, 660/7]
> Q[coords];                             ⟶      [-110/7, 55/7, 275/7]
```

SEE ALSO: projgeom[lccutc2p] and projgeom[lccutr]

6.3.12   projgeom[lccutr] – find the intersection of a conic and a line in a real extended plane

Calling Sequence:
    lccutr(c, l, P, Q)

Parameters:

> c –   a conic
>
> l –   a line
>
> $P,\ Q$ –   name of the intersections

Synopsis:

- There must be four arguments, with the last two arguments being the names of the intersection points.

- Use `op(P)`, `op(Q)`, `P[coords]`, and `Q[coords]` for more details of the intersections `P` and `Q`.

Examples:

```
> with (projgeom):
  conic(c, [0, 0, 0, -7, -11, 75/2]),
  line(l, [7, -11, 5]):
  lccutr(l, c, P, Q);                    ⟶            P, Q

> P[form];                              ⟶            point

> P[coords];                            ⟶       [220/7, 440/7, 660/7]

> Q[form];                              ⟶            point

> Q[coords];                            ⟶       [-110/7, 55/7, 275/7]
```

SEE ALSO: `projgeom[lccutr2p]` and `projgeom[lccutc]`

6.3.13   projgeom[lccutr2p] – find the intersection of a conic and a line joining two given points in a real extended plane

Calling Sequence:

> `lccutr2p(c, P, Q, A1, A2)`

Parameters:

> c –   a conic
>
> $P,\ Q$ –   two points which determine a line
>
> $A1,\ A2$ –   name of the intersections

Synopsis:

- There must be five arguments in the above order.

- There may be zero, one, or two intersections.

- Use `op(A1)`, `op(A2)`, `A1[coords]`, and `A2[coords]` for more details of the intersections `A1` and `A2`.

Examples:

```
> with (projgeom):
  conic(c, [0, 0, 0, -7, -11, 75/2]);       ⟶                c
```

> point(P, [1,2,3]), point(Q, [-2,1,5]);	$\longrightarrow$	P, Q
> lccutr2p(c, P, Q, A1, A2);	$\longrightarrow$	A1, A2
> A1[form];	$\longrightarrow$	point
> A1[coords];	$\longrightarrow$	[1, 2, 3]
> A2[form];	$\longrightarrow$	point
> A2[coords];	$\longrightarrow$	[-2, 1, 5]

SEE ALSO: `projgeom[lccutr]` and `projgeom[lccutc]`

6.3.14   projgeom[line] – define the lines

Calling Sequence:

    line(l, [lx, ly, lz]);

Parameters:

> l –   name of the line
>
> px, py, pz –   coefficients of the homogeneous equation of l

Synopsis:

- The coordinates are the coefficients of the homogeneous equation of the line.

- The coordinates must always be given in a list.

- For more details of the line l, use `l[coords]`.

Examples:

> with (projgeom):		
line(l,[0,0,1]), line(t,[2,0,3]);	$\longrightarrow$	l, t
> l[form];	$\longrightarrow$	line
> t[form];	$\longrightarrow$	line
> l[coords];	$\longrightarrow$	[0, 0, 1]
> t[coords];	$\longrightarrow$	[2, 0, 3]

SEE ALSO: `projgeom[point]`, `projgeom[conic]`

6.3.15   projgeom[linemeet] – find a line concurrent to two given lines

Calling Sequence:

    linemeet(l1, l2, s, l);

Parameters:

s	–	a parameter
l	–	name of the new line
l1, l2	–	two given lines

Synopsis:

- `l1[coords] + k l2[coords] = l[coords]`. This means the line `l` divides `l1` and `l2` by the ratio `k`.

- There must be four arguments in the above order.

- Use `op(l)` or `l[coords]` for detail of the line `l`.

- Use `help(projgeom, line)` for the definition of lines.

Examples:
```
> with (projgeom):
  linemeet(line(l1, [1,0,1]), line(l2, [0,
  0,1]), k, l);                              ⟶              l
> l[form];                                   ⟶             line
> l[coords];                                 ⟶         [1, 0, 1 + k]
```

SEE ALSO: `projgeom[onsegment]`

6.3.16   projgeom[midpoint] – find the midpoint of the line joining two given points

Calling Sequence:

    midpoint(A, B, M);

Parameters:

A, B	–	any two points
M	–	name of the midpoint

Synopsis:

- The midpoint of two given points is the point which is harmonic to the intersection of the line at infinity and the line joining two above points.

- Use `op(M)` or `M[coords]` for the detail of the midpoint M.

- Use `help(projgeom, point)` for the definition of points.

Examples:
```
> with (projgeom):
  midpoint(point(A, [1,2,3]), point(B, [2,
  3,4]), M);                                 ⟶                M
```

> M[form];	$\longrightarrow$	point
> M[coords];	$\longrightarrow$	[-4, -7, -10]

6.3.17   projgeom[onsegment] – find a point that is collinear to two given points and divides these points by a constant ratio

Calling Sequence:
    onsegment(P, Q, s, M)

Parameters:
    P, Q   –   two given points
    s   –   a constant ratio
    M   –   name of the new point

Synopsis:
- Point M divides P and Q by the ratio k.
- There must be four arguments in the above order.
- Use op(M) or M[coords] for detail of point M.
- Use help(projgeom, point) for the definition of points.

Examples:

> with (projgeom):
   onsegment(point(P, [1,0,1]), point(Q,
   [0,0,1]), k, M);
	$\longrightarrow$	M

> M[form];	$\longrightarrow$	point
> M[coords];	$\longrightarrow$	[1, 0, 1 + k]

SEE ALSO: projgeom[linemeet]

6.3.18   projgeom[point] – define the points

Calling Sequence:
    point(P, [px, py, pz]);

Parameters:
    P   –   name of the point
    px, py, pz   –   the homogeneous coordinate of P

Synopsis:
- The coordinates must always be given in a list.

- For more details of the point P, use `P[coords]`.

Examples:
```
> with (projgeom):
  point(A,[0,0,1]), point(B,[2,0,3]);
```
          $\longrightarrow$ A, B

```
> A[form];
```
          $\longrightarrow$ point

```
> B[form];
```
          $\longrightarrow$ point

```
> A[coords];
```
          $\longrightarrow$ [0, 0, 1]

```
> B[coords];
```
          $\longrightarrow$ [2, 0, 3]

SEE ALSO: `projgeom[line]`, `projgeom[conic]`

6.3.19   projgeom[polarp] – determine the polar line of a given point with respect to a given conic

Calling Sequence:
```
polarp(A, c, l);
```

Parameters:

    A –   a point

    c –   a conic

    l –   name of the polar line

Synopsis:

- There must be four arguments in the above order.

- Use `op(l)` or `l[coords]` for more details of the polar line l.

Examples:
```
> with (projgeom):
  point(A, [1, 1, 1]), conic(c, [-516, 0,
  0, 350, 518, -402]):
  polarp(A, c, l);
```
          $\longrightarrow$ l

```
> l[form];
```
          $\longrightarrow$ line

```
> l[coords];
```
          $\longrightarrow$ [-400, -52, 868]

SEE ALSO: `projgeom[poleline]`

6.3.20   projgeom[poleline] – determine the pole of a given line with respect to a given conic

Calling Sequence:
```
poleline(l, c, P)
```

Parameters:

> l – a line
>
> c – a conic
>
> P – name of the pole of the line l

Synopsis:

- There must be four arguments in the above order.

- Use P[coords] or op(P) for details of the pole P of the line 1.

Examples:
```
> with (projgeom):
  line(l, [-400, -52, 868]), conic(c,
  [-516, 0, 0, 350, 518, -402]):
  poleline(l, c, P);                    ⟶                    P
> P[form];                             ⟶                  point
> P[coords];                           ⟶   [-82555200, -82555200, -82555200]
```

SEE ALSO: projgeom[polarp]

6.3.21   projgeom[ptangent] – find the tangent line to the conic at any point on the conic

Calling Sequence:

>   ptangent(P, c, 1);

Parameters:

> P – a point on a conic
>
> c – a conic
>
> l – name of the tangent line

Synopsis:

- There must be three arguments in the above order.

- The point must be on the conic.

- Use 1[coords] or op(1) for more details of the tangent line 1.

Examples:
```
> with (projgeom):
  point(A, [1, 1, 2]), conic(c, [0, 0, 0,
  2, 0, -4]):
  ptangent(A, c, 1);                    ⟶                    1
> 1[form];                             ⟶                  line
> 1[coords];                           ⟶             [-4, 0, 2]
```

SEE ALSO: projgeom[ctangent] and projgeom[rtangent]

6.3.22   projgeom[rtangent] – find the tangent line to the conic at any point in the real extended plane

Calling Sequence:

```
rtangent(P, c, l1, l2)
```

Parameters:

P	–	a point
c	–	a conic
l1, l2	–	name of the tangent lines

Synopsis:

- There must be four arguments in the above order.

- There might be zero, one, or two tangent lines.

- Use op or `l1[coords]`, `l2[coords]` for details of the tangent lines `l1` and `l2`.

Examples:

```
> with (projgeom):
  point(A, [1,2,3]), conic(c,[1,0,0,2,0,-4]
  );
```
$\longrightarrow$ A, c

```
> rtangent(A, c, l1, l2);
```
$\longrightarrow$ l1, l2

```
> l1[form];
```
$\longrightarrow$ line

```
> l2[form];
```
$\longrightarrow$ line

```
> l1[coords];
```
$\longrightarrow$ $[\dfrac{240}{49}, \dfrac{24}{49}, -\dfrac{96}{49}]$

```
> l2[coords];
```
$\longrightarrow$ $[\dfrac{156}{49}, \dfrac{156}{49}, -\dfrac{156}{49}]$

SEE ALSO: `projgeom[ctangent]` and `projgeom[ptangent]`

6.3.23   projgeom[tangentte] – check if a given line touches a given conic

Calling Sequence:

```
tangentte(l, c)
```

Parameters:

l	–	a line
c	–	a conic

Synopsis:

- The arguments must be given in the above order.

- The output will be **true**, **false**, or a **condition**.

Examples:
```
> with (projgeom):
> line(l, [-4, 0, 2]), conic(c, [0, 0, 0, 2, 0, -4]):
> tangentte(l, c);
```
$$\text{true}$$
```
> line(l1, [x,y,1]), conic(c1, [0, 0, 0, 2, 0, -4]):
> tangentte(l1, c1);
```
$$\text{A LINE TOUCHES A CONIC IF :}$$
$$16 + 16\ x + 4\ x^2 = 0$$

6.3.24 projgeom[tharmonic] – check whether a pair of points is harmonic conjugate to another pair of points

Calling Sequence:
>   tharmonic(A, B, C, D)

Parameters:
>   *A, B, C, D* – four given points

Synopsis:
- The output will be **true**, **false**, or a **condition**.

- The points should be collinear first.

Examples:
```
> with (projgeom):
> point(A,[1,-2,3]), point(B,[0,0,1]), point(C,[1,-2,5]),
> point(D,[1,-2,1]):
> tharmonic(A, B, C, D);
```
$$\text{true}$$

SEE ALSO: projgeom[harmonic]

7

Miscellaneous Packages

7.1 The Orthogonal Polynomial Package

7.1.1 Introduction to the orthopoly package

Calling Sequence:

orthopoly[*function*] (*args*)

function (*args*)

Synopsis:

- To use an orthopoly function, either define that function alone using `with(orthopoly, `*function*`)`, or define all orthopoly functions using `with(orthopoly)`.

- The functions available are:

  G   H   L   P
  T   U

- `G(n, a, x)` generates the nth Gegenbauer polynomial.

- `H(n, x)` generates the nth Hermite polynomial.

- `L(n, x)` generates the nth Laguerre polynomial.

- `L(n, a, x)` generates the nth generalized Laguerre polynomial.

- `P(n, x)` generates the nth Legendre polynomial.

- `P(n, a, b, x)` generates the nth Jacobi polynomial.

- `T(n, x)` generates the nth Chebyshev polynomial of the first kind.

- `U(n, x)` generates the nth Chebyshev polynomial of the second kind.

- For a more detailed description of a particular function see `orthopoly[`*function*`)]`.

- The package functions are always available without applying `with`, using the long-form notation orthopoly[*function*](*args*). This notation is necessary whenever there is a conflict between a package function name and another function used in the same session.

SEE ALSO: with

7.1.2   orthopoly[G] – Gegenbauer (ultraspherical) polynomial

Calling Sequence:

   G(n, a, x)

   orthopoly[G] (n, a, x)

Parameters:

   n –   a non-negative integer

   a –   any non-rational algebraic expression or a rational number greater than -1/2

   x –   any algebraic expression

Synopsis:

- The function G computes the nth ultraspherical (Gegenbauer) polynomial, with parameter a, evaluated at x.

- These polynomials are orthogonal on the interval [-1,1], with respect to the weight function w(x) = (1-x^2)^(a-1/2). Thus

$$
\int_{-1}^{1} w(x)\ G(n,\ a,\ x)\ G(m,\ a,\ x)\ dx\ =\ 0
$$

if m is not equal to n.

- These ultraspherical polynomials satisfy the following recurrence relation, for a<>0.

```
G(0,a,x) = 1
G(1,a,x) = 2*a*x
G(n,a,x) = 2*(n+a-1)/n*x*G(n-1,a,x)
              - (n+2*a-2)/n*G(n-2,a,x), for n>1.
```

- When a=0, G(0,0,x) = 1, and G(n,0,x) = 2/n*T(n,x) for n>0.

- This function is part of the **orthopoly** package, and so can be used in the form G(..) only after performing the command with(orthopoly) or with(orthopoly, G). The function can always be accessed in the long form orthopoly[G](..).

Examples:

> with(orthopoly); $\longrightarrow$ [G, H, L, P, T, U]

```
> G(2,5,x);
```
$\longrightarrow$
$$60\ x^2 - 5$$

```
> G(3,2,x);
```
$\longrightarrow$
$$32\ x^3 - 12\ x$$

7.1.3   orthopoly[H] – Hermite polynomial

Calling Sequence:

H(n, x)

orthopoly[H](n, x)

Parameters:

n – a non-negative integer

x – any algebraic expression

Synopsis:

- The function H will compute the nth Hermite polynomial, evaluated at x.

- These polynomials are orthogonal on the interval (-infinity, infinity), with respect to the weight function w(x) = exp(-x^2). Thus

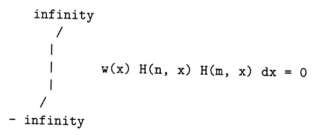

if **m** is not equal to **n**.

- They satisfy the following recurrence relation:

```
H(0,x)  = 1,
H(1,x)  = 2*x,
H(n,x)  = 2*x*H(n-1,x) - 2*(n-1)*H(n-2,x), for n>1.
```

- This function is part of the orthopoly package, and so can be used in the form H(..) only after performing the command with(orthopoly) or with(orthopoly, H). The function can always be accessed in the long form orthopoly[H](..).

Examples:

```
> with(orthopoly);
```
$\longrightarrow$
$$[G,\ H,\ L,\ P,\ T,\ U]$$

```
> H(2,x);                          ⟶                    2
                                                      4 x  - 2

> H(3,x);                          ⟶                    3
                                                      8 x  - 12 x
```

SEE ALSO: linalg[hermite]

7.1.4   orthopoly[P] – Legendre and Jacobi polynomials

Calling Sequence:

> P(n, a, b, x)
> P(n, x)
> orthopoly[P](n, a, b, x)
> orthopoly[P](n, x)

Parameters:

n –   a non-negative integer

x –   any algebraic expression

a, b –   any non-rational algebraic expressions or rational numbers greater
            than -1

Synopsis:

- The function P will compute the nth Jacobi polynomial, with parameters a and b, evaluated at x. In the case of only two arguments, P(n, x) will find the nth Legendre (spherical) polynomial, which is equal to P($n,0,0,x$).

- These polynomials are orthogonal on the interval [-1,1] with respect to the weight function w(x) = (1-x)^a*(1+x)^b when a and b are greater than -1. Thus

$$
\int_{-1}^{1} w(x)\ P(n,\ a,\ b,\ x)\ P(m,\ a,\ b,\ x)\ dx = 0
$$

if **m** is not equal to **n**. They are undefined for negative integer values of **a** or **b**.

- They satisfy the following recurrence relation:

```
P(0,a,b,x)  = 1,
P(1,a,b,x)  = (a/2-b/2) + (1+a/2+b/2)*x,
P(n,a,b,x)  = (2*n+a+b-1-x) * (a^2 b^2+(2*n+a+b-2)
    * (2*n+a+b)*x) / (2*n*(n+a+b)*(2*n+a+b-2)) * P(n-1,a,b,x)
```

```
       - 2*(n+a-1)*(n+b-1)*(2*n+a+b)
             / (2*n*(n+a+b)*(2*n+a+b-2)) * P(n-2,a,b)
    for n>1.
```

- This function is part of the `orthopoly` package, and so can be used in the form P(..) only after performing the command `with(orthopoly)` or `with(orthopoly, P)`. The function can always be accessed in the long form `orthopoly[P](..)`.

Examples:

```
> with(orthopoly);                    ⟶        [G, H, L, P, T, U]

> P(3,x);                             ⟶             3
                                             5/2 x  - 3/2 x

> P(2,1,3/4,x);                       ⟶       91     19         437  2
                                            - --- + ---- x + --- x
                                              128    64        128
```

SEE ALSO: `numtheory[legendre]`, `numtheory[jacobi]`

7.1.5 orthopoly[L] – Laguerre polynomial

Calling Sequence:

$L(n, a, x)$

$L(n, x)$

$orthopoly[L](n, a, x)$

$orthopoly[L](n, x)$

Parameters:

n – a non-negative integer

a – any non-rational algebraic expression or a rational number greater than -1

x – any algebraic expression

Synopsis:

- The function L will compute the nth generalized Laguerre polynomial with parameter a, evaluated at x.

- In the two argument case, $L(n, x)$ will compute the nth Laguerre polynomial, which is equal to $L(n, 0, x)$.

- These polynomials are orthogonal on the interval [0, infinity), with respect to the weight function w(x) = exp(-x)*x^a. Thus

```
              infinity
                 /
```

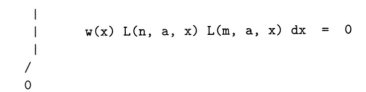

if m is not equal to n.

- Laguerre polynomials satisfy the following recurrence relation:

```
L(0,a,x) = 1,
L(1,a,x) = -x+1+a,
L(n,a,x) = (2*n+a-1-x)/n*L(n-1,a,x) - (n+a-1)/n*L(n-2,a,x)
for n > 1.
```

- This function is part of the `orthopoly` package, and so can be used in the form `L(..)` only after performing the command `with(orthopoly)` or `with(orthopoly, L)`. The function can always be accessed in the long form `orthopoly[L](..)`.

Examples:

```
> with(orthopoly);
```
$\longrightarrow$ [G, H, L, P, T, U]

```
> L(3,x);
```
$\longrightarrow$ $1 - 3 x + 3/2 x^2 - 1/6 x^3$

```
> L(2,1,x);
```
$\longrightarrow$ $3 - 3 x + 1/2 x^2$

7.1.6   orthopoly[T] – Chebyshev polynomial (of the first kind)

Calling Sequence:

   T(n, x)

   orthopoly[T](n,x)

Parameters:

   n –   a non-negative integer

   x –   any algebraic expression

Synopsis:

- The function T computes the nth Chebyshev polynomial of the first kind, evaluated at x.

- These polynomials are orthogonal on the interval [-1, 1], with respect to the weight function `w(x) = (1-x^2)^(-1/2)`. Thus

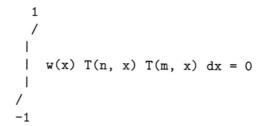

$$\int_{-1}^{1} w(x)\ T(n,\ x)\ T(m,\ x)\ dx = 0$$

if **m** is not equal to **n**.

- They satisfy the following recurrence relation:

```
T(0,x) = 1,
T(1,x) = x,
T(n,x) = 2*x*T(n-1,x) - T(n-2,x),   for n>1.
```

- This function is part of the **orthopoly** package, and so can be used in the form T(..) only after performing the command with(orthopoly) or with(orthopoly, T). The function can always be accessed in the long form orthopoly[T](..).

Examples:

> with(orthopoly);	⟶	[G, H, L, P, T, U]
> T(2,x);	⟶	$2x^2 - 1$
> T(3,x);	⟶	$4x^3 - 3x$

SEE ALSO: chebyshev

7.1.7   orthopoly[U] – Chebyshev polynomial of the second kind

Calling Sequence:

    U(n, x)

    orthopoly[U](n, x)

Parameters:

    n – a non-negative integer

    x – any algebraic expression

Synopsis:

- The function U will compute the nth Chebyshev polynomial of the second kind, evaluated at x.

- These polynomials are orthogonal on the interval [-1, 1], with respect to the weight function w(x) = (1-x^2)^(1/2). Thus

$$\int_{-1}^{1} w(x)\ U(n,\ x)\ U(m,\ x)\ dx = 0$$

if **m** is not equal to **n**.

- They satisfy the following recurrence relation:

```
U(0,x) = 1,
U(1,x) = 2*x,
U(n,x) = 2*x*U(n-1,x) - U(n-2,x),  for n>1.
```

- This function is part of the `orthopoly` package, and so can be used in the form U(..) only after performing the command `with(orthopoly)` or `with(orthopoly, U)`. The function can always be accessed in the long form `orthopoly[U](..)`.

Examples:

`> with(orthopoly);`	$\longrightarrow$	`[G, H, L, P, T, U]`
`> U(2,x);`	$\longrightarrow$	$4 x^2 - 1$
`> U(3,x);`	$\longrightarrow$	$8 x^3 - 4 x$

7.2 The Formal Power Series Package

7.2.1 Introduction to the `powseries` package

Calling Sequence:

> *function*(**args**)
>
> `powseries`[*function*](**args**)

Synopsis:

- The `powseries` package contains functions to create and manipulate formal power series represented in general form.

- The functions available are:

add	compose	evalpow	inverse
multconst	multiply	negative	powcreate
powdiff	powexp	powint	powlog
powpoly	powsolve	quotient	reversion
subtract	tpsform		

- For more information on a particular function see **powseries**[*function*], where *function* is taken from the above list.

- For example, to multiply two power series **A** and **B**, use the command **with(powseries, multiply)** followed by the command **multiply(A, B)**.

- To use a **powseries** function, either define that function alone using **with(powseries, *function*)**, or define all **powseries** functions by typing **with(powseries)**. The package functions are always available without applying **with**, using the long-form notation: powseries[*function*](*args*).

- This latter notation is necessary whenever there is a conflict between a package function name and another function used in the same session.

- For more information on the general representation of formal power series, see the information on **powseries**[powseries].

SEE ALSO: **with, series, asympt**

7.2.2   powseries[add] – addition of formal power series

Calling Sequence:
    add(*A, B, ..., Z*)

Parameters:
    A, B, ..., Z – an arbitrary number of formal power series

Synopsis:
- The function **add**(*A, B, ..., Z*) returns the formal power series which is the sum of *A, B, ..., Z*.

- This function is part of the **powseries** package, and so can be used in the form **add(..)** only after performing the command **with(powseries)** or **with(powseries, add)**. The function can always be accessed in the long form **powseries[add](..)**.

Examples:
```
> with(powseries):
> powcreate(t(n)=t(n-1)/n,t(0)=1):
> powcreate(v(n)=v(n-1)/2,v(0)=1):
> s := add(t, v):
> tpsform(s, x, 7);
```

$$2 + 3/2\ x + 3/4\ x^2 + 7/24\ x^3 + 5/48\ x^4 + \frac{19}{480}\ x^5 + \frac{49}{2880}\ x^6 + O(x^7)$$

7.2.3   powseries[compose] – composition of formal power series

Calling Sequence:

compose(*a*, *b*)

Parameters:

  a, b – formal power series

Synopsis:

- The function compose(*a*, *b*) computes and returns the formal power series which is *a(b)*, the composition of *a* and *b*. This power series is: *a(0) + a(1)*b + a(2)*b^2 + a(3)*b^3 + ...*

- Note that *b(0)* must be *0* to ensure that the composition is well defined. Otherwise, compose returns an error message.

- This function is part of the powseries package, and so can be used in the form compose(..) only after performing the command with(powseries) or with(powseries, compose). The function can always be accessed in the long form powseries[compose](..).

Examples:
```
> with(powseries):
> powcreate(t(n)=t(n-1)/n,t(0)=1):
> powcreate(v(n)=v(n-1)/2,v(0)=0,v(1)=1):
> s := compose (t, v):
> tpsform(s, x);
```

$$1 + x + x^2 + \frac{11}{12}\ x^3 + \frac{19}{24}\ x^4 + \frac{157}{240}\ x^5 + O(x^6)$$

7.2.4   powseries[evalpow] – general evaluator for power series expressions

Calling Sequence:

evalpow(*expr*)

Parameters:

    expr –   any arithmetic expression involving formal power series and
                  constants

Synopsis:

- The function **evalpow** evaluates any arithmetic expression involving power series and constants. It returns an unnamed power series.

- The following operators can be used: **+, -, *, /, ^** .

- Also, functions may be composed with each other. For example, **f(g)** can be used.

- The other functions that can be used in **evalpow** are:

powexp	powinv	powlog
powneg	powrev (reversion)	powdiff (first derivative)
powint (first integral)	powquo (quotient)	powsub (subtract)

- This function is part of the **powseries** package, and so can be used in the form **evalpow(..)** only after performing the command **with(powseries)** or **with(powseries, evalpow)**. The function can always be accessed in the long form **powseries[evalpow](..)**.

Examples:
```
> with(powseries):
> powcreate(f(n)=f(n-1)/n,f(0)=1):
> powcreate(g(n)=g(n-1)/2,g(0)=0,g(1)=1):
> powcreate(h(n)=h(n-1)/5,h(0)=1):
> k:=evalpow(f^3+g-pquo(h,f)):
> tpsform(k,x,5);
                      233  2    7273  3    52171  4       5
            24/5 x + --- x  + ---- x  + ----- x  + O(x )
                      50       1500      15000

> powcreate(t(n)=1/(n-5), t(5)=0):
> powcreate(u(n)=u(n-1)+u(n-2),u(0)=0,u(1)=1):
> v := evalpow((t+3*u)/t):
> tpsform(v,x);
                         2   155  3   105  4   6775  5       6
            1 - 15 x + 15/4 x  - --- x  - --- x  + ---- x  + O(x )
                                  16       64       768
```

SEE ALSO: `powseries[powcreate]`

7.2.5   powseries[inverse] – multiplicative inverse of a formal power series

Calling Sequence:

    inverse(*p*)

Parameters:

   p – formal power series

Synopsis:

- The function `inverse(`p`)` returns a formal power series which is the inverse of p with respect to `multiply`.

- That is, `multiply(`p`, inverse(`p`))` will return the formal power series equivalent to 1.

- This function is part of the **powseries** package, and so can be used in the form `inverse(..)` only after performing the command `with(powseries)` or `with(powseries, inverse)`. The function can always be accessed in the long form `powseries[inverse](..)`.

Examples:
```
> with(powseries):
> powcreate(t(n)=t(n-1)/n,t(0)=1):
> s := inverse(t):
> tpsform(s, x, 7);
                 2       3       4        5        6      7
       1 - x + 1/2 x  - 1/6 x  + 1/24 x  - 1/120 x  + 1/720 x  + O(x )
```

7.2.6   powseries[multconst] – multiplication of a power series by a constant

Calling Sequence:

   `multconst(`p`,` *const*`)`

Parameters:

   p – formal power series

   const – scalar constant

Synopsis:

- The function `multconst(`p`,` *const*`)` returns the formal power series where each coefficient is *const* times the corresponding coefficient in p.

- This function is part of the **powseries** package, and so can be used in the form `multconst(..)` only after performing the command `with(powseries)` or `with(powseries, multconst)`. The function can always be accessed in the long form `powseries[multconst](..)`.

Examples:
```
> with(powseries):
> powcreate(t(n)=t(n-1)/n,t(0)=1):
> s := multconst(t, -42):
> tpsform(s, x);
                     2      3       4        5        6
        - 42 - 42 x - 21 x  - 7 x  - 7/4 x  - 7/20 x  + O(x )
```

7.2.7   powseries[multiply] – multiplication of formal power series

Calling Sequence:
    multiply(a, b)

Parameters:
    a, b –   formal power series

Synopsis:
- The function multiply(a, b) returns the formal power series which is the product of a and b.

- This function is part of the powseries package, and so can be used in the form multiply(..) only after performing the command with(powseries) or with(powseries, multiply). The function can always be accessed in the long form powseries[multiply](..).

Examples:
```
> with(powseries):
> powcreate(t(n)=t(n-1)/n,t(0)=1):
> powcreate(v(n)=v(n-1)/2,v(0)=1):
> s := multiply(t, v):
> tpsform(s, x);
```
$$1 + 3/2\ x + 5/4\ x^2 + \frac{19}{24}\ x^3 + 7/16\ x^4 + \frac{109}{480}\ x^5 + O(x^6)$$

7.2.8   powseries[negative] – negation of a formal power series

Calling Sequence:
    negative(p)

Parameters:
    p –   formal power series

Synopsis:
- The function negative(p) returns a formal power series where each coefficient is the negative of the corresponding coefficient in p.

- This function is part of the powseries package, and so can be used in the form negative(..) only after performing the command with(powseries) or with(powseries, negative). The function can always be accessed in the long form powseries[negative](..).

Examples:
```
> with(powseries):
> powcreate(t(n)=t(n-1)/n,t(0)=1):
> s := negative(t):
> tpsform(s, x);
```
$$- 1 - x - 1/2\ x^2 - 1/6\ x^3 - 1/24\ x^4 - 1/120\ x^5 + O(x^6)$$

7.2.9 powseries[powcreate] – create formal power series

Calling Sequence:

powcreate(*eqns*)

Parameters:

eqns – expression sequence of equations

Synopsis:

- The function powcreate creates a formal power series whose coefficients are specified by *eqns*.

- The expression sequence of equations passed to powcreate should consist of one equation, possibly recursive, that specifies the nth coefficient, followed by zero or more equations that specify initial conditions.

- More specifically, powcreate($t(n)=expr$, $t(0)=a_1$, ..., $t(m)=a_m$) creates a formal power series, t, whose nth coefficient is $expr$, and whose initial terms are $t(0) = a_1$,..., $t(m) = a_m$.

- Any initial conditions among *eqns* override the general expression. Thus, it is possible to include particular values of coefficients in *eqns*.

- This function is part of the powseries package, and so can be used in the form powcreate(..) only after performing the command with(powseries) or with(powseries, powcreate). The function can always be accessed in the long form powseries[powcreate](..).

Examples:
```
> with(powseries):
> powcreate(t(n)=1/n!,t(0)=1):
> tpsform(t, x, 7);
                  2         3          4           5           6          7
      1 + x + 1/2 x  + 1/6 x  + 1/24 x  + 1/120 x  + 1/720 x  + O(x )

> powcreate(v(n)=(v(n-1)+v(n-2))/4,v(0)=4,v(1)=2):
> tpsform(v, x);
                        2         3    19    4    47   5         6
      4 + 2 x + 3/2 x  + 7/8 x  + ---- x  + --- x  + O(x )
                                   32        128
```

7.2.10 powseries[powdiff] – differentiation of a formal power series

Calling Sequence:

powdiff(*p*)

Parameters:

p – formal power series

Synopsis:

- The function powdiff(p) returns the formal power series which is the derivative of p with respect to the variable of p.

- This function is part of the `powseries` package, and so can be used in the form `powdiff(..)` only after performing the command `with(powseries)` or `with(powseries, powdiff)`. The function can always be accessed in the long form `powseries[powdiff](..)`.

Examples:
```
> with(powseries):
> t := powpoly(x+1, x):
> s := powlog(t):
> tpsform(s, x, 6);
```
$$x - 1/2\ x^2 + 1/3\ x^3 - 1/4\ x^4 + 1/5\ x^5 + O(x^6)$$
```
> r := powdiff(s):
> tpsform(r, x, 10);
```
$$1 - x + x^2 - x^3 + x^4 - x^5 + x^6 - x^7 + x^8 - x^9 + O(x^{10})$$

7.2.11 powseries[powexp] – exponentiation of a formal power series

Calling Sequence:

  `powexp(p)`

Parameters:

  p – formal power series

Synopsis:

- The function `powexp(p)` returns the formal power series which is `exp(p)`.

- This function is part of the `powseries` package, and so can be used in the form `powexp(..)` only after performing the command `with(powseries)` or `with(powseries, powexp)`. The function can always be accessed in the long form `powseries[powexp](..)`.

Examples:
```
> with(powseries):
> t := powpoly(x, x):
> s := powexp(t):
> tpsform(s, x, 7);
```
$$1 + x + 1/2\ x^2 + 1/6\ x^3 + 1/24\ x^4 + 1/120\ x^5 + 1/720\ x^6 + O(x^7)$$

7.2.12 powseries[powint] – integration of a formal power series

Calling Sequence:

  `powint(p)`
  `powseries[powint](p)`

Parameters:

  p – formal power series

Synopsis:

- The function `powint(`p`)` returns the formal power series which is the formal anti-derivative of p with respect to the variable of the power series.

- This function is part of the **powseries** package, and so can be used in the form `powint(..)` only after performing the command `with(powseries)` or `with(powseries, powint)`. The function can always be accessed in the long form `powseries[powint](..)`.

Examples:
```
> with(powseries):
> powcreate(t(n)=1/n,t(0)=1):
> tpsform(t, x, 5);
```
$$1 + x + 1/2\ x^2 + 1/3\ x^3 + 1/4\ x^4 + O(x^5)$$
```
> r := powint(t):
> tpsform(r, x, 7);
```
$$x + 1/2\ x^2 + 1/6\ x^3 + 1/12\ x^4 + 1/20\ x^5 + 1/30\ x^6 + O(x^7)$$

7.2.13   powseries[powlog] – logarithm of a formal power series

Calling Sequence:

  `powlog(`p`)`

Parameters:

  p – formal power series

Synopsis:

- The function `powlog(`p`)` returns the formal power series which is the natural logarithm of p.

- The power series p must have a non-zero first term $(p(0) <> 0)$ for its logarithm to be well-defined.

- This function is part of the **powseries** package, and so can be used in the form `powlog(..)` only after performing the command `with(powseries)` or `with(powseries, powlog)`. The function can always be accessed in the long form `powseries[powlog](..)`.

Examples:
```
> with(powseries):
> t := powpoly(1 + x, x):
> s := powlog(t):
> tpsform(s, x, 7);
```
$$x - 1/2\ x^2 + 1/3\ x^3 - 1/4\ x^4 + 1/5\ x^5 - 1/6\ x^6 + O(x^7)$$

7.2.14 powseries[powpoly] – create a formal power series from a polynomial

Calling Sequence:

powpoly(*poly, var*)

Parameters:

poly – a polynomial in *var*

var – a variable

Synopsis:

- The function powpoly(*poly, var*) returns a formal power series equivalent to *poly*.
- Thus, for all *i*, the *i*th coefficient of the power series returned is coeff(*poly, var, i*).
- This function is part of the powseries package, and so can be used in the form powpoly(..) only after performing the command with(powseries) or with(powseries, powpoly). The function can always be accessed in the long form powseries[powpoly](..).

Examples:

```
> with(powseries):
  t:=powpoly(2*x^5+4*x^4-x+5, x):
  tpsform(t, s, 5);                  ⟶                4      5
                                              5 - s + 4 s  + O(s )

> tpsform(t, s, 4);                  ⟶                       4
                                                 5 - s + O(s )
```

7.2.15 Representation of formal power series

Calling Sequence:

function(args)

Synopsis:

- Formal power series are procedures which return the coefficients of the power series they represent. Thus name(i) is the coefficient of x^i in the series called name.
- The values of the coefficients are saved using a remember table, so all computed coefficients can be seen via op(4, op(name)) .
- The actual procedure is identical for all power series; the only differences are the general term and the values that each remembers.
- The general term of the power series can be obtained via name(_k) .
- Note: that each intermediate power series created in a calculation should be named.

Examples:
```
> with(powseries):
> powcreate(e(n)=1/n!,e(0)=1):
> powcreate(f(n)=f(n-1)/(n^2*f(n-2)),f(0)=1,f(1)=5,f(2)=2):
> tpsform(e, x, 6);
```
$$1 + x + 1/2\ x^2 + 1/6\ x^3 + 1/24\ x^4 + 1/120\ x^5 + O(x^6)$$

```
> tpsform(f, x, 6);
```
$$1 + 5\ x + 2\ x^2 + 2/45\ x^3 + 1/720\ x^4 + 1/800\ x^5 + O(x^6)$$

```
# Compute the series  f^e .
> logf := powlog(f):
> elogf := multiply(e, logf):
> result := powexp(elogf):
> tpsform(result, x, 6);
```
$$1 + 5\ x + 7\ x^2 + \frac{767}{45}\ x^3 + \frac{2351}{240}\ x^4 + \frac{39231}{800}\ x^5 + O(x^6)$$

7.2.16   powseries[powsolve] – solve linear differential equations as power series

Calling Sequence:

> powsolve(*sys*)

Parameters:

> *sys* – a set or an expression sequence containing a linear differential
> equation and optional initial conditions

Synopsis:

- The function `powsolve` solves a linear differential equation which may or may not have initial conditions.

- All the initial conditions must be at zero.

- Derivatives are denoted by applying D to the function name. For example, the second derivative of y at 0 is D(D(y))(0) .

- The solution returned is a formal power series which represents the infinite series solution.

- This function is part of the `powseries` package, and so can be used in the form `powsolve(..)` only after performing the command `with(powseries)` or `with(powseries, powsolve)`. The function can always be accessed in the long form `powseries[powsolve](..)`.

Examples:
```
> with(powseries):
> a:=powsolve(diff(y(x),x)=y(x),y(0)=1):
> tpsform(a, x);
```
$$1 + x + 1/2\ x^2 + 1/6\ x^3 + 1/24\ x^4 + 1/120\ x^5 + O(x^6)$$

```
# second system
> v:=powsolve({diff(y(x),x$4)=y(x),y(0)=3/2,D(y)(0)=-1/2,
>               D(D(y))(0)=-3/2,D(D(D(y)))(0)=1/2}):
> tpsform(v, x);
```

$$3/2 - 1/2\ x - 3/4\ x^2 + 1/12\ x^3 + 1/16\ x^4 - 1/240\ x^5 + O(x^6)$$

7.2.17   powseries[quotient] – quotient of two formal power series

Calling Sequence:

>   quotient(a, b)

Parameters:

> a, b – formal power series

Synopsis:

- The function quotient(a, b) returns the formal power series equal to a/b .

- This function yields the same result as multiply(a, inverse(b)), but is faster.

- This function is part of the powseries package, and so can be used in the form quotient(..) only after performing the command with(powseries) or with(powseries, quotient). The function can always be accessed in the long form powseries[quotient](..).

Examples:
```
> with(powseries):
  powcreate(t(n)=t(n-1)/n,t(0)=1):
  powcreate(v(n)=v(n-1)/5,v(0)=1):
  s := quotient(t, v):
  tpsform(s, x, 4);
```
$$\longrightarrow \quad 1 + 4/5\ x + 3/10\ x^2 + 1/15\ x^3 + O(x^4)$$

7.2.18   powseries[reversion] – reversion of formal power series

Calling Sequence:

>   reversion(a, b)
>   reversion(a)

Parameters:

> a – formal power series
> b – (optional) formal power series

Synopsis:

- The function reversion(a, b) returns the formal power series which is the reversion of a with respect to b. If b is not specified then it is assumed to be the formal power series with one nonzero coefficient, $b(1)=1$.

- Since reversion is the inverse of composition, composition of the result into a will give b.

- Note that $a(0)$ must be 0, $a(1)$ must be 1, and $b(0)$ must be 0. If not, the reversion is not well defined and **reversion** returns an error message.

- This function is part of the **powseries** package, and so can be used in the form **reversion(..)** only after performing the command **with(powseries)** or **with(powseries, reversion)**. The function can always be accessed in the long form **powseries[reversion](..)**.

Examples:
```
> with(powseries):
> powcreate(t(n)=t(n-1)/n,t(0)=0,t(1)=1):
> powcreate(v(n)=v(n-1)/2,v(0)=0,v(1)=1):
> s := reversion(t,v):
> tpsform(s,x,11);
```
$$x + 1/12\ x^3 + 1/80\ x^5 + 1/448\ x^7 + 1/2304\ x^9 + O(x^{11})$$

```
> ts := compose(t,s):
> tpsform(ts,x);
```
$$x + 1/2\ x^2 + 1/4\ x^3 + 1/8\ x^4 + 1/16\ x^5 + O(x^6)$$

```
> tpsform(v,x);
```
$$x + 1/2\ x^2 + 1/4\ x^3 + 1/8\ x^4 + 1/16\ x^5 + O(x^6)$$

7.2.19   powseries[subtract] – subtraction of two formal power series

Calling Sequence:

   subtract(a, b)

Parameters:

   a, b –   formal power series

Synopsis:

- The function **subtract**(a, b) returns the formal power series in which each coefficient is the corresponding coefficient of a minus the corresponding coefficient of b.

- This function yields the same result as **add**(a, **negative**(b)), but is faster.

- This function is part of the **powseries** package, and so can be used in the form **subtract(..)** only after performing the command **with(powseries)** or **with(powseries, subtract)**. The function can always be accessed in the long form **powseries[subtract](..)**.

Examples:
```
> with(powseries):
> powcreate(t(n)=t(n-1)/n,t(0)=1):
> powcreate(v(n)=v(n-1)/2,v(0)=1):
> s := subtract(v, t):
> tpsform(s, x, 6);
```
$$- 1/2\ x - 1/4\ x^2 - 1/24\ x^3 + 1/48\ x^4 + \frac{11}{480}\ x^5 + O(x^6)$$

7.2.20 powseries[tpsform] – truncated power series form of a formal series

Calling Sequence:
> tpsform(*p, var, order*)
> tpsform(*p, var*)

Parameters:

p	–	name of a formal power series
var	–	variable to be used in the truncated series
order	–	(optional) number of terms to which resulting series should be accurate (this defaults to the global variable `Order`)

Synopsis:

- The function `tpsform(`*p, var, order*`)` creates a truncated power series (a series data structure) in *var*. This truncated series is equal to *p* up to, but not including, the term containing $x\hat{\ }order$.

- In some cases `tpsform` can determine when the series is exact and so it drops the $O(x\hat{\ }order)$ term. Formal power series created using `powpoly` fall into this category.

- This function is part of the `powseries` package, and so can be used in the form `tpsform(..)` only after performing the command `with(powseries)` or `with(powseries, tpsform)`. The function can always be accessed in the long form `powseries[tpsform](..)`.

Examples:
```
> with(powseries):
> powcreate(t(n)=t(n-1)/n,t(0)=1):
> t(_k);
```
$$\frac{t(\_k - 1)}{\_k}$$

```
> tpsform(t, x, 6);
```
$$1 + x + 1/2\ x^2 + 1/6\ x^3 + 1/24\ x^4 + 1/120\ x^5 + O(x^6)$$

SEE ALSO: `type[series]`

7.3 The Total Ordering of Names Package

7.3.1 Introduction to the `totorder` package

Calling Sequence:

    assume(r)
    is(r)
    forget(r)
    ordering()

Parameters:

    r – sequence of relations

Synopsis:

- The `assume` function allows a total ordering to be defined on a sequence of names.

- Having defined such an ordering, the `is` function can be used to make queries with respect to the ordering.

- The `forget` function deletes relationships within the ordering. To delete all relationships use the command `forget(everything)`.

- The `ordering` function prints out the current total ordering.

- This package should be defined by the command `with(totorder)` before it is used.

Examples:

```
> with(totorder):
  assume(a<b, b<c, c=d, d<f);                 ⟶      assumed, a < b, b < c, c = d, d < f

> assume(u>b);                                ⟶               assumed, b < u

> is(u<f);                                    ⟶                    true

> forget(u);                                  ⟶                forgotten, u

> ordering();                                 ⟶        a < b, b < c, c = d, d < f

> assume(u>d);                                ⟶               assumed, d < u

> is(2*c-c<f);                                ⟶                    true

> is(c<g);                                ⟶   Error, (in is) unknown symbol, g
```

Index